A Problem-Solving Approach to Aquatic Chemistry

James N. Jensen

University at Buffalo

John Wiley & Sons, Inc.

ACQUISITIONS EDITOR   Wayne Anderson

SENIOR MARKETING MANAGER   Katherine Hepburn

COVER PHOTOGRAPH   W.S. Keller/National Park Service

ISBN 0-471-41386-0
ISBN 13: 978-0-471-41386-8

To

Anne, Robert, and Sarah

Preface

For the Student

Introduction

You are about to embark on a journey into the world of aquatic chemistry. As with all journeys, there will be moments of joy and surprise, frustration and disappointment, and discovery and growth. My goal in writing this book was to help you in your exploration of aquatic chemistry concepts and to help convert the challenges into understanding.

What is in store for you? In the summer before I took my first course in water chemistry, I was sharing my choice of fall classes with my relatives. One of them feigned shock: "Water chemistry? What do you need to know about the chemistry of water other than H two O?" This question may be on your mind as well. After all, isn't water just water? In a word: No. Natural waters contain enough dissolved gases (such as oxygen), dissolved minerals (such as sodium ion), and organic matter to sustain a thriving ecosystem. Wastewater is a rich mixture of chemical species. Even the purest waters that you encounter every day (drinking water and perhaps bottled water) are influenced by their chemical makeup. Water is a fertile amalgamation of chemical species and not a barren desert of H_2O molecules. Soon you will find it difficult to look at rain or a stream or tap water without thinking about the wealth of chemistry in every drop. Why it is necessary to learn about water chemistry? Perhaps you are motivated by understanding and maintaining the quality of the aquatic environment. If so, then you may appreciate that the chemical makeup of water influences and is influenced by the system ecology. For example, water quality models start with nutrients and nutrients undergo some of the transformations described in this text. If your interests lie in treating wastewater and drinking water, then you may appreciate that treatment processes will fail unless the chemistry of the water to be treated is taken into account.

How to Use This Book for Learning

This book contains many devices to assist in communicating the main ideas of each chapter. First, each chapter begins with an introduction and ends with a summary. The chapter introductions set the stage for each chapter and provide a guide to the chapter material. Chapter summaries are concise reviews of the main points in each chapter.

Second, the text contains the definition of important terms in the left margin. Terms to be defined in the margin are indicated in the text by a boldface italic font. The left-margin definitions provide good study aids for reviewing the important new terms in each chapter.

Third, the text contains key ideas in the left margin. Key ideas are indicated by the icon in the left margin. As with the definitions, reading the key idea summaries provides a good review of the important new concepts in each chapter. Key ideas are listed for your convenience at the end of every chapter.

Fourth, boxes containing a *Thoughtful Pause* interrupt the text frequently. Please try to answer the questions in each Thoughtful Pause before continuing to read. The questions in each Thoughtful Pause are always answered in the following text. The Thoughtful Pause is intended to simulate classroom interaction between instructor and student. At other places in the text, you will encounter phrases such as: "You may wish to try this yourself before proceeding." Please stop at these signposts and take the time to work through the question at hand before continuing to read.

Fifth, examples are used to illustrate key quantitative concepts in the text. Some examples are found in the left margin of the page. Longer examples usually are embedded in the main text. You may want to read through some of the examples before attempting to work them.

Last, each of the five parts of this book contains a road map in an introductory chapter. The road map highlights the appropriate paths through the part for students with varying backgrounds in aquatic chemistry.

Other portions of this text also will enrich your learning experience. Each part of the book contains a case study. The case studies allow you to apply the main ideas for each part of the book to interesting and complex problems in environmental science and engineering. Portions of the case study are analyzed in each chapter. In addition, this text emphasizes computer solution methods. Numerous computer files illustrating the practical uses of computer solution techniques may be found on the accompanying website, www.wiley.com/college/jensen. Computer files are indicated in the left margin by a CD icon ⊕. Special computer files, called animations and indicated by the 🎞 icon, allow you to see how species concentrations change with independent variables. The website also contains a software package for solving chemical equilibrium (NanoQL) that was developed specifically for this text

For the Instructor

Introduction

Several excellent textbooks in aquatic chemistry are available. Your challenge is to select a textbook that matches the background of your students, the course objectives, and your teaching style. This text was developed for advanced undergraduates or first-year graduate students in environmental engineering and science. It demands little background from the student other than freshman chemistry, aptitude in algebra, and basic computer skills.

The text can be used in a one-semester course in water chemistry or in an environmental chemistry course that emphasizes aquatic systems. I have designed the text to cover equilibrium calculation techniques and equilibrium types, along with a chapter on chemical kinetics. In addition, most chapters contain some material of a more advanced nature. In this way, the text can be used either to quickly "hit the highlights" in a general environmental chemistry course or to allow exploration of more complex systems in a water chemistry course.

How To Use This Book for Teaching

Most of us unconsciously judge textbooks as being good if the author thinks the same way that we do. I have made several choices in this text that may be different from the way you teach the material. First, I postpone a quantitative discussion of activity coefficients until the final part of the text. The thermodynamic development in Chapter 3 is based on activity. In other early chapters, it is assumed frequently that activity and concentration are interchangeable and that the temperature of interest is the temperature at which the thermodynamic data were collected. I have found it useful in my own teaching to separate equilibrium calculation fundamentals from activity coefficient and temperature corrections. If you prefer to teach activity coefficient and temperature corrections up front, you may wish to cover Sections 21.2 to 21.4 earlier in your course.

Second, I teach the concept of alkalinity in the context of natural waters first and then generalize to systems with other weak acids. If your applications usually contain weak acids other than the carbonate system, you may wish to rearrange the material in Chapter 13. Third, most water chemistry texts struggle with how to sequence complexation and precipitation. If complexation is presented first, then the discussion of the important effects of complexation on solubility must be delayed. If precipitation is presented first, then metal precipitation examples must include some extra discussion of metal hydrolysis. I have chosen to present complexation first, using complexation examples with homogeneous systems only. The effects of complexation on solubility then are presented in the precipitation chapter.

Finally, this text reflects my own teaching style for my version of this course. I employ the discussion method in my lectures, with frequent opportunities for student input. Each Thoughtful Pause in the text come from questions I have asked my classes over the years during lecture. Professionals wishing to remind themselves of water chemistry principles also can use this text for self-study. These readers should use the index and table of contents carefully to select their particular topics of interest.

Acknowledgments

I have been taught and inspired by several excellent teachers of aquatic chemistry. My first experiences with water chemistry came in the fall of 1979 in the classroom of Professor James J. Morgan. In addition to Jim Morgan's long-lasting contributions to the field through research, leadership, and textbook writing, I can attest to his excellence as a teacher and a role model for students. I had the opportunity to sit in on Professor Philip C. Singer's classroom on numerous occasions in the early 1980s. His command of the material and patience with students has inspired me ever since. I often prepare for lectures by asking: "What would Jim and Phil do?" Their imprint can be seen throughout this text.

I want to acknowledge the many other people who have contributed to this text. My colleagues at the University at Buffalo have been very supportive of the effort to give birth to the book you hold in your hands. In particular, my conversations with Professors John E. Van Benschoten, A. Scott Weber, and Mark R. Matsumoto (currently at University of California, Riverside) have shaped and improved the text. I am also very thankful to the students with whom I have shared ideas over the years. They more than anyone have taught me how to teach.

Wayne Anderson and his staff and colleagues at John Wiley & Sons have shown patience beyond reason with me. This book is better for their input and guidance. The comments of the reviewers are also very much appreciated. I am thankful to you, the reader, as well. Writing this text has allowed me to rediscover and renew my appreciation of the richness of the aquatic world.

Finally, I want to thank my wife Anne and my family for indulging me in the "long, strange trip" of writing this text.

James N. Jensen
Buffalo, NY
September 2002

Brief Table of Contents

Table of Contents

Part III: Acid-Base Equilibria in Homogenous Aqueous Systems   191

Fundamental Concepts

The fundamental things apply
As time goes by.
Herman Hupfeld

The beginning sets the rules.
Mason Cooley

CHAPTER 1

Getting Started with the Fundamental Concepts

1.1 INTRODUCTION

The first part of this text reviews fundamental concepts that must be mastered prior to learning how to calculate and interpret species concentrations in aquatic systems. In this chapter, the motivation for studying chemical species and a few general principles concerning aquatic systems are presented.

In Section 1.2, the motivation for why engineers and scientists are interested in individual chemical species concentrations at equilibrium will be discussed. Important water quality parameters, called *master variables*, are introduced in Section 1.3. It is impossible to study water chemistry without a little knowledge of the structure of water. A few of the unique properties of water will be explored in Section 1.4, especially as they relate to the chemical reactions that occur in water. In Section 1.5, a road map for Part I of the text is presented and discussed. Finally, the Part I case study is presented at the conclusion of this chapter. Before beginning Part I of the text, you are urged to review the chemistry background material in Appendix A (Section A.2).

1.2 WHY CALCULATE CHEMICAL SPECIES CONCENTRATIONS AT EQUILIBRIUM?

1.2.1 Overview

hydrolysis reactions: reactions with water

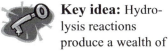 **Key idea:** Hydrolysis reactions produce a wealth of dissolved chemical species

The bulk of this book is dedicated to the calculation of species concentrations at equilibrium. The focus here is on chemical species that undergo chemical reactions; in other words, reactive species. More specifically, the emphasis here is on chemical species which react with water. Reactions with water are called **hydrolysis reactions** (from the Greek *hydōr* water + *lyein* to loosen). When substances react with water, numerous other compounds can be formed. Indeed, the richness of aquatic chemistry stems from the large number of substances that react not only with water but also with the products of myriad other hydrolysis reactions.

This richness is illustrated in Figure 1.1. Inputs of chemical species (from aqueous discharges, runoff, atmospheric deposition, dissolution from

3

the atmosphere, and dissolution from sediments) react with water to form hydrolysis products. The hydrolysis products and input chemicals react further to increase the complexity of aquatic systems.

So why so much interest in calculating the equilibrium concentrations of chemical species? This question is really two questions. First, why calculate the concentrations of *individual* chemical species? Second, why calculate species concentrations at *equilibrium*?

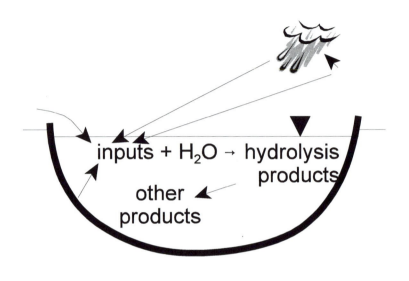

inputs + H$_2$O → hydrolysis products

other products

Figure 1.1: Complexity of Aquatic Systems

1.2.2 The importance of individual chemical species

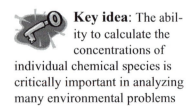

Key idea: The ability to calculate the concentrations of individual chemical species is critically important in analyzing many environmental problems

Throughout this text, you shall be reminded again and again that knowing the concentrations of *individual chemical species* is critically important in analyzing many environmental problems. At first glance, this statement may not make sense. After all, many environmental regulations are based on *total* concentrations of classes of compounds rather than on the concentrations of individual species. Should you be more concerned about the total amount of mercury or phenol or ammonia than about individual species stemming from the hydrolysis of mercury or phenol or ammonia?

In fact, you will find that *individual* species frequently are more important. Three general examples will illustrate this point. First, adverse impacts on human health and ecosystem viability may be due to only one or several of a large number of related hydrolysis products. A prime example is the transition metals (such as mercury, copper, zinc, cadmium, iron, and lead), in which toxicity varies dramatically among the hydrolysis products. Another example is cyanide. Hydrogen cyanide (HCN) is much more toxic to humans than cyanide ion (CN^-).

Second, the success of engineered treatment systems may depend on knowledge of the concentrations of key individual species. Since hydrolysis products vary in their physical, chemical, and biochemical properties, the design and operation of treatment processes depend on quantitative models for the concentrations of individual chemical species. For example, the addition of gaseous chlorine to wastewater for disinfection results in the formation of many chemical species (including HOCl, OCl^-, NH_2Cl, and $NHCl_2$), each of which differs in its ability to inactivate (i.e., kill) microorganisms.

Third, individual species vary greatly in how readily they cross cell membranes or cell walls and are assimilated by aquatic biota. Thus, understanding the cycling of trace nutrients in the aquatic environment (and humankind's impact on nutrient cycling) requires knowledge of the concentrations of individual chemical species.

As an example of the importance of the concentrations of individual chemical species, consider the soup created when copper sulfate crystals, $CuSO_4(s)^\dagger$, are added to a reservoir for algae control. The $CuSO_4(s)$ dissolves in water to form a copper-containing ion (called aquo cupric ion) and sulfate. The structure of the aquo cupric ion usually is abbreviated as Cu^{2+}. The Cu^{2+} ions thus formed react very quickly with water to form a number of hydrolysis products, including $CuOH^+$, $Cu(OH)_2(aq)^\dagger$, $Cu(OH)_3^-$, $Cu(OH)_4^{2-}$, and $Cu_2(OH)_2^{2-}$. Under certain chemical conditions, copper may precipitate as CuO(s). As you spread the copper sulfate from the back of a boat, carbon dioxide in the atmosphere is equilibrating with the reservoir water to form its own hydrolysis products. The hydrolysis products of carbon dioxide are H_2CO_3, HCO_3^-, and CO_3^{2-}. The aquo cupric ion will react to some extent with the hydrolysis products of carbon dioxide to form $CuCO_3(aq)$, $Cu(CO_3)_2^{2-}$, and perhaps even solids containing copper and carbonate (CO_3^{2-}). By adding one copper compound to a natural water body, you may be faced with accounting for as many as ten copper-containing species even in a relatively simple chemical model.

$\dagger$ Here, (s) represents a solid species and (aq) represents a dissolved species in water. See Appendix A (Section A.2.2) for a review of this nomenclature.

Of course, the real world is even more complex. The reservoir water contains many more species that can react with copper than just hydroxide (OH^-) and carbonate. A realistic model for copper in the reservoir would have to include the reactions of Cu^{2+} with (among other chemical species) chloride, amino acids, ammonia, particulates, and microorganisms. In reality, the act of throwing copper sulfate crystals into the reservoir will produce dozens of chemical species containing copper.

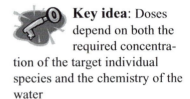

Key idea: Doses depend on both the required concentration of the target individual species and the chemistry of the water

Why should you care that copper sulfate forms many copper-containing species in a lake? Remember that copper is added to kill algae. It is well-established that copper toxicity to algae is due almost entirely to *one* chemical species: Cu^{2+} (Jackson and Morgan, 1978). Thus, to determine the copper sulfate dose, you must be able to calculate the concentration of Cu^{2+} after a certain amount of copper sulfate is added to the reservoir. Since the Cu^{2+} concentration usually is exceedingly small, this is akin to counting needles of Cu^{2+} in this haystack of copper-containing species. In practice, you would back-calculate the copper sulfate dose required to achieve a required level of Cu^{2+}. Even if two reservoirs had the same amounts of algae, different water chemistries in the reservoirs may lead to very different copper sulfate doses to achieve the same Cu^{2+} concentration. The chemistry of the water determines how the required concentration of one species (Cu^{2+}) is translated back into a copper sulfate dose.

The process of relating a dose to a required concentration of an individual chemical species is illustrated in Figure 1.2. The arrows in Figure 1.2 indicate chemical reactions that must be included in a mathematical model to allow for the determination of the copper sulfate dose. In this text, you will learn the tools to make *quantitative* decisions to solve similar problems in the aquatic environment.

1.2.3 The importance of equilibrium

An entire chapter of this book (Chapter 3) is devoted to developing the thermodynamic basis of equilibrium. For the present, you can think of the equilibrium state as the condition in which the concentrations of all chemical species do not change with time. To impose equilibrium on a chemical system, the interesting and important time-dependent nature of chemical concentrations are excluded. The study of the rates of chemical reaction is called **chemical kinetics** and is covered briefly in Chapter 22. Why constrain the discussion mainly to the equilibrium state here, with shorter coverage of chemical kinetics?

chemical kinetics: the study of chemical reaction rates

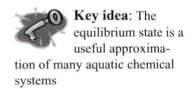

Key idea: The equilibrium state is a useful approximation of many aquatic chemical systems

There are two reasons for focusing on equilibrium. First, many of the chemical reactions you shall examine in this text are fast. For example, the reaction of H^+ and OH^- to form water occurs on the time scale of 10^{-5} s at natural water conditions (Morgan and Stone, 1985). Second, equilibrium models give insight into chemical systems, even when kinetics are known to be important. As an example, equilibrium chemistry provides a good framework to understand coagulation chemistry in drinking water treatment. Chemical kinetics also play a role in coagulation, but equilibrium models remain useful. Thus, the focus on equilibrium is somewhat constraining but not overly restrictive.

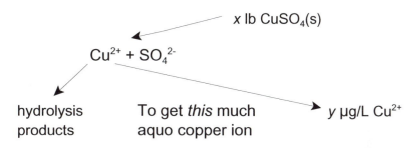

You must add *this* much copper sulfate...

x lb $CuSO_4(s)$

$Cu^{2+} + SO_4^{2-}$

hydrolysis products

To get *this* much aquo copper ion

y µg/L Cu^{2+}

Figure 1.2: Qualitative Relationship Between the Dose Required and End Species Concentrations Desired

1.3 MASTER VARIABLES: THE IMPORTANCE OF pH AND pe

master variable: a species concentration (or function of a species concentration) that controls the chemistry of a system

pH: a master variable controlling H^+ (proton) transfer and defined by $-\log(H^+$ activity)

pe: a master variable controlling electron transfer and defined by $-\log(e^-$ activity)

As discussed in Section 1.2, it is important to know the concentrations of individual chemical species in aquatic systems. Some species may be of greater or lesser importance, but *a small number of chemical species often control the chemistry of an aquatic system.* "Controlling the chemistry" means that certain chemical species play a dominant role in determining the concentrations of other chemical species.

The concentrations of the controlling species are expressed in parameters called *master variables*.[†] The two most important master variables in water are *pH* and *pe*. The master variable pH is defined as the negative of the log of the activity of the ion H^+.[††] Similarly, pe is defined as the negative of the log of the activity of the electron, e^-. Since the atom H contains one proton and one electron in its most abundant state, H^+ is a proton. Thus, the transfer of H^+ sometimes is called *proton transfer* or *acid-base chemistry*. Electron transfer is also called *oxidation-reduction chemistry* or *redox chemistry*. Acid-base chemistry is discussed in more detail in Chapter 11, and redox chemistry is discussed in more detail in Chapter 14.

[†] More formally, a master variable is the single variable that determines the ratios of species concentrations (see Sillén, 1959).
[††] Activity is an idealized concentration. It is discussed further in Chapter 3. The characteristics of the *p* function are explored in Appendix A (Section A.3).

You will note in Appendix A that a low pH corresponds to high activity of H^+. Conditions of high H^+ activity are called *acidic*. Thus, low pH (high H^+ activity) means acidic conditions. High pH (low H^+ activity) corresponds to *basic* conditions. Low pe (high e^- activity) corresponds to *reducing* conditions, and high pe (high e^- activity) corresponds to *oxidizing* conditions.

1.4 PROPERTIES OF WATER

1.4.1 Introduction

Water is a unique substance. In freshman chemistry courses, you probably learned of the periodic nature of the properties of the elements. Elements in the same column of the periodic table share similar chemical properties. Yet water, H_2O, has very different physical and chemical properties than other dihydrogen complexes of elements in oxygen's column of the periodic table. For example, water is a liquid under standard temperature and pressure conditions, but H_2S is a gas.

The unique properties of water stem from the large differences between the affinity of oxygen and hydrogen for electrons. Oxygen has a much higher affinity for electrons than hydrogen, resulting in a highly polarized bond between O and H. In fact, the oxygen atoms in water are partially negatively charged and the hydrogen atoms are partially positively charged, as illustrated in Figure 1.3. As a result of this polar bond and the difference in partial charge, oxygen atoms in water have the ability to form weak, but important, chemical bonds with more that two hydrogen atoms. These weak bonds are called ***hydrogen bonds***. Without exaggeration, hydrogen bonds influence *every* molecular interaction in the aquatic environment.[†]

Hydrogen bonds affect the three-dimensional shape of water. The bond distance between oxygen in water and the hydrogen-bonded hydrogen is about twice the bond distance between the oxygen in water and one of its own hydrogen atoms. Hydrogen bonding allows for the formation of very large clusters of water molecules (about 400 molecules per cluster; Luck, 1998). The clusters are shifting constantly since the average lifetime of a hydrogen bond is only a few picoseconds (10^{-12} s). The shifting clusters of water molecules contribute to the solubilization of ions in water, as will be discussed in Section 1.4.2.

Example 1.1: **Solubilization of Sodium Chloride in Water**

Why does NaCl dissolve in water?

Solution:
One reason NaCl dissolves in water is that the potential energy of interaction between Na^+ and Cl^- is reduced in water. To illustrate this point, calculate the potential energy of interaction between Na^+ and Cl^- separated by 1 nm in water.

The potential energy of two particles of charge q_1 and q_2 separate by a distance r is given by Coulomb's Law:

$$\text{potential energy} = q_1 q_2 / (4\pi r \epsilon_0 \epsilon)$$

where ϵ_0 is the permittivity of a vacuum ($= 8.854 \times 10^{-12}$ $J^{-1}C^2m^{-1}$) and ϵ is the dielectric constant ($= 1$ for a vacuum and 80 for water). The unit charge on Na^+ is $+1.602 \times 10^{-19}$ C and -1.602×10^{-19} C on Cl^-.

[†] Consider, for example, the influence of hydrogen bonds on ice. The structure of ice is different than the structure of water. In ice, all the water molecules participate in four hydrogen bonds arranged in a tetrahedronal shape. The shape results in an open structure for ice and a corresponding density of 0.92 g/cm^3. The density of ice is less than the density of liquid water at 0°C, so ice floats. The length of the hydrogen bond is directly responsible for the open structure and low density of ice. If the hydrogen bond was 5% shorter, ice would sink and water bodies would have frozen solid during the ice ages. If hydrogen bonds were slightly shorter, life on Earth may not have survived long enough to allow you to read this book.

Plugging in the values, you will find that the potential energy is -139 kJ per mole in a vacuum and -1.7 kJ per mole in water (the negative sign indicates an energy of attraction and one mole is 6.022×10^{23} ions).

The attractive potential energy of interaction between Na^+ and Cl^- in water is very small (only about one-tenth of the energy of a hydrogen bond).

Sodium chloride is soluble in water mainly because the attractive energy between Na^+ and Cl^- is reduced by water.

(Note: The energy of the system also is altered as the atoms in the salt become more disordered and the water molecules become more ordered around the ions.)

hydration: the process of forming compounds with water

polar: a polar species contains atoms or groups exhibiting different charges (or partial charges)

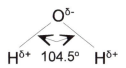

Figure 1.3: Charge Distribution Among the Atoms in Water

1.4.2 Solubility of ionic species

Throughout this text, it shall be assumed that salts containing sodium, potassium, or chloride at low concentrations dissolve nearly completely in water to form ions. Why are salts so soluble in water? Hydrogen bonding allows for the orientation of a large number of molecules simultaneously. This orientation reduces the field strength of an applied field significantly. More specifically, the orientation reduces the attractive forces between pairs of anions and cations. As the attractive forces are reduced, the salt can be solubilized. For example, you can show that water significantly reduces the attractive forces between Na^+ and Cl^- (see Example 1.1).

1.4.3 Hydration

Once dissolved, the ions become **_hydrated_**. In other words, the ions form molecules (called *complexes*) with water. Water is adept at hydrating both cations and anions. Why? Recall from Section 1.41 that water has both a partially positive pole on H and partially negative pole on O. Water is said to be very **_polar_**. For example, when alum is added to water at low pH, Al^{3+} forms and is quickly hydrated to $Al(H_2O)_6^{3+}$. Such structures are frequently abbreviated. For example, sometimes Cu^{2+} is written to represent $Cu(H_2O)_4^{2+}$.

Even the proton is hydrated. Free H^+ probably does not exist in solution. The free proton is hydrated to form $H(H_2O)^+ = H_3O^+$, $H(H_2O)_2^+ = H_5O_2^+$, and others. For convenience, the collection of hydrated protons is abbreviated as H^+ or H_3O^+.

1.5 PART I ROAD MAP

Part I of this book contains the fundamental concepts that must be mastered before calculating the equilibrium concentrations of chemical species. Readers with a chemistry background limited to freshman chemistry only should study each Section of each chapter and work the problems assigned.

In Chapter 2, concentration units are reviewed. The main take-home lesson from Chapter 2 is the definition of molar concentration units and why we use molar concentration units for equilibrium calculations.

Readers with a strong chemistry background may wish to focus on Sections 2.3, 2.4, 2.6, and 2.7.

Chapter 3 develops the thermodynamic basis of equilibrium. It is critical for you to understand the definitions of the main thermodynamic properties in Sections 3.2 through 3.5. A key thermodynamic property, the equilibrium constant, is developed in Section 3.9.

In Chapter 4, the rules for interpreting and manipulating equilibrium expressions are reviewed. The manipulation of equilibrium expressions is a skill that must be honed before attempting equilibrium calculations. Careful review of Chapter 4 and practice with manipulating equilibria is required of all readers of this text.

1.6 SUMMARY

In this chapter, you learned that hydrolysis reactions lead to the formation of a number of chemical species in natural waters. Individual chemical species vary greatly in their importance in natural and engineered systems. Thus, you must learn to calculate the concentrations of *individual* chemical species, rather than just measuring total elemental compositions in water. Chemical species concentrations may be controlled by rates (kinetics) or thermodynamics (equilibrium). The focus is on systems at equilibrium in this text because many aqueous phase reactions are fast and equilibrium models are useful in many applications.

The importance of pH [$= -\log(H^+$ activity)] and pe [$= -\log(e^-$ activity)] were discussed. Both pH and pe are called *master variables*. In many cases, pH controls H^+ transfer (also called proton transfer or acid-base chemistry), and pe controls electron transfer (also called redox chemistry).

Water is a unique chemical species. Because of the charge distribution in its constituent atoms, water participates in weak chemical bonds called hydrogen bonds. As a result, salts are very soluble in water and many species become hydrated (i.e., form compounds with water). A road map to Part I of the text also was provided in this chapter.

1.7 PART I CASE STUDY: CAN METHYLMERCURY BE FORMED CHEMICALLY IN WATER?

1.7.1 Background[†]

Mercury is another example of a pollutant for which knowledge of chemical speciation is very important. Of particular concern is the chemical species methylmercury, CH_3Hg^+. Methylmercury can be formed in the aquatic environment by bacterial transformation of mercury.

[†] Background material was obtained from NRC (2000) and the National Institute for Minamata Disease web site (http://www.nimd.go.jp).

Consumption of methylmercury can lead to severe neurological damage and even death. Infants exposed to methylmercury *in utero* may be born with devastating disabilities. The primary route of exposure of adults to methylmercury is through the consumption of fish. In environments with significant mercury concentrations, methylmercury can accumulate through the food web, leading to high levels in game fish and shellfish.

From the mid-1950s through the 1970s, symptoms of neurological damage were noticed in the population residing near Minamata Bay in southern Japan. The collection of symptoms was named *Minamata disease*. Investigations revealed that mercury from a catalyst used in the manufacture of acetaldehyde was migrating to the bay.

Another tragic example of methylmercury poisoning occurred in the same time period in Iraq. In this case, people were exposed to methylmercury through bread. The bread was made with grain that had been treated with a methylmercury-containing fungicide.

1.7.2 Case study question

We know that methylmercury can be formed by bacterial transformation of mercury. It is important to know whether other sources of methylmercury are significant. Since this text focuses on chemical reactions, it is of interest to ask whether methylmercury can be synthesized in water by non-microbial mechanisms. The question posed in the Part I case study is as follows: *Can methylmercury form in significant concentrations from the chemical reaction of methane and mercury in water?*

1.7.3 Lessons from Chapter 1

At the conclusion of each chapter in Part 1, important chapter concepts will be applied to the case study. In this chapter, you learned that speciation is important. Thus, you can see that the case study question is poorly posed.

Thoughtful Pause

Why is the case study question poorly posed?

The reactive species have not been specified. The case study question can be refined by identifying the species of interest in the chemical synthesis of methylmercury. Although many species can be considered, this case study will focus on the reaction of Hg^{2+} and dissolved methane, $CH_4(aq)$, to form CH_3Hg^+. Recall also that methane can exist in the gas phases as methane gas, $CH_4(g)$. Methane gas may dissolve into water to form $CH_4(aq)$. Thus, from Chapter 1, a refined case study question has been forwarded: Can methylmercury form in significant concentrations from the chemical reaction of dissolved methane and Hg^{2+} in water?

SUMMARY OF KEY IDEAS

- Hydrolysis reactions produce a wealth of dissolved chemical species

- The ability to calculate the concentrations of individual chemical species is critically important in analyzing many environmental problems

- Doses depend on both the required concentration of the target individual species and the chemistry of the water

- The equilibrium state is a useful approximation of many aquatic chemical systems

CHAPTER 2

Concentration Units

2.1 INTRODUCTION

concentration: the quantity of the material per volume (or mass) of the surrounding environment

Use of the correct and consistent units is key to solving problems in the sciences and engineering. Environmental engineers and scientists often are interested in the concentrations of chemical species. Thus, you must decide which of the many possible sets of concentration units makes sense for a given application.

Concentration refers to the quantity of the material per volume or mass of the surrounding environment. There are many measures of the quantity of material in the environment and thus many different concentration units. Some concentration units make chemical sense, whereas others persist because they are perceived to be more convenient or have the inertia of common use behind them.

In this chapter, the rules for tracking units through calculations will be reviewed. Common units of concentration for chemical species in the environment will be discussed in Section 2.2. The common molar, mass, and dimensionless concentration units for dissolved species are reviewed in Sections 2.3, 2.4, and 2.5, respectively. Equivalents units are introduced in Section 2.6. In Section 2.7, a table for interconverting units is presented. Common units for gas phase and solid phase species are investigated in Section 2.8 and 2.9, respectively. In Section 2.10, the concept of activity is introduced.

The focus of this chapter is on *concentration units*. Units are the currency by which amounts of the material and volumes of the surroundings are quantified. In addition to units, you also may choose between several *concentration scales*. Concentration scales have their own standard states and will be discussed more thoroughly in Section 3.7.2. Common concentration scales are molarity, molality, and mole fraction.

2.2 UNITS ANALYSIS

The tracking or accounting system for units is called *units analysis*. The rules of units analysis are simple:

- Addition and subtraction
  Add or subtract only terms with identical units.

- Multiplication, division, and raising to a power

  When multiplying or dividing terms (or raising terms to a power), also multiply or divide (or raised to a power) the units associated with the terms.

- Other functions

  Most other common mathematical functions (e.g., logarithms, exponentiation, and the trigonometric functions) can operate only on terms with no units (i.e., *unitless* or *dimensionless* terms). In other words, the arguments of logarithmic, exponential, and trigonometric functions must be dimensionless.

Units are associated with each value you calculate. However, you can manipulate units independent of values. For example, you can use units alone to find conversion factors between units (see Example 2.1).

2.3 MOLAR CONCENTRATION UNITS

Example 2.1: **Units Conversion**

How do you convert the concentration units of milligrams per liter (mg/L) to the concentration units of moles per liter (mol/L)?

Solution:
To answer this question, you do not require values. In fact, you do not even need to know what the words milligrams and moles mean. Simply treat the units as terms in an equation:

$$(mg/L)(X) = (mol/L)$$

Solving for X: X = mol/mg

Thus, to convert from mg/L to mol/L, **multiply by the number of moles per milligram.**

---

mole (abbr. *mol*): one mole is 6.022×10^{23} atoms, molecules, or ions

2.3.1 Introduction

Environmental chemistry often concerns the *combinations* of chemical species to form other species. Thus, it makes sense to choose concentration units that reflect *the proportions in which chemicals combine.* One logical choice of units is the number of atoms, molecules, or ions of a given substance per volume (or mass) of the system. After all, if one H^+ ion combines with one OH^- ion to form one H_2O molecule, why not express each concentration in units of the number of ions (or molecules) per liter? This choice of units makes great chemical sense. However, it is unwieldy because the numbers of atoms, molecules, and ions are so large. For example, 1 milliliter of water contains about 3×10^{22} molecules of water. (To put this large number in perspective, 3×10^{22} cans of soft drink would fill the world's oceans eight times over.) If you chose the *number per liter* concentration units, then the concentrations of the major ions in seawater are $[Na^+] = 2.82 \times 10^{23}$ ions/L and $[Cl^-] = 3.28 \times 10^{23}$ ions/L; not very practical numbers. (Note: As usual, denote the concentration of A by [A].)

2.3.2 The mole

You can create a more practical system of concentration units by assigning a name to a large, but arbitrary, number of atoms. The name ***mole*** (abbreviation: mol) has been assigned to 6.022×10^{23} atoms, molecules, or ions. The number 6.022×10^{23} is called *Avogadro's number.* The value of Avogadro's number is arbitrary and evolved through a tortuous history (see the *Historical Note* at the end of this chapter). The concentration units based on the mole are called ***molar units***, typically in moles/L (abbreviated M, with mM = millimoles per liter and μM = micromoles per liter).

molar units: concentration units of moles per liter (1 molar = 1 M = 1 mol/L)

Thoughtful Pause

What are the sodium and chloride ion concentrations in seawater in units of moles per liter?

The major ions in seawater in molar units are

$$[Na^+] = (2.82 \times 10^{23} \text{ ions/L})/(6.022 \times 10^{23} \text{ ions/mol})$$
$$= 0.468 \text{ mol/L} = 0.468 \text{ M, and:}$$

$$[Cl^-] = (3.28 \times 10^{23} \text{ ions/L})/(6.022 \times 10^{23} \text{ ions/mol})$$
$$= 0.545 \text{ mol/L} = 0.545 \text{ M}$$

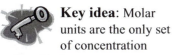

Key idea: Molar units are proportional to the number of atoms, molecules, or ions in a given volume

It is important to remember that molar units are *proportional to the number of atoms, molecules, or ions in a given volume*. The proportionality constant relating the number of atoms (or molecules or ions) to the number of moles is the same for every substance.

Thoughtful Pause

In molar units, what is the constant relating moles to the number of atoms, molecules, or ions and what are the units of this constant?

Key idea: Molar units are the only set of concentration units with the same constant relating the concentration units to the number of atoms, molecules, or ions for all chemical species

In molar units, the constant relating moles to numbers is Avogadro's number (6.022×10^{23} atoms, molecules, or ions per mole). The molar system is the only set of concentration units with the *same constant relating the concentration units to the number of atoms, molecules, or ions for all chemical species*. Another way to express this idea is to take the approach of Example 2.1. For any set of concentration units, you can write (something/L)(X) = (number/L). Molar units (where: something = moles) is the only set of concentration units where X is constant for all chemical species.

To summarize, molar units are the only concentration units with the same constant relating the concentration units to the number of atoms, molecules, or ions for all chemical species. Thus, molar units are especially useful in equilibrium calculations, where combining ratios are critical. You should *always use molar units in equilibrium calculations* for the concentrations of dissolved species.

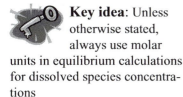

Key idea: Unless otherwise stated, always use molar units in equilibrium calculations for dissolved species concentrations

2.3.3 Molarity

As will be discussed in the *Historical Note*, Avogadro's number was set by defining the mass of 1 mole of the ^{12}C isotope of carbon atoms equal to exactly 12 grams. (The ^{12}C isotope of carbon has 6 neutrons in addition to

atomic weight: the mass of 1 mole of an element

molecular weight: the mass of 1 mole of a molecule or ion

stoichiometric coefficient: here, the number of occurrences of an atom in a molecule or ion

gram molecular weight: molecular weight in grams per mole

***Example 2.2:* Calculation of Molecular Weight**

What is the molecular weight of ferrous ammonium sulfate?

Solution:
The molecular formula is $Fe(NH_4)_2(SO_4)_2$. One mole contains 1 mole of Fe, 2 moles of N, 2 moles of S, 8 moles of O, and 8 moles of H. The molecular weight is

$$
\begin{aligned}
MW = {} & 1(AW_{Fe}) + 2(AW_N) + \\
& 2(AW_S) + 8(AW_O) + \\
& 8(AW_H) \\
= {} & 1(55.85 \text{ g/mol}) + \\
& 2(14.01 \text{ g/mol}) + \\
& 2(32.06 \text{ g/mol}) + \\
& 8(16.00 \text{ g/mol}) + \\
& 8(1.01 \text{ g/mol}) \\
= {} & 284.07 \text{ g/mol}
\end{aligned}
$$

The molecular weight of $Fe(NH_4)_2(SO_4)_2$ is 284.07 g/mol.

---

analytical concentration: number of moles of starting material added per liter of water

6 protons and 6 electrons.) With Avogadro's number fixed, you can use the combining ratios of the elements to calculate the mass of 1 mole of each element. The mass of 1 mole of an element (i.e., the mass of 6.022×10^{23} atoms of an element) is called its **atomic weight**. The **molecular weight** of a molecule is calculated by summing the atomic weights of each atom in the molecule (or ion) multiplied by the number of times the atom occurs in the molecule or ion. The number of times the atom occurs is called its **stoichiometric coefficient** in the molecule or ion. The concept of stoichiometry (from the Greek *stoicheion* element + *metrein* to measure) shall be used throughout this text. The molecular weight[†] of a molecule or ion containing n different atoms is given by $\sum_{i=1}^{n} v_{ij} AW_i$, where v_{ij} = stoichiometric coefficient of atom i in molecule (or ion) j, and AW_i = atomic weight of atom i.

The calculation of molecular weight is illustrated in Example 2.2. In the older literature, the term ***gram molecular weight*** is used to refer to the molecular weight expressed in grams (or, more properly, g/mol). Thus, the statements "The molecular weight of water is 18 g/mol" and "The gram molecular weight of water is 18 g/mol" are identical.

2.3.4 Analytical concentrations and formality

In some applications, molar units may seem inappropriate. For example, how would you make up a 1 M NaCl solution? When NaCl is added to water, it dissociates nearly completely to Na^+ and Cl^- ions (see Section 1.4.2). Thus, although you might add 1 mole (about 58.44 g) of NaCl molecules to 1 liter of water, the actual number of moles of NaCl molecules in solution after a short period of time is quite small. The basic question is: Does a 1 M NaCl solution contain 1 M of NaCl molecules *before* any chemical reactions occur or *after* chemical reactions occur?

The most common terminology in environmental applications is to label solutions by the number of moles per liter of starting material *added*, regardless of the final composition of the mixture. For example, a solution made up by diluting 1 mole of HCl to 1 L of water shall be referred to as a 1 M HCl solution (even though the actual HCl concentration at equilibrium is only about 1×10^{-3} M[††]). The number of moles per liter of starting material added sometimes is called the ***analytical concentration***.

In the older literature, a solution obtained by adding 1 mole of NaCl molecules to 1 L of water was called a 1 formal (or 1 F) NaCl solution. The use of *formal concentration units* (or *formality*) is uncommon in environmental applications and will not be used again in this text.

[†] The term *molecular weight* is commonly used to refer to the mass per mole of ions as well as molecules. More precisely, you can use the term *formula weight* to indicate the mass of any species per mole of that species. The formula weight uses the stoichiometry given in the chemical formula of the species.

[††] After Chapter 7, you will be able to calculate the equilibrium HCl concentration.

2.3.5 Molality

Another potential problem with molar concentrations is that they depend on the temperature and pressure of the system. Why? Molar concentrations depend on temperature and pressure because the *solution volume depends on temperature and pressure.*[†] You can create a temperature- and pressure-independent concentration scale by dividing the number of moles of material by the solvent *mass* rather than the solvent *volume*. The resulting concentration units are called the **molal concentration units**:

molal concentration units: concentration units of moles per kg of solvent (1 molal = 1 m = 1 mol/kg solvent)

> 1 molal (abbreviated 1 m or 1 *m*) = 1 mol/kg of solvent

The difference between the molar and molal scales is small for dilute aqueous solutions near 25°C and 1 atm of pressure. To convert units:

$$\text{mol/L solution} = (\text{mol/kg solvent})(\text{kg solvent/L solution})$$
$$= (\text{mol/kg solvent})(\text{kg solvent/kg solution})(\text{kg solution/L solution})$$

Thus:

$$\text{molarity} = (\text{molality})(\text{kg solvent/kg solution})\rho$$

where ρ = density of the solution in kg/L. For aqueous solutions, the solvent is pure water.

Thoughtful Pause

Why is the difference between molar and molal units small for dilute aqueous solutions at room temperature and pressure?

For *dilute* aqueous solutions, the mass of the dissolved species is small and 1 kg of solution contains very close to 1 kg of water. In addition, near 25°C and 1 atm of pressure, the density of *dilute* solutions is near 1 kg/L. Thus:

$$\text{molarity} = (\text{molality})(\text{kg solvent/kg solution})\rho$$
$$= (\text{molality})(\approx 1 \text{ kg solvent/kg solution})(\approx 1 \text{ kg solution/L solution})$$
$$\approx \text{molality}$$

An example of conversion between molality and molarity is shown in Example 2.3.

Example 2.3: **Conversion of Molal and Molar Units**

What is the molarity of a 1 m NaCl solution (at 25°C and 1 atm)?

Solution:
From the text:
molarity = (molality)(kg solvent/kg solution)ρ

For a 1 m NaCl solution: mass of solvent = 1 kg water, mass of solution = mass of water + mass of solute = 1.05844 kg, and ρ = 1.0405 kg/L (at 25°C and 1 atm).

Thus, 1 M NaCl = (1 m)(1 kg/1.05844 kg)(1.0405 kg/L) = **0.98 M**.

[†] By way of justification, assume for a moment that dilute aqueous species behave like ideal gases. The ideal gas law states that (pressure)(volume) = (constant)(number of moles)(temperature); see Section 2.8. In other words, the molar concentration (= number of moles/volume) is proportional to the ratio of the pressure and temperature. Thus, molar concentrations vary with temperature and pressure.

2.4 MASS CONCENTRATION UNITS

2.4.1 Introduction

The amount of material added to the environment typically is quantified by its mass. As a result, mass concentration units are very common. Typical units are milligrams per liter ($= 10^{-3}$ g/L = mg/L), micrograms per liter ($= 10^{-6}$ g/L = μg/L), and nanograms per liter ($= 10^{-9}$ g/L = ng/L).

Units analysis reveals that you should multiply molar concentration (in mol/L) by the molecular weight (MW, in g/mol) to convert to the mass scale (in g/L) (see also Example 2.1):

$$\text{concentration in g/L} = (\text{MW in g/mol})(\text{concentration in mol/L})$$

The conversion factor relating mass and molar units is the molecular weight. Thus, the conversion factor between mass units and the number of atoms (or moles) is *different* for every substance having a different molecular weight.

This observation leads to two problems with mass units. First, mass concentration units do not give you the combining proportions directly. For example, 1 mole of silver ions (Ag^+) reacts with 1 mole of chloride ions (Cl^-) to form silver chloride. One mole of silver ions also can react with 1 mole of bromide ions (Br^-) to form silver bromide. In molar units, the 1:1 stoichiometry between silver and chloride or silver and bromide is clear. In mass units, a solution containing 1 mg/L of Ag^+ requires 0.33 mg/L of Cl^- or 0.74 mg/L of Br^- to satisfy the 1:1 stoichiometry. Thus, the combining proportion is not obvious when concentrations are expressed in mass units. Another example of conversion between molar stoichiometry and mass units is given in Example 2.4.

Second, mass concentration units are difficult to compare. For example, you may wish to verify that a 1 mg/L solution of nitrite (NO_2^-) contains *more* nitrogen than a 1 mg/L solution of nitrate (NO_3^-) (see Example 2.5).

Key idea: The conversion factor relating mass and molar units is the molecular weight: concentration in g/L = (MW in g/mol)(concentration in mol/L)

Example 2.4: **Stoichiometry in Mass Units (AgI Example)**

What mass concentration of iodide ions is required to react with 1 mg/L silver ions if Ag^+ and I^- react with 1:1 stoichiometry?

Solution:
1 mg/L Ag^+ = 1×10^{-3} g Ag^+/L = $(1 \times 10^{-3}$ g Ag^+/L)/(107.868 g Ag/mol Ag) = 9.27×10^{-6} M Ag^+

This is equivalent to $(9.27 \times 10^{-6}$ mol I^-/L)(126.905 g I^-/mol I^-) = 1.18×10^{-3} g/L or **1.18 mg/L.**

2.4.2 Mass concentrations as other species

To allow for comparison of mass concentration units, the concentrations of related species sometimes are calculated by using the *molecular weight of a common species*. If the common species is Y, you say that the concentration of species X is "1 mg/L as Y" or "1 mg Y/L" or "1 mg X-Y/L." *This notation means that you are using the molecular weight of Y for the molecular weight of X.* For example, the concentration of the 1 mg/L NO_2^- solution discussed above can be expressed in terms of mg/L as N (see Example 2.5). You write that the solution concentration is "0.30 mg/L NO_2^- as N" or "0.30 mg N/L" or "0.30 mg NO_2^--N/L." In doing so, you are using that the molecular weight of atomic nitrogen (14 g/mol) for the molecular weight of nitrate.

***Example 2.5:* Stoichiometry in Mass Units (Nitrogen Example)**

What is the nitrogen content of a 1 mg/L nitrite solution and a 1 mg/L nitrate solution?

Solution:
1 mg/L NO_2^- = [$(1\times10^{-3}$ g NO_2^-/L)/(46 g NO_2^-/mol)](14 g N/mol) = **0.30 mg of N per L**

1 mg/L NO_3^- = [$(1\times10^{-3}$ g NO_3^-/L)/(62 g NO_2^-/mol)](14 g N/mol) = **0.23 mg of N per L**

***Example 2.6:* Mass Concentrations as Another Species**

A rinse water from an electro-plating bath has a chromate (CrO_4^{2-}) concentration of 10 mg/L. What is the chromate concentration expressed as Cr?

Solution:
The bath is
$(10$ mg/L$)(10^{-3}$ g/mg)/(116 g chromate/mol chromate)
$= 8.62\times10^{-5}$ M chromate
Expressed as Cr, the concentration is
$(8.62\times10^{-5}$ mol chromate/L)(1 mol Cr/mol chromate)(52 g Cr/mol)(1000 mg/g)
= **4.48 mg/L as Cr**

Key idea: Use the *mass as* notation to convert the mass concentrations of several species to the same set of units to compare or add them

When expressing the concentration as another species, it is useful to first convert the mass concentration to molar units, then multiply by the molecular weight of the *as* species. For example, suppose the personnel at a wastewater treatment plant wish to determine if the measured effluent ammonium (NH_4^+) concentration of 2.3 mg/L NH_4^+ violates the discharge limit of 1.8 mg NH_4^+-N/L. The ammonium concentration is

$$[NH_4^+] = (2.3 \text{ mg/L } NH_4^+)(10^{-3} \text{ g/mg})/(17 \text{ g } NH_4^+/\text{mol } NH_4^+)$$
$$= 1.4\times10^{-4} \text{ mol/L } NH_4^+$$

This is equivalent to $(1.4\times10^{-4}$ mol/L $NH_4^+)(14$ g N/mol N)(10^3 mg/g) = 2.0 mg NH_4^+-N/L; in violation of the discharge limit. See also Example 2.6.

In performing these calculations, be sure to account for stoichiometry; that is, the number of times the *as* fragment appears in the original substance. For example, suppose 25 mg/L of alum ($Al_2(SO_4)_3\cdot18H_2O$) is added at a drinking water treatment plant. What is the alum dose expressed as Al? The alum dose is:

$$\text{dose} = (25 \text{ mg alum/L})(10^{-3} \text{ g/mg})/(666.43 \text{ g alum/mol alum})$$
$$= 3.75\times10^{-5} \text{ mol alum/L}$$

There are 2 moles of Al per mole of alum. Thus, the dose is:

$$\text{dose} = (2 \text{ mol Al/mol alum})(3.75\times10^{-5} \text{ mol alum/L})$$
$$= 7.50\times10^{-5} \text{ mole Al/L, or:}$$
$$= (7.50\times10^{-5} \text{ mol Al/L})(26.98 \text{ g Al/mol})(10^3 \text{ mg/g})$$
$$= 2.02 \text{ mg/L as Al}$$

Another example is shown in Example 2.7. To summarize the procedure for converting a concentration in mg X/L into mg Y/L:

concentration in mg as Y/L =
(concentration in mg X/L)(mol Y/mol X)(MW Y)/(MW X)

Why use the *mass as* notation? The advantage of expressing mass concentrations as other species is that species concentrations can be compared easily. If a natural water is said to contain 0.5 mg/L NO_3^- and 0.2 mg/L NH_4^+, it is difficult to see at first that ammonium is the larger source of nitrogen. However, when you express the nitrate and ammonium concentrations *in the same units* (i.e., 0.11 mg/L NO_3^- as N and 0.16 mg/L NH_4^+ as N; please verify these values), it becomes clear that ammonium is the larger nitrogen source. In a similar fashion, economists may express currencies in common units (euros or U.S. dollars) to compare prices in different countries.

***Example 2.7:* Mass Concentrations as Another Species with Stoichiometry Correction**

An industrial waste stream contains 5 mg/L of phenol (C_6H_6O). The discharge limit for the industry is 3 mg/L total organic carbon (TOC). TOC is the mass concentration of carbon from organic compounds. Does the waste stream meet the discharge limit?

Solution:
The molecular weight of phenol is:

$$(6 \text{ C})(12 \text{ g/mol C}) + (6 \text{ H})(1 \text{ g/mol H}) + (1 \text{ O})(16 \text{ g/mol O})$$
$$= 94 \text{ g/mol}$$

The phenol concentration is:
$$[\text{phenol}] = (5 \text{ mg/L})(10^{-3} \text{ g/mg})/(94 \text{ g phenol/mol phenol})$$
$$= 5.32 \times 10^{-5} \text{ mol/L}$$

There are 6 moles of C per mole of phenol. Thus, the TOC concentration is:

$$(6 \text{ mol C/mol phenol}) \times (5.32 \times 10^{-5} \text{ M phenol})$$
$$= 3.19 \times 10^{-4} \text{ mole C/L}$$

Or:
$$= (3.19 \times 10^{-4} \text{ mol C/L})(12 \text{ g C/mol})(1000 \text{ mg/g})$$
$$= 3.8 \text{ mg/L as C}$$

The TOC is 3.8 mg/L, exceeding the discharge permit.

 Key idea: When adding species concentrations in mass units, convert all concentrations to the same units by expressing each value as a common species (exceptions: TDS and salinity)

The *mass as* notation is particularly useful for compounds that are measured as a group because it allows you to express the concentration of two different species by using the same mass units. Recall from Section 2.2 that you can add or subtract only terms with the same units. Thus, the *mass as* approach allows you to add and subtract species concentrations when using mass units. For example, you may wish to know how much of the total soluble cyanide in a waste stream is present as the species cyanide (CN^-) and how much is present as cadmium cyanide (assuming for the moment that cyanide and cadmium cyanide are the only forms of cyanide in the waste). Some common cyanide measurement methods measure the *sum* of the two species. Expressing the total soluble cyanide as CN eliminates any confusion introduced by the difference in molecular weights between cyanide and cadmium cyanide. Thus, if the CN^- concentration was 1.1 mg/L as CN and the cadmium cyanide concentration was 4.2 mg/L as CN, the total cyanide concentration can be calculated as 1.1 mg CN/L + 4.2 mg CN/L = 5.3 mg/L as CN.

Before leaving the *mass as* notation, please note that the stoichiometry is not always clear from inspection. As an example of the confusion in stoichiometry, consider the case of chlorine. Species containing *active chlorine* (usually meaning chlorine in the +I oxidation state) often are expressed in units of mg/L as Cl_2. The pertinent reactions for several chlorine-containing species of environmental significance are as follows:

$$Cl_2 + H_2O = HOCl + Cl^- + H^+ \qquad \text{eq. 2.1}$$
$$HOCl = OCl^- + H^+ \qquad \text{eq. 2.2}$$
$$HOCl + NH_3 = NH_2Cl + H_2O \qquad \text{eq. 2.3}$$

It is clear from these reactions that 1 mole of HOCl forms from 1 mole of Cl_2 (eq. 2.1), 1 mole of OCl^- forms from 1 mole of HOCl (eq. 2.2), and 1 mole NH_2Cl forms from 1 mole of HOCl (eq. 2.3). Thus, each mole of HOCl, OCl^-, and NH_2Cl is equivalent to 1 mole of Cl_2. As a result: 1 mg/L Cl_2 as Cl_2 = 1 mg/L HOCl as Cl_2 = 1 mg/L OCl^- as Cl_2 = 1 mg/L NH_2Cl as Cl_2.

2.4.3 Unusual mass concentration units

Having stressed that you can add only terms with the same units, it now must be noted that there are a few common water quality parameters where you add species concentrations expressed in *different* mass units. One example is ***total dissolved solids*** (TDS), operationally defined as the mass of material that passes through a specified filtration operation. The TDS is *measured* gravimetrically (i.e., by weight). It also can be *calculated* by adding the masses of the individual components in the water. This approach violates the rules of units analysis because it requires that you add terms with different units. For example, the TDS of a solution containing 30 mg/L Na^+ and 40 mg/L Cl^- is 70 mg/L. The sodium ion and chloride ion

total dissolved solids (*TDS*): sum of the mass concentrations of species passing through a specified filter

concentrations have different units (mg/L as Na^+ and mg/L as Cl^-, respectively). Nonetheless, you add these terms of different units when calculating the TDS from species concentrations.

Another example in which species concentrations with different mass units are added is *salinity*, S(‰). The salinity of seawater is related to the sum of the masses of dissolved inorganic species per kg of seawater. Thus, *TDS and salinity are exceptions to the general rule that masses are not additive when expressed in different units*.

2.5 DIMENSIONLESS CONCENTRATION UNITS

It is common to use dimensionless units to express the concentrations of chemical species in the environment. Several examples are illustrated below.

mole fraction: concentration units equal to the number of moles of the substance divided by the total number of moles in the system (dimensionless; mole fraction of A = x_A)

2.5.1 Mole fraction

The ***mole fraction*** is the number of moles of the substance divided by the total number of moles in the system. The total number of moles in the system is the sum of the moles of solvent and moles of all dissolved species, where the solvent is pure water in dilute aqueous solutions. The mole fraction (denoted x) is used frequently in chemical engineering and is useful in some thermodynamic derivations (see Chapter 3).

2.5.2 Parts per X

Pollutant concentrations are frequently expressed in parts per some number of parts. Examples are parts per million (ppm), parts per billion (ppb), and parts per trillion (ppt or pptr). Using the example of ppm, the notation 10 ppm refers to 10 parts of something per million parts of something else. The word *parts* can refer to masses or volumes (denoted, for example, ppm_m and ppm_v, respectively) but typically refers to masses for aqueous or solid systems.

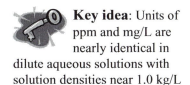 **Key idea**: Units of ppm and mg/L are nearly identical in dilute aqueous solutions with solution densities near 1.0 kg/L

In dilute aqueous systems at environmentally important temperatures and pressures, the density of water is very close to 1.0 kg/L. Thus, 1 liter of water weighs approximately 1 kg or 10^6 mg. A concentration of 1 mg/L means 1 mg of material in 1 liter of water = 1 mg in 10^6 mg = 1 part in 1 million parts = 1 ppm. For dilute aqueous systems near 25°C and 1 atm, the units mg/L and ppm are nearly identical. Similarly, ppb and µg/L are nearly interchangeable in dilute aqueous systems (near 25°C and 1 atm), as are ppt and ng/L.

In the more general case:

ppm = mg/kg solution
    = (mass concentration in mg/L)/(solution density in kg/L)

For example, consider a waste stream consisting of 1 mg of lead in 1 L of water (density = 1.0 kg/L at 25°C, 1 atm). The lead concentration is 1 mg/L or (1 mg Pb/L)/(1.0 kg/L) = 1.0 mg Pb/kg = 1 ppm. If 1 mg of lead was placed in 1 L of a 25% sodium chloride solution (density = 1.19 kg/L at 25°C, 1 atm), the solution still would be 1 mg/L. However, in ppm, the lead concentration is:

$$[\text{Pb in ppm}] = (1 \text{ mg Pb/L solution})/(1.19 \text{ kg solution/L solution})$$
$$= 0.84 \text{ mg Pb/kg solution}$$
$$= 0.84 \text{ ppm.}$$

There are a few caveats to remember when using the parts per something units. First, make sure you know whether the parts are mass or volume. Second, remember that ppm and mg/L (or ppb and µg/L or ppt and ng/L) are nearly identical *only* in dilute solutions when the density of water is very close to 1.0 kg/L.

2.5.3 Percentage
The concentrations of more concentrated solutions are often expressed as a percentage. This dimensionless unit refers to the mass (or volume) of the substance per mass (or volume) of the system. In aqueous solution, the percentage concentration is the mass of the species per mass of solution, expressed as a percentage. Thus:

$$\text{conc. (in \% by mass)} = 10^{-4} (\text{conc. in ppm})$$

For dilute aqueous solutions near 25°C and 1 atm:

$$\text{conc. (in \% by mass)} = 10^{-4} (\text{conc. in ppm})$$
$$\approx 10^{-4} (\text{conc. in mg/L})$$

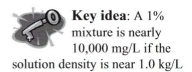 **Key idea**: A 1% mixture is nearly 10,000 mg/L if the solution density is near 1.0 kg/L

In general:

$$\text{conc. (in \% by mass)} = 10^{-4} (\text{conc. in ppm})$$
$$= 10^{-4} (\text{conc. in mg/L})/(\text{density in kg/L})$$

Thus, a 1% sludge stream is very close to 10,000 mg/L solids if its density is near 1.0 kg/L.

2.6 EQUIVALENTS

Another common set of concentration units in aquatic chemistry is equivalents units. The term *equivalents* is confusing at first blush because it appears to carry a different meaning with each use. In fact, converting from molar to equivalents units is trivial. Analogous to Example 2.1, units analysis will reveal how to convert from molar to equivalents units:

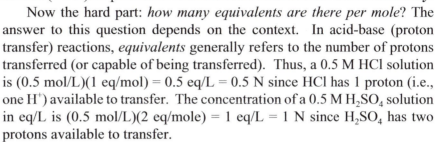

$$\text{equivalents/L} = \text{eq/L} = (\text{conversion factor})(\text{mol/L})$$

normality: concentration units
in equivalents per liter (1
equivalent/L = 1 eq/L = 1
normal = 1 N)

Thus, the conversion factor that multiplies mol/L to get eq/L must have units of equivalents/mol. A solution containing 1 eq/L is said to be 1 normal (= 1 N). Equivalence concentration units also are called *normality*.

Now the hard part: *how many equivalents are there per mole*? The answer to this question depends on the context. In acid-base (proton transfer) reactions, *equivalents* generally refers to the number of protons transferred (or capable of being transferred). Thus, a 0.5 M HCl solution is (0.5 mol/L)(1 eq/mol) = 0.5 eq/L = 0.5 N since HCl has 1 proton (i.e., one H^+) available to transfer. The concentration of a 0.5 M H_2SO_4 solution in eq/L is (0.5 mol/L)(2 eq/mole) = 1 eq/L = 1 N since H_2SO_4 has two protons available to transfer.

Similarly, in redox (electron transfer reactions), *equivalents* generally refers to the number of electrons transferred. Thus, a 0.4 mM O_2 solution is: (0.4 mmol/L)(4 eq/mol) = 1.6 meq/L = 1.6 mN, since each molecule of O_2 can accept four electrons to become two molecules of H_2O (formally: $O_2 + 4e^- + 4H^+ \rightarrow 2H_2O$; more on redox reactions in Chapter 16).

You also can use equivalents to express species concentrations in terms of important fragments. For example, a 1.2×10^{-5} M $Cu(CN)_3^-$ solution made be said to contain $(1.2\times10^{-5}$ mol/L)(3 eq of cyanide/mol) = 3.6×10^{-5} eq/L of cyanide.

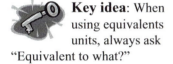

Key idea: When using equivalents units, always ask "Equivalent to what?"

As can been seen by these examples, there are many uses of equivalents units. It is critical to know the basis of comparison. *When using equivalents units, always ask, "Equivalent to what?"*

2.7 REVIEW OF UNITS INTERCONVERSION

The conversion factors for changing from one set of units to another are listed in Table 2.1. Remember that each conversion factor has units.

2.8 COMMON CONCENTRATION UNITS IN THE GAS PHASE

Gas phase concentrations usually are expressed in the partial pressure scale. Thus, the concentration of a gas is defined as its partial pressure, P_i, where P_i = pressure exerted by the gas/total pressure. The most common unit in aquatic chemistry for expressing the concentration of a gas is the *atmosphere* (atm).[†] Other concentration units for trace air pollutants are mass concentration (typically $\mu g/m^3$), parts per X (especially ppm_v = parts by volume per million parts by volume), and percentage units. Gas concentration units are dependent on pressure and temperature.

[†] The formal (Système International d'Unités, or SI) unit of pressure is the pascal (abbreviated Pa), where 1 Pa = 1 N/m^2. One atmosphere is 101,325 Pa or 101.325 kPa. One bar of pressure is 100,000 Pa or 0.9869 atm.

Table 2.2: Interconversion Factors for Concentration Units in Aqueous Systems
[Multiply units in row by table entry to get units in the column.
Example: To convert mol/L to mg/L, multiply by (1000)(molecular weight)]

	Molar (mol/L = M)	Mole fraction (x)	Mass (mg/L)	Mass (mg/L as Y)	Parts per million (ppm$_m$)	Percentage (mass basis)
Molar (mol/L = M)	1	$\dfrac{1}{T}$	$1000(MW)$	$1000(MW_Y)$	$\dfrac{1000(MW)}{\rho}$	$\dfrac{10^{-4}(1000)(MW)}{\rho}$
Mole fraction (x)	T	1	$1000(MW)(T)$	$1000(MW_Y)(T)$	$\dfrac{1000(MW)(T)}{\rho}$	$\dfrac{10^{-4}(1000)(MW)(T)}{\rho}$
Mass (mg/L)	$\dfrac{1}{1000(MW)}$	$\dfrac{1}{1000(MW)(T)}$	1	$\dfrac{MW_Y}{MW}$	$\dfrac{1}{\rho}$	$\dfrac{10^{-4}}{\rho}$
Mass (mg/L as Y)	$\dfrac{1}{1000(MW_Y)}$	$\dfrac{1}{1000(MW_Y)(T)}$	$\dfrac{MW}{MW_Y}$	1	$\dfrac{MW}{(MW_Y)\rho}$	$\dfrac{10^{-4}(MW)}{(MW_Y)\rho}$
Parts per million (ppm$_m$)	$\dfrac{\rho}{1000(MW)}$	$\dfrac{\rho}{1000(MW)(T)}$	ρ	$\dfrac{(\rho)MW_Y}{MW}$	1	10^{-4}
Percentage (mass basis)	$\dfrac{10^4\rho}{1000(MW)}$	$\dfrac{10^4\rho}{1000(MW)(T)}$	$10^4\rho$	$\dfrac{(10^4\rho)MW_Y}{MW}$	10^4	1

Notes: MW = molecular weight of species of interest (g/mol)   MW$_Y$ = molecular weight of Y (g/mol)
        T = total mol/L in solution, including the solvent   The number 1000 has units of mg/g
        ρ = solution density (kg/L)

The concentration unit ppm$_v$ is defined for a gas as:

$$= \dfrac{10^6 \left(\dfrac{\text{moles of species}}{\text{volume}} \right)}{\dfrac{\text{moles of air}}{\text{volume}}}$$

The ideal gas law states that:

$$PV = nRT$$

where P = pressure, V = volume, n = number of moles of gas, R = ideal gas constant, and T = absolute temperature ($^{\circ}$K). Thus: moles of air per unit volume = $n/V = P/RT$. In addition, the number of moles of a species per unit volume is given by:

$$\text{number of moles per volume} = 10^{-6}(\text{conc. in } \mu g/m^3)/MW$$

Substituting:

$$\text{concentration (ppm}_v = \frac{RT}{P}\left(\frac{\text{concentration in } \mu g/m^3}{MW}\right)$$

2.9 COMMON CONCENTRATION UNITS IN THE SOLID PHASE

In solid environments (e.g., soils, sludges, sediments, and other sorbents), common mass concentration units are mg/kg (i.e., milligrams of pollutant per kilogram of solid), parts per X (i.e., parts per million), and percentages. As with aqueous systems, the units of ppm and mg/kg are equivalent: 1 ppm = 1 mg/kg. For equilibrium calculations, the concentration of a pure solid phase is given by the mole fraction.

2.10 ACTIVITY

In the first Section of this chapter, concentration was defined as the quantity of the material per volume or mass of the surrounding environment. Now you might ask: Is mass or molar concentration the *best* measure of how a substance actually *behaves* in the environment? When seeking to describe the behavior of substances in the environment, mass or molar concentration may not be the most appropriate indicator.

To understand the limitations on concentration, consider the problem of determining the behavior of a group of children in a swimming pool. One might start by counting the children. Say there are 20 children in a small pool of volume 15 m^3. Is the number concentration (also called the *bather load*, which equals 20 per 15 m^3 = 1.3×10^{-3}/L) the best measure of the sprightliness of the children?

Thoughtful Pause
What factors (besides the number per volume) affect the liveliness of the children?

The activity of the children may depend on several factors other than their number concentration. For example, the children may be less active if the water temperature is cooler. Also, the activity of the children may change if adults are present in the pool.

Similarly, chemical species may behave differently under different environmental conditions, *even at the same mass or molar concentration*. As with the children in the pool, the behavior of chemical species depends on the water temperature and concentration of other species. In addition, the system pressure may influence the way in which chemical species behave.

As an example, consider three glasses of water, each with the same small amount of salt dissolved in the water. One glass is at room temperature, one at elevated temperature, and one at room temperature with a lot of Epsom salts (a form of $MgSO_4$) also dissolved in the water. The mass (and molar) concentrations of sodium and chloride ions will be the same for each glass. However, the ions will be *more active* in the warmer water and *less active* with the Epsom salts than in the first glass. The ions are less active in the presence of Epsom salts because the sodium ions interact electrostatically with sulfate ions and the chloride ions interact electrostatically with the magnesium ions. This is analogous to the children in the pool: the children may be less active (i.e., less rambunctious) if they are interacting with the adults in the pool.

Scientists have recognized for nearly 100 years that temperature, pressure, and the presence of other chemical species affect the behavior of dissolved chemicals in water. As a result, it has been necessary to develop a new approach in quantifying the amount of material in a system. This approach is an idealized concentration (sometimes called, in the older literature, a *thermodynamic concentration*), which takes into account the effects of temperature, pressure, and the presence of other chemical species on the behavior of dissolved chemicals. The idealized concentration is called **activity**. The activity of species A is denoted as {A}.

Activity is the proper way to express the quantity of species in thermodynamic calculations. As you shall see in Chapter 3, equilibrium is a thermodynamic concept. Thus, you *should* use activity (and *not* concentration) as the sole indicator of species quantity. However, you shall find that activity and concentration are nearly identical in dilute solutions (where the concentrations of other chemical species are small) at near standard temperature and pressure. In other words, the conversion factor relating activity and concentration (called the *activity coefficient*) is nearly 1 in dilute solution near 25°C and 1 atm of pressure.

In this text, the effects of other inert chemical species will be ignored for most of the book, and molar concentration units will be used. In Chapter 3, activity will be used in developing thermodynamic relationships. In Chapter 21, a detailed analysis will be made of the effects of other species on activity.

activity: an idealized concentration, used in thermodynamic calculations, which takes into account the effects of temperature, pressure, and the presence of other chemical species on the behavior of dissolved chemicals

2.11 SUMMARY

Units are key to engineering and science calculations, and concentration units are critical when manipulating chemical species concentrations. In equilibrium calculations, molar units (mol/L = M) are useful because they capture the combining proportions inherent in chemical reactions. In other cases, you may wish to express mass concentrations in the same units so they can be added. Thus, concentrations sometimes are expressed as some other species. In expressing the concentration of species X as Y, use the molecular weight of Y as the molecular weight of X and take into account the number of moles of Y per mole of X. Dimensionless concentration units (mole fraction; parts per million, billion, or trillion; and percentages) also are in common use. For aqueous system with a density near 1 kg/L, 1 part per million (1 ppm) is very close to 1 mg/L.

Another way to use the same units (or currency) when writing concentrations is to use units of normality (or equivalence/L). In this case, always ask yourself: Equivalent to what? In other words, the *basis of the equivalence* (i.e., the number of equivalence per mole) must be established clearly. Common equivalences are the number of H^+ or electrons accepted or donated. Gas phase and solid phase concentration units also were reviewed in this chapter.

Mass or molar concentration units may not be the most appropriate way to describe the *behavior* of chemical species in solution. In particular, temperature, pressure, and the presence of other chemical species are known to affect the behavior of dissolved chemicals. This observation necessitates a new look at concentration and has lead to the development of an *idealized concentration* called *activity*. Activity is the most appropriate choice of expressing the quantity of material in thermodynamic calculations. In this text, the effects of other inert chemicals, temperature, and pressure on the behavior of selected chemical species will be ignored until Chapter 21.

2.12 PART I CASE STUDY: CAN METHYLMERCURY BE FORMED CHEMICALLY IN WATER?

In Chapter 1, four species were identified as being important in determining whether methylmercury can be formed at significant concentrations from the reaction of dissolved methane and Hg^{2+} in water. The four species are CH_3Hg^+, Hg^{2+}, $CH_4(aq)$, and $CH_4(g)$. Based on the lessons of this chapter, it should be possible to select units for the four species in the chemical system.

Thoughtful Pause

What concentration units would you recommend for the species
in the Part I case study?

In reality, any consistent set of concentration units can be used.[†] However, some concentration units make more sense than others. In this case study, the concentration units should capture the combining proportions of Hg^{2+} and $CH_4(aq)$. In addition, the water involved is considered to be very dilute. Thus, the dependencies of solution volume on temperature and pressure can be ignored and molal units are not necessary. As a result, *molar concentration units appear to be appropriate for the dissolved species*. For the gaseous species, *concentration units of atmospheres will be used*. They are the most commonly used gas phase concentration units.

SUMMARY OF KEY IDEAS

- Molar units are proportional to the number of atoms, molecules, or ions in a given volume

- Molar units are the only set of concentration units with the same constant relating the concentration units to the number of atoms, molecules, or ions for all chemical species

- Unless otherwise stated, always use molar units in equilibrium calculations for dissolved species concentrations

- The conversion factor relating mass and molar units is the molecular weight: concentration in g/L = (MW in g/mol)(concentration in mol/L)

- Use the *mass as* notation to convert the mass concentrations of several species to the same set of units to compare or add them

- When adding species concentrations in mass units, convert all concentrations to the same units by expressing each value as a common species (exceptions: TDS and salinity)

- Units of ppm and mg/L are nearly identical in dilute aqueous solutions with solution densities near 1.0 kg/L

[†] For thermodynamic calculations, activities make the most sense for the dissolved species (see Section 2.10). However, in the calculations to be considered in this case study, activities and concentrations shall be assumed to be interchangeable. Thus, suitable concentration units are sought here.

- A 1% mixture is nearly 10,000 mg/L if the solution density is near 1.0 kg/L

- When using equivalents units, always ask: "Equivalent to what?"

HISTORICAL NOTE: AMADEO AVOGADRO AND AVOGADRO'S NUMBER

Amadeo Avogadro

The first half of the nineteenth century saw a great deal of progress in the development of atomic theory. The ancients developed the concept that all matter was built from fundamental, indivisible particles (called *atoms*, from the Greek *a* not + *temnein* to cut). The philosophy of atomism was abandoned by the Greeks by 40 B.C. because the idea of a simple building block was at odds with the apparent complexity of the natural world. A consistent atomic theory was lacking until the work of the influential chemists and physicists of the early 1800s.

John Dalton (1766-1844) contributed greatly to the development of modern atomic theory. He struggled to establish a consistent framework for assigning molecular formulas and atomic weights. The experimental work of Joseph Louis Gay-Lussac (1778-1850) became the basis for determining molecular formulas. His work implied that equal volumes of gases should contain equal numbers of atoms or molecules.[†]

Enter Lorenzo Romano Amadeo Carlo Avogadro, the count of Quaregna and Ceretto (1776-1856). Avogadro allowed for the acceptance of the *equal number for equal volumes* idea (later called *Avogadro's Law*) by proposing in 1811 that some gases were diatomic. Avogadro's ideas were not appreciated until after his death. Stanislao Cannizzaro (1826-1910) fought for the acceptance of Avogadro's ideas at the influential Karlruhe Conference of 1860. He used Avogadro's Law to determine atomic weights.

Avogadro's Law led to the need to express a large number of atoms or molecules by a convenient unit to describe reasonable gas volumes (say, 1 L). As stated in Section 2.3.2, a large number of atoms or molecules (Avogadro's number, or N_A) is defined to be 1 mole of atoms or molecules. Avogadro's number can be calculated[††] by a number of methods. For example, consider a crystal containing one mole of NaCl. The crystal contains $2N_A$ atoms (N_A atoms of Na and N_A atoms of Cl) and has a volume

[†] Dalton could not accept this, in part because he held the belief that all gases were monoatomic. For example, Gay-Lussac found that 1 L of CO combined with 0.5 L of oxygen to form 1 L of CO_2. To conclude that equal volumes of gas contain equal numbers of molecules from this observation, you must accept that oxygen is O_2, not O. Dalton did not think of simple substances as being polyatomic and thus rejected the idea of equal numbers for equal volumes (Pauling, 1970).

[††] It is perhaps more accurate to say that Avogadro's number can be *back-calculated* by the approach shown. After all, the value of $N_A = 6.022 \times 10^{23}$ mol^{-1} was used to establish the molecular weight of NaCl as 58.448 g/mol. If a different value of Avogadro's number had been agreed to, then the molecular weight of NaCl (and every other substance) would be different.

equal to $2N_Ad^3$, where d = interatomic distance. The volume also equals the mass divided by the density (or, for one mole, the molecular weight divided by the density). Combining, you get $N_A = MW/(2\rho d^3)$. For NaCl, MW = 58.448 g/mol, ρ = 2.165 g/cm^3, and d = 2.819×10^{-8} cm. Thus: $N_A \approx$ 6.026×10^{23} mol^{-1}.

Avogadro's number is related to the atomic weight scale. In 1961, the mass of 1 mole of ^{12}C (containing six neutrons, six protons, and six electrons) was defined to be exactly 12 g. This definition set the value of Avogadro's number at 6.022×10^{23} mol^{-1}.

PROBLEMS

2.1 What system of concentration units is most appropriate for following the progress of chemical reactions? Why?

2.2 A drinking water treatment plant adds lime (CaO) at a dose of 200 mg/L as CaCO$_3$. Determine the lime dose in mg/L (as Ca) and molar units.

2.3 In concentrated ferric iron solutions, numerous species may appear as intermediates before iron precipitates as Fe(OH)$_3$(s). What is the total ferric iron concentration in mg/L as Fe if the concentrations of Fe^{3+}, Fe(OH)$^{2+}$, Fe(OH)$_2^+$, and Fe$_2$(OH)$_2^{4+}$ are, respectively, 1.0×10^{-3} M, 8.9×10^{-4} M, 4.9×10^{-7} M, and 1.2×10^{-3} M.

2.4 What is the solids concentration (in mg/L) in a sludge with density 1.1 g/cm^3 and 5.6% solids?

2.5 Seawater typically contains 19.354 g Cl$^-$/kg and 0.0673 g Br$^-$/kg. The *chlorinity*, Cl(‰), of seawater is defined as the number of grams of silver required to precipitate chloride and bromide in 328.5233 g of seawater. Calculate Cl(‰) in typical seawater if 1 mole of silver reacts with 1 mole of chloride or bromide.

2.6 Calculate the number of grams per liter required to make the following solutions:

   A. 0.5 M NaCl (recall from Section 2.3.3 that seawater is about 0.47 M in Na$^+$ and 0.55 M in Cl$^-$)

   B. 0.1 M K$_2$Cr$_2$O$_7$ (a reagent used in the chemical oxygen demand test)

   C. 10^{-2} M Fe(NH$_4$)$_2$(SO$_4$)$_2$ (a reagent used in chlorine measurement)

2.7 Calculate the molarity and normality of the following solutions:

A. 20 g/L NaOH (assume 1 eq/mol)

B. 25 g/L H_2SO_4 (assume 2 eq/mol)

2.8 The major ions in a groundwater sample were found to be as follows: $[Ca^{2+}] = 75$ mg/L, $[Mg^{2+}] = 40$ mg/L, $[Na^+] = 10$ mg/L, $[HCO_3^-] = 300$ mg/L, $[Cl^-] = 10$ mg/L, and $[SO_4^{2-}] = 109$ mg/L (values from Benefield and Morgan, 1990).

A. Calculate the concentration of each ion in molar units and equivalents per liter, where equivalents is defined here as the number of charges per mole.

B. If the water is electrically neutral, the sum of the concentration of cations should equal the sum of the concentration of anions. Is the groundwater electrically neutral?

C. Calculate the TDS of the water sample.

2.9 From eq. 2.3, 1 mole of hypochlorous acid (HOCl) reacts with 1 mole of ammonia (NH_3) to form 1 mole of monochloramine (NH_2Cl). In practice, chlorine concentrations are expressed in mg/L as Cl_2 and ammonia concentrations are expressed in mg/L as N. Determine the HOCl concentration (in mg/L as Cl_2) and ammonia concentration (in mg/L as N) required to form 1 mg/L NH_2Cl as Cl_2.

2.10 The term *hardness* refers to the sum of the concentrations of the divalent cations (species with charge +2) in water. For many waters, the most important divalent cations are Ca^{2+} and Mg^{2+}. Hardness is typically expressed in units of mg/L as $CaCO_3$. In this approach, each mole of Ca^{2+} and Mg^{2+} is equivalent to 1 mole of $CaCO_3$.

A. Calculate the hardness (in mg/L as $CaCO_3$) of a water containing 80 mg/L of Ca^{2+}.

B. Calculate the hardness (in mg/L as $CaCO_3$) of the groundwater described in Problem 2.8.

C. Water with hardness ranging from 150 to 300 mg/L as $CaCO_3$ is considered hard. How much calcium ion (in mg/L as Ca^{2+}) would be required to account for a hardness of 150 mg/L as $CaCO_3$?

CHAPTER 3

Thermodynamic Basis of Equilibrium

3.1 INTRODUCTION

thermodynamics: the study of transformations of energy

3.1.1 Chapter overview

Thermodynamics (from the Greek *therme* heat + *dynamis* power) is the study of transformations of energy. Thermodynamics is a far-reaching and beautifully self-consistent field of study. Thermodynamics seeks to explain phenomena as diverse as high jumping (where kinetic energy is converted into potential energy) and thermonuclear weapons. This text is concerned with reactions between chemical species. Thus, the following discussion of thermodynamics will focus on chemical reactions. This is a slice of thermodynamics sometimes called *chemical thermodynamics*.

The intent of this chapter is to show that the concept of equilibrium has thermodynamic underpinnings. To accomplish this task, we must start with some basic thermodynamic principles and definitions. The thermodynamic functions Gibbs free energy and chemical potential will be introduced, along with their dependence on species concentrations. Equilibrium will be defined from these thermodynamic functions. Thus, you shall find that the equilibrium state of a chemical system is reflected in the concentrations of the species comprised in the system. The concept of an equilibrium constant will be developed. Further thermodynamic concepts will be introduced in Chapter 21.

3.1.2 Scope

As stated in section 3.1.1, the goal of this chapter is to develop expressions for the equilibrium state of a system based on thermodynamic concepts and the concentrations of chemical species. There are many ways to derive equilibrium expressions from thermodynamic laws. Which thermodynamic concepts should be employed?

In this chapter, the discussion will be guided by three principles. First, thermodynamic properties will be developed that are *convenient*. In some cases, this may mean combining simple properties into groupings to reach the chapter goal in an expedient fashion. Second, thermodynamic properties will be sought that are *reusable*. This means that we will take advantage of tabulated thermodynamic properties. Third, thermodynamic properties will be developed which describe the *system* of interest. Thus, the focus will be on the properties of the *system*, not properties of the matter surrounding the system.

3.1.3 Motivation

Before diving into definitions and calculations, it is informative to step back and decide if a discussion of thermodynamics is really necessary. If your goal (from Chapter 1) is to be able to determine the concentrations of individual chemical species at equilibrium, then thermodynamics is absolutely required.

Thermodynamics will allow you to determine quickly whether certain chemical reactions will proceed as written. More significantly, thermodynamics will empower you to calculate species concentrations under specified conditions. You will find that thermodynamic properties of chemical reactions will let you determine, for example:

- The extent to which ammonia partitions between the atmosphere and a lake

- The effect of acid rain on national monuments

- The dose of soda ash required to neutralize an acid spill

- The effect of pH on the disinfection strength of chlorine

- The speciation of phosphorous in a drinking water reservoir

In short, thermodynamics is the key to many applications of aquatic chemistry.

3.2 THERMODYNAMIC PROPERTIES

3.2.1 Introduction

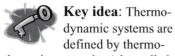

Key idea: Thermodynamic systems are defined by thermodynamic properties (also called *state variables*)

The state of a system can be defined by a number of *thermodynamic properties*. Thermodynamic properties sometimes are called *state variables* because their values depend on the state of the system and not on the manner in which the state was achieved. Thermodynamic properties are divided into two types: extensive and intensive properties. Property types are discussed in Sections 3.2.2 and 3.2.3. Some thermodynamic properties are *conserved*; that is, they are additive even after transformations of a system or when two systems are combined. Conservation of thermodynamic properties is discussed in Section 3.2.4. Finally, since the state of a system is defined by thermodynamic properties, it is useful to ask: how many thermodynamic properties does it take to define a system? This question is answered in Section 3.2.5.

3.2.2 Extensive properties

extensive property: a thermodynamic property dependent on
the amount of material in the
system

As stated in Section 3.2.1, thermodynamic properties can be extensive
properties or intensive properties. ***Extensive properties*** are *dependent on
the amount of material in the system*. For example, mass is an extensive
property because it is dependent on the amount of the substance present.
Examples of extensive properties are listed with their common units in
Table 3.1.

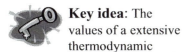

Key idea: The
values of a extensive
thermodynamic
property for each portion of a
system add up to the value of
the extensive thermodynamic
property for the whole system

The extensive thermodynamic properties share a unique feature: the
sum of the values of an extensive property for each portion of the system
equals the value of the extensive property for the system as a whole. For
example, if you empty a glass of water one drop at a time, the sum of the
volumes (or masses) of each drop will be equal to the volume (or mass) of
the original water in the glass.[†]

Table 3.1: Extensive and Intensive Properties
(common units in parentheses)

Type	Extensive Property	Intensive Property
Mass or mole	mass (g or kg)	concentration (mg/L) and density (kg/L)
	number of moles	concentration (mol/L)
Volume	volume (L)	specific volume (L/kg) and molar volume (L/mol)
Thermal	heat capacity (J/$^{\circ}$K)	specific heat (J/g-$^{\circ}$K)
	energy (kJ)	molar energy (kJ/mol)
	enthalpy (kJ)	molar enthalpy (kJ/mol)
	entropy (kJ/$^{\circ}$K)	molar entropy (kJ/mol-$^{\circ}$K)
	free energy (kJ)	molar free energy (kJ/mol)
Other		pressure (atm)
		temperature ($^{\circ}$K)

3.2.3 Intensive properties

intensive property: a thermodynamic property independent
of the amount of material

Intensive properties are *independent* of the amount of material. In other
words, intensive properties are the same for each packet of material in the
system. Temperature is an example of an intensive property. Concentration
also is an intensive property since concentration is a mass (or number of
moles) per unit volume (or mass). Examples of intensive properties are
listed with their common units in Table 3.1.

[†] This definition of extensive properties is valid for systems having constant
thermodynamic properties at every point in space.

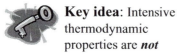 **Key idea**: Intensive thermodynamic properties are *not* additive within a system

---

***Example 3.1:* Balancing Thermodynamic Properties**

What is the sodium ion concentration in a solution formed when one drop (0.05 mL) of seawater ($[Na^+] = 10.8$ g/L) at 25°C is added to 1 L of river water ($[Na^+] = 6$ mg/L) at 25°C?

Solution:
Concentration is not an extensive property and therefore is not additive. Mass is additive and mass = $m = VC$, where V = volume and C = concentration.

$$m_s = m_{sw} + m_{rw}$$
$$= V_{sw}C_{sw} + V_{rw}C_{rw}$$

(The subscripts s, sw, and rw refer here to solution, seawater, and river water, respectively.)

$$m_s = (5 \times 10^{-5} \text{ L})(10.8 \text{ g/L}) +$$
$$(1 \text{ L})(6 \times 10^{-3} \text{ g/L})$$
$$= 6.54 \times 10^{-3} \text{ g}$$

Also: $C_s = m_s/V_s$

Since the density of the final solution is expected to be very close to the density of the river water, you can approximate: $V_s = V_{sw} + V_{rw} = 1 \text{ L} + 5 \times 10^{-5} \text{ L} = 1.00005$ L. Thus:

$$C_s = m_s/V_s$$
$$= (6.54 \times 10^{-3} \text{ g})/(1.00005 \text{ L})$$
$$= \textbf{6.5 mg/L}$$

Some intensive properties are formed from extensive properties by normalizing for the amount of material. In such cases, the intensive property is called a *specific* or *molar* property. For example, the specific volume is calculated as the volume divided by the mass, and the molar energy is the energy per mole. (Related intensive functions, the *partial molar* properties, will be discussed in Section 3.6.3.)

Recall that intensive properties are the same for each packet of material in the system. Thus, unlike extensive properties, *intensive properties are not additive within a system*. For example, the temperature of the water in a glass is *not* equal to the sum of the temperatures of each drop of water in the glass.

3.2.4 Conservation of some thermodynamic properties

Some thermodynamic properties are *conserved*; that is, they are additive even after transformations of a system or when two systems are combined. The general concept of the conservation laws will be explored in Section 3.4.1. In this section, the thermodynamic properties that are candidates for conservation will be identified.

Thoughtful Pause
Which thermodynamic properties are candidates for conservation: extensive or intensive properties?

You know that intensive properties are not conserved. Intensive properties are not additive even *inside* a system, so it does not make sense that they should be additive when systems are transformed or combined. Only the extensive properties are candidates for conservation.

Here you must tread carefully. Although only extensive properties are *candidates* for conservation, it does not follow that *every* extensive property is conserved. Hundreds of years of observation has led to the conclusion (see Section 3.4.1) that the extensive properties mass and energy (along with momentum) are conserved. Thus, you can balance mass, energy, and momentum as systems are transformed or combined.

It is very important to perform balances only on mass, energy, or momentum and *not* on intensive properties. One important implication in environmental engineering and science is that we must perform balances on the mass (or mass flux) of chemicals, not on chemical concentrations. This principle is illustrated in Example 3.1.

3.2.5 How many thermodynamic properties are enough?
Thermodynamic properties are used to define the state of a system. You need to know when to stop adding properties to your description of

systems. In other words, we would like to know how many thermodynamic properties are required to define a system.

The number of thermodynamic properties required to define the state of a system depends on the complexity of the system. More specifically, the number of thermodynamic properties required depends on the number of phases and number of *components* of the system. The concept of the component is used throughout this text. A **component** is a chemical species (or fragment of a chemical species) that can be varied independently in a system. A system with P phases and C components requires $C - P + 2$ thermodynamic properties. This is called the **Gibbs Phase Rule** (after J. Willard Gibbs, 1839-1903; see the *Historical Note* at the end of the chapter). The simplest possible aqueous system is a known volume of water. This is a one-phase, one-component system ($P = C = 1$). Therefore, two thermodynamic properties ($C - P + 2 = 1 - 1 + 2 = 2$) are required to define the system. In aqueous systems, it is convenient to use temperature and pressure as the two thermodynamic properties. Thus, a system of 1 L of pure water is completely and uniquely defined by specifying the temperature and pressure.

Now, thermodynamics would be pretty useless if it told you how to describe systems of only pure water. In fact, the Gibbs Phase Rule is very powerful. It will assist you in describing the equilibrium state of systems. In particular, it will assist in determining the species concentrations at equilibrium. For example, suppose you want to determine the equilibrium concentrations of every species in a system containing P phases and C components. The Gibbs Phase Rule tells you that you need to specify $C - P + 2$ thermodynamic properties from Table 3.1. In most cases, those properties will be temperature, pressure, and $C - P$ species concentrations (or combinations of species concentrations).

component: a chemical species (or species fragment) that can be varied independently in a system

Gibbs Phase Rule: a system with P phases and C components requires $C - P + 2$ thermodynamic properties

3.3 WHY DO WE NEED THERMODYNAMICS TO CALCULATE SPECIES CONCENTRATIONS?

Key idea: Thermodynamics can tell you which chemical reactions are possible (under a given set of conditions) and whether species concentrations are not time-dependent

We seek to determine the concentrations of each chemical species involved in a set of chemical reactions. To accomplish this, two features of chemical reactions must be determined. First, you must have a way of deciding *whether a chemical reaction will proceed as written*. If you can determine that a reaction will ***not*** occur under a given set of conditions (e.g., temperature, pressure, and species concentrations), then clearly it is of little interest to you.

Second, we would like to determine if the chemical species concentrations are changing with time. In particular, the goal for most of this text is confined to calculating species concentrations when they do **not** change with time. The time-dependent nature of species concentrations will be considered in Chapter 22.

How do you know whether a reaction proceeds as written or whether species concentrations are not time-dependent? We shall devise thermodynamic properties specifically for determining whether reactions occur and whether species concentrations change with time. Before developing these thermodynamic properties, it is necessary to examine the concepts of whether reactions occur and the time-independence of species concentrations in more detail. To do so, we must introduce several related concepts, namely spontaneity, equilibrium, and reversibility.

3.3.1 Spontaneity

A reaction that occurs as written without energy added is called a ***spontaneous reaction***. Spontaneous reactions also are called *possible* reactions. In other words, reactions that are *not spontaneous* as written are said to be *not possible under the given set of thermodynamic conditions*. Identifying spontaneous reactions will allow the division of all reactions into two types: possible (i.e., spontaneous) reactions and impossible reactions.

It is important to remember that reactions are possible (spontaneous) or impossible under a *specific set of thermodynamic conditions*. In other words, you must specify the temperature, pressure, and chemical species concentrations when you determine whether or not a reaction is spontaneous. A reaction could occur spontaneously at, say, one temperature, but not occur spontaneously at another temperature. As an example, the evaporation of water to form water vapor is spontaneous at temperatures greater than 100°C (at 1 atm of pressure) but not spontaneous at temperatures less than 100°C (at 1 atm of pressure).

In addition, reactions may occur spontaneously at one set of species concentrations, but not spontaneously at another set of species concentrations. For example, the dissolution of sodium chloride to form sodium and chloride ions is spontaneous at 25°C and 1 atm if sodium chloride is the only source of Na^+ and Cl^-. However, in a saturated brine solution, the dissolution of sodium chloride is not spontaneous.

3.3.2 Equilibrium

Systems in which the species concentrations do not change with time are said to be at ***equilibrium*** (from the Latin *aequi* equal + *libra* weight). The formal thermodynamic definition of equilibrium is that *the thermodynamic properties of the system at equilibrium do not change with time and the system does not have a net loss or gain of heat or mass with its surroundings.*

spontaneous reaction: a reaction occurring as written without energy added

equilibrium: a state where the thermodynamic properties of the system do not change with time (for a system that does not have a net loss or gain of heat or mass with its surroundings)

Thoughtful Pause

Is the formal definition of equilibrium more or less restrictive than the requirement that species concentrations be independent of time?

Note that the formal definition of equilibrium is *very* restrictive. It says that ***no*** thermodynamic properties (not just the property of species concentration) change with time at equilibrium. A system at equilibrium also does not have a net loss or gain of heat or mass with its surroundings. In addition, the chemical composition of the system is independent of time. Why? At equilibrium, all thermodynamic properties (including the species concentrations) are independent of time. Identifying the equilibrium state will allow the division of all spontaneous reactions into two types: reactions at equilibrium and reactions not at equilibrium.

The definitions of spontaneity and equilibrium allow you to divide all chemical reactions into groups. One classification scheme is shown in Figure 3.1. Note that all reactions at equilibrium are spontaneous, but not all spontaneous reactions are at equilibrium.

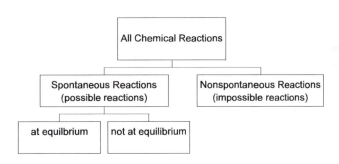

Figure 3.1: Classification of Chemical Reactions
(spontaneous reactions are possible and nonspontaneous reactions are impossible under a specified set of thermodynamic conditions)

3.3.3 Reversibility

The concept of equilibrium is closely aligned with the idea of *reversibility*. A reaction is said to be ***reversible*** if the reverse reaction (i.e., conversion of products into reactants) can occur spontaneously with an infinitesimal increase in the product concentration. Consider the reaction: $H^+ + Cl^- \rightarrow$ HCl. The reaction is reversible if the reverse reaction (HCl $\rightarrow$ $H^+ + Cl^-$) also occurs spontaneously when the HCl concentration is increased slightly. In general, a process is said to be reversible if it returns to its initial state when the mass, heat, and energy flows are reversed.

reversible reaction: a reaction where the reverse reaction (i.e., conversion of products into reactants) can occur spontaneously with an infinitesimal increase in the product concentration

What is the relationship between equilibrium and reversibility? Reversible systems are said to be near equilibrium, whereas irreversible systems are far from the equilibrium state. *For a chemical reaction to be at equilibrium, the reaction must be reversible.*

3.3.4 Summary

The connection between spontaneity, equilibrium, and reversibility casts a new light on equilibrium. *For a reaction to be at equilibrium, the reaction*

Key idea: For a reaction to be at equilibrium, both the reaction and its reverse reaction must be spontaneous

must be both spontaneous and reversible. In other words, both the reaction and its reverse reaction must be spontaneous. For example, it is known that hydrochloric acid equilibrates quickly with H^+ and Cl^-. Thus, both the reactions $HCl \rightarrow H^+ + Cl^-$ and $H^+ + Cl^- \rightarrow HCl$ must be spontaneous. This leads to the expression of equilibrium as reactions that proceed in both directions, denoted $HCl \rightleftharpoons H^+ + Cl^-$, or more commonly in aquatic chemistry: $HCl = H^+ + Cl^-$.

At equilibrium, the concentrations of all chemical species do not change with time. This fact sometimes conjures up the image that no reactions are occurring at equilibrium. However, you now know that all chemical reactions and their reverse reactions proceed spontaneously at equilibrium. Overall, the species concentrations are not changing with time. However, reactants are converted to products and products to reactants continuously at equilibrium in such a way that the reactant and product concentrations do not change over time. In the hydrochloric acid example, HCl is being formed continuously from reaction of H^+ and Cl^-. In addition, HCl is continuously dissociating to form H^+ and Cl^-. The system is not static at equilibrium, but no net change in species concentration occurs.

An analogy may make this clearer. At many college campuses and shopping malls, parking space is at a premium.[†] During the day, the number of empty parking spaces remains nearly constant (at some very small number), and the number of filled parking spaces remains nearly constant as well. In spite of this observation, the actual cars parked in the parking spaces change throughout the day. Parking is reversible. The processes of parking and vacating a parking space (analogous to a reaction and its reverse reaction) are spontaneous. The number of empty and full spaces (analogous to species concentrations) does not change much over time, even though new empty parking spaces are becoming available and being filled continuously. As with chemical systems at equilibrium, the system is not static, but no net change in the number of empty parking spaces occurs.

[†] This leads to the cynical definition of a university as a place where faculty, students, and staff gather together to complain about parking.

3.4 THERMODYNAMIC LAWS

3.4.1 Conservation laws

In the context of thermodynamics, laws are concepts held to be true without derivation or proof. The entire codex of thermodynamics is built upon the assumption of three conservation statements: in any process, mass, energy, and momentum are conserved. (An important exception is nuclear reactions, where mass and energy are interconverted according to Einstein's famous equation: $E = mc^2$.) The conservation laws are logical but unproven. In fact, they are unprovable; you can verify the conservation of mass, energy, and momentum only up to the accuracy of the measurements. The law of conservation of energy is given a special name. It is called the First Law of Thermodynamics and will be explored further in Section 3.4.2.

3.4.2 The First Law of Thermodynamics

First Law of Thermodynamics:
energy is conserved, so any change in the heat of a system or work done by the system must be accounted for in the internal energy of a system

The *First Law of Thermodynamics* states that energy is conserved. Energy may be present in the form of heat or work. Thus, the First Law is written in terms of work and heat. The First Law is a statement that *any energy added to the system* (i.e., any change in the heat of a system or work done by a system) *must be reflected by a change in the internal energy of a system*. In other words:

$$dE = dQ + dW$$

where: dE = differential change in the internal energy (E) of the system, dQ = differential change in the heat absorbed (Q) by the system, and dW = differential work done (W) by the system. In this presentation, *work done by the system is considered to be negative*.

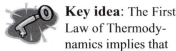 **Key idea**: The First Law of Thermodynamics implies that energy is conserved and that the internal energy changes only through heat exchange or work

Note that the First Law is really two statements in one. First, it is truly a statement that energy is conserved; the difference between energy added to the system (by heat absorbed and work) and energy released from the system (by heat released and work done) must show up as a change in the internal energy of the system. Second, the First Law states that the only ways that the internal energy changes in a system is through heat exchange (heat absorbed or released) and work.

Heat is a fairly common form of energy. The work done by the system is a little more complicated because work can take many forms. For the moment, consider only work done by changes in the pressure or volume of the system. This is called *pressure-volume work* (or *P-V* work).

Thoughtful Pause

How is *P-V* work related to pressure and volume?

By units analysis, work (= force × distance) has units of force units × length units = $(ML/T^2)(L) = ML^2/T^2$, where M, L, and T are units of mass, length, and time, respectively. (Recall that: force = mass × acceleration, so the units of force are ML/T^2.) Pressure is force per unit area (units: $ML/T^2 \times 1/L^2 = M/L\text{-}T^2$) and volume has units of L^3. Hence, by units analysis, work is proportional to the product of pressure and volume. In fact: $W = PV$.

If the process is reversible (in other words, if the process can proceed infinitesimally in either direction), then the pressure change must be zero ($dP = 0$) and:

$$dW = VdP - PdV = -PdV \text{ (constant pressure; } dP = 0\text{)}$$

The sign convention can be confusing, but be assured that it is consistent. If a gaseous system expands, then it does work and therefore dW must be negative. Expansion means $dV > 0$, so the negative sign is needed to insure that dW is negative. Combining the First Law ($dE = dQ + dW$) and the change in work at constant pressure ($dW = -PdV$), you can show:

$$dE = dQ_P - PdV \text{ (constant pressure)} \qquad \text{eq. 3.1}$$

where dQ_P = differential heat absorbed at constant pressure (subscript P indicates constant pressure).

3.4.3 Enthalpy

Equation 3.1 was developed for systems involving P-V work. As a result, eq. 3.1 is very useful for describing changes in internal energy of gaseous systems. However, eq. 3.1 is not very useful for aqueous systems.

Thoughtful Pause

Why is eq. 3.1 not very applicable to aqueous systems?

Equation 3.1 is not very useful for aqueous systems because $dV \approx 0$ for most transformations in aqueous systems. As discussed in Section 3.1.2, perhaps we can combine thermodynamic properties to form a new property that is more convenient to use.

A more convenient way to discuss energy changes for chemical reactions in aqueous systems is the extensive thermodynamic function *enthalpy* (H). Enthalpy is defined as:

enthalpy (H): the sum of the internal energy of the system and the product of pressure and volume ($H = E + PV$)

$$H = E + PV$$

What does enthalpy mean? Enthalpy is the sum of internal energy, E, and work (since $W = PV$). This formal definition probably does not help you understand enthalpy or how to use it. A more meaningful view of enthalpy is found after a little effort. For constant pressure conditions ($dP = 0$):

$$dH = dE + VdP + PdV = dE + PdV \text{ (constant pressure)}$$

Rearranging:

$$dE = dH - PdV \text{ (constant pressure)} \qquad \text{eq. 3.2}$$

Comparing eqs. 3.1 and 3.2:

$$dH = dQ_P \qquad \text{eq. 3.3}$$

Equation 3.3 gives a better physical feeling for enthalpy: *the change in enthalpy is the heat absorbed at constant pressure.*

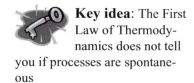

Key idea: The change in enthalpy is the heat absorbed at constant pressure

3.4.4 The Second Law of Thermodynamics
Common experience tells you that the First Law is not adequate to explain all observations of the natural world. In particular, the First Law does ***not*** tell you which processes are *spontaneous*. (Recall that spontaneous processes occur without energy added.) For example, if you place two blocks of identical material at different temperatures in contact with each other, you expect that the temperatures will equilibrate after some time. The First Law does not predict this change in temperature.

Key idea: The First Law of Thermodynamics does not tell you if processes are spontaneous

---

Thoughtful Pause

Two blocks of identical material, each at 20°C, are brought in contact. Does the First Law *prevent* the temperatures of the blocks from changing spontaneously to 10°C and 30°C?

In fact, as demonstrated in the *Thoughtful Pause* above, the First Law even allows for spontaneous events to occur that you know are impossible. Clearly, a new property is needed to account for spontaneous processes.

In 1875, Rudolph Clausius (1822-1888) proposed a thermodynamic function related to spontaneous change called ***entropy*** (from the Greek *en* in + *trepein* to turn, i.e., to direct). Entropy, S, usually is described as a measure of disorder. However, focus for a moment on its original intent as an extensive property required to account for the spontaneity of processes.

entropy (S): a property related to spontaneous change and a measure of disorder

Second Law of Thermodynamics: entropy increases for spontaneous processes and is related to heat exchange ($dS_{total} = dS_{sys} + dS_{surr} \geq 0$ and $dS_{sys} \geq dQ/T$)

The **Second Law of Thermodynamics** describes two properties of entropy. First, entropy increases for spontaneous processes (but does not change for reversible processes). Second, the entropy of a system is related to the change in the heat absorbed by the system. The Second Law can be summarized by:

$$dS_{total} = dS_{sys} + dS_{surr} \geq 0 \text{ (for spontaneous processes)} \qquad \text{eq. 3.4a}$$
$$(dS_{total} = 0 \text{ only for reversible processes})$$

$$dS_{sys} \geq dQ/T \text{ (for spontaneous processes)} \qquad \text{eq. 3.4b}$$
$$(dS_{sys} = dQ/T \text{ only for reversible processes})$$

(Here, the subscripts sys and surr refer to the system and surroundings, respectively.)

At first glance, these properties of entropy appear to be arbitrary. However, it can be shown that defining $dS_{sys} = dQ/T$ for reversible processes leads to the conservation statement that $dS_{total} = dS_{sys} + dS_{surr} = 0$ for reversible processes (see, for example, Eisenberg and Crothers, 1979). Rearranging eq. 3.4b, note that for reversible processes:

$$dQ = TdS_{sys} \text{ (for reversible processes)} \qquad \text{eq. 3.5}$$

3.5 GIBBS FREE ENERGY

Gibbs free energy (G): $G = H - TS_{sys}$

The concept of enthalpy combines internal energy and pressure-volume work. You now know that a more descriptive energy term should include entropy as well to account for reaction spontaneity. The motivation here is to develop a convenient energy term that describes reversibility and spontaneity. Such an extensive property, **Gibbs free energy (G)**, was developed by the ubiquitous J. Willard Gibbs. Gibbs free energy is defined as:

$$G = E + PV - TS = H - TS_{sys} \qquad \text{eq. 3.6}$$

Why is this apparently arbitrary collection of thermodynamic properties so useful? Gibbs free energy is useful for four reasons, as discussed in Sections 3.5.1 through 3.5.4.

3.5.1 Gibbs free energy as the link between the First and Second Laws

The Gibbs free energy combines enthalpy and entropy, two important extensive properties of chemical systems. Enthalpy embodies internal energy and work and is related to heat absorbed by the system (Section 3.4.3). Entropy is related to spontaneity and also has a connection to changes in heat (Section 3.4.4). Thus, Gibbs free energy serves as a link between the First and Second Laws.

3.5.2 Gibbs free energy as a function of temperature and pressure

For aqueous systems, temperature and pressure are natural intensive properties and G can be expressed in terms of temperature and pressure. For reversible processes involving only pressure-volume work (see Example 3.2 for the derivation):

$$dG = VdP - S_{sys}dT \text{ (for reversible } P\text{-}V \text{ work)} \qquad \text{eq. 3.7}$$

3.5.3 Gibbs free energy as a measure of reversibility and spontaneity

Gibbs free energy is a convenient way to determine if a system is reversible. From Example 3.2, for any system involving only P-V work, $dG \leq VdP - S_{sys}dT$. For reversible processes involving only P-V work: $dG = VdP - S_{sys}dT$. At constant temperature and pressure ($dP = dT = 0$):

> $dG \leq 0$ always, and:
> $dG = 0$ for reversible processes

Thus, the Gibbs free energy tells you whether or not reactions are *reversible*. The test: $dG \leq 0$ for all reactions but $dG = 0$ only for reversible processes.

The property dG also is a measure of whether a reaction is *spontaneous*. You can show that for spontaneous reactions at constant temperature and pressure, $dG \leq 0$. A derivation of this useful criterion for spontaneity is given in Example 3.3.

3.5.4 Advantages of Gibbs free energy compared to entropy

Why bother relating dG to reversibility and spontaneity when we already know that entropy itself is related to reversibility and spontaneity? After all, we know already that $dQ = TdS_{sys}$ for reversible processes (eq. 3.5) and $dQ \leq TdS_{sys}$ for a spontaneous process (from eq. 3.4b).

The condition $dG = 0$ is a much more convenient measure of reversibility for many chemical systems than $dQ = TdS_{sys}$. Similarly, $dG < 0$ is a more convenient measure of spontaneity than $dQ \leq TdS_{sys}$. Why? The term dS_{sys} includes contributions from the surroundings to the system, not just internal changes to the system entropy from chemical reaction. On the other hand, dG can be calculated by examining the chemical reactions of the system only. Nonetheless, both Gibbs free energy and entropy can be related to reversibility and spontaneity. The thermodynamic conditions required for reaction reversibility and spontaneity are reviewed in Table 3.2.

Example 3.2: **Gibbs Free Energy as a Function of T and P**

Derive: $dG = VdP - S_{sys}dT$

Solution:
From eq. 3.6:
$$dG = dH - TdS_{sys} - S_{sys}dT$$

Or, from the definition of H:
$$dG = (dE + PdV + VdP) - TdS_{sys} - S_{sys}dT \text{ (*)}$$

Now: $dE = dQ + dW$

For only reversible P-V work ($dP = 0$), $dQ = TdS_{sys}$ (eq. 3.5) and $dW = -PdV$. So:

$$dE = TdS_{sys} - PdV \text{ (**)}$$

Substituting (**) into (*):

$$\mathbf{dG = VdP - S_{sys}dT \text{ (for reversible } P\text{-}V \text{ work)}}$$

Note: For irreversible systems, $dQ < TdS_{sys}$. Thus: $dG < VdP - S_{sys}dT$

 Key idea: Gibbs free energy is a measure of reversibility: $dG \leq 0$ for all reactions, but $dG = 0$ only for reversible processes

 Key idea: Gibbs free energy is a measure of spontaneity: $dG \leq 0$ for spontaneous reactions at constant T and P

3.6 PROPERTIES OF THERMODYNAMIC FUNCTIONS

 Key idea: Thermodynamic functions are independent of path

3.6.1 Path independence

Thermodynamic functions have a number of important characteristics. First, as discussed in Section 3.2.1, they are *path-independent*. This means that the value of a thermodynamic function in a given thermodynamic state does not depend on the process taken to reach that state. As an example, consider potential energy. A ball of mass m located at a distance h above the floor has potential energy equal to mgh (where g = gravitational acceleration). The potential energy is mgh regardless of how the ball got to the height h. The potential energy at height h is the same if the ball was held at height h, bounced to height h, or dropped from some higher height to height h. In other words, the potential energy is *independent of the path* used to achieve the state of the system when the ball is at the height h.

***Example 3.3:* Gibbs Free Energy as a Measure of Spontaneity**

How is G a measure of reaction spontaneity?

Solution:

Imagine a system which exchanges only heat at constant pressure with the surroundings (i.e., $dS_{surr} = -dQ_p/T$).

From eq. 3.3: $dS_{surr} = -dH/T$
Thus:
$$dS_{total} = dS_{sys} + dS_{surr}$$
$$= dS_{sys} - dH/T \geq 0$$

Or: $TdS_{total} = TdS_{sys} - dH \geq 0$ (*)

From the definition of G:
$$dG = dH - TdS_{sys} - SdT$$
$$= dH - TdS_{sys}$$
$$(T, P \text{ constant})$$

Recall that G, H, and S_{sys} are values of the system

So: $TdS_{sys} - dH = -dG$ (**)

Comparing (*) and (**):

$$TdS_{total} = -dG$$

Thus, the usual condition for spontaneity ($dS_{total} \geq 0$) is equivalent to $-dG \geq 0$ (since T is always ≥ 0).

Table 3.2: Conditions of Reversibility and Spontaneity

Function	Reversible	Irreversible	Spontaneous	Impossible
		Reaction Type:		
dS_{total} $(= dS_{sys} + dS_{surr})$	0	>0	≥ 0	<0
dS_{sys}	dQ/T	$>dQ/T$	$\geq dQ/T$	$<dQ/T$
$dG_{T,P}$	0	<0	≤ 0	>0

Why should you care that thermodynamic functions are path-independent? The property of path independence greatly simplifies the manipulation of thermodynamic functions. It allows you to consider only the starting state and ending state of a system. For chemical reactions, where reactants react to form products, you can focus only on the reactants and products without worrying about *how* reactants are converted to products.[†]

3.6.2 Additivity

Second, extensive thermodynamic functions are *additive within the system* (see Section 3.2.2). Additivity means that the sum of the E, H, S, or G for each component of a system should add up to the corresponding thermodynamic function for the system as a whole.

[†]The process of how reactants are converted to products, called the reaction mechanism, is related to chemical kinetics (see Chapter 22).

For spontaneous reactions at constant T and P, $TdS_{total} \geq 0$ (eq. 3.4b). Thus: $dG \leq 0$.

Example 3.4: **Thermodynamic Properties at Standard State**

Which thermodynamic properties are equal to zero at standard state for $O_2(g)$, $O_2(l)$, $O_2(aq)$, $Hg(l)$, and $Hg(g)$?

Solution:
At standard state (25°C and 1 atm), oxygen is most stable as a gas and mercury is most stable as a liquid. Thus: $H^o = G^o = 0$ **for $O_2(g)$ and $Hg(l)$**. Note that $S^o \neq 0$ for each, since the species are not perfectly ordered.

Third Law of Thermodynamics: entropy of a perfectly ordered material approaches zero as the temperature approaches 0°K

A corollary of path-independence and additivity is that the total change in an extensive property for a process consisting of multiple steps is the sum of the changes in the property for each step. For enthalpy, this concept is expressed by *Hess's Law of Heat Summation* (after Germain Henri Hess, 1802-50), which states that the total enthalpy change is equal to the sum of the enthalpy changes of each step.

How can you use rules like Hess's Law of Heat Summation to your advantage? It is possible to take a chemical reaction (i.e., reactants → products) and convert it into any number of convenient steps. We shall find it convenient to break reactions into two steps: (1) conversion of reactants to a state in which energy levels are zero, and (2) conversion from the zero energy state to products. To accomplish this trickery, it is necessary to define zero levels.

3.6.3 Defining zero levels
Before investigating how to add thermodynamic functions, we must define zero levels for the thermodynamic functions. Recall that all energy scales are measured relative to some zero level. In the ball example of Section 3.6.1, the zero level of potential energy was defined to be the floor.

Values of zero enthalpy and zero Gibbs free energy are assigned to elements in their *standard states*. An element in its standard state is an element at a standard set of temperature and pressure conditions. Typically, the standard state temperature and pressure are 25°C and 1 atm, respectively.

Define $H = 0$ and $G = 0$ for each element in its standard state when the element is in the *physical state* (i.e., gas, liquid, or solid) at which it is most stable under the standard conditions of temperature, pressure, and concentration. Thermodynamic functions at standard state are indicated with a superscript o; thus: $H^o = G^o = 0$ for elements in their most stable physical state at standard state. Some standard states are given in Example 3.4.

For entropy, it is not necessary to define an arbitrary zero level. The entropy (here best thought of as the degree of disorder) approaches zero for a perfectly ordered material (such as a perfect crystal) as the temperature approaches 0°K. Thus: $S \to 0$ as $T \to 0$ for perfectly ordered materials. This statement is sometimes called the *Third Law of Thermodynamics*. Note that unlike H^o and G^o, S^o is *not* equal to zero for elements in their standard state. Rather, S^o is nearly zero for well-ordered materials. For example, S^o is only 2.5 J/mol-°K for carbon as diamond at 25°C.

3.6.4 Intensive equivalents of extensive properties
It will be convenient later to work with the intensive equivalents of enthalpy and Gibbs free energy. In other words, we would like to normalize extensive properties to the number of moles of material.

There are two reasonable ways to normalize thermodynamic properties to the number of moles of material. The simplest approach would be to divide the thermodynamic property by the number of moles. The resulting thermodynamic functions are called *molar properties*. Examples include molar enthalpy and molar entropy (see Table 3.1). Molar properties are simply the property divided by the number of moles of the species and are usually denoted by an underbar. For example, the molar enthalpy, $\underline{H}$, is H/n, where n = number of moles of a chemical species. Molar properties are used in pure systems consisting of one species (e.g., pure water or pure toluene).

partial molar enthalpy $\left(\overline{H}_i\right)$:

$\overline{H}_i = \dfrac{\partial H}{\partial n_i}$ (at constant T, P, and $n_{j \neq i}$)

In environmental application, mixtures of species are much more important. For mixtures, the appropriate intensive equivalents of the extensive properties are called *partial molar properties* and are indicated by an overbar. Define: $\overline{H}_i = \dfrac{\partial H}{\partial n_i}$ and $\overline{G}_i = \dfrac{\partial G}{\partial n_i}$, where: $\overline{H}_i$ = **partial molar enthalpy of species *i*, $\overline{G}_i$ = partial molar Gibbs free energy of species *i*,** n_i = number of moles of species *i*, and the partial derivatives are taken while holding constant the temperature, pressure, and the number of moles of all other species in the system. With the notation introduced in eq. 3.1:

partial molar Gibbs free energy

$\left(\overline{G}_i\right): \overline{G}_i = \dfrac{\partial G}{\partial n_i}$ (at constant T, P, and $n_{j \neq i}$)

$$\overline{H}_i = \left(\frac{\partial H}{\partial n_i}\right)_{T,P,n_{j \neq i}} \quad \text{and} \quad \overline{G}_i = \left(\frac{\partial G}{\partial n_i}\right)_{T,P,n_{j \neq i}}$$

Note that $\overline{H}_i = \overline{G}_i = 0$ for elements at standard state since $H^o = G^o = 0$ for elements at standard state.

3.6.5 Chemical potential

The partial molar Gibbs free energy, $\overline{G}_i$, is given a special name. It is called the ***chemical potential***. The chemical potential of species *i* is denoted by

μ_i: $\mu_i = \overline{G}_i = \dfrac{\partial G}{\partial n_i}$ (at constant T, P, and $n_{j \neq i}$).

chemical potential (μ_i): $\mu_i = \overline{G}_i = \dfrac{\partial G}{\partial n_i}$ (at constant T, P, and $n_{j \neq i}$)

Why introduce a new name for the partial molar Gibbs free energy? The term *chemical potential* is very descriptive. The name emphasizes that *differences in Gibbs free energy drive chemical systems to do work*.

To see the link between chemical potential and work, return to the concept of work. Work takes many forms, all of which are responses to driving forces. By using the concepts of intensive and extensive variables, all work can be described as the integral of an intensive variable multiplied by the differential of an extensive variable. For example, pressure-volume work is $\int P dV$, where pressure is the driving force for accomplishing work. Since $dE = P dV$ in the case of no heat absorption (see eq. 3.1), you could express pressure-volume work as $\int \frac{\partial E}{\partial V} dV$, where E = internal energy. Work in a gravitational field is $\int mgdh$, where g = gravitational acceleration, m = mass, and h = height in the gravitational field. This work could

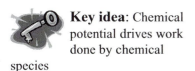

Key idea: Chemical potential drives work done by chemical species

be written as $\int \frac{\partial(PE)}{\partial h} dh$, where: PE = potential energy. Similarly, electrical work is $\int$(potential difference)(dq), where q = charge and the potential difference is the driving force. By analogy, the work done by a chemical species is equal to $\int \frac{\partial G}{\partial n_i} dn_i = \int \mu_i dn_i = \int$(chemical potential)$dn_i$.

Chemical potential drives work done by chemical species.

3.7 CHANGES IN THERMODYNAMIC PROPERTIES DURING CHEMICAL REACTIONS

3.7.1 Thermodynamic properties and chemical reactions

As thermodynamic systems change (e.g., ice melts or two chemical species react to form a third species), the thermodynamic properties may change. How can you quantify the changes in thermodynamic functions such as H, S, and G (and their partial molar values) as the system moves from one thermodynamic state to another? For the change from state 1 to state 2, it is logical to define: $\Delta H = H_{\text{state 2}} - H_{\text{state 1}}$ (with analogous definitions of ΔS and ΔG). However, the problem of calculating $H_{\text{state 2}}$, $H_{\text{state 1}}$, $S_{\text{state 2}}$, and so on remains.

To aid in the calculation, recall from Section 3.6.2 that the pathway of moving from state 1 to state 2 does not affect the values of state variables such as H, S, and G. Thus, rewrite the process:

$$\text{state 1} \rightarrow \text{state 2}$$

as:

$$\text{state 1} \rightarrow \text{standard state} \rightarrow \text{state 2} \qquad \text{eq. 3.8}$$

Recalling additivity and Hess's law of heat summation, you can show that for the process in eq. 3.8:

$$\Delta H = \Delta H_{\text{state 1} \rightarrow \text{ss}} + \Delta H_{\text{ss} \rightarrow \text{state 2}} \qquad \text{eq. 3.9}$$

where ss = standard state.

This text is concerned primarily with the reactions of chemical species; that is, with the transformation of reactants to products. Thus, we wish to calculate changes in thermodynamic properties during chemical reactions. Clearly, the definition of ΔH as $\Delta H = H_{\text{state 2}} - H_{\text{state 1}}$ must be expanded. In addition, you will find it useful to define ΔH (and ΔS and ΔG) for reactions as being *intensive* properties. It is logical to define:

$$\Delta H_{rxn} = \sum_{products} v_i \overline{H}_i - \sum_{reactants} v_i \overline{H}_i$$

reaction stoichiometric coefficient: a dimensionless number representing the number of moles of a species lost or gained per mole of reaction (> 0 for products and < 0 for reactants)

where v_i = ***reaction stoichiometric coefficient*** of species i. The functions ΔS_{rxn} and ΔG_{rxn} can be defined in a similar fashion. If we adopt the common convention of treating *the stoichiometric coefficients of reactants as always negative and stoichiometric coefficients of products as always positive*, then:

$$\Delta H_{rxn} = \sum v_i \overline{H}_i$$
$$\Delta S_{rxn} = \sum v_i \overline{S}_i$$
$$\Delta G_{rxn} = \sum v_i \overline{G}_i = \sum v_i \mu_i$$

The left-hand terms are called the enthalpy (or entropy or Gibbs free energy) of reaction.

A word about units: the stoichiometric coefficients, v_i, are dimensionless. Stoichiometric coefficients represent a sort of normalized number of moles: the units of v_i are moles of species i gained or lost per mole of reaction.[†] As a result, ΔH_{rxn}, ΔS_{rxn}, and ΔG_{rxn} have the same units as their partial molar quantities, typically kJ/mol; J/$^\circ$K-mol; and kJ/mol for ΔH_{rxn}, ΔS_{rxn}, and ΔG_{rxn}, respectively.

3.7.2 Calculation of ΔH_{rxn}, ΔS_{rxn}, and ΔG_{rxn}

To calculate changes in thermodynamic properties during a reaction, write two new reactions in which reactants are converted to elements at their standard states first and then the elements are transformed into products. (Remember that such manipulation is allowed because the values of the thermodynamic properties are path-independent.)

As an example, consider the reaction of aqueous silver ions and aqueous chloride ions to form solid silver chloride: $Ag^+(aq) + Cl^-(aq) \rightarrow AgCl(s)$. Using the approach of Section 3.7.1, divide this reaction into three reactions: silver ion $\rightarrow$ silver in its standard state, chloride ion $\rightarrow$ chlorine in its standard state, and silver and chlorine in their standard states $\rightarrow$ solid silver chloride. The standard state of silver is solid silver, and the standard state of chlorine is gaseous chlorine. Thus, we write the following three reactions:

[†] Formally: $dn_i = v_i d\xi$, where: ξ is the extent of the reaction. Thus, a reactant with $v_i = -2$ loses two moles per mole of reaction and a product with $v_i = +1$ gains one mole per mole of reaction.

$$Ag^+(aq) \rightarrow Ag(s) \qquad\qquad\qquad \text{reaction 1}$$
$$Cl^-(aq) \rightarrow \tfrac{1}{2}Cl_2(g) \qquad\qquad\qquad \text{reaction 2}$$
$$Ag(s) + \tfrac{1}{2}Cl_2(g) \rightarrow AgCl(s) \qquad\qquad \text{reaction 3}$$

What is the advantage of breaking up the reaction into three reactions? Reaction 3 represents the formation of a compound, AgCl(s), from elements in their standard states. Reactions 1 and 2 represent the reverse of the formation of two ions from elements in their standard states. Thus, by breaking up the reaction, you have three reactions which relate chemical species to elements in their standard states.

Why the focus on elements in their standard state? You know from Section 3.6.3 that the partial molar enthalpy, entropy, and Gibbs free energy are fixed for elements in their standard state at a given temperature and pressure (and equal to zero for $\overline{H}_i$ and $\overline{G}_i$). The thermodynamic properties for the formation of chemical species from elements at their standard states have been tabulated. These properties are called the partial molar enthalpy of formation ($\overline{H}_{f,i}$), partial molar entropy of formation ($\overline{S}_{f,i}$), and partial molar enthalpy of formation ($\overline{G}_{f,i}$). The subscript f indicates formation.

Returning to the example of sodium chloride formation, you can write the thermodynamic properties of the *reaction* in terms of the partial molar thermodynamic properties of *formation*. For enthalpy:

$$\Delta H_{rxn} = \Delta H_{rxn1} + \Delta H_{rxn2} + \Delta H_{rxn3}$$
$$= \left(-\overline{H}_{Ag^+(aq)} + \overline{H}_{Ag(s)}\right) + \left(-\overline{H}_{Cl^-(aq)} + \tfrac{1}{2}\overline{H}_{Cl_2(g)}\right)$$
$$+ \left(-\overline{H}_{Ag(s)} - \tfrac{1}{2}\overline{H}_{Cl_2(g)} + \overline{H}_{AgCl(s)}\right)$$
$$= -\overline{H}_{f,Ag^+(aq)} - \overline{H}_{f,Cl^-(aq)} + \overline{H}_{f,AgCl(s)}$$

Evaluating under standard-state conditions (usually 25°C and 1 atm):

$$\Delta H^o_{rxn} = \overline{H}^o_{f,AgCl(s)} - \left(\overline{H}^o_{f,Ag^+(aq)} - \overline{H}^o_{f,Cl^-(aq)}\right) \qquad \text{eq. 3.10}$$

The thermodynamic function ΔH^o_{rxn} is called the ***standard enthalpy of reaction***. Note that it is a partial molar quantity and thus has units of kJ/mol. In general, for the reaction: aA + bB $\rightarrow$ cC + dD:

standard enthalpy of reaction
(ΔH^o_{rxn}): $\Delta H^o_{rxn} = \sum v_i \overline{H}^o_{f,i}$

$$\Delta H^o_{rxn} = \sum v_i \overline{H}^o_{f,i} = c\overline{H}^o_{f,C} + d\overline{H}^o_{f,D} - a\overline{H}^o_{f,A} - b\overline{H}^o_{f,B}$$

standard Gibbs free energy of reaction (ΔG^o_{rxn}): ΔG^o_{rxn}
$= \sum v_i \overline{G}^o_{f,i}$

standard entropy of reaction (ΔS^o_{rxn}): $\Delta S^o_{rxn} = \sum v_i \overline{S}^o_{f,i}$

___

***Example 3.5:* Calculation of ΔH^o_{rxn}, ΔS^o_{rxn}, and ΔG^o_{rxn}**

Calculate ΔH^o_{rxn}, ΔS^o_{rxn}, and ΔG^o_{rxn} for the reaction:
$CO_2(g) + H_2O(l) \rightarrow H_2CO_3(aq)$

Solution:
The pertinent thermodynamic data are:

Species	$\Delta \overline{H}_{f,i}$	$\Delta \overline{S}_{f,i}$	$\Delta \overline{G}_{f,i}$
$CO_2(g)$	−393.5	213.6	−394.4
$H_2O(l)$	−285.8	69.9	−237.2
$H_2CO_3(aq)$	−699.6	187.0	−623.2

(Units: kJ/mol, J/°K-mol, and kJ/mol for $\Delta \overline{H}^o_{f,i}$, $\Delta \overline{S}^o_i$, and $\Delta \overline{G}^o_{f,i}$, respectively. All values at 25°C = 298°K)

Here, all $v_i = 1$. Thus:

$\Delta H^o_{rxn} = -699.6 - (-393.5) - (-285.8)$
$\quad = \textbf{−20.3 kJ/mol}$
$\Delta S^o_{rxn} = 187.0 - 213.6 - 69.9$
$\quad = \textbf{−96.5 J/°K-mol}$
$\Delta G^o_{rxn} = -623.2 - (-394.4) - (-237.2)$
$\quad = \textbf{+8.4 kJ/mol}$

You can verify that: $\Delta G^o_{rxn} = \Delta H^o_{rxn} - T\Delta S^o_{rxn}$.

___

A similar definition can be written for $\Delta G^o_{rxn} = $ ***standard Gibbs free energy of reaction*** $= \sum v_i \overline{G}^o_{f,i} = \sum v_i \mu^o_i$ and $\Delta S^o_{rxn} = $ ***standard entropy of reaction*** $= \sum v_i \overline{S}^o_{f,i}$.

3.7.3 Example

The standard enthalpy, entropy, and Gibbs free energy of formation have been tabulated for many chemical species. This greatly facilitates the calculation of the standard enthalpy, entropy, and Gibbs free energy of reaction. Using the language of Section 3.1.2, the standard properties make thermodynamic properties *reusable*. Consider again the example of the formation of silver chloride from aqueous silver ions and aqueous chloride ions. You can calculate ΔH^o_{rxn}, ΔS^o_{rxn}, and ΔG^o_{rxn} as follows. The pertinent standard partial molar properties of formation are listed in Table 3.3.

Table 3.3: Thermodynamic Data for Ag⁺(aq), Cl⁻(aq), and AgCl(s)

Species	$\Delta \overline{H}^o_{f,i}$ (kJ/mol)	$\Delta \overline{S}^o_{f,i}$ (J/mol-°K)	$\Delta \overline{G}^o_{f,i}$ (kJ/mol)
$Ag^+(aq)$	+105.6	+73.4	+77.12
$Cl^-(aq)$	−167.2	+56.5	−131.3
$AgCl(s)$	−127.1	+96	−109.8

Using equations analogous to eq. 3.10:

$\Delta H^o_{rxn} = -127.1 \text{ kJ/mol} - (+105.6 \text{ kJ/mol} - 167.2 \text{ kJ/mol})$
$\quad = -65.5 \text{ kJ/mol}$
$\Delta S^o_{rxn} = 96 \text{ J/mol-°K} - (+73.4 \text{ J/mol-°K} + 56.5 \text{ J/mol-°K})$
$\quad = -33.9 \text{ J/mol-°K}$
$\Delta G^o_{rxn} = -109.8 \text{ kJ/mol} - (+77.12 \text{ kJ/mol} - 131.3 \text{ kJ/mol})$
$\quad = -55.6 \text{ kJ/mol}$

___

Thoughtful Pause
How are ΔH^o_{rxn}, ΔS^o_{rxn}, and ΔG^o_{rxn} related?

___

From the definition of G, you expect: $\Delta G^o_{rxn} = \Delta H^o_{rxn} - T\Delta S^o_{rxn}$. For the formation of silver chloride at 25°C:

$\Delta H^o_{rxn} - T\Delta S^o_{rxn} = -65.5 - (298°K)(-33.9 \text{ J/mol-°K})(10^{-3} \text{ kJ/J})$
$\quad = -55.4 \text{ kJ/mol}$

The value of -55.4 kJ/mol is about equal to ΔG^o_{rxn}. Another example of calculating standard thermodynamic properties of reaction is given in Example 3.5.

3.7.4 Interpretation of ΔH^o_{rxn}, ΔS^o_{rxn}, and ΔG^o_{rxn}

What do ΔH^o_{rxn}, ΔS^o_{rxn}, and ΔG^o_{rxn} mean? For ΔH^o_{rxn}, recall that the change in enthalpy is equal to the change in heat at constant pressure (eq. 3.3). Thus: $\Delta H^o_{rxn} = \Delta Q_P$ under standard-state conditions. When heat is generated by a reaction (i.e., $\Delta Q_P = \Delta H^o_{rxn} < 0$; see Section 3.4.2 for the sign convention for heat), the reaction is called ***exothermic*** (from the Greek *exo*, meaning out or outside). When heat is absorbed during a reaction (i.e., $\Delta Q_P = \Delta H^o_{rxn} > 0$), the reaction is called ***endothermic*** (from the Greek *endon*, meaning in or within).

exothermic reaction: a reaction in which heat is generated

endothermic reaction: a reaction in which heat is absorbed

Thoughtful Pause

Are the reactions in Section 3.7.3 and Example 3.5 exothermic or endothermic?

Note that the reactions in Section 3.7.3 and Example 3.5 are exothermic since $\Delta H^o_{rxn} < 0$.

The sign of ΔS^o_{rxn} tells one whether the degree of disorder is increasing, decreasing, or remaining constant for a reaction. When reactions involve transitions from less ordered phases to more highly ordered states, the degree of disorder decreases and $\Delta S^o_{rxn} < 0$. In predicting the sign of the entropy change, it is important to keep in mind the change in entropy of water. For the sodium chloride example, ΔS^o_{rxn} is negative since order is increased (disorder decreased) when the aqueous species form the crystalline product. For the reaction in Example 3.5, ΔS^o_{rxn} is very negative, reflecting the large increase in order (decrease in disorder) as gaseous carbon dioxide dissolves in water to form carbonic acid (H_2CO_3); the gas is disordered, whereas the dissolved species are constrained by water molecules.

As discussed in Section 3.5, the change in G at constant temperature and pressure is a measure of spontaneity. You expect $\Delta G^o_{rxn} < 0$ for spontaneous reactions and $\Delta G^o_{rxn} > 0$ for reactions that will not occur spontaneously as written under standard-state conditions. The dissolution of carbon dioxide gas is not spontaneous under standard conditions. Note that ΔG^o_{rxn} is positive in this case in spite of the fact that heat is given off ($\Delta H^o_{rxn} < 0$) because of the large decrease in entropy under standard conditions.

3.8 RELATING GIBBS FREE ENERGY TO SPECIES CONCENTRATIONS

3.8.1 Ideal gases and ideal gas mixtures

In addition to the fundamental laws of thermodynamics, the study of heat transformation is governed by a series of empirical formulas called *equations of state*. One common equation of state is the ***ideal gas law***:

$$PV = nRT$$

ideal gas law: $PV = nRT$, where: R = ideal gas constant = 0.0205 L-atm/mol-°K = 8.314 J/mol-°K

where R = ideal gas constant = 0.082057 L-atm/mol-°K = 8.314 J/mol-°K or 8.314×10^{-3} kJ/mol-°K. (The ideal gas constant is equal to Boltzmann's constant divided by Avogadro's number.) The ideal gas law is valid only for gases at low pressure, where gas molecules do not interact. Similarly, in an ***ideal gas mixture***, each gas behaves as an ideal gas and thus does not interact with other gaseous molecules.

ideal gas mixture: a mixture where each constitutive gas behaves as an ideal gas and does not interact with other gases

What is the Gibbs free energy of an ideal gas? From eq. 3.7: $dG = VdP - SdT$ (for reversible P-V work). Thus, at constant temperature: $dG = VdP$.[†] From the ideal gas law: $V = nRT/P$. So: $dG = (nRT/P)dP$. Integrating from $P = P_0$ to P at constant temperature:

$$\int_{P_0}^{P} dG = nRT \int_{P_0}^{P} \frac{dP}{P}$$

Solving:

$$G = G_{P_0} + nRT \ln \frac{P}{P_0} \qquad \text{eq. 3.11}$$

Unfortunately, eq. 3.11 is valid only as $P \to 0$ since the ideal gas law is only valid as $P \to 0$. In mixtures, the concentration of the gas (represented by the partial pressure) will be influenced by other gases.

Physical chemists have circumvented this problem by defining an idealized pressure called the ***fugacity***, f. For species i:

fugacity: an idealized pressure

$$G_i = G_{i,P_0} + n_i RT \ln \frac{f_i}{f_{i,P_0}} \qquad \text{eq. 3.12}$$

If you take the derivative of eq. 3.12 with respect to n_i and recall that $\partial G/\partial n_i = \mu_i$ = chemical potential of species i (Section 3.6.3), then:

[†] More accurately, in a mixture undergoing reversible P-V work at constant temperature: $\bar{V} = \left(\dfrac{\partial \bar{G}}{\partial P} \right)_T$.

$$\mu_i = \mu_{i,P_0} + RT \ln \frac{f_i}{f_{i,P_0}} \qquad \text{eq. 3.13}$$

3.8.2 Application to dissolved species

Equation 3.13 can be applied to dissolved species in water. The term μ_{i,P_0} represents the chemical potential at some standard pressure, P_0. Thus, replace it with μ_i^o, the standard chemical potential. In addition, replace the fugacity (an idealized pressure) with an idealized concentration called *activity* (see Section 2.10 for an introduction to activity). The activity of species i will be denoted $\{i\}$. Thus, replace f_{i,P_0} (the fugacity of species i at standard pressure) with $\{i\}_o$ (the activity of species i at standard state). Now, eq. 3.13 becomes:

$$\mu_i = \mu_i^o + RT \ln \frac{\{i\}}{\{i\}_o} \qquad \text{eq. 3.14}$$

The value of μ_i^o and $\{i\}_o$ depend on the concentration scale. For most of this text, the molarity scale will be used for the concentrations of dissolved species. In this scale, we define $\{i\}_o = 1$ M. Thus:

$$\mu_i = \mu_i^o + RT \ln \{i\} \qquad \text{eq. 3.15}$$

At first glance, eq. 3.15 may appear troublesome. It looks as if we are trying to take the natural logarithm of a term that has units. Of course, the ln function can operate only on dimensionless terms. However, the term to the right of the ln function actually is dimensionless. It is the activity divided by the activity at standard state (see also eq. 3.14).

3.8.3 Application to chemical reactions

In Section 3.5, it was noted that Gibbs free energy has many important characteristics. Thus, it is desirable to rewrite eq. 3.15 in terms of G rather than μ. Also, it is desirable to apply eq. 3.15 to chemical reactions. From eq. 3.10: $\Delta G_{rxn} = \Sigma v_i \mu_i$. Combining this with eq. 3.15:

$$\begin{aligned}\Delta G_{rxn} &= \Sigma v_i(\mu_i^o + RT\ln\{i\}) \\ &= \Sigma v_i \mu_i^o + \Sigma v_i RT\ln\{i\} \quad \text{or:}\end{aligned}$$

$$\Delta G_{rxn} = \Delta G_{rxn}^o + RT \ln\{i\}^{v_i} \qquad \text{eq. 3.16}$$

where $\Delta G_{rxn} = \sum v_i \mu_i^o = \sum v_i \overline{G}_{f,i}^o$ (see Section 3.7.2).

Equation 3.16 can be applied to chemical reactions. Remember that the reaction stoichiometric coefficients (v_i) are considered to be negative for reactants and positive for products. Thus, for the general reaction aA + bB = cC + dD:

$$\Delta G_{rxn} = \Delta G_{rxn}^{o} + RT \ln\left(\frac{\{C\}^c \{D\}^d}{\{A\}^a \{B\}^b}\right) \qquad \text{eq. 3.17}$$

3.8.4 Minimization of free energy

Key idea: ΔG_{rxn} is related to species activities

As summarized in Table 3.2, dG is less than or equal to zero for spontaneous reactions and less than zero for reversible reactions (at constant temperature and pressure). At equilibrium, the thermodynamic properties (including the activities of the chemical species) do not change with time. Thus, at equilibrium, the reaction must be both spontaneous and reversible, and dG = 0. In other words, *equilibrium corresponds to a minimum in the Gibbs free energy*. For a reaction, this means $\Delta G_{rxn} = 0$ at equilibrium. Once again, the Gibbs free energy is revealed to be a powerful tool to apply to equilibria.

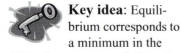

Key idea: Equilibrium corresponds to a minimum in the Gibbs free energy or $\Delta G_{rxn} = 0$ at equilibrium

For simple reactions, it is easy to track G during the course of the reaction and identify the equilibrium state as the system composition that minimizes G. Consider the dissolution of oxygen into water.[†] Imagine bringing 1 L of air (partial pressure of oxygen = P_{O_2} = 0.209 atm) at 25°C into contact with 1 L of water. For the overall system:

$$G = n_{H_2O} \mu_{H_2O} + n_{O_2(aq)} \mu_{O_2(aq)} + n_{O_2(g)} \mu_{O_2(g)}$$

Since we are seeking a minimum in G and the number of moles of water does not change during oxygen dissolution, it makes sense to track:

$$G - G_{initial} = n_{O_2(aq)} \mu_{O_2(aq)} + n_{O_2(g)} \mu_{O_2(g)} \qquad \text{eq. 3.18}$$

To find the minimum in $G - G_{initial}$, it is necessary to calculate all variables on the right side of this equation. Let ξ = extent of the reaction = number of moles of oxygen dissolved = $n_{O_2(aq)}$. The initial amount of oxygen in the air (from rearrangement of the ideal gas law) is:

$$n_{O_2(g),initial} = \frac{P_{O_2} V}{RT}$$
$$= (0.209 \text{ atm})(1 \text{ L})/[(0.082057 \text{ L-atm/mol-°K})(298.16°K)]$$
$$= 8.54 \times 10^{-3} \text{ mol}$$

[†] This example was inspired by a similar illustration for carbon dioxide in Stumm and Morgan (1996).

Thus:

$$n_{O_2(g)} = n_{O_2(g),initial} - n_{O_2(aq)} = 8.54 \times 10^{-3} - \xi \, \text{mol} \qquad \text{eq. 3.19}$$

The chemical potential of dissolved oxygen is given by:

$$\mu_{O_2(aq)} = \mu^o_{O_2(aq)} + RT \ln\{O_2(aq)\} \qquad \text{eq. 3.20}$$

where: $\mu^o_{O_2(aq)} = \overline{G}^{\,o}_{f,O_2(aq)} = 16.32$ kJ/mol. For 1 L of water, $\{O_2(aq)\}$ $\approx n_{O_2(aq)} = \xi$. The chemical potential of gaseous oxygen is given by:

$$\mu_{O_2(g)} = \mu^o_{O_2(g)} + RT \ln P_{O_2} \qquad \text{eq. 3.21}$$

where $\mu^o_{O_2(g)} = \overline{G}^{\,o}_{f,O_2(g)} = 0$. From the ideal gas law:

$$P_{O_2} = \frac{n_{O_2(g)} RT}{V} = (8.54 \times 10^{-3} - \xi) RT/(1 \, \text{L}) \qquad \text{eq. 3.22}$$

Combining eqs. 3.18-3.22:

$$G - G_{initial} \, (\text{kJ}) = \xi(16.32 + RT\ln\xi) + $$
$$(8.54 \times 10^{-3} - \xi) RT\ln[R'T(8.54 \times 10^{-3} - \xi)]$$

(Note: To calculate G in kJ, use $R = 8.314 \times 10^{-3}$ kJ/mol-^{o}K and $R' = 0.082057$ L-atm/mol-^{o}K .)

The change in Gibbs free energy for the system as a function of the extent of the reaction is shown in Figure 3.2. The minimum in Gibbs free energy occurs when only a small amount of oxygen has dissolved. The portion of free energy diagram at very small extents of reaction is shown in Figure 3.3.

Thoughtful Pause
From Figure 3.3, what is the equilibrium state of the system?

The equilibrium state is given by the minimum in G. This occurs when 2.79×10^{-4} mol of oxygen has dissolved (3.27% of available oxygen transferred) and represents a dissolved oxygen concentration of 8.9 mg/L (see dotted lines in Figure 3.2). The minimum $G - G_{initial}$ is -7.0×10^{-4} kJ.

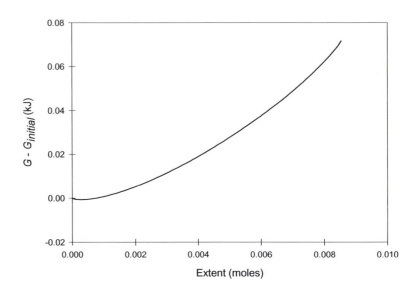

Figure 3.2: $G - G_{initial}$ **as a Function of the Extent of O$_2$ Dissolution**

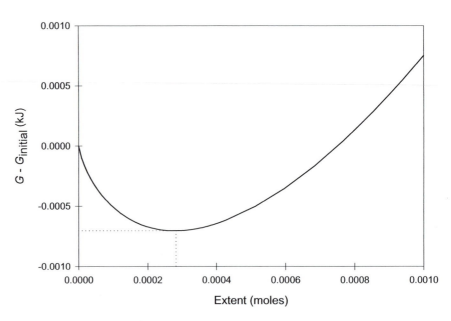

Figure 3.3: Detail from Figure 3.2
(equilibrium indicated by dotted lines)

3.9 CHEMICAL EQUILIBRIUM AND THE EQUILIBRIUM CONSTANT

3.9.1 Equilibrium constants

From Section 3.7.4, you know that $\Delta G_{rxn} = 0$ at equilibrium. This means that there must be some set of species activities that allows ΔG_{rxn} to be zero. At equilibrium, these activities are called the *equilibrium activities* (denoted with a subscript eq), so that eq. 3.17 becomes:

$$0 = \Delta G_{rxn}^{o} + RT \ln\left(\frac{\{C\}_{eq}^{c} \{D\}_{eq}^{d}}{\{A\}_{eq}^{c} \{B\}_{eq}^{c}}\right) \qquad \text{eq. 3.23}$$

Since ΔG_{rxn}^{o} is a constant for a given reaction, it follows that the term in parentheses in eq. 3.23 must be constant at equilibrium. The term in parentheses is called the *equilibrium constant* and is denoted K or K_{eq}. For the reaction aA + bB = cC + dD, the equilibrium constant is:

Key idea: The equilibrium constant is the product of the reactant activities divided by the product of the product activities, each raised to the power of their reaction stoichiometric coefficient

$$K = \frac{\{C\}_{eq}^{c} \{D\}_{eq}^{d}}{\{A\}_{eq}^{a} \{B\}_{eq}^{b}}$$

This text is focused on chemical reactions at equilibrium. Therefore, the subscript eq will be understood and generally not written. The activity of species A, $\{A\}$, will be assumed to be the equilibrium activity unless otherwise indicated.

So far, activity has not been defined very precisely. Activity is an idealized concentration. As a thermodynamic property, activity depends on the temperature, pressure, and number of moles of material in the system. The dependence of activity on the temperature, pressure, and number of moles of material in the system will be examined in Chapter 21. For now, assume you are working with dilute solutions (i.e., the number of moles of material in the system is small) at standard and constant temperature and pressure. The dilute solution assumption is valid for many freshwaters. Under these conditions, the concentration of a chemical species is nearly equal to its activity. Thus, the activity of species A, $\{A\}$, can be replaced

Key idea: In dilute solution, the equilibrium constant is approximately equal to the product of the reactant concentrations divided by the product of the product concentrations, each raised to the power of their reaction stoichiometric coefficient

with the concentration of species A, $[A]$. You can write $K = \dfrac{[C]^{c}[D]^{d}}{[A]^{a}[B]^{b}}$.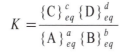

Again, the concentrations are assumed here to be equilibrium concentrations unless otherwise noted. In words, *the equilibrium constant in dilute solution is the product of the reactant concentrations divided by the product of the product concentrations, each raised to the power of their reaction stoichiometric coefficient.*

3.9.2 Properties of the equilibrium constant

All equilibrium constants show some important properties. First, equilibrium constants are thermodynamic functions. As such, they may depend on the temperature, pressure, and concentrations of other species in the system. These relationships will be quantified in Chapter 21.

Second, equilibrium constants can be calculated from ΔG^o_{rxn}. From eq. 3.18 and the definition of the equilibrium constant:

***Example 3.6:* Calculation of ΔG^o_{rxn} from K**

Calculate ΔG^o_{rxn} for the reaction $H_2O(l) \rightarrow OH^-(aq) + H^+(aq)$ if $K = 1.0 \times 10^{-14}$ M at 25°C

Solution:
$\Delta G^o_{rxn} = -RT\ln K = (8.314 \times 10^{-3}$ kJ/mol-°K)(298°K)ln(1.0×10^{-14}) = **+79.9 kJ**

You can verify this value using:
$\Delta G^o_{rxn} = \sum \nu_i \Delta \overline{G}^o_{f,i}$, where the $\Delta \overline{G}^o_{f,i}$ values for $H_2O(l)$, $OH^-(aq)$, and $H^+(aq)$ are −237.2, 0, and −157.3 kJ/mol, respectively.

$$K = e^{-\frac{\Delta G^o_{rxn}}{RT}}$$

Thus, you can calculate K if you know ΔG^o_{rxn}. As an example, the value of $\Delta G^o_{rxn} = +8.7$ kJ/mol was calculated for the reaction $CO_2(g) + H_2O(l) \rightarrow H_2CO_3(aq)$ in Example 3.5. For this reaction at 25°C (= 298°K):

$$
\begin{aligned}
K &= \exp[-(8.7 \text{ kJ/mol})/(8.314 \times 10^{-3} \text{ kJ/mol-°K})(298°K)] \\
&= 4.35 \times 10^{-10}
\end{aligned}
$$

You also can calculate ΔG^o_{rxn} if you know K (see Example 3.6).

Third, it is necessary to think carefully about the units of equilibrium constants. In reality, equilibrium constants have no units. Each term is a concentration (or activity) divided by the concentration (or activity) of the standard state (see Section 3.8.2). Thus, all K values are unitless, and you should have no qualms about taking the natural logarithm of K in calculating ΔG^o_{rxn}. However, as discussed in Section 4.3, it is common to associate units with equilibrium constants to indicate the concentration units of the species involved in the equilibrium.

3.10 SUMMARY

This chapter provided a brief review of the thermodynamic basis of chemical equilibrium. Equilibrium is defined as the state in which the thermodynamic functions of the system (such as temperature, pressure, and species concentrations) do not change with time. Three links were made in this chapter to allow for calculation of species concentrations at equilibrium. First, the concept of equilibrium was related to the change in one particular thermodynamic function, Gibbs free energy. Gibbs free energy combines internal energy, pressure-volume work, and entropy (a measure of disorder in the system). The change in the Gibbs free energy of a reaction is zero at equilibrium.

Second, Gibbs free energy was related to chemical potential. Differences in chemical potential drive chemical reactions, just as differences in potential energy (i.e., elevation) drive some mechanical processes.

Third, the chemical potential (and thus the change in Gibbs free energy for a reaction) was related to species activities. Activity is an idealized concentration, which takes into account the effects of temperature, pressure, and other species concentrations on the behavior of a given species.

Also in this chapter, the concept of the equilibrium constant was introduced. The equilibrium constant is a thermodynamic function, related to both the standard Gibbs free energy of a reaction and to species activities. The equilibrium constant in dilute solutions is the product of the reactant concentrations divided by the product of the product concentrations, each raised to the power of their reaction stoichiometric coefficient.

3.11 PART I CASE STUDY: CAN METHYLMERCURY BE FORMED CHEMICALLY IN WATER?

Recall from Sections 1.7 and 2.11 that the Part I case study involves deciding whether methylmercury (CH_3Hg^+) can be formed in significant concentration from the chemical reaction of aqueous methane and Hg^{2+}. Based on the knowledge of thermodynamics acquired in this chapter, we should be able to make good progress towards answering the question posed in the case study.

The best course of action would be to write a reaction with $CH_4(aq)$ and Hg^{2+} as reactants and CH_3Hg^+ as a product. Unfortunately, the topic of balancing chemical reactions has not been covered yet in this text. Thus, the case study cannot be completed until Chapter 4. However, it is still possible to select an approach to analyzing the case study and calculate the necessary thermodynamic values.

To address the case study question, you must calculate a thermodynamic property that will reveal whether or not the reaction between $CH_4(aq)$ and Hg^{2+} to form CH_3Hg^+ proceeds spontaneously.

Thoughtful Pause

What is the appropriate thermodynamic property to test whether the reaction between $CH_4(aq)$ and Hg^{2+} to form CH_3Hg^+ is spontaneous?

The thermodynamic variable we seek is the Gibbs free energy of reaction. Recall that a reaction is spontaneous as written under standard-state conditions if $\Delta G^o_{rxn} < 0$. We cannot calculate ΔG^o_{rxn} yet because we cannot write the balanced reaction yet. However, you know that you will need $\Delta \overline{G}^o_f$ for at least $CH_4(aq)$, Hg^{2+}, and CH_3Hg^+ to calculate ΔG^o_{rxn}. Thus, to make progress on the case study, we need $\Delta \overline{G}^o_f$ for $CH_4(aq)$, Hg^{2+}, and CH_3Hg^+. The $\Delta \overline{G}^o_f$ values for $CH_4(aq)$, Hg^{2+}, and CH_3Hg^+ are -34.4,

+164.4, and +112.9 kJ/mol, respectively (Stumm and Morgan, 1996). These thermodynamic data will be used to calculate ΔG^o_{rxn} when the balanced reaction is written in Section 4.9.

SUMMARY OF KEY IDEAS

- Thermodynamic systems are defined by thermodynamic properties (also called *state variables*)

- The values of a extensive thermodynamic property for each portion of a system add up to the value of the extensive thermodynamic property for the whole system

- Intensive thermodynamic properties are ***not*** additive within a system

- Thermodynamics can tell you which chemical reactions are possible (under a given set of conditions) and whether species concentrations are not time-dependent

- For a reaction to be at equilibrium, both the reaction and its reverse reaction must be spontaneous

- The First Law of Thermodynamics implies that energy is conserved and that the internal energy changes only through heat exchange or work

- The change in enthalpy is the heat absorbed at constant pressure

- The First Law of Thermodynamics does not tell you if processes are spontaneous

- Gibbs free energy is a measure of reversibility: $dG \leq 0$ for all reactions, but $dG = 0$ only for reversible processes

- Gibbs free energy is a measure of spontaneity: $dG \leq 0$ for spontaneous reactions at constant T and P

- Chemical potential drives work done by chemical species

- ΔG_{rxn} is related to species activities

- Equilibrium corresponds to a minimum in the Gibbs free energy or $\Delta G_{rxn} = 0$ at equilibrium

- The equilibrium constant is the product of the reactant activities divided by the product of the product activities, each raised to the power of their reaction stoichiometric coefficient

- In dilute solution, the equilibrium constant is approximately equal to the product of the reactant concentrations divided by the product of the product concentrations, each raised to the power of their reaction stoichiometric coefficient

HISTORICAL NOTE: Josiah Willard Gibbs

J. Willard Gibbs

J. Willard Gibbs was one of the most influential American scientists of the nineteenth century. His pioneering work in thermodynamics, vector analysis, electromagnetics, and statistical mechanics set the mathematical groundwork for modern physical chemistry.

Gibbs lived most of life in one house in New Haven, Connecticut. He received a Ph.D. in engineering from Yale University in 1863 for his work on gear design. His was the first Ph.D. in engineering granted in the United States. After studying in Europe from 1866 to 1869, Gibbs returned to Yale in 1869 and was appointed professor of mathematical physics at Yale in 1871.

Scientific minds of Gibbs's caliber usually make their mark on the scientific community at an early age. It is interesting to note that Gibbs did not publish his first paper until 1873, when he was 34 years old. His contributions to thermodynamics were many, but his most influential work was "On the Equilibrium of Heterogeneous Substances," published in two parts in 1876 and 1878.

J. Willard Gibbs died in New Haven in 1903. His lasting impact on science is indicated by the range of items bearing his name, among them Gibbs free energy (G), the Wilbraham-Gibbs constant ($G = \int_0^\pi \frac{\sin\theta}{\theta} d\theta =$ 1.81937052...), the American Mathematical Society Gibbs Lecture, and Crater Gibbs (a 76-km-wide crater on the moon, named for Gibbs in 1964).

PROBLEMS

3.1 For reactions at constant pressure involving liquids and solids, it is commonly assumed that $\Delta H \approx \Delta E$. Why?

3.2 To illustrate the relationship in Problem 3.1, calculate ΔH and ΔE when 100 g of ice melts at 25°C and

1 atm. How much heat is absorbed by the system? For ice: $\overline{H}_f^o = -292.80$ kJ/mol and molar volume = 0.0196 L/mol; for water: $\overline{H}_f^o = -285.83$ kJ/mol and molar volume = 0.0180 L/mol.

3.3 What pressure would be required to convert 1 g of graphite into 1 g of diamond at 25°C and 1 atm? Assume graphite and diamond are incompressible. The $\overline{G}_f^o$ values are 0 and -2.59 kJ/mol and densities are 2.25 and 3.51 g/cm³ for graphite and diamond, respectively. Hint: Start with eq. 3.7 and realize that temperature is constant here.

[Before you quit your day job to make 1 g (= 5 carat) diamonds, realize that artificial diamonds are made at much higher temperatures and pressures (> 100,000 atm) than you calculated. Graphite is dissolved in a molten metal catalyst and diamond precipitates. Why are the temperatures and pressures employed much higher than calculated here?]

3.4 Consider the reaction: $HSO_4^- \rightarrow SO_4^{2-} + H^+$. Is the reaction endothermic or exothermic if all concentrations are 1 M? Does the reaction proceed spontaneously as written if all concentrations are 1 M? For HSO_4^-: $\overline{H}_f^o = -887.3$ kJ/mol, $\overline{S}_f^o = 132$ J/mol-°K, and $\overline{G}_f^o = -756.0$ kJ/mol. For SO_4^{2-}: $\overline{H}_f^o = -909.2$ kJ/mol, $\overline{S}_f^o = 20.1$ J/mol-°K, and $\overline{G}_f^o = -744.6$ kJ/mol. For H^+: $\overline{H}_f^o = 0$ kJ/mol, $\overline{S}_f^o = 0$ J/mol-°K, and $\overline{G}_f^o = 0$ kJ/mol.

3.5 Consider the reaction: $NH_4^+ \rightarrow NH_3 + H^+$. Is the reaction endothermic or exothermic if all concentrations are 0.01 M? Does the reaction proceed spontaneously as written if all concentrations are 0.01 M? For NH_4^+: $\overline{H}_f^o = -132.5$ kJ/mol, $\overline{S}_f^o = 113.4$ J/mol-°K, and $\overline{G}_f^o = -79.37$ kJ/mol. For NH_3: $\overline{H}_f^o = -80.3$ kJ/mol, $\overline{S}_f^o = 111$ J/mol-°K, and $\overline{G}_f^o = -26.57$ kJ/mol. For H^+: $\overline{H}_f^o = 0$ kJ/mol, $\overline{S}_f^o = 0$ J/mol-°K, and $\overline{G}_f^o = 0$ kJ/mol.

3.6 Calculate K for the reaction in Problem 3.4 at equilibrium. (At equilibrium, you cannot assume all species concentrations are 1 M.) At what pH are the equilibrium activities of H_2SO_4 and HSO_4^- equal?

3.7 Calculate K for the reaction in Problem 3.5 at equilibrium. (At equilibrium, you cannot assume all species concentrations are 0.01 M.) At what range of pH is $\{NH_4^+\} > \{NH_3\}$ at equilibrium?

3.8 What is the criterion for equilibrium in terms of G? ΔG_{rxn}? ΔG_{rxn}^o?

3.9 From the example discussed in Section 3.8.4, calculate the equilibrium constant for the equilibrium $O_2(g)$ = $O_2(aq)$ from the equilibrium concentrations of $O_2(g)$ and $O_2(aq)$. How does your value compare to the accepted value of 1.26×10^{-3} mol/L-atm? (Be careful about units.) How does your value compare to that calculated from ΔG^o_{rxn}?

3.10 From the example discussed in Section 3.7.3, calculate the equilibrium constant for $AgCl(s)$ = Ag^+ + Cl^-. How does your value compare to the accepted value of 2.8×10^{-10}?

CHAPTER 4

Manipulating Equilibrium Expressions

4.1 INTRODUCTION

You learned in Chapter 3 that the concept of chemical equilibrium has a thermodynamic basis. At this point, you can write equilibrium expressions and calculate equilibrium constants from Gibbs free energy of formation values. Given enough thermodynamic data, you could calculate the equilibrium constant for any equilibrium.

However, there are many instances in environmental engineering and science where you do *not* have sufficient thermodynamic data to calculate the equilibrium constant for an equilibrium of interest to you. However, you may know equilibrium constants for *related* equilibria. How can information about *similar* equilibria be used to determine the concentrations of chemical species at equilibrium? To calculate species concentrations efficiently, you must be able to exploit known equilibria. Throughout this text, you shall be manipulating systems of equilibrium expressions to calculate species concentrations at equilibrium. In this chapter, some simple rules for the manipulation of equilibrium expressions will be presented.

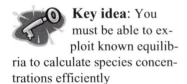
Key idea: You must be able to exploit known equilibria to calculate species concentrations efficiently

Three points will be emphasized throughout this chapter. First, as discussed in Section 4.2, chemical equilibria can be interpreted as both *chemical expressions* and *mathematical statements*. Although the chemical and mathematical forms of equilibria are used for different purposes, they should convey consistent information about the way in which species and species concentrations relate to the equilibrium expression. Second, equilibrium constants can have units associated with them (Section 4.3). Third, knowledge of an equilibrium constant for one reaction allows for the calculation the equilibrium constants for a number of related reactions. To calculate additional equilibrium constant, you must master the skills of reversing (Section 4.4), changing the stoichiometric constants (Section 4.5), and adding equilibria (Section 4.6). With these skills, new equilibria and equilibrium constants can be created, as discussed in Section 4.7.

4.2 CHEMICAL AND MATHEMATICAL FORMS OF EQUILIBRIA

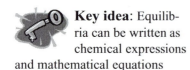

Key idea: Equilibria can be written as chemical expressions and mathematical equations

4.2.1 Chemical expressions

Chemical equilibria can be written in two ways: chemical expressions and mathematical expressions. *Chemical expressions* are idealized ways of writing the chemical reactions that occur at equilibrium. As an example, consider the dissociation of ammonium (NH_4^+) to ammonia (NH_3) and H^+. For this chemical equilibrium, you can write the chemical expression

$$NH_4^+ \;=\; NH_3 + H^+$$

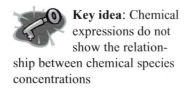

Key idea: Chemical expressions show the reactants, products, and stoichiometric relationship between chemical species

Chemical expressions have several important characteristics. First, chemical expressions describe the relationships between *chemical species*. In particular, chemical expressions show the reactants, products, and stoichiometry of the chemical reactions. For the ammonium/ammonia example, the chemical expression shows that each mole of ammonium that dissociates produces 1 mole of ammonia and 1 mole of H^+. By convention, reaction stoichiometric coefficients equal to 1 are omitted in the chemical expressions.

Second, the $=$ symbol in the chemical expression signifies that the reaction is at equilibrium. Thus, the chemical expression tells us that the reaction is reversible and spontaneous. This means that the reactions proceed in each direction. In this way, the symbol $=$ is not a mathematical sign of equality, but rather a shorthand notation for $\rightleftharpoons$. In the example, the chemical expression reveals that ammonium dissociates to form 1 mole of ammonia and 1 mole of H^+ and also that for each mole of ammonia and H^+ that combine, 1 mole of ammonium is formed.

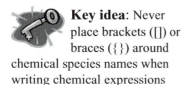

Key idea: Chemical expressions do not show the relationship between chemical species concentrations

Third, chemical expressions have a serious limitation. They do **not** tell you the relationship between chemical species *concentrations*. In the ammonium example, the chemical expression does *not* mean that the concentration of ammonium equals the sum of the concentrations of ammonia and H^+ (i.e., the chemical express does *not* imply that $[NH_4^+] = [NH_3] + [H^+]$). When writing chemical expressions, *never place brackets or braces around the names of the chemical species*.

Key idea: Never place brackets ([]) or braces ({}) around chemical species names when writing chemical expressions

4.2.2 Mathematical equations

Equilibria also can be written as *mathematical equations*. To write an equilibrium as a mathematical equation, set the equilibrium constant equal to the product of the product species concentrations raised to their stoichiometric coefficients divided by the product of the reactant species concentrations raised to their stoichiometric coefficients (as in Section 3.9.1).[†] For the example above:

[†] In this chapter, the term *stoichiometric coefficient* will be used to mean the stoichiometric coefficients in reactions (i.e., reaction stoichiometric coefficients).

Write the mathematical form of the equilibrium given by
$$FeOH_3(s) = Fe^{3+} + 3OH^-$$
Write the chemical expression for the equilibrium given by:
$$K = \frac{Fe_3(PO_4)_2(s)}{[Fe^{2+}]^3[PO_4^{3-}]^2}$$

Solution:
For the first equilibrium, set K equal to the product of the species concentrations raised to their signed stoichiometric coefficients (assuming concentrations are nearly equal to activities). Thus

$$K = [Fe^{3+}][OH^-]^3/[FeOH_3(s)]$$

For the second equilibrium, interpret the numerator term as a product and the denominator terms as reactants. The chemical expression is
$$3Fe^{2+} + 2PO_4^{3-} = Fe_3(PO_4)_2(s)$$

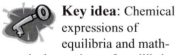 **Key idea**: Chemical expressions of equilibria and mathematical equations of equilibria contain the same information and must lead to consistent conclusions

Le Chatelier's principle: a change to a thermodynamic function controlling the equilibrium will cause the system to adjust to minimize the change

$$K = [NH_3][H^+]/[NH_4^+]$$

The equation $K = [NH_3][H^+]/[NH_4^+]$ is the proper *mathematical* relationship between species concentrations.

In Chapter 3, the symbol v_i for the stoichiometric coefficient of species i in an equilibrium was introduced. Recall that stoichiometric coefficients of reactants are considered negative and stoichiometric coefficients of reactants are considered positive. Thus, you can write in general:

$$K = \prod_i \{i\}^{v_i} \approx \prod_i [i]^{v_i}$$

where $\{i\}$ and $[i]$ are the activity and concentration of species i, respectively.[†]

The ammonium example demonstrates the relationship between chemical expressions of equilibrium and mathematical statements of equilibrium. In forming the mathematical equations, you *are translating the chemical expressions into mathematical equations*. Two other illustrations are shown in Example 4.1. It is important to be able to translate quickly between chemical expressions and mathematical equations.

4.2.3 An example of chemical expressions and mathematical equations: Le Chatelier's principle

Chemical expressions and mathematical equations are two ways to write the information contained in chemical equilibria. As you have seen, each approach has its uses. However, the two ways of writing equilibria *must lead to consistent conclusions*. Throughout this chapter, you will see that chemical expressions and mathematical equations must be consistent in the trends they predict in species concentrations.

One example in which predictions from chemical expressions and mathematical equations can be compared is *Le Chatelier's principle* (after Henry-Louis Le Chatelier, 1850-1936; see the *Historical Note* at the end of the chapter). For a system initially at equilibrium, Le Chatelier's principle states that a change to a thermodynamic function controlling the equilibrium (e.g., a change in the system temperature, system pressure, or species concentration) will cause the system to adjust to minimize the change. Le Chatelier's principle usually is applied to chemical expressions. For the ammonium/ammonia example, assume the system is at equilibrium. Now, imagine that the H^+ concentration is increased.

[†] In this chapter, it will be assumed that the solutions are dilute. Thus, activities and concentrations are nearly equal, and you can use concentrations as approximations for activities in equilibrium expressions (see Section 3.9.1).

Example 4.2: **Le Chatelier's Principle**

Using Le Chatelier's principle, explain why bleach manufacturers bubble chlorine gas into sodium hydroxide solutions (rather than into pure water) when making bleach. The pertinent equilibria are

$$Cl_2(g) + H_2O = HOCl + H^+ + Cl^-$$

$$H_2O = H^+ + OH^-$$

Solution:
From the first equilibrium, the bleach (here, HOCl or hypochlorous acid) concentration is increased when $Cl_2(g)$ is increased and/or when $[H^+]$ is decreased. From the second equilibrium, $[H^+]$ is decreased when $[OH^-]$ is increased (since the activity of water, a pure liquid, is fixed; see Section 4.3.2). Thus, bleach manufacturers add NaOH to increase the HOCl concentration (thus saving money by reducing the mass of water shipped per kg of bleach).

Thoughtful Pause
What will happen to the $NH_3/H^+/NH_4^+$ equilibrium if the concentration of H^+ is increased?

An increase in the H^+ concentration will shift the system away from equilibrium. According to Le Chatelier's principle, the system will adjust to minimize the impact of the increase in the H^+ concentration. To accomplish this, the equilibrium ($NH_4^+ = NH_3 + H^+$) will shift to the left, with a resulting increase in the ammonium concentration.

Is Le Chatelier's principle consistent with the mathematical version of equilibrium information? In the mathematical expression ($K = [NH_3][H^+]/[NH_4^+]$), K is constant at constant temperature and pressure. Think of this equation as purely a mathematical construct divorced from the chemistry: K = constant = xy/z. If you make y larger but keep K constant, then one of the following statements must be true: (1) x decreases, (2) z increases, or (3) both x decreases and z increases. Back to the chemistry: to maintain a constant amount of N in the system, you cannot decrease the ammonia concentration without increasing the ammonium concentration by an equal amount. Similarly, you cannot increase $[NH_4^+]$ without decreasing $[NH_3]$ by an equal amount. Thus, the ammonia concentration must decrease *and* the ammonium concentration must increase. In the symbols used previously, x decreases *and* z increases. The mathematical statement of equilibrium tells you that the ammonium concentration must increase (and the ammonia concentration must decrease) as H^+ is increased. Thus, Le Chatelier's principle (usually applied to the chemical expression) is consistent with the mathematical version of equilibrium information. Another instance of Le Chatelier's principle is shown in Example 4.2.[†]

4.3 UNITS OF EQUILIBRIUM CONSTANTS

 Key idea: Equilibrium constants are dimensionless, but units are associated with equilibrium constants as a reminder of the units of the species in the equilibrium

4.3.1 Introduction
What are the units of equilibrium constants? Recall from Section 3.9.2 that *equilibrium constants are dimensionless*, because each term in the equilibrium constant is normalized to a standard state having the same concentration (or activity) units as the species concentration. You can write mathematical expressions such as $\log(K)$ and $\Delta G_{rxn}^o = -RT\ln(K)$ without concern about the units of K.

[†] Le Chatelier's principle is a powerful tool for understanding the *qualitative* behavior of simple chemical systems. The great American chemist and double Nobel laureate Linus Pauling wrote (Pauling, 1964): "When you have obtained a grasp of Le Chatelier's principle, you will be able to think about any problem of chemical equilibrium that arises, and, by use of a simple argument, to make a qualitative statement about it.... Some years after you have finished your college work, you may ... have forgotten all the mathematical equations relating to chemical equilibrium. I hope, however, that you have not forgotten Le Chatelier's principle."

Although equilibrium constants are inherently dimensionless, *it is convenient to associate units with equilibrium constants to reflect the concentration units of the species in the equilibrium expression.* This is an accounting trick to aid in manipulating equilibria and will be used occasionally throughout this text. In addition, writing units with equilibrium constants allows one another way to check whether the manipulation of the equilibria and equilibrium constants was performed properly; the units associated with the final equilibrium constant should match the final equilibrium expression.

Example 4.3: Units Associated with Equilibrium Constants

Using typical concentration units, what units should be associated with the first equilibrium in Example 4.2? The equilibrium is

$$Cl_2(g) + H_2O = HOCl + H^+ + Cl^-$$

Solution:
Typical concentration units for gases, pure liquids (such as water), and dissolved species are atm, none, and mol/L, respectively. Thus, you would associate the units of **mol³/L³-atm** (or **M³/atm**) with the first equilibrium in Example 4.2.

4.3.2 Concentration scales
To be able to associate units with equilibrium constants, you must choose a consistent set of concentration scales. Four types of species are found commonly in equilibria in aqueous systems: dissolved species, pure solids, pure liquids, and gases. As discussed in Chapter 2, dissolved species concentrations in dilute solution usually are expressed using the molarity concentration scale. Thus, dissolved species concentrations commonly are written in mol/L = M. For example, the equilibrium constant for ammonium dissociation ($K = [NH_3][H^+]/[NH_4^+]$) may be written with units of $M^2/M = M$ to indicate the units of $[NH_3]$, $[H^+]$, and $[NH_4^+]$.

Pure solids and pure liquids usually employ the mole fraction concentration scale. Pure solids and pure liquids have a mole fraction of unity. As a result, pure solids and pure liquids (e.g., water) do not affect the units of equilibrium constants. Thus, for the reaction $CaCO_3(s) = Ca^{2+} + CO_3^{2-}$, you can write

$$K = [Ca^{2+}][CO_3^{2-}]/[CaCO_3(s)] = [Ca^{2+}][CO_3^{2-}]$$

The equilibrium constant K usually has associated units of M^2. You shall frequently make use of the following reaction: $H_2O = H^+ + OH^-$. For this reaction, you can write

$$K = [H^+][OH^-]/[H_2O] = [H^+][OH^-]$$

where K usually has associated units of M^2.

Gases usually employ the partial pressure concentration scale. Thus, gases commonly have concentrations of atm. For example, for the reaction $O_3(aq) = O_3(g)$, you can write $K = [O_3(g)]/[O_3(aq)]$ with typical associated units of atm/M = atm-L/mol. Another illustration of associating units with equilibrium constants may be found in Example 4.3.

4.3.3 Using units with equilibrium constants
Recall that a general rule of quantitative analysis is to avoid comparing values with different units. For example, it is improper to conclude that an automobile with a gas mileage of 30 miles per gallon is more fuel efficient

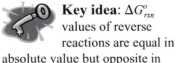

Key idea: Take great care when comparing equilibrium constants with different units

than an automobile with a gas mileage of 12.8 km/L. In fact, the gas mileage is about the same for the two vehicles. Similarly, *you cannot compare the values of equilibrium constants with different associated units*. For example, K for the equilibrium $H_3PO_4 = PO_4^{3-} + 3H^+$ is 3.2×10^{-22} and K for the equilibrium $HCN = CN^- + H^+$ is 6.3×10^{-11}. Can we use this information to compare the ability of phosphoric acid (H_3PO_4) and hydrocyanic acid (HCN) to produce H^+? The answer is no: the equilibrium constant for the first equilibrium has units of M^3 ($= mol^3/L^3$) associated with it, whereas the equilibrium constant for the second equilibrium has units of M associated with it. The equilibrium constants have different units and cannot be compared. The units associated with equilibrium constants are a reminder that you should not compare values with different units.

4.4 REVERSING EQUILIBRIA

reversing an equilibrium: an equilibrium is reversed when its reactants and products are interchanged

One of the take-home lessons from this chapter is that knowledge of one equilibrium constant allows you to calculate easily the equilibrium constants of related equilibria. In this Section, you will see that knowing an equilibrium constant allows for the calculation of the equilibrium constant for the reaction formed when reactants and products in the original reaction are interchanged. Interchanging reactants and products is called *reversing an equilibrium*.

For the example equilibrium of this chapter ($NH_4^+ = NH_3 + H^+$), the equilibrium constant, K, is about 5×10^{-10} M $= 10^{-9.3}$ M.[†] What is the equilibrium constant for the reverse reaction ($NH_3 + H^+ = NH_4^+$)? You can answer this question in two ways: by examining the chemical expression and by examining the mathematical equation for K. From the chemical expression, it is clear that ΔG^o_{rxn} for the ammonia/H^+ association reaction is -1 times ΔG^o_{rxn} for the ammonium dissociation reaction.

Thoughtful Pause
Why are the ΔG^o_{rxn} of reverse reactions equal in absolute value but opposite in sign?

Key idea: ΔG^o_{rxn} values of reverse reactions are equal in absolute value but opposite in sign and K values of reverse reactions are reciprocals of one another

---

[†] Equilibrium constants can take values over many orders of magnitude. In addition, you know that $\log(K)$ values are proportional to ΔG^o_{rxn} and therefore have special meaning. Thus, we usually express K values as the antilog base 10; in other words, as 10^x.

Example 4.4: **Reversing Equilibria**

Find the equilibrium constant for $HgCl^+ = Hg^{2+} + Cl^-$ if $\Delta G^o_{rxn} = -38.54$ kJ/mol for the equilibrium:
$$Hg^{2+} + Cl^- = HgCl^+$$

Solution:

If ΔG^o_{rxn} for the equilibrium $Hg^{2+} + Cl^- = HgCl^+$ is -38.54 kJ/mol, then ΔG^o_{rxn} for the reverse reaction ($HgCl^+ = Hg^{2+} + Cl^-$) is $+38.54$ kJ/mol. Thus, K for $HgCl^+$ dissociation at 25°C is

$$K = \exp(-\Delta G^o_{rxn}/RT)$$
$$= \mathbf{1.76 \times 10^{-7}}$$

Alternatively, you could calculate the equilibrium constant for the equilibrium $Hg^{2+} + Cl^- = HgCl^+$ from ΔG^o_{rxn} ($K = 5.70 \times 10^6$) and note that the equilibrium constant for the equilibrium $HgCl^+ = Hg^{2+} + Cl^-$ is $1/5.70 \times 10^6 = 1.76 \times 0^{-7}$.

For any reaction, $\Delta G^o_{rxn} = \sum v_i \overline{G}^o_{f,i} = \sum_{products} |v_i| \overline{G}^o_{f,i} - \sum_{reactants} |v_i| \overline{G}^o_{f,i}$. Thus, ΔG^o_{rxn} of reverse reactions are equal in absolute value but opposite in sign, since the reactants and products are simply interchanged for reverse reactions. Now, since $K = e^{-\frac{\Delta G^o_{rxn}}{RT}}$, K for the reverse reaction must be equal to the *reciprocal* of K for the original reaction. For the example, K for the $NH_3 + H^+$ association reaction is $1/(K$ for the ammonium dissociation reaction) $= 1/10^{-9.3}$ M $= 10^{+9.3}$ M^{-1}. Thus, examination of the chemical expression tells you that K *for one reaction is the reciprocal of K for the reverse reaction.*

The mathematical form of the equilibrium also should yield that K for one reaction is the reciprocal of K for the reverse reaction. This is obvious by inspection. For example, the ammonium dissociation reaction has an equilibrium constant equal to $[NH_3][H^+]/[NH_4^+]$, whereas the equilibrium constant for the ammonia/H^+ association reaction is $[NH_4^+]/([NH_3][H^+])$: K for the $NH_3 + H^+$ association reaction is the reciprocal of K for the ammonium dissociation reaction.

You can show that K for one reaction is the reciprocal of K for the reverse reaction in the general case by using the product notation of Section 4.2.2. For any equilibrium (where activity and concentration are nearly interchangeable): $K = \prod_i [i]^{v_i}$. For the reverse reaction, $v_{i,rev} = -v_i$ since reactants and products are reversed in the reverse reaction. Thus:

$$K_{rev} = \prod_i [i]^{v_{i,rev}} = \prod_i [i]^{-v_i} = \frac{1}{K}.$$ See also Example 4.4.

4.5 EFFECTS OF STOICHIOMETRY

4.5.1 Linear independence

Compare the following two reactions: $NH_4^+ = NH_3 + H^+$ with $K = K_1$ and $2NH_4^+ = 2NH_3 + 2H^+$ with $K = K_2$. The chemical expressions convey the same information: for each mole of ammonium dissociated, 1 mole of ammonia and 1 mole of H^+ are formed. (In addition, for each mole of ammonia associating with 1 mole of H^+, 1 mole of ammonium is formed.)

Since the chemical expressions $NH_4^+ = NH_3 + H^+$ and $2NH_4^+ = 2NH_3 + 2H^+$ contain the same information, it would be improper to use both equilibria in describing a chemical system. Why? You cannot use more than one expression that contains the same information in *any* physical system. For example, if you were solving an algebraic system for the values of x and y, you could not use *both* the equation $x + 2y = 5$ and the equation $3x + 6y = 15$; the two equations convey the same information.

When two equilibria differ only by a constant multiple of their stoichiometric coefficients, we say the equilibria are *linearly dependent*. (The definition of linear dependence is expanded in Section 4.6.2.) You will need to seek a set of linearly independent equilibria to describe a chemical system.

4.5.2 Free energy and equilibrium constants

Return to two reactions: $NH_4^+ = NH_3 + H^+$ with $K = K_1$ and $2NH_4^+ = 2NH_3 + 2H^+$ with $K = K_2$. What is ΔG^o_{rxn} for each equilibrium? Since $\Delta G^o_{rxn} = \sum_i v_i \overline{G}^o_{f,i}$, the standard Gibbs free energy of the second equilibrium should be twice the standard Gibbs free energy of the first equilibrium: doubling the stoichiometric coefficients values doubles ΔG^o_{rxn}. What about K? You know that

$$\begin{aligned} K_2 &= \exp(-\Delta G^o_{rxn,2}/RT) \\ &= \exp(-2\Delta G^o_{rxn,1}/RT) \\ &= [\exp(-\Delta G^o_{rxn,1}/RT)]^2 \\ &= K_1^2 \end{aligned}$$

Thus, K for $2NH_4^+ = 2NH_3 + 2H^+$ is the square of K for $NH_4^+ = NH_3 + H^+$.

Is this result consistent with the mathematical form of the equilibrium information? For the first equilibrium, $K_1 = [NH_3][H^+]/[NH_4^+]$, whereas for the second equilibrium $K_2 = [NH_3]^2[H^+]^2/[NH_4^+]^2 = K_1^2$. As expected, the mathematical form of the equilibria also reveals that K for the second equilibrium is the square of K for the first equilibrium. In general, let K be the equilibrium constant for an equilibrium. If you multiply each of the stoichiometric coefficients by c, then the equilibrium constant for the new equilibrium is $\prod_i [i]^{cv_i} = \prod_i \left([i]^{v_i}\right)^c = K^c$.

The effects of stoichiometry can be summarized as follows. Multiplying each stoichiometric coefficient in an equilibrium expression by a constant does not convey any new information about the equilibrium. However, *multiplying stoichiometric coefficients by a constant, c, results in an equilibrium with ΔG^o_{rxn} equal to c times the ΔG^o_{rxn} for the original equilibrium and K equal to K for the original equilibrium raised to the c power.* You shall find that it is sometimes convenient to track $\log(K)$ instead of K. Multiplying stoichiometric coefficients by a constant, c, multiplies ΔG^o_{rxn} by c and multiplies $\log(K)$ by c [since $c\log(K) = \log(K^c)$]. The effects of changing stoichiometric constants also are shown in Example 4.5.

It should be noted that reversing equilibria (discussed in Section 4.3) is really a special case of changing reaction stoichiometry. Recall that we usually think of stoichiometric coefficients for reactants as negative and stoichiometric coefficients for products as positive. Therefore, reversing an equilibrium (i.e., switching reactants and products) is equivalent to

---

Example 4.5: **Changing Stoichiometry**

Find the equilibrium constant for $HCN = H^+ + CN^-$ if ΔG^o_{rxn} for the reaction $\frac{1}{2}HCN = \frac{1}{2}H^+ + \frac{1}{2} CN^-$ is +27 kJ/mol.

Solution:

If ΔG^o_{rxn} for the reaction $\frac{1}{2}HCN = \frac{1}{2}H^+ + \frac{1}{2}CN^-$ is +27 kJ/mol, then ΔG^o_{rxn} for the reaction $HCN = H^+ + CN^-$ is 2(+27 kJ/mol) = +54 kJ/mol. Thus, K for $HCN = H^+ + CN^-$ at 25^oC is

$$\begin{aligned} K &= \exp(-\Delta G^o_{rxn}/RT) \\ &= \exp\{(-54\ kJ/mol)/ \\ &\quad [(8.314\times10^{-3}kJ/mol- \\ &\quad {}^oK)(298^oK)]\} \\ &= \mathbf{3.42\times10^{-10}} \end{aligned}$$

You also could calculate the equilibrium constant for $\frac{1}{2}HCN = \frac{1}{2}H^+ + \frac{1}{2}CN^-$ from ΔG^o_{rxn} ($K = 1.85\times10^{-5}$) and note that the equilibrium constant for $HCN = H^+ + CN^-$ is $(1.85\times10^{-5})^2 = 3.42\times10^{-10}$.

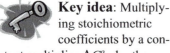 **Key idea**: Multiplying stoichiometric coefficients by a constant multiplies ΔG^o_{rxn} by the constant and raises K to the constant

multiplying each stoichiometric coefficient by -1. Thus, in reversing an equilibrium, the sign of the Gibbs free energy of reaction is multiplied by -1 and the new equilibrium constant is the reciprocal of the original equilibrium constant (i.e., K is raised to the -1 power).

4.6 ADDING EQUILIBRIA

4.6.1 Free energy and equilibrium constants

You can form new equilibria by adding existing equilibria. For example, how would you create a new equilibrium to show the relationship between ammonium, ammonia, hydroxide ion (OH^-), and water? You can accomplish this by adding the following two equilibria:

$$NH_4^+ = NH_3 + H^+ \qquad\qquad \text{reaction 1}$$
$$H^+ + OH^- = H_2O \qquad\qquad \text{reaction 2}$$

Adding:

$$NH_4^+ + H^+ + OH^- = NH_3 + H^+ + H_2O$$

Eliminating the common species on both sides of the equilibrium (here, only H^+ appears on both sides):

$$NH_4^+ + OH^- = NH_3 + H_2O \qquad\qquad \text{reaction 3}$$

How do ΔG^o_{rxn} and K for the new equilibrium (reaction 3) compare to ΔG^o_{rxn} and K for the original equilibria (reactions 1 and 2)? You know from Section 3.2.2 that free energies are additive. Thus, you might expect that $\Delta G^o_{rxn,3} = \Delta G^o_{rxn,1} + \Delta G^o_{rxn,2}$. For the equilibrium constant:

$$
\begin{aligned}
K_3 &= \exp(-\Delta G^o_{rxn,3}/RT) \\
&= \exp[-(\Delta G^o_{rxn,1} + \Delta G^o_{rxn,2})/RT] \\
&= \exp(-\Delta G^o_{rxn,1}/RT)\exp(-\Delta G^o_{rxn,2}/RT) \\
&= K_1 K_2
\end{aligned}
$$

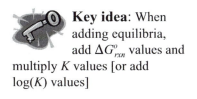

Key idea: When adding equilibria, add ΔG^o_{rxn} values and multiply K values [or add $\log(K)$ values]

Therefore, *adding equilibria results in a new equilibrium with Gibbs free energy equal to the sum of the Gibbs free energies of the individual equilibria and with an equilibrium constant equal to the product of the individual equilibrium constants.* Recall that if $K_3 = K_1 K_2$, then $\log(K_3) = \log(K_1) + \log(K_2)$. Thus, when adding equilibrium expressions, add ΔG^o_{rxn} values and multiply K values or, equivalently, add ΔG^o_{rxn} values and add $\log(K)$ values. This analysis points a nice trick involving $\log(K)$ values: *log(K) values are additive when equilibria are added.* With this fact, you often can calculate equilibrium constants for new equilibria in your head.

The mathematical form of the equilibria should give the same result. Indeed:

$$K_1 = [NH_3][H^+]/[NH_4^+]$$
$$K_2 = [H_2O]/[H^+][OH^-], \text{ and:}$$
$$K_3 = [NH_3][H_2O]/[NH_4^+][OH^-]$$

It is clear that $K_3 = K_1 K_2$. In general, if $K_1 = \prod_i [i]^{v_i}$ and $K_2 = \prod_i [i]^{v_i'}$ then

K_3, the equilibrium constant formed by adding reactions 1 and 2, is given by $\prod [i]^{v_i + v_i'} = K_1 K_2$.

4.6.2 Linear independence revisited

It is important to note that reaction 3 does *not* convey any information in addition to the information conveyed by reactions 1 and 2. In fact, given any two of reactions 1, 2, or 3, it is possible to create the other reaction by judiciously adding and/or reversing the reactions in hand. It is possible to extend the concept of linearly independent equilibria developed in Section 4.4: *a set of equilibria is **linearly independent** if no equilibrium expression can be formed by linearly combining some or all of the other equilibria in the set*. Here, *linearly combining* means reversing equilibria, multiplying stoichiometric coefficients by a constant, and/or adding equilibria. Another illustration of linear independence is shown in Example 4.6.

linearly independent equilibria: a set of equilibria where no equilibrium expression can be formed by linearly combining some or all of the other equilibria in the set

4.7 CREATING EQUILIBRIA

***Example 4.6:* Linear Independence**

Is the following set of equilibria linearly independent?

$$Al(OH)_3(s) = Al^{3+} + 3OH^-$$
$$Al^{3+} + OH^- = Al(OH)^{2+}$$
$$Al(OH)^{2+} + 2OH^- =$$
$$Al(OH)_3(s)$$

Solution:
The set is not linearly independent. **The third equilibrium can be formed by reversing and adding the first two equilibria**:

$$Al^{3+} + 3OH^- = Al(OH)_3(s)$$
$$\underline{+ Al(OH)^{2+} = Al^{3+} + OH^-}$$
$$Al(OH)^{2+} + 2OH^- = Al(OH)_3(s)$$

In performing equilibrium calculations, you often have to manipulate known equilibria to write a target equilibrium in the form desired. Two skills are required: balancing chemical reactions and manipulating equilibria.

4.7.1 Balancing chemical reactions

Conservation of mass (Section 3.2.2) requires that chemical reactions are balanced. Balancing reactions is a five-step process (see also Appendix C, Section C.1.2). The process will be illustrated with the equilibrium between hypochlorous acid (HOCl, a common disinfectant) and chloride (Cl^-).

Step 1: Write the known reactants on the left and known products on the right

In the example, write: HOCl = Cl^-

Step 2: Adjust stoichiometric coefficients to balance all elements except H and O

Key idea: To balance a chemical reaction, balance all elements except H and O, add water to balance O, add H^+ to balance H, and add e^- to balance the charge

---

Example 4.7: **Balancing Chemical Reactions**

Balance the equilibrium between thiosulfate ($S_2O_3^{2-}$) and sulfate (SO_4^{2-}).

Solution:
Follow the procedure in the text:

Step 1: Write the known species
$$S_2O_3^{2-} = SO_4^{2-}$$
Step 2: Balance all but O, H
$$S_2O_3^{2-} = 2SO_4^{2-}$$
Step 3: Add H_2O to balance O
$$S_2O_3^{2-} + 5H_2O = 2SO_4^{2-}$$
Step 4: Add H^+ to balance H
$$S_2O_3^{2-} + 5H_2O =$$
$$2SO_4^{2-} + 10H^+$$
Step 5: Add e^- to balance the charge
$$S_2O_3^{2-} + 5H_2O =$$
$$2SO_4^{2-} + 10H^+ + 8e^-$$

The balanced reaction is

$$S_2O_3^{2-} + 5H_2O =$$
$$2SO_4^{2-} + 10H^+ + 8e^-$$

In the example, all elements except H and O already are balanced and you still have: $HOCl = Cl^-$. If you were balancing the equilibrium between molecular chlorine (Cl_2) and chloride, you would adjust the stoichiometric coefficients to balance the element Cl. You could accomplish this in either one of two equivalent ways: $Cl_2 = 2Cl^-$ or $\frac{1}{2}Cl_2 = Cl^-$.

Step 3: Add water (H_2O) to balance the element O

In the example, oxygen will be balanced if you add water as a product: $HOCl = Cl^- + H_2O$.

Step 4: Add H^+ to balance the element H

In the example, hydrogen will be balanced if you add one H^+ as a reactant: $HOCl + H^+ = Cl^- + H_2O$. At this point in the process, all elements should be balanced.

Step 5: Add electrons (e^-) to balance the charge

In the example, the left-hand side has a net charge of +1 and the right-hand side has a net charge of −1. You must add two e^- as reactants: $HOCl + H^+ + 2e^- = Cl^- + H_2O$. The equilibrium is now balanced.

Although it may seem slow and unwieldy at first, it is recommended that you follow this process step by step when balancing reactions. Another illustration of balancing reactions is given in Example 4.7. The balancing procedure is extended in Problem 4.7 at the end of the chapter.

4.7.2 Manipulating equilibria
After balancing the desired reaction, it may be possible to relate it to other equilibria by reversing equilibria, multiplying stoichiometric coefficients by a constant, and/or adding equilibria. It is important to master these skills before you proceed to the mechanics of equilibrium calculations in Part II of this text.

Consider the following example. In developing remediation strategies for a landfill leachate, you wish to know the equilibrium constant for the dissolution of cadmium carbonate solid to form divalent cadmium and carbon dioxide gas. You search tables of equilibrium constants and find the following information:

$$CdCO_3(s) = Cd^{2+} + CO_3^{2-} \qquad K_1 = 10^{-11.66}\ M^{2\dagger}$$

[†] Recall from Section 4.3 that the typical units of solids and liquids are the mole fraction. The mole fractions of pure solids [e.g., $CdCO_3(s)$] and pure liquids (e.g., H_2O) are unity. Thus, the units of pure solids and pure liquids do not appear in the equilibrium constants.

$$HCO_3^- = CO_3^{2-} + H^+ \qquad\qquad K_2 = 10^{-10.3} \text{ M}$$
$$H_2CO_3 = HCO_3^- + H^+ \qquad\qquad K_3 = 10^{-6.3} \text{ M}$$
$$H_2CO_3 = CO_2(g) + H_2O \qquad\qquad K_4 = 10^{1.5} \text{ atm-L/mol}$$

You can find the equilibrium constant for the target equilibrium by (1) writing the balanced equilibrium, and (2) manipulating the known equilibria to match the target equilibrium. First, balance the reaction by using the five-step process outlined in Section 4.6. (You may wish to try this first on your own.)

Step 1: Write the known reactants on the left and known products on the right

$$CdCO_3(s) = Cd^{2+} + CO_2(g)$$

Step 2: Adjust stoichiometric coefficients to balance all elements except H and O

All elements except H and O are balanced in Step 1

Step 3: Add water (H_2O) to balance the element O

$$CdCO_3(s) = Cd^{2+} + CO_2(g) + H_2O$$

Step 4: Add H^+ to balance the element H

$$CdCO_3(s) + 2H^+ = Cd^{2+} + CO_2(g) + H_2O$$

Step 5: Add electrons (e^-) to balance the charge

The charges are balanced in Step 4. The balanced reaction is:
$$CdCO_3(s) + 2H^+ = Cd^{2+} + CO_2(g) + H_2O$$

Now you need to combine the existing equilibria to obtain $CdCO_3(s) + 2H^+ = Cd^{2+} + CO_2(g) + H_2O$. Start with the K_1 equilibrium. It has $CdCO_3(s)$ and Cd^{2+} in the desired places as a reactant and product, respectively. You now have:

$$CdCO_3(s) = Cd^{2+} + CO_3^{2-} \qquad\qquad K_1 = 10^{-11.66} \text{ M}^2$$

You do not want carbonate (CO_3^{2-}) as a product. Thus, you need to add an equilibrium with carbonate as a reactant to cancel the carbonate as a product in the equilibrium you are building. This can be accomplished by reversing the K_2 equilibrium and adding:

$$
\begin{array}{ll}
CdCO_3(s) \;=\; Cd^{2+} + CO_3^{2-} & K_1 = 10^{-11.66}\ M^2 \\
+\ CO_3^{2-} + H^+ \;=\; HCO_3^- & K_5 = 1/K_2 = 10^{10.3}\ M^{-1} \\
\hline
CdCO_3(s) + H^+ \;=\; Cd^{2+} + HCO_3^- & K_6 = K_1 K_5 = 10^{-1.36}\ M
\end{array}
$$

Notice that you take the reciprocal of K_2 when you reverse the K_2 equilibrium. Also, you multiply the equilibrium constants when adding reactions. Again, you do not want bicarbonate (HCO_3^-) as a product. To eliminate HCO_3^-, reverse the K_3 equilibrium and add:

$$
\begin{array}{ll}
CdCO_3(s) + H^+ \;=\; Cd^{2+} + HCO_3^- & K_6 = 10^{-1.36}\ M \\
+\ HCO_3^- + H^+ \;=\; H_2CO_3 & K_7 = 1/K_3 = 10^{6.3}\ M^{-1} \\
\hline
CdCO_3(s) + 2H^+ \;=\; Cd^{2+} + H_2CO_3 & K_8 = K_6 K_7 = 10^{4.94}
\end{array}
$$

Finally, add the K_4 equilibrium to eliminate H_2CO_3 and have $CO_2(g)$ as a product:

$$
\begin{array}{ll}
CdCO_3(s) + 2H^+ \;=\; Cd^{2+} + H_2CO_3 & K_8 = 10^{4.94} \\
+\ H_2CO_3 \;=\; CO_2(g) + H_2O & K_4 = 10^{1.5}\ atm\text{-}L/mol \\
\hline
CdCO_3(s) + 2H^+ \;=\; Cd^{2+} + CO_2(g) + H_2O & \\
& K_9 = K_8 K_4 \\
& \quad = 10^{6.44}\ atm\text{-}L/mol
\end{array}
$$

Thus, the target equilibrium (cadmium carbonate solid in equilibrium with divalent cadmium and carbon dioxide gas) has an equilibrium constant equal to $10^{6.44}$ atm-L/mol.

4.7.3 Why learn how to manipulate equilibria?
In Section 4.7.2, you calculated an equilibrium constant for a target equilibrium by linearly combining other equilibria.

Thoughtful Pause
From Chapter 3, is there another way to calculate the
equilibrium constant for the target equilibrium?

Key idea: Equilibrium constants can be calculated by linearly combining other equilibria or from ΔG^o_{rxn} values

You know from Chapter 3 that you can calculate K from $\exp(-\Delta G^o_{rxn}/RT)$. The standard Gibbs free energy of formation values for $CdCO_3(s)$, H^+, Cd^{2+}, $CO_2(g)$, and H_2O are, respectively, -669.4, 0, -77.58, -394.37, and -237.18 kJ/mol. These values give $\Delta G^o_{rxn} = -39.73$ kJ/mol or $K = 10^{6.96}$ at $25^{\circ}C$. This is about three times larger than the value of $K = 10^{6.44}$ atm-L/mol determined in Section 4.7.2. Such discrepancies are not unusual, given the uncertainty in the thermodynamic data and errors in the equilibrium constants of the constituent equilibria. It is important to become adept at calculating equilibrium constants both from linearly combining other equilibria and from thermodynamics.

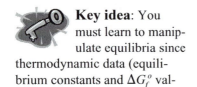

Key idea: You must learn to manipulate equilibria since thermodynamic data (equilibrium constants and ΔG_f^o values) are not always available

Why should you learn how to calculate equilibrium constants by manipulating known equilibria when you can always calculate equilibrium constants from thermodynamics? In a reasonably complex aqueous system, it is unlikely you will be able to look up all required equilibria in the form you want. In addition, the ΔG_f^o values may not be available for the species of interest. If ΔG_f^o values are not available, then ΔG_{rxn}^o cannot be calculated. Thus, it is extremely important to be able to manipulate equilibria to calculate equilibrium constants.

4.7.4 A systematic approach to manipulating equilibria: an advanced concept

The process described in Section 4.7.2 of linearly combining equilibria to create a target equilibrium may seem somewhat haphazard at first. For simple systems, it is often expedient to play with the equilibria as jigsaw puzzle pieces until the target equilibrium emerges (as was done in the cadmium carbonate example).

For more complex systems, you may wish to take a more systematic approach as follows. List the n constituent equilibria that are to be linearly combined to form the target equilibrium. Let m be the total number of chemical species in the n constituent equilibria and in the target equilibrium. Write the stoichiometric coefficients for each species in each equilibrium, using the usual notation that stoichiometric coefficients of reactants are negative and the stoichiometric coefficients of products are positive. Let ν_{ij} be the stoichiometric coefficient of species j in constituent equilibrium i. Now, multiply the stoichiometric coefficients of constituent equilibrium i by a constant c_i. For each species j, sum the values of $\nu_{ij}c_i$ over all the equilibria and set the sum equal to the stoichiometric coefficient of species j for the target equilibrium $(= \nu_{target,i})$: $\sum_{i=1}^{n} c_i \nu_{ij} = \nu_{tj}$. This will yield a linear set of m equations (one for each species) in n unknowns $(c_1, ..., c_n)$ that can be solved to find the c_i values. From the c_i values, you can determine the linear operations to perform on the constituent equilibria to form the target equilibrium and thus the target equilibrium K.

As an example, consider the $CdCO_3(s)$ problem of Section 4.7.2. The target equilibrium is:

$$CdCO_3(s) + 2H^+ = Cd^{2+} + CO_2(g) + H_2O \quad K_{target} = ?$$

The $n = 4$ constituent equilibria are:

$$\begin{array}{llll} \text{Eq. 1:} & CdCO_3(s) = Cd^{2+} + CO_3^{2-} & K_1 = 10^{-11.66} \\ \text{Eq. 2:} & HCO_3^- = CO_3^{2-} + H^+ & K_2 = 10^{-10.3} \\ \text{Eq. 3:} & H_2CO_3 = HCO_3^- + H^+ & K_3 = 10^{-6.3} \\ \text{Eq. 4:} & H_2CO_3 = CO_2(g) + H_2O & K_4 = 10^{1.5} \end{array}$$

The system has $m = 8$ chemical species: $CdCO_3(s)$, Cd^{2+}, $CO_2(g)$, H_2CO_3, HCO_3^-, CO_3^{2-}, H^+, and H_2O. Writing the matrix of stoichiometric coefficients:

Equil.	$CdCO_3(s)$	Cd^{2+}	$CO_2(g)$	H_2CO_3	HCO_3^-	CO_3^{2-}	H^+	H_2O
1	−1	1	0	0	0	1	0	0
2	0	0	0	0	−1	1	1	0
3	0	0	0	−1	1	0	1	0
4	0	0	1	−1	0	0	0	1
Target	−1	1	1	0	0	0	−2	1

Multiplying the stoichiometric coefficients of constituent equilibrium i by unknowns c_i:

Equil.	$CdCO_3(s)$	Cd^{2+}	$CO_2(g)$	H_2CO_3	HCO_3^-	CO_3^{2-}	H^+	H_2O
1	$-c_1$	c_1	0	0	0	c_1	0	0
2	0	0	0	0	$-c_2$	c_2	c_2	0
3	0	0	0	$-c_3$	c_3	0	c_3	0
4	0	0	c_4	$-c_4$	0	0	0	c_4
Target	−1	1	1	0	0	0	−2	1

Summing the columns for each chemical species and setting each sum equal to the stoichiometric coefficient for the target equilibrium yields eight linear equations:

$$-c_1 = -1 \qquad c_4 = 1 \qquad -c_2 + c_3 = 0 \qquad c_2 + c_3 = -2$$
$$c_1 = 1 \qquad -c_3 - c_4 = 0 \qquad c_1 + c_2 = 0 \qquad c_4 = 1$$

This system of linear equations can be solved (by inspection in this simple case) to reveal: $c_1 = 1$, $c_2 = -1$, $c_3 = -1$, and $c_4 = 1$. In other words, to obtain the target reaction, add equilibrium 1, the reverse of equilibrium 2 (i.e., multiply the stoichiometric coefficients of the species in equilibrium 2 by -1), the reverse of equilibrium 3, and equilibrium 4. Since free energies are additive: $\Delta G^o_{rxn,target} = \Delta G^o_{rxn,1} - \Delta G^o_{rxn,2} - \Delta G^o_{rxn,3} + \Delta G^o_{rxn,4}$ and $K_{target} = K_1 K_4 / (K_2 K_3)$. In general, the n linear and independent equations can be solved by the methods of linear algebra to determine the n values of c_1, ..., c_n. In matrix notation:

$$
\begin{bmatrix} v_{11} & v_{12} & \cdots & v_{1m} \\ v_{21} & v_{22} & \cdots & v_{2m} \\ \cdot & \cdot & \cdots & \cdot \\ \cdot & \cdot & \cdots & \cdot \\ v_{n1} & v_{n2} & \cdots & v_{nm} \end{bmatrix} \begin{pmatrix} c_1 \\ c_2 \\ \cdot \\ \cdot \\ c_m \end{pmatrix} = \begin{pmatrix} v_{t\,arg\,et1} & v_{t\,arg\,et2} & \cdots & v_{t\,arg\,etn} \end{pmatrix}
$$

where, again, v_{ij} is the stoichiometric coefficient of species j in constituent equilibrium i and v_{tj} is the stoichiometric coefficient of species j in the target equilibrium. In general for this approach: $K_{t\,arg\,et} = \prod_{i=1}^{n} K_i^{c_i}$.

4.8 SUMMARY

Chemical equilibria can be interpreted as both *chemical expressions* and *mathematical statements*. Chemical expressions show relationships between chemical species and mathematical statements show relationships between species concentrations. The chemical and mathematical forms of equilibria convey similar information about the way in which species and species concentrations relate to the equilibrium expression.

Knowing the equilibrium constant of one reaction allows you to calculate the equilibrium constants for a number of related reactions. Reversing an equilibrium results in a new equilibrium with K equal to the reciprocal of the K of the original equilibrium ($K_{new} = 1/K_{old}$). Multiplying the stoichiometric coefficients of an equilibrium by a constant (say, c) results in a new equilibrium with K equal to the K of the original equilibrium raised to the constant ($K_{new} = K_{old}^c$). Finally, adding two equilibria results in a new equilibrium with K equal to the product of the equilibrium constants of the original equilibria ($K_{new} = K_{old,1} K_{old,2}$).

4.9 PART I CASE STUDY: CAN METHYLMERCURY BE FORMED CHEMICALLY IN WATER?

Recall that the Part I case study involves deciding whether methylmercury, CH_3Hg^+, can be formed in water from $CH_4(aq)$ and Hg^{2+}. To answer this question, you must write a balanced reaction between reactants and products. Following the procedures introduced in this chapter, you can show that:

$$
CH_4(aq) + Hg^{2+} = CH_3Hg^+ + H^+
$$

From Section 3.11, the $\Delta \overline{G}_f^o$ values for $CH_4(aq)$, Hg^{2+}, and CH_3Hg^+ are, respectively, -34.4, $+164.4$, and $+112.9$ kJ/mol. The $\Delta \overline{G}_f^o$ value for H^+ is 0 kJ/mol. Thus, $\Delta G_{rxn}^o = -17.1$ kJ/mol, and $K = \exp(-\Delta G_{rxn}^o/RT) = 10^{3.0}$.

To see if methylmercury formation is significant, you will have to do a small equilibrium calculation. If concentrations can be used for activities, recall that:

$$K = [CH_3Hg^+][H^+]/([CH_4(aq)][Hg^{2+}])$$

Assume for the moment that the only methane-containing species are $CH_4(aq)$ and CH_3Hg^+ and the only Hg-containing species are Hg^{2+} and CH_3Hg^+. (This is a weak assumption: other Hg-containing species likely exist in this system.) If $[CH_3Hg^+]$ is given the symbol x, then: $[CH_4(aq)]$ = total methane $- x$ and $[Hg^{2+}]$ = total Hg $- x$. Thus:

$$K = x[H^+]/[(\text{total methane} - x)(\text{total Hg} - x)]$$

For $K = 10^{3.0}$, pH 7, total methane = 10^{-5} M, and total Hg = 10^{-7} M, you can calculate that the CH_3Hg^+ concentration is about 10^{-7} M. Thus, under these conditions, the CH_3Hg^+ concentration is small but much larger than the concentration of Hg^{2+}.

SUMMARY OF KEY IDEAS

- You must be able to exploit known equilibria to calculate species concentrations efficiently

- Equilibria can be written as chemical expressions and mathematical equations

- Chemical expressions show the reactants, products, and stoichiometric relationship between chemical species but not the relationship between chemical species concentrations

- Chemical expressions do not show the relationship between chemical species concentrations

- Never place brackets ([]) or braces ({}) around chemical species names when writing chemical expressions

- Chemical expressions of equilibria and mathematical equations of equilibria contain the same information and must lead to consistent conclusions

- Equilibrium constants are dimensionless, but units are associated with equilibrium constants as a reminder of the units of the species in the equilibrium

- Take great care when comparing equilibrium constants with different units

- ΔG^o_{rxn} values of reverse reactions are equal in absolute value but opposite in sign, and K values of reverse reactions are reciprocals of one another

- Multiplying stoichiometric coefficients by a constant multiplies ΔG^o_{rxn} by the constant and raises K to the constant

- When adding equilibria, add ΔG^o_{rxn} values and multiply K values [or add $\log(K)$ values]

- To balance a chemical reaction, balance all elements except H and O, add water to balance O, add H^+ to balance H, then add e^- to balance the charge

- Equilibrium constants can be calculated by linearly combining other equilibria or from ΔG^o_{rxn} values

- You must learn to manipulate equilibria since thermodynamic data (equilibrium constants and ΔG^o_f values) are not always available

HISTORICAL NOTE: HENRI-LOUIS LE CHATELIER AND LE CHATELIER'S PRINCIPLE

Henri-Louis Le Chatelier

Who was the man behind Le Chatelier's Principle? Henri Le Chatelier was a contemporary of J. Williard Gibbs (see the *Historical Note* in Chapter 3). Le Chatelier, son of an engineer, was trained as an engineer at École des Mines in Paris. He worked for the French government as a mining engineer in the Corps des Mines for two years. Rather unexpectedly, he was offered a professorship in chemistry at his alma mater. This was the beginning of a career in academia that would span 30 years at Parisian universities, including École des Mines, École Polytechnique, College de France, and the Sorbonne.

In contrast with Gibbs's theoretical brilliance, Le Chatelier was interested in the applications of chemistry. His training as a problem-solving engineer was reflected in his accomplishments. Le Chatelier's work on cements, thermodynamics, combustion, ammonia synthesis, and metallurgy was almost always motivated by practical problems.

The principle for which his name is linked was refined by Le Chatelier through a series of papers published in the 1880s. This version of the principle, published in 1880, echoes Newton's Third Law of Motion: "Every change of one of the factors of an equilibrium occasions a rearrangement of the system in such a direction that the factor in question experiences a change in a sense opposite to the original change." (Le Chatelier, 1888).

Le Chatelier devoted the later years of his career to social issues related to his science and engineering background, including educational reform and the scientific management of industry (called Taylorism, after Frederick W. Taylor). As a teacher, he changed chemical education by emphasizing general principles over memorization. It is not difficult to imagine the hundreds of students in *fin de siècle* lecture halls learning about Le Chatelier's Principle from Henri Le Chatelier himself.

PROBLEMS

4.1 If you add two equilibria (with equilibrium constants K_1 and K_2) to get a third equilibrium (with equilibrium constant K_3), you know from Section 4.6 that $K_3 = K_1 K_2$. Use this information to "prove" that Gibbs free energies are additive.

4.2 The equilibrium constant for the equilibrium $H_2O = H^+ + OH^-$ is 10^{-14} M^2. What is the equilibrium constant for the equilibrium $3H^+ + 3OH^- = 3H_2O$? What units are associated with this equilibrium constant?

4.3 An environmental engineer is seeking to find a way to recover silver from a metal finishing operation. She finds in a textbook an equilibrium constant of $10^{6.3}$ for the reaction $\frac{1}{2}Ag_2O(s) + H^+ = Ag^+ + \frac{1}{2}H_2O$. What is the equilibrium constant for the equilibrium containing the same information but with all stoichiometric coefficients integers less than 3?

4.4 An equilibrium expression is added to its reverse reaction to form a new equilibrium. What is the equilibrium constant of the new equilibrium? What is the Gibbs free energy of reaction for the new equilibrium? Do your answers make sense?

4.5 Balance the equilibrium between chromate ($Cr_2O_7^{2-}$) and trivalent chromium (Cr^{3+}).

4.6 Write a balanced equilibrium for the reaction with $Al(OH)^{2+}$ as a reactant and $Al(OH)_2^+$ as a product.

4.7 In balancing reactions, you sometimes may wish to write a final equilibrium in terms of OH^- rather than H^+. How would you modify the balancing procedure of Section 4.7.1 to produce an equilibrium in terms of OH^-? Illustrate your procedure with the chlorine example found in Section 4.7.1 or an equilibrium of your choosing.

4.8 Using your answer to Problem 4.6 and the constituent equilibria below, what is the equilibrium constant for the balanced equilibrium between $Al(OH)^{2+}$ and $Al(OH)_2^+$?

$$Al(OH)_3(s) + 3H^+ = Al^{3+} + 3H_2O \quad K_1 = 10^{10.8} \text{ M}^{-2}$$
$$Al(OH)_3(s) = Al(OH)^{2+} + 2OH^- \quad K_2 = 10^{-22.2} \text{ M}^3$$
$$Al^{3+} + 2OH^- = Al(OH)_2^+ \quad K_3 = 10^{18.5} \text{ M}^{-2}$$
$$(H_2O = H^+ + OH^-) \quad K_4 = 10^{-14} \text{ M}^2$$

4.9 Are the four constituent equilibria in Problem 4.8 linearly independent? Explain your answer.

4.10 Are the five equilibria in Problem 4.8 [four constituent equilibria plus the target equilibrium between $Al(OH)^{2+}$ and $Al(OH)_2^+$] linearly independent? Explain your answer.

4.11 By trial and error, determine the equilibrium constant for the equilibrium between $FeCO_3(s)$ as a reactant and $Fe(OH)_2(s)$ and $CO_2(g)$ as products. Use some or all of the constituent equilibria below.

$$FeCO_3(s) = Fe^{2+} + CO_3^{2-} \quad K_1 = 10^{-10.68} \text{ M}^2$$
$$Fe(OH)_2(s) = Fe^{2+} + 2OH^- \quad K_2 = 10^{-15.1} \text{ M}^3$$
$$HCO_3^- = CO_3^{2-} + H^+ \quad K_3 = 10^{-10.3} \text{ M}$$
$$H_2CO_3 = HCO_3^- + H^+ \quad K_4 = 10^{-6.3} \text{ M}$$
$$H_2CO_3 = CO_2(g) + H_2O \quad K_5 = 10^{1.5} \text{ atm-L/mol}$$
$$H^+ + OH^- = H_2O \quad K_6 = 10^{14} \text{ M}^{-2}$$

4.12 Repeat Problem 4.11 using the systematic method of Section 4.7.3.

PART II

Solution of Chemical Equilibrium Problems

A problem clearly stated is a problem half solved.
Dorothea Brande

When I'm working on a problem, I never think about beauty. I think only how to solve the problem. But when I have finished, if the solution is not beautiful, I know it is wrong.
R. Buckminster Fuller

CHAPTER 5

Getting Started with Chemical Equilibrium Calculations

5.1 INTRODUCTION

The second part of this text concerns the techniques used to calculate species concentrations at equilibrium. Part II is the heart of a quantitative investigation of chemical equilibria. It is critical that you master the material in Chapters 5 through 9 before moving on to Part III. In this chapter, the general principles behind the calculation of chemical species concentrations at equilibrium will be presented. In addition, a road map for Part II of the text is presented and discussed.

5.2 A FRAMEWORK FOR SOLVING CHEMICAL EQUILIBRIUM PROBLEMS

5.2.1 Analysis approaches

It is likely that you have some experiences in the analysis of systems from course work in thermodynamics, fluid mechanics, hydraulics, or water quality. These experiences will help you to understand the framework for the calculation of species concentrations at equilibrium. In all cases, the goal is the same: to determine the values of state variables in a defined space over a defined time (see Section 3.2 for a review of state variables). The space element and time interval that is being studied is called a ***control volume***.

control volume: the space element and time interval about which inferences are to be made

 Key idea: To analyze systems in general, define the control volume and state variables; identify mass, energy, and momentum fluxes crossing the control volume boundaries; and identify processes inside the control volume that affect the state variables

For the analysis problems in thermodynamics, fluid mechanics, hydraulics, and water quality, the same approach is used: define the control volume; enumerate the state variables; identify the fluxes of mass, energy, and momentum that *cross the boundaries* of the control volume; and identify processes occurring *inside* the control volume that affect the state variables of the system. Expressions for the state variables as functions of the independent variables (usually spatial dimensions, x, y, and z, and time) are then developed. The expressions must take into account the control volume geometry (through boundary conditions), control volume history (through initial conditions), fluxes, and processes. The resulting expressions constitute a mathematical model of the real world. As engineers and scientists, we attempt to solve the mathematical model to make statements about the state variables in the control volume.

5.2.2 Application to chemical equilibrium modeling

For chemical equilibrium modeling, a similar approach is used. However, the characteristics of chemical equilibria allow for simplification of the analysis.

Thoughtful Pause

How do the characteristics of chemical equilibria allow us to simplify the analysis process?

Knowing that the system is at equilibrium means that the *state variables (e.g., temperature, pressure, and species concentrations) do not change with time* (Section 3.3.2). The time-invariant nature of systems at equilibrium eliminates the need to consider changes in species concentrations with time. How do you know that a system is at equilibrium? This can be a difficult question to answer. For now, assume that the decision has been made to use an equilibrium model. Nonequilibrium models are discussed in more detail in Chapter 22.

The general analysis approach discussed in Section 5.2.1 has four steps: (1) define the control volume, (2) enumerate the state variables, (3) identify pertinent fluxes that cross the boundaries of the control volume, and (4) identify processes affecting the state variables that occur inside the control volume. Each of these steps will be discussed in more detail below.

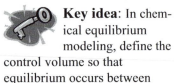

Key idea: In chemical equilibrium modeling, define the control volume so that equilibrium occurs between phases in the control volume

Control volume and fluxes. The choice of the control volume is influenced by fluxes. A flux means that mass, energy, and/or momentum are transported across the control volume boundary. At equilibrium, no transport of mass occurs across the boundary of the control volume. This means that the control volume must be defined carefully to include material exchange with the phases of interest. Chemical equilibrium modeling involves defining the control volume so that equilibrium occurs between *phases in the control volume.*[†]

State variables. The choice of state variables is narrowed in most chemical equilibrium models. In many environmental applications, the temperature and pressure are known.[††] Thus, the state variables to be solved for are the equilibrium concentrations of the chemical species. The remaining question is: which species? In chemical equilibrium calculations, an important step is to *identify the species of interest*.

[†] There are important differences between a mass flux analysis at steady-state and an equilibrium model. The steady-state and equilibrium models are compared in Appendix B.

[††] In this text, it is assumed that the system temperature and pressure are known and constant. Thus, it is assumed that solutions are sufficiently dilute so that heat generated or absorbed by chemical reactions does not affect the temperature and pressure of the system. For some problems (especially gas or nonaqueous phase reactions in low heat capacity media), the effects of reaction on system temperature and pressure must be included.

Processes affecting the state variables. Finally, the standard analysis approach includes the identification of processes affecting the state variables. In chemical equilibrium modeling, you know that *species concentrations are constrained by thermodynamics*. In particular, the equilibrium expressions must be satisfied. Additional constraints on species concentrations will be discussed in Section 5.5 and Chapter 6.

The standard analysis approach now can be applied to chemical equilibrium modeling. First, *define the system*. In particular, you must decide if the system is allowed to equilibrate with gas and/or solid phases. This decision is very important and is discussed in more detail in Section 5.3. Second, *identify the chemical species to be considered*. Techniques for enumerating chemical species are discussed in Section 5.4. Last, you must *identify the equilibrium and other constraints* on the species concentrations. This aspect of chemical equilibrium modeling is discussed in Section 5.5.

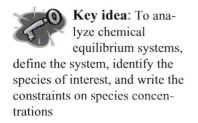

Key idea: To analyze chemical equilibrium systems, define the system, identify the species of interest, and write the constraints on species concentrations

5.3 INTRODUCTION TO DEFINING THE CHEMICAL SYSTEM

How do you define the control volume for a chemical system? In many instances, the choice is dictated by (1) the physical system to be modeled, and (2) how closely you wish the mathematical model to describe the physical system in question. For example, an equilibrium model for groundwater chemistry would likely include equilibria between dissolved species and minerals. Dissolution/precipitation equilibria should be included since groundwater generally is in intimate contact with the surrounding soil. An equilibrium model for groundwater might exclude gas/liquid exchange with the atmosphere and still be a reasonable approximation of reality.

As the groundwater example shows, aqueous systems may (or may not) be in equilibrium with gas or solid phases. Do gas or solid phases exist in a given chemical system? This question shall be explored separately for gas and solid phases. Typically, a gas phase exists if we allow it to exist. In other words, we usually *define* the system to include or exclude a gas phase. What is meant by a gas phase? Gases, at environmentally important temperatures and pressures, exist as uncharged molecules (or atoms). Each molecule (or atom) may exist as two distinct chemical species: a *dissolved gas* species and a gaseous species. For example, molecular oxygen may be found in the environment as dissolved oxygen, denoted $O_2(aq)$, and gaseous oxygen, $O_2(g)$. If the system is defined to include a gas phase, then we allow for the existence of gaseous oxygen. Allowing for gaseous oxygen to exist has two ramifications. First, the system now has an additional chemical species, $O_2(g)$. Second, an equilibrium expression that describes the equilibrium partitioning of oxygen between gas and water must be included. In other words, an equilibrium between $O_2(g)$ and $O_2(aq)$ must be included in the system model.

Solid phases are more problematic. First, it is sometimes difficult to know if a solid phase will exist at equilibrium. For example, if you add a small amount of NaCl to water, common experience tells you that the salt likely will dissolve completely. What if you add some calcium carbonate to water? It is difficult to know whether the calcium carbonate will dissolve completely or whether some solid calcium carbonate will remain at equilibrium without performing the calculation (or experiment). Second, it is possible for more than one solid phase to exist in a system. (In environmental applications, we rarely consider more than one gas phase.) The complexities involved when more than one solid phase exists are discussed in Section 19.4.2.

5.4 INTRODUCTION TO ENUMERATING CHEMICAL SPECIES

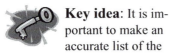 **Key idea**: It is important to make an accurate list of the chemical species in the equilibrium model

To determine whether you have enough information to calculate species concentrations at equilibrium, you must first count the number of species in the control volume. Again, the choice of which species to include in the model depends on both the physical system and how accurately you wish your model to describe the physical system. In fact, it is quite easy to miss important species. For example, suppose you tried to predict the effects of equilibration with the atmosphere on the pH (and thus, the H^+ activity) of a carbonated cola drink. If your model of the cola drink included only dissolved carbon dioxide and associated chemical species, the effects of opening the bottle on the pH would be greatly overestimated. Why? The pH of the soda is in fact controlled by the large amount of phosphoric acid in the mixture. Without phosphoric acid (and related species) in the list of important species, an equilibrium model cannot predict the pH of the cola drink. On the other hand, carbon dioxide exchange alone *can* be used to predict accurately the pH of clean rainwater. However, other chemical species (such as sulfate, nitrate, and chloride) must be included to calculate the pH of acid rain.

species list: the list of chemical species to be evaluated in an equilibrium calculation

Throughout this text, the roster of chemical species to be included in the equilibrium model shall be referred to as the *species list*. Techniques for developing a species list are discussed in Section 6.3.

5.5 INTRODUCTION TO DEFINING THE CONSTRAINTS ON SPECIES CONCENTRATIONS

So far, chemical systems have been described from the standpoint of aquatic chemistry. Now it is time to think about the translation of chemical systems into mathematical expressions to be solved. Suppose your species list contains n species.

Thoughtful Pause
How many unknowns are there in a chemical system consisting of n species?

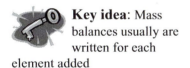

Key idea: To determine the equilibrium concentrations of n species, n linearly independent equations are required and may be found through mathematical versions of equilibrium expressions, mass balances, a charge balance, and equations from the definition of the standard state of mixtures

If the temperature and pressure are known, a system of n chemical species has n unknowns. The unknowns are the n equilibrium concentrations of the n species. How many equations are required to solve a mathematical system of n unknowns? You need n equations.[†] Now you must put your chemistry hat on again and come up with n equations to solve the system of n unknowns.

The process of finding n equations in the n equilibrium concentrations will be described in more detail in Section 6.4. There are four types of equations: mathematical versions of equilibrium expressions, mass balances, a charge balance, and equations from the definition of the standard state of mixtures.

5.5.1 Equilibrium expressions

From Section 4.2.2, equilibria can be written as both chemical expressions and mathematical equations. To calculate the concentrations at equilibrium, all pertinent equilibria must be translated into mathematical equations. How many equilibrium expressions will result? The number of equilibrium expressions will vary with the size and nature of the chemical system. However, *every aqueous system has at least one equilibrium*: the equilibrium between H_2O, H^+ and OH^-.

5.5.2 Mass balance expressions

Mass is, of course, conserved. The mass of certain elements is added to the system through the starting materials. The added mass can be redistributed through equilibria into many chemical species. For example (see Section 1.2.2 for details), the addition of copper sulfate to a reservoir results in the formation at equilibrium of at least 10 copper-containing species. For a closed system of constant volume without solids, conservation of mass means that the sum of the masses of each element at equilibrium must be equal to the added mass of the element.[††]

Key idea: Mass balances usually are written for each element added

As an example, suppose you add enough cyanide to water to generate an initial cyanide concentration of 1×10^{-3} M. Imagine that, through equilibria, the added cyanide is partitioned into two species, each of which is formed from 1 mole of cyanide per mole of species. The mass balance constraint requires that the sum of the concentrations of the two cyanide-containing species must be 1×10^{-3} M.

[†] Please note that to solve a system of n unknowns, n *linearly independent* equations are required. See Section 4.6.2 for a discussion of the application of linear independence to chemical equilibria.

[††] Remember that mass, *not concentration*, is conserved. In the case where the volume of the system and the density of the fluid are constant, conservation of mass can be expressed through concentration since mass = concentration × volume. Mass balances usually are expressed as mole balances. Although the *total* number of moles of material in a system are not necessarily conserved, the number of moles of, say, carbon are conserved. Mass balances sometimes are called *material balances*.

How many mass balance equations are required? Again, the answer depends on the complexity of the system. In general, mass balance equations are written for each element added (see Section 6.4.2 for a more detailed discussion). If electron transfer reactions are not considered, you write mass balance equations for each oxidation state of each element added. For all but the simplest system (i.e., pure water), there is at least one mass balance equation.

5.5.3 Charge balance expression

Water is electrically neutral in the environment. This fact constrains the possible values of species concentrations and will be referred to as the *electroneutrality constraint*. The electroneutrality constraint means that the sum of the positive charges in solution (contributed by cations) must be equal to the sum of the negative charges in solution (contributed by anions). For each chemical system, you write *one* charge balance equation to ensure compliance with the electroneutrality constraint.

5.5.4 Other constraints on species concentrations

There are two other important constraints on the concentrations of chemical species (and therefore constraints on the unknowns). First, species concentrations cannot be less than zero. Thus, you can reject all negative roots of equations involving species concentrations. Second, the choice of concentration units and standard states of mixtures[†] leads to defined activities of certain species. For most systems of environmental interest, *the mole fractions of pure liquids (such as water) and pure solids are taken to be unity* (see also Section 4.3). This choice of the standard state of mixtures greatly simplifies many equilibrium calculations.

 **Key idea**: Species concentrations are constrained by the fact that the sum of the positive charges in solution must be equal to the sum of the negative charges in solution (called the *electroneutrality constraint*)

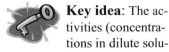 **Key idea**: The activities (concentrations in dilute solution) of pure liquids (such as water) and pure solids are taken to be unity

5.6 PART II ROAD MAP

Part II of this book contains the techniques needed to calculate the equilibrium concentrations of chemical species. The examples provided in Part II are fairly simple. More complex systems will be encountered later in the text. The solution techniques used to solve complex systems are identical to the methods developed in Part II. Thus, it is imperative that you master each chapter of this part of the text.

In Chapter 6, the governing equations constraining species concentrations at equilibrium are presented in more detail. By the conclusion of Chapter 6, a complete mathematical description of a chemical system at equilibrium can be written.

[†] The standard state of a mixture is different from the standard state of an element (described in Section 3.6.2). The standard state of a mixture defines when the activity is equal to the concentration and when the activity (in the chosen concentration scale) is unity.

Chapters 7 through 9 provide techniques to solve the system of mathematical equations developed in Chapter 6. Three approaches are presented: manual, graphical, and computer calculations. For a given chemical system, one solution technique may be superior to the others. Thus, it is important that you master each solution method. Developing a facility with at least one computer method described in Chapter 9 is especially important. Computer methods allow for the solution of more complex systems comprising many chemical species.

5.7 SUMMARY

In this chapter, you learned about a framework for the analysis of chemical systems at equilibrium. To analyze chemical equilibrium systems, three tasks must be accomplished. First, the system is defined as in equilibrium with gas or solid phases or not in equilibrium with gas or solid phases. Second, a species list is generated to identify the species of interest. Third, the constraints on species concentrations are written. Constraints stem from equilibrium expressions, mass balances, and electroneutrality. In addition, species concentrations must be nonnegative and comply with the thermodynamic standard states (i.e., pure liquids and solids have activities of unity). Techniques to complete the three tasks are presented in Chapter 6.

5.8 PART II CASE STUDY: HAVE YOU HAD YOUR ZINC TODAY?

The case study for Part II of this text concerns a dietary supplement containing a metal. Trace metals are necessary for human life. They play vital roles in oxygen transport, enzyme function, and endocrine system activity. If trace metal concentrations in food and water are insufficient, mineral supplements may be indicated.

The chemical form of the supplement and water chemistry of the mouth and stomach often dictate the speciation of the metal in the digestive system. The form of the metal may be critical in determining whether or not the metal is absorbed by the body and transported to the appropriate site. Formation of the "wrong" species can result in excretion of the metal and loss if its beneficial effects.

This case study will explore the chemical speciation of zinc in a zinc acetate supplement. Consider the consumption of a zinc acetate tablet containing 20 mg of zinc as Zn. Assume the tablet is dissolved in 100 mL of water and consumed. The zinc-containing water will pass through the mouth (saliva pH = 6.5-8.5) and into the stomach (gastric juice pH = 1-3). The case study question is: How do the dominant forms of zinc change as th pH of the water is reduced from the pH 6.5-8.5 range to the pH 1-3 range when zinc acetate is consumed?

SUMMARY OF KEY IDEAS

- To analyze systems in general, define the control volume and state variables; identify mass, energy, and momentum fluxes crossing the control volume boundaries; and identify processes inside the control volume that affect the state variables

- In chemical equilibrium modeling, define the control volume so that equilibrium occurs between phases in the control volume

- To analyze chemical equilibrium systems, define the system, identify the species of interest, and write the constraints on species concentrations

- It is important to make an accurate list of the chemical species in the equilibrium model

- To determine the equilibrium concentrations of n species, n linearly independent equations are required and may be found through mathematical versions of equilibrium expressions, mass balances, a charge balance, and equations from the definition of the standard state of mixtures

- Mass balances usually are written for each element added

- Species concentrations are constrained by the fact that the sum of the positive charges in solution must be equal to the sum of the negative charges in solution (called the *electroneutrality constraint*)

- The activities (concentrations in dilute solution) of pure liquids (such as water) and pure solids are taken to be unity

CHAPTER 6

Setting Up Chemical Equilibrium Problems

6.1 INTRODUCTION

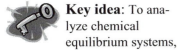 **Key idea**: To analyze chemical equilibrium systems, define the system, identify the species of interest, and write the constraints on species concentrations

In this chapter, a unified approach to setting up chemical equilibrium problems will be developed. By the end of the chapter, you will be able to generate a species list and write the equations controlling the species concentrations at equilibrium. This chapter expands on the principles of chemical systems at equilibrium that were developed in Chapter 5. The general approach from Chapter 5 is repeated in the **Key idea** at the left.

It may seem a little pedantic to devote an entire chapter to developing the governing equations for species concentrations at equilibrium. The equally important tasks of setting up the equations (Chapters 5 and 6) and solving them (Chapters 7 through 9) have been separated in this text on purpose. It is very important to master the *procedure* for setting up the equations before you attempt to *solve* the equations. The steps necessary to build a mathematical model for a chemical system at equilibrium are developed in sections 6.2 through 6.4 and summarized in section 6.5.

6.2 DEFINING THE CHEMICAL SYSTEM

open system: an aqueous system in equilibrium with a gas phase

closed system: an aqueous system not in equilibrium with a gas phase

As discussed in section 5.2, you must define the system to include or exclude gas phases. A system that includes a gas phase is called an *open system.* If equilibration with a gas phase is not allow, the system is *closed*. The labels "open" and "closed" refer only to the presence of a gas phase and should not be interpreted as indicating the presence or absence of solid phases.

Systems may undergo transitions from closed to open states (or from open to closed states). If groundwater is pumped to the surface as a drinking water source, some chemical constituents of the system may equilibrate with the atmosphere. As another example, consider an unopened bottle of soda. You could calculate the equilibrium concentrations of key chemical species (e.g., dissolved carbon dioxide) in the soda. Your experience tells you that the concentration of dissolved carbon dioxide will be significantly less after the bottle is opened and the carbon dioxide equilibrates with the atmosphere. At equilibrium (i.e., after the soda goes flat), the total pressure and the dissolved carbon dioxide concentration inside the bottle change. It is important to note that the systems (i.e, the control volumes) for the closed-bottle case and open-bottle case are different. The control volume is the soda bottle only in the first case and the soda bottle plus the atmosphere in the second case (see Figure 6.1).

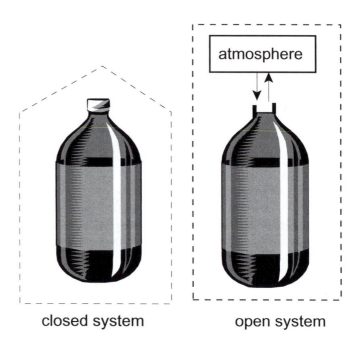

Figure 6.1: Control Volumes for Open and Closed Systems
(control volumes indicated by dashed lines)

6.3 ENUMERATING CHEMICAL SPECIES

 Key idea: To enumerate species, identify the starting materials, identify the initial hydrolysis products, enumerate subsequent hydrolysis products, and enumerate other reaction products

As discussed in Section 5.4, it is necessary to develop a list of the species to be considered in the equilibrium model. Four steps are necessary to generate an accurate species list: identification of the starting materials, identification of the initial hydrolysis products, enumeration of subsequent hydrolysis products, and enumeration of other reaction products. Each step will be discussed in more detail below.

6.3.1 Step 1: Identify the starting materials

starting materials: chemicals added to water before equilibration reactions begin

In all environmental applications, you know something about the chemical composition of the system prior to starting the equilibrium calculations. The chemicals added to water initially are called *starting materials* (also called *the recipe*; Morel and Hering, 1993). Frequently, the total concentrations of some elements are known because the starting materials are known. In the copper sulfate example of Section 1.2.2, the total mass (and, hence, the total concentration at constant system volume) of copper was known.[†]

It is important to correctly identify starting materials for two reasons. First, you will use the total concentrations of starting materials as constraints on the concentrations of individual species (see Section 6.4.2). Second, knowledge of the starting materials is critical in generating a list of species in the system. The focus in this section is on this second reason for identifying starting materials.

The chemical composition of the starting materials often affects the equilibrium composition of the system. For example, equilibrium concentrations may be different if copper is added in the form of copper chloride or copper carbonate instead of in the form of copper sulfate. For the ammonia example in Chapter 4, you need to know if the source of ammonia is ammonia gas or ammonium chloride or ammonium hydroxide prior to performing equilibrium calculations.

You will find it useful to adopt the convention that water is always a starting material. This makes sense since aquatic systems obviously always contain water. Although it is easy to forget, *always include water as a starting material.*

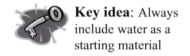

Key idea: Always include water as a starting material

6.3.2 Step 2: Identify the initial hydrolysis products

As stated in Section 1.2, the primary focus of this text is on chemical species that react with water. To generate a species list, you must ascertain how the added material will dissociate to form the initial hydrolysis products. Dissociation (from the Latin *dis* not + *sociare* to join) refers here to the initial bond-cleaving reactions that form multiple chemical species from the added material. For example, when sodium chloride is added to water, it dissociates at least in part to form sodium ions (Na^+) and chloride ions (Cl^-). Note that dissociation *does not have to be complete*. We are **not** saying at this point that *all* the NaCl dissociates to form Na^+ and Cl^- but only that some (as yet unknown) concentrations of Na^+ and Cl^- are formed by the addition of NaCl to water. The *complete dissolution* of added materials is discussed in Section 6.4.4. Recall from Section 1.4 that salts such as NaCl dissociate in water because water can hydrate the dissolved ions and reduce the attractive forces between the ions.

In most cases, *the initial dissociation of starting materials is very fast.* You can identify the initial hydrolysis products by performing a thought experiment where the first dissociation processes are isolated. For now, you do not care what happens to the initial dissociation products. The job at hand is merely to identify them.

Example 6.1: **Dissociation**

What are the initial dissociation products for the dissolution of $Fe(NH_4)_2(SO_4)_2$ (ferrous ammonium sulfate), $K_2Cr_2O_7$ (potassium dichromate), and ClO_2 (chlorine dioxide) in water?

[†] As your chemical intuition grows, it will no longer be necessary to be restricted to problems in which the starting materials are known. In the copper example, you can solve for equilibrium concentrations if you know (say, by measurement) the total concentrations of dissolved copper, dissolved chloride, dissolved carbonate, and so on.

Solution:

From $Fe(NH_4)_2(SO_4)_2$, common ions can be formed through initial dissociation to Fe^{2+}, NH_4^+, and SO_4^{2-}. Note that the iron is in the +II oxidation state (see Appendix A).

The starting material $K_2Cr_2O_7$ contains potassium and would likely dissociate to form K^+. This leaves $Cr_2O_7^{2-}$ initially.

For ClO_2, no common ions can be formed. The Cl in ClO_2 is in the +IV oxidation state, so dissociation to Cl^- is not allowed (see text).

In all cases, H_2O is also a starting material. Thus, H^+ and OH^- also are initial dissociation products.

Thus:

$Fe(NH_4)_2(SO_4)_2$ and water dissociate initially to Fe^{2+}, NH_4^+, SO_4^{2-}, H^+, and OH^-; $K_2Cr_2O_7$ and water dissociate initially to form K^+, $Cr_2O_7^{2-}$, H^+, and OH^-; and ClO_2 likely does not dissociate initially, so only H^+ and OH^- are formed as initial dissociation products of water.

 Key idea: When considering whether dissociation occurs, look for the formation of simple ions (Table 6.1) without a change in the oxidation state

The sodium chloride example was very simple. How do you know what chemical species are formed when the starting materials dissociate? How do you even know if the starting material dissociates at all? As you work through this text, you will develop an intuition about dissociation. A few rules will help you get started in developing an intuitive feel for dissociation.

Look for the formation of simple ions. To decide whether dissociation occurs, see if the starting materials form simple, common ions. For example, in most aqueous systems of environmental interest, compounds containing sodium or potassium dissociate to form Na^+ ions and K^+ ions. A list of common ions is provided in Table 6.1. Other examples of the formation of simple ions as initial dissociation products are given in Example 6.1.

Table 6.1: Ions Frequently Formed as Initial Dissociation Products

Ion Type	Examples
Cations	
monoatomic[1]	H^+, K^+, Na^+, Ca^{2+}, Mg^{2+}
polyatomic	NH_4^+
Anions	
monoatomic	Cl^-, Br^-, S^{2-}
polyatomic[2]	OH^-, SO_4^{2-}, CO_3^{2-}, HCO_3^-, NO_3^-, PO_4^{3-}

Notes: 1. Also any other metal ion of charge $+n$, M^{n+}
　　　　　2. Called hydroxide, sulfate, carbonate, bicarbonate, nitrate, and phosphate ion, respectively

You must be careful about using the list of ions in Table 6.1. It is *incorrect* to change the oxidation state when dissociation occurs (see Appendix A for a review of the concept of the oxidation state). A change in oxidation state means that electron transfer has occurred. Electron transfer (or redox) equilibria will be discussed in Chapter 16. For example, it is proper to allow HCl to dissociate into H^+ and Cl^-, because both the chlorine in HCl and the chlorine in Cl^- are in the $-I$ oxidation state. In addition, the hydrogen in HCl and the hydrogen in H^+ are both in the +I oxidation state. However, it is **not** proper to allow hypochlorous acid (HOCl) to dissociate to Cl^-.

Thoughtful Pause

Why is it incorrect to allow HOCl to dissociate to Cl^-?

Hypochlorous acid cannot be allowed to dissociate into Cl because the chlorine in HOCl is in the +I oxidation state and the chlorine in Cl⁻ is in the −I oxidation state.

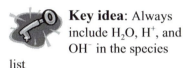

Key idea: An H^+ usually is available for transfer to water if it is O or N in the starting material

Look for H⁺ transfer. As you will see in Chapter 11, the transfer of H^+ to water (also called proton transfer or acid/base chemistry) is an important part of aquatic chemistry. Always look for H^+ transfer to water. In general, *an H^+ is usually available for transfer to water if it is bonded to O or N in the starting material*. Thus, you might think to dissociate HNO_3 to H^+ and NO_3^- and you might think to dissociate NH_4^+ to NH_3 and H^+, but you would **not** dissociate CH_4 to H^+ and CH_3^-. Why? The H in methane (CH_4) is bonded to C in the starting material, not to O or N.[†]

Always include H₂O, H⁺, and OH⁻ in the species list. Aqueous systems, by definition, always contain water. Recall from Section 6.3.1 that water should be considered a starting material in all aquatic equilibrium systems. As you will see later in this chapter, water dissociates to H^+ and OH^-. Therefore, H_2O, H^+, and OH^- are always in the species list.

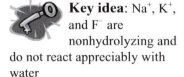

Key idea: Always include H₂O, H⁺, and OH⁻ in the species list

6.3.3 Step 3: Enumerate subsequent hydrolysis products

After the starting materials have been dissociated, expand the species list by adding the chemical species formed upon hydrolysis of the initial dissociation products. A useful technique for determining subsequent hydrolysis products is to ask *repeatedly* whether a given dissociation product reacts with water. If a given initial dissociation product does not react with water, abandon it for the moment and move to the next dissociation product. Some chemical species, called ***nonhydrolyzing species***, do not react appreciably with water. Common examples include Na^+, K^+, and F^-.

nonhydrolyzing species: chemical species that do not react appreciably with water

If a given initial dissociation product *does* react with water, then write the hydrolysis products. Question whether the new hydrolysis products react with water. This process continues with each new generation of species until no new hydrolysis products are generated.

Key idea: Na^+, K^+, and F^- are nonhydrolyzing and do not react appreciably with water

---

[†] In general, H^+ will dissociate from a species (and be available for transfer) if H and the atom it is bonded to have very different affinities for electrons. A difference in electron affinity leads to a highly polar bond (see Section 1.4) called an *ionic bond*. Examples include the H-N bond, H-O bond, and H-Cl bond. If H and the atom it is bonded to have similar affinities for electrons, the bond is called a *covalent bond* and H^+ dissociation is much less likely.

6.3.4 Step 4: Enumerate other reaction products

In the final step of generating a species list, you must assess whether any species on the list equilibrates with other species on the list. This process again continues with each new generation of species until no new species are generated. How do you know whether two or more species react? You will develop an intuition for conjuring up equilibria and reaction products as you work through this text. A useful tool is to examine lists of chemical equilibria and look for reactions involving the species on the species list. If the species list contains all the reactants (or all the products) in an equilibrium, then include all the products (or all the reactants) from the equilibrium in the species list. For example, if the species list contains HCO_3^- and H^+, then H_2CO_3 should be added to the list because of the equilibrium: $H_2CO_3 = HCO_3^- + H^+$. This approach also helps generate a list of equilibria to be considered for inclusion in the model. Examples of equilibria in water are found in Appendix D. If the system is open, be sure to include species resulting from equilibria partitioning of mass between gas and aqueous phases.

Key idea: Use equilibria to determine whether species on the species list react to form other species

6.3.5 An example of generating a species list

An example will make the process of generating a species list a bit clearer. Let us generate a species list for the addition of ammonium chloride (NH_4Cl) to water in a closed system. (You may wish to try this on your own before reading further.) The starting materials are $NH_4Cl(s)$ and H_2O (don't forget water as a starting material!). Now allow each starting material to dissociate. Ammonium chloride will dissociate to form the common ions ammonium and chloride. The other starting material (water) will dissociate to form H^+ and OH^-. After allowing only the starting materials to dissociate, the chemical system is

Starting materials: $NH_4Cl(s)$, H_2O
Species list: $NH_4Cl(s)$, H_2O, NH_4^+, Cl^-, H^+, and OH^-

You now have four species (NH_4^+, Cl^-, H^+, and OH^-) that came from the dissociation of starting material. You must ask whether any of these species hydrolyze. To answer this question, *look for equilibria for which you already have all the reactants or all the products in the species list.* Recall the following equilibrium from Chapter 4:

$$NH_4^+ = NH_3 + H^+$$

The species list contains all the reactants (only ammonium in this case), so the products must be added to the species list. The species list now is

Species list: $NH_4Cl(s)$, H_2O, NH_4^+, Cl^-, H^+, OH^-, and NH_3

CO_2 & H_2O CO_2 HCO_3^-
H, OH^-, H_2CO_3,

***Example 6.2:* Generating a Species List**

Write a species list for an open system in equilibrium with $CO_2(g)$ in the atmosphere.

Solution:
The starting materials are $CO_2(g)$ and H_2O. The first dissociation products of water, as usual, are H^+ and OH^-. To decide on the hydrolysis products of carbon dioxide, refer to the pertinent equilibria in Problem 4.8 (repeated here):

$$HCO_3^- = CO_3^{2-} + H^+$$
$$H_2CO_3 = HCO_3^- + H^+$$
$$H_2CO_3 = CO_2(g) + H_2O$$

The first hydrolysis product of $CO_2(g)$ is H_2CO_3. To determine subsequent reaction products, note that the species list contains all the reactants (only H_2CO_3) of the middle equilibrium. Thus, the products (HCO_3^-) should be added to the species list. Similarly, CO_3^{2-} is added because all the reactants from the first equilibrium (only HCO_3^-) are on the species list.

Thus, the species list is $CO_2(g)$, H_2CO_3, HCO_3^-, CO_3^{2-}, H_2O, H^+, and OH^-.

What about Cl^-? From Section 3.8.2, you are aware of the equilibrium: $HCl = H^+ + Cl^-$. All the products in this equilibrium are in the species list, and so the reactant, HCl, must added to the species list. The system becomes

Starting materials: $NH_4Cl(s)$, H_2O
Species list: $NH_4Cl(s)$, H_2O, NH_4^+, Cl^-, H^+, OH^-, NH_3, and HCl

The process of generating a species list for this example is summarized in Figure 6.2. Diagrams such as Figure 6.2 are useful in several ways. They allow you to look for possible interacts between species in a partially completed species list. In addition, they show clearly from which starting material each species comes. For example, in Figure 6.2, it is clear that NH_4^+ and NH_3 both come from ammonium chloride. This information will be helpful in generating mass balances. Another illustration of generating a species list is given in Example 6.2.

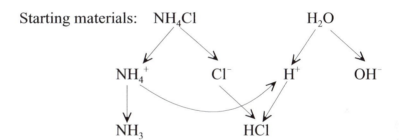

Figure 6.2: Species List Diagram for the Addition of Ammonium Chloride to Water

6.3.6 When is the species list complete?
It is sometimes difficult to know if you have included *all* pertinent species in a species list. How do you know if the species list is complete? For many chemical equilibrium calculations, you can use existing lists of equilibria to make sure that all pertinent species and equilibria have been included. Lists of equilibria are provided in Appendix D of this text.

Proceed with caution when using existing lists of equilibria. A given species may be easily ignored in some applications but may be important in others. From the copper example in Chapter 1, Cu^{2+} is important for algal toxicity, even though its concentration in the aquatic environment at equilibrium is *very* small. On the other hand, we usually ignore the species $N_2(aq)$ in environmental systems because it is unreactive, even though its concentration is relatively large (about 5.2×10^{-4} M in equilibrium with the

Na_2CO_3

H_2O

OH^-

Na

CO_3

H^+, HCO_3^-,

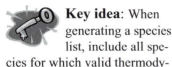

Key idea: When generating a species list, include all species for which valid thermodynamic data are available

atmosphere at 25°C). When in doubt, *include all species for which valid thermodynamic data are available.*

As an example of when to stop adding species, consider whether the species NH_2^- should be included in the example of Section 6.3.5. After all, you can write the equilibrium:

$$NH_3 \; = \; NH_2^- + H^+$$

This equilibrium is not commonly listed as important in aqueous systems. You can show (upon completion of Part II of this text) that the equilibrium concentration of NH_2^- in a solution formed by adding 1×10^{-3} M NH_4Cl to a closed system of pure water is about 10^{-23} M, or about 6 ions per liter of water. Is NH_2^- important? The answer depends on the application, of course. However, it is unlikely that a species present at several ions per liter will influence the chemistry of the system significantly.

6.4 DEFINING THE CONSTRAINTS ON SPECIES CONCENTRATIONS

The species list leads to a mathematical system with n unknowns; namely, the equilibrium concentrations of the n species. As discussed in Section 5.5, the n equations necessary to solve for the n species concentrations at equilibrium are of four types: equilibrium expressions, mass balances, electroneutrality, and constraints based on the standard state of mixtures. (In addition, the species concentrations must be nonnegative.) Each of these constraints will be discussed in more detail below.

6.4.1 Equilibrium constraints

Key idea: Always include the self-ionization of water as an equilibrium

self-ionization of water: the equilibrium $H_2O \; = \; H^+ + OH^-$

ion product of water (K_W): the equilibrium constant for the self-ionization of water (= 1.01×10^{-14} at 25°C)

Every aquatic chemical system has at least one equilibrium. The ubiquitous equilibrium is the expression relating H_2O, H^+, and OH^-. This equilibrium is called the ***self-ionization of water*** and is given by:

$$H_2O \; = \; H^+ + OH^- \qquad K = K_W$$

The equilibrium constant for this reaction is called the ***ion product of water*** and is given a special symbol, K_W. At 25°C, $K_W = 1.01 \times 10^{-14}$. *Always include the self-ionization of water in aqueous equilibrium calculations.*

What other equilibria should be included in an equilibrium model? You should include all equilibria populated solely by the species in the species list. In fact, you may have compiled this list of equilibria already as you were generating the species list. Thus, the list of pertinent species and list of pertinent equilibria are intertwined.

The slate of equilibria developed during the generation of the species list may have to be modified before inclusion in the equilibrium model. Recall that you need n linearly *independent* equations to solve a system of n unknowns. Thus, *the equilibria to be included in the model must be linearly independent.* Remember from Section 4.6.2 that a set of equilibria is linearly independent if no equilibrium expression can be formed by linearly combining the other equilibria. In this context, "linearly combining" means reversing equilibria, multiplying stoichiometric coefficients by a constant, and/or adding equilibria. *Always check to see that the equilibria in the equilibrium list are linearly independent.*

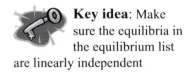

Key idea: Make sure the equilibria in the equilibrium list are linearly independent

Returning to the ammonium chloride example, a list of *potential* equilibria is as follows (from Section 4.6.1, 6.3.5, and the self-ionization of water):

$$NH_4^+ = NH_3 + H^+ \qquad K_1$$
$$NH_4^+ + OH^- = NH_3 + H_2O \qquad K_2$$
$$HCl = H^+ + Cl^- \qquad K_3$$
$$Cl^- + H_2O = HCl + OH^- \qquad K_4$$
$$H_2O = H^+ + OH^- \qquad K_W$$

(The subscripts on the equilibrium constants in this list, except for K_W, are arbitrary.)

Thoughtful Pause

Is this a legitimate list of equilibria for the NH_4Cl/H_2O system?

The equilibria in the list are not linearly independent and thus the list cannot be used as is. For example, the K_2 equilibrium can be formed by reversing the K_W reaction and adding it to the K_1 equilibrium. Similarly, the K_4 equilibrium can be formed by reversing the K_3 reaction and adding it to the K_W equilibrium. A proper (and linearly independent) set of equilibria would be K_W, either K_1 or K_2, and either K_3 or K_4. From the equilibria listed, the four sets of linearly independent equilibria are (K_W, K_1, K_3); (K_W, K_1, K_4); (K_W, K_2, K_3); and (K_W, K_2, K_4).[†] For this example, assume you have selected the set K_W, K_1, and K_3. This choice may seem arbitrary now (and it is somewhat arbitrary), but it will appear to be more natural after you learn about acid-base chemistry in Chapter 11.

You now need to translate these four equilibrium expressions into mathematical equations. (Please try this on your own before continuing.) The chemical statement of the system is:

[†] Four other sets of equilibria also describe the system: (K_1, K_2, K_3); (K_1, K_2, K_4); (K_1, K_3, K_4); and (K_2, K_3, K_4). In this text, we shall emphasize an approach in which the self-ionization of water always is included in the equilibrium list.

Starting material: $NH_4Cl(s)$, H_2O
Species list: $NH_4Cl(s)$, H_2O, NH_4^+, Cl^-, H^+, OH^-, NH_3, and HCl
Equilibria:

$$NH_4^+ = NH_3 + H^+ \qquad K_1$$
$$HCl = H^+ + Cl^- \qquad K_3$$
$$H_2O = H^+ + OH^- \qquad K_W$$

If the solution is dilute enough so that activities and concentrations are interchangeable, the mathematical equivalent of the chemical system becomes:

Unknowns: $[NH_4Cl(s)]$, $[H_2O]$, $[NH_4^+]$, $[Cl^-]$, $[H^+]$, $[OH^-]$, $[NH_3]$, and $[HCl]$

Equilibria: $K_1 = [NH_3][H^+]/[NH_4^+]$
$\qquad\qquad K_3 = [H^+][Cl^-]/[HCl]$
$\qquad\qquad K_W = [H^+][OH^-]/[H_2O]$

You now have eight unknowns and three equations - five equations to go.

🔑 **Key idea**: Translate the equilibrium expressions into mathematical equations (as described in Chapter 4)

6.4.2 Mass balance constraints

The next equations are provided by mass balances. Recall from Section 5.5.2 that mass balances are written for each element represented in the species list (ignoring H and O and writing different mass balances for each oxidation state if electron transfer is allowed). In one-phase systems (here, a closed system without solids), mass balances are true constraints on species concentrations: the mass added must be equal to the mass in the closed system at equilibrium. The use of mass balances in multiphase systems will be considered in Part IV of this text.

As you approach mass balances for the first time, it is probably easiest to write mass balances on each element or oxidation state. For example, if you add sulfuric acid (H_2SO_4) to water, you may write a mass balance on sulfur: the sum of the mass of sulfur in all species must equal the mass of sulfur added. However, mass balances do not have to be written on elements. Mass balances can be written on non-dissociating fragments of starting materials. For the example of sulfuric acid, note that sulfuric acid equilibrates with bisulfate (HSO_4^-) and sulfate (SO_4^{2-}). Each species contains the fragment SO_4. Thus, instead of writing a mass balance on S, you could write a mass balance on the fragment SO_4, called *total sulfate*. Here: total sulfate = $[H_2SO_4] + [HSO_4^-] + [SO_4^{2-}]$. As another example, acetic acid (formally, ethanoic acid, CH_3COOH) hydrolyzes to acetate (CH_3COO^-), but the CH_3COO group remains intact. It would make sense to perform a mass balance on total acetate = $[CH_3COOH] + [CH_3COO^-]$. Formally, any element, oxidation state, or fragment of starting materials about which a mass balance is written is called a ***component*** of the system.

component: an element, oxidation state, or fragment of starting materials about which a mass balance is written

Four aspects of the mass balances of the system must be ascertained. First, the number of mass balance equations must be determined. One mass balance must be created for each element, oxidation state, or nondissociating fragment in the system.

Thoughtful Pause

How many mass balance equations are required for the NH_4Cl/H_2O system?

In the NH_4Cl/H_2O system, two elements other than H and O (namely, N and Cl) must be balanced. Thus, you need two mass balance equations. In this example, mass balances will be performed on total soluble N(–III) (= N_T) and total soluble Cl(–I) (= Cl_T).[†]

Second, the total concentration of each element, oxidation state, or nondissociating fragment must be listed. The total concentration can be determined by the starting materials or measured values. For example, you may know the *total* amount of iron in the +II oxidation state through measurements, but you may not know the concentrations of *individual* species (Fe^{2+}, $Fe(OH)^+$, etc.) making up the total Fe(+II). In the current example $N_T = Cl_T$ since all the N(–III) and all the Cl(–I) in the system come from ammonium chloride (see Figure 6.2). For example, if you add 1×10^{-3} mole of NH_4Cl to 1 L of water, then $N_T = Cl_T = 1\times10^{-3}$ M. Note that **both** N_T and Cl_T are 1×10^{-3} M since the system contains 1×10^{-3} M of N(–III) and 1×10^{-3} M of Cl(–I).

Third, almost all the soluble species in the species list must be associated with at least one mass balance. In assigning species to mass balances, ignore the species containing only H and O (since mass balances are not written for H and O). Typically, this means you do not assign H^+, OH^-, or H_2O to mass balances. Species list diagrams (such as Figure 6.2) can assist in assigning species to components. Thus, the generation of the species list and mass balance equations are intertwined.

Thoughtful Pause

In the chapter example, to which mass balance should each species in the species list be assigned?

[†] In this text, the subscript *T* will be used to denote total concentrations, as in Cl_T. Another common approach is to label the total concentration with the prefix *TOT*, as in *TOT*Cl. A special kind of mass balance on H^+, called the *proton condition*, will be introduced in Section 8.5.2.

Key idea: Remember that each species concentration in a mass balance equation is multiplied by a stoichiometric coefficient equal to the number of equivalents of the total mass per mole of the species

---

***Example 6.3*: Mass Balance**

A metal plating wastewater contains cyanide, iron, nickel, and zinc. Develop a mass balance equation for total cyanide in the wastewater if the species list contains the following cyanide-bearing species: HCN, CN^-, $Fe(CN)_6^{3-}$, $Ni(CN)_4^{2-}$, $Zn(CN)_2(aq)$, $Zn(CN)_3^-$, and $Zn(CN)_4^{2-}$.

Solution:
It makes sense to conduct a mass balance on the fragment CN (called total cyanide = CN_T) since each species contains the CN fragment. We must determine how many times the fragment CN appears in each species. This is easy to determine from the molecular formulas. Thus:

CN_T = (1 eq/mole)[HCN]
   + (1 eq/mole)[CN^-]
   + (6 eq/mole) [$Fe(CN)_6^{3-}$]
   + (4 eq/mole)[$Ni(CN)_4^{2-}$]
   + (2 e-
q/mole)[$Zn(CN)_2(aq)$]
   + (3 eq/mole)[$Zn(CN)_3^-$]
   + (4 eq/mole)[$Zn(CN)_4^{2-}$]

Or:

CN_T = **[HCN] + [CN^-] + 6[$Fe(CN)_6^{3-}$] + 4[$Ni(CN)_4^{2-}$] + 2[$Zn(CN)_2(aq)$] + 3[$Zn(CN)_3^-$] + 4[$Zn(CN)_4^{2-}$]**

In the NH_4Cl/H_2O system, the species NH_4^+ and NH_3 would be assigned to N_T and the species Cl^- and HCl would be assigned to Cl_T. [The species $NH_4Cl(s)$ is not assigned to a component because it is a solid and not a soluble species.]

Third, determine the stoichiometry of the element (or oxidation state or fragment) in each species associated with a mass balance. In the example, NH_4^+ and NH_3 each contain one N (the basis of N_T) and the species Cl^- and HCl each contain one Cl (the basis of Cl_T). In determining the stoichiometry, you are using the concept of equivalents (see Section 2.6). In other words, NH_4^+ and NH_3 each contain one equivalent relative to N_T and the species Cl^- and HCl each contain one equivalent relative to Cl_T.

To perform the mass balance, add the products of the mass balance stoichiometric coefficients and the species concentrations and set the sum equal to the total concentration of the element, oxidation state, or fragment. In the example:

$$N_T = (1 \text{ eq/mole})[NH_4^+] + (1 \text{ eq/mole})[NH_3]$$
$$Cl_T = (1 \text{ eq/mole})[Cl^-] + (1 \text{ eq/mole})[HCl]$$

The coefficients mean that NH_4^+ contains 1 mole (or one equivalent) of N per mole of NH_4Cl, NH_3 contains 1 mole (or one equivalent) of N per mole of NH_3, and so on. We commonly leave off the stoichiometric coefficients if they are equal to 1. *Do not forget that the stoichiometric coefficients are present.* The stoichiometric coefficients are really there for units conversion, to convert the units of mole/L of species to eq/L of total mass.

The mathematical equivalent of the chemical system becomes:

Unknowns: [$NH_4Cl(s)$], [H_2O], [NH_4^+], [Cl^-], [H^+], [OH^-], [NH_3], and [HCl]

Equilibria: $K_1 = [NH_3][H^+]/[NH_4^+]$
       $K_3 = [H^+][Cl^-]/[HCl]$
       $K_W = [H^+][OH^-]/[H_2O]$

Mass balances: $N_T = [NH_4^+] + [NH_3]$
        $Cl_T = [Cl^-] + [HCl]$

Another illustration of the development of mass balance equations is shown in Example 6.3. In the NH_4Cl/H_2O system, you now have five equations to describe the eight unknowns - three more equations to go.

6.4.3 Electroneutrality (charge balance) constraint
The starting materials, including water, almost always are uncharged. If you perform a number balance on electrons, the system at equilibrium also must have a net charge of zero. This is expressed in the ***electroneutrality constraint*** or ***charge balance***.

(Note that the coefficients have units of eq CN/mole of species = number of CN fragments per molecule or ion of species.)

---

electroneutrality constraint (*charge balance*): the requirement that the sum of the positive charges is equal to the sum of the negative charges

The electroneutrality constraint requires that the sum of the positive charges is equal to the sum of the negative charges. For a system of constant water volume, the number of equivalents of positive charge per liter must be equal to the equivalents of negative charge per liter. You calculate the equivalents of positive or negative charge per liter by summing the products of the charge on each cation (or anion) and the cation (anion) concentration.

Thoughtful Pause
For the NH_4Cl/H_2O system, what is the charge balance equation?

The charge balance in the ammonium chloride example is:

$$(1 \text{ eq/mole})[NH_4^+] + (1 \text{ eq/mole})[H^+] =$$
$$(1 \text{ eq/mole})[Cl^-] + (1 \text{ eq/mole})[OH^-]$$

Here, "1 eq/mole" means "one equivalent of charge per mole" (i.e., the charge on the species is +1 or −1). We usually write the charge balance without the units on the coefficients:

$$[NH_4^+] + [H^+] = [Cl^-] + [OH^-]$$

Key idea: Remember that each species concentration in the charge balance equation is multiplied by a stoichiometric coefficient equal to the number of equivalents (charges) per mole of the species

Again, do not forget that each species in the charge balance is multiplied by a coefficient. The coefficient represents the number of charges on the species and has units of equivalents of charge/mole (or charges/ion).

There are several important features of the charge balance equation. First, each chemical system has *only one* charge balance equation. Second, neutral species do not appear in the charge balance equation. Third, $[H^+]$ always appears on the cation side of the charge balance and $[OH^-]$ always appears on the anion side of the charge balance in aqueous systems since H^+ and OH^- always are in the species list.

The mathematical equivalent of the NH_4Cl/H_2O chemical system becomes:

Unknowns: $[NH_4Cl(s)]$, $[H_2O]$, $[NH_4^+]$, $[Cl^-]$, $[H^+]$, $[OH^-]$, $[NH_3]$, and $[HCl]$

Equilibria: $K_1 = [NH_3][H^+]/[NH_4^+]$
$K_3 = [H^+][Cl^-]/[HCl]$
$K_W = [H^+][OH^-]/[H_2O]$

Mass balances: $N_T = [NH_4^+] + [NH_3]$
$Cl_T = [Cl^-] + [HCl]$

Charge balance: $[NH_4^+] + [H^+] = [Cl^-] + [OH^-]$

You now have six equations and eight unknowns and need two more equations.

6.4.4 Other constraints

The final equations in the model come from the activities of pure liquids and solids in the standard state for mixtures adopted in Sections 5.5.4 and 4.3. Water in freshwater systems is considered a pure liquid. Thus, we have a new equation: activity of water = 1.

What about $NH_4Cl(s)$? To understand the constraints on $[NH_4Cl(s)]$, think about the solubility of solids. You will take up the case of the solubility of solids in Chapter 19. For now, you should recognize that solids either exist or do not exist. A solid does not exist if it dissolves completely when added to water. If the solid does not exist, its concentration is zero. If a pure solid exists, it has unit activity. It is fairly common to have salts as starting materials. A *salt* is a compound in which the transferable H^+ has been replaced by another cation. If the cation is K^+, Na^+, or NH_4^+, then the salt is very soluble in water. For example, the solubility of ammonium chloride in water at 25°C is about 370 g/L (almost 7 M). Thus, a small amount of ammonium chloride added to water is expected to dissolve completely. The assumption of complete dissolution (or complete dissociation) is appropriate for many common salts containing K^+, Na^+, or NH_4^+ (e.g., NaOH, NaCl, NaHCO$_3$, and KOH). With the assumption of complete dissolution, you have generated another equation: $[NH_4Cl(s)] = 0$.

You now have a complete mathematical model as follows:

salt: a compound in which the transferable H^+ is replaced by another cation

Unknowns: $[NH_4Cl(s)]$, $[H_2O]$, $[NH_4^+]$, $[Cl^-]$, $[H^+]$, $[OH^-]$, $[NH_3]$, and $[HCl]$

Equilibria: $K_1 = [NH_3][H^+]/[NH_4^+]$
$K_3 = [H^+][Cl^-]/[HCl]$
$K_W = [H^+][OH^-]/[H_2O]$

Mass balances: $N_T = [NH_4^+] + [NH_3]$
$Cl_T = [Cl^-] + [HCl]$

Charge balance: $[NH_4^+] + [H^+] = [Cl^-] + [OH^-]$

Other: activity of $H_2O = 1$ (standard state)
$[NH_4Cl(s)] = 0$ (complete dissolution)

This is a system of eight equations in eight unknowns. Techniques for solving this mathematical system will be discussed in Chapters 7 through

9. To solve the system, you require numerical values of K_1, K_3, and K_W and numerical values for the total masses of N_T and Cl_T.

6.5 REVIEW OF PROCEDURES FOR SETTING UP EQUILIBRIUM SYSTEMS

It is imperative that you become very familiar with the process of setting up equilibrium systems described in Sections 6.2-6.4. Several shortcuts are possible and will be discussed later. For now, it is recommended that the procedures presented here be followed closely. For convenience, the procedures are summarized below and in Appendix C.2:

Step 1: Define the chemical system
    Goal: Decide if an open or closed model is to be used
    Tool: Match the model to the system to be modeled

Step 2: Generate a species list
    Goal: Create a list of n unknowns (equilibrium concentrations
        of n species)
    Tools: Identify the starting materials
        Always include H_2O as a starting material
        Identify the initial hydrolysis products
            Look for the formation of simple ions
                (as shown in Table 6.1)
            Look for H^+ transfer
            Always include H_2O, H^+, and OH^-
        Enumerate subsequent hydrolysis products
        Enumerate other reaction products
        Generate a species list diagram (see Figure 6.2)

Step 3: Define the constraints on species concentrations
    Goal: Create a list of n equations
    Tools: Equilibrium constraints
        Always include the self-ionization of water
        Mass balance constraints
            Identify elements, oxidation states, or fragments
                to serve as the bases of mass balances
            Find the total concentration of mass to be balanced
            Associate each soluble species with at least one
                mass balance (exceptions: H_2O, H^+, and OH^-)
            Determine the stoichiometry of each soluble
                species with respect to its component(s)
        Electroneutrality constraint (charge balance)
        Other constraints

Activity of pure liquids = 1
Activity of pure solids = 1 (if solid exists)
Activity of solids = 0 (if solid dissolves completely)

6.6 CONCISE MATHEMATICAL FORM FOR EQUILIBRIUM SYSTEMS: AN ADVANCED TOPIC

It is possible to write the mathematical form of a closed equilibrium system in a very concise format. Let C_i be the equilibrium concentration of species i. Assume the system has n_s species, n_{eq} equilibria, and n_{mb} components. For each equilibrium, let v_{ij}^{eq} be the stoichiometric coefficient of species i in equilibrium j.[†] For a given equilibrium j, v_{ij}^{eq} will be zero for any species not involved in the reaction. Recall from Chapters 3 and 4 that v_{ij}^{eq} is negative for reactants and positive for products.

For each component, let v_{ij}^{mb} be the stoichiometric coefficient of species i in mass balance j.[††] Again, for a given component, v_{ij}^{mb} will be zero for any species that is not a part of the mass balance. Let TOT_j be the total concentration of total mass j.

Finally, let v_{ij}^{ch} be the signed charge on species i (< 0 for anions, > 0 for cations, and equal to zero for neutral species). The mathematical form of a closed equilibrium system can be written as:

$$\text{Equilibria: } K_j = \prod_{i=1}^{n_s} C_i^{v_{ij}^{eq}}, \text{ for } j = 1 \text{ to } n_{eq} \qquad \text{eq. 6.1}$$

$$\text{Mass balances: } TOT_j = \sum_{i=1}^{n_s} v_{ij}^{mb} C_i, \text{ for } j = 1 \text{ to } n_{mb} \qquad \text{eq. 6.2}$$

$$\text{Charge balance: } 0 = \sum_{i=1}^{n_s} v_i^{ch} C_i \qquad \text{eq. 6.3}$$

Other constraints:

$$C_i \geq 0 \text{ for all } i \qquad \text{eq. 6.4}$$

activity of i = 1 if species i is a pure liquid or   eq. 6.5
a pure solid

activity of i = 0 if species i is a solid that   eq. 6.6
dissolves completely

[†] Previously (Sections 3.7.1 and 4.4), the stoichiometric coefficients for chemical reactions were given the symbol v_i. At the risk of confusion, symbols here are changed to emphasize that there are several types of stoichiometric coefficients in chemical equilibrium systems. The superscripts indicate the type of stoichiometric coefficients: *eq* for equilibrium, *mb* for mass balances, and *ch* for charge balances.

[††] This is analogous to the stoichiometric coefficient used in Section 3.7.2 to calculate standard enthaplies of formation (= v_i = number of occurrences of element j in species i).

This mathematical form is the basis of the computer calculation methods described in Chapter 9. To test your understanding of these equations, you may wish to confirm that the mathematical form of the NH_4Cl/H_2O system can be written by using equations 6.1 through 6.6.

6.7 SUMMARY

A detailed protocol for setting up chemical equilibrium problems was developed in this chapter. The first step is to define the chemical system; that is, to decide if an open or closed model is to be used.

The second step is to generate a species list. This can be accomplished by identifying the starting materials (always include H_2O as a starting material), identifying the initial hydrolysis products (by looking for the formation of simple ions and H^+ transfer), enumerating subsequent hydrolysis products, and enumerating other reaction products. It is useful to draw a species list diagram to identify the source of each hydrolysis species.

The third step is to define the constraints on species concentrations. The constraints stem from equilibria (including the self-ionization of water), mass balances, electroneutrality (charge balance), and other considerations. Mass balance constraints can be determined by identifying components, finding the total concentration of each component, associating each soluble species with a component (except for H_2O, H^+, and OH^-), and determining the stoichiometry of each soluble species with respect to its component(s). Only one charge balance equation exists and it insures that the sum of the signed charges is zero. Other common constraints are that the activity of a pure liquid or solid is unity, the activity of a solid is zero if it dissolves completely, and all species concentrations must be nonnegative.

6.8 PART II CASE STUDY: HAVE YOU HAD YOUR ZINC TODAY?

Recall from Section 5.8 that the Part II case study involves determining the form of zinc when a zinc acetate tablet dissolves in your mouth and is swallowed. Having competed this chapter, you should be able to set up the chemical system for zinc acetate in water.

The procedure reviewed in Section 6.5 will be followed. Please try each step on your own before reading the answer. The first step is to define the chemical system. For describing the water chemistry in your mouth and stomach, a closed system appears to be reasonable. Thus, the control volume will not include a gas phase.

The second step is to make a species list. First, the starting materials need to be identified. For the problem at hand, the starting materials are zinc acetate solid [$Zn(CH_3COO)_2(s)$, abbreviated here $Zn(Ac)_2(s)$] and water.

Second, identify the initial hydrolysis products. It is logical to assume that $Zn(Ac)_2(s)$ will dissolve in water to form Zn^{2+} and Ac^-. Water, of course, forms H^+ and OH^- as initial hydrolysis products.

Third, enumerate subsequent hydrolysis products. To generate a list of subsequent hydrolysis products, it is useful to ask whether each of the initial hydrolysis products reacts with water. You know that H^+ and OH^- do not react further with water, so you need only examine whether Zn^{2+} and Ac^- react with water. By examining the equilibria in Table D.1, it is apparent that Zn^{2+} reacts with water (or the water hydrolysis product OH^-) to form $ZnOH^+$. Soluble zinc hydrolysis products, formed from subsequent reactions with OH^- (from Table D.1), include $Zn(OH)_2^0$, $Zn(OH)_3^-$, and $Zn(OH)_4^{2-}$. From Table D.3, it appears that the insoluble hydrolysis product $Zn(OH)_2(s)$ also should be included in the species list. How about Ac^-? It is logical that acetate (Ac^-) might accept a proton from water. The hydrolysis product (acetic acid, or HAc) also can be seen in Table D.1.

Fourth, enumerate other reaction products. To accomplish this task, examine lists of equilibria to see if all the products or all the reactants of a given equilibrium are on the preliminary species list. Table D.2 lists several soluble species that can be formed by the reaction between Zn^{2+} and Ac^-. The new species include $ZnAc^+$, $Zn(Ac)_2^0$, and $Zn(Ac)_3^-$. The resulting species list is as follows:

Species list: $Zn(Ac)_2(s)$, Zn^{2+}, Ac^-, H_2O, H^+, OH^-, $ZnOH^+$, $Zn(OH)_2^0$, $Zn(OH)_3^-$, $Zn(OH)_4^{2-}$, $Zn(OH)_2(s)$, HAc, $ZnAc^+$, $Zn(Ac)_2^0$, and $Zn(Ac)_3^-$

Fifth, generate a species list diagram. Although this step is not required, it does provide a nice summary of the species formed and their origin. Also, species list diagrams are a good way to check visually whether other species may be formed. A species list diagram for the $Zn(Ac)_2(s)/H_2O$ system is shown in Figure 6.3.

The third step is to define the constraints on the species concentrations. First, write a list of pertinent equilibria. From the tables in Appendix D, a suitable list of equilibria can be found (subscripts on the equilibrium constants, except for K_W, are arbitrary):

$$\begin{aligned}
H_2O &= H^+ + OH^- & K_W \\
Zn^{2+} + OH^- &= ZnOH^+ & K_1 \\
Zn^{2+} + 2OH^- &= Zn(OH)_2^0 & K_2 \\
Zn^{2+} + 3OH^- &= Zn(OH)_3^- & K_3 \\
Zn^{2+} + 4OH^- &= Zn(OH)_4^{2-} & K_4 \\
Zn^{2+} + 2OH^- &= Zn(OH)_2(s) & K_5 \\
HAc &= H^+ + Ac^- & K_6 \\
Zn^{2+} + Ac^- &= ZnAc^+ & K_7
\end{aligned}$$

$$Zn^{2+} + 2Ac^- = Zn(Ac)_2^0 \qquad K_8$$
$$Zn^{2+} + 3Ac^- = Zn(Ac)_3^- \qquad K_9$$

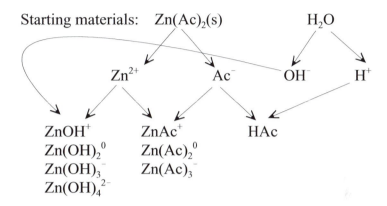

Starting materials: $Zn(Ac)_2(s)$ H_2O

Figure 6.3: Species List Diagram for the $Zn(Ac)_2(s)/H_2O$ System

Equilibrium constraints are given by the mathematical forms of the equilibria and will be listed at the end of this section.

The fourth step is to write the mass balance constraints. You can verify that two mass balances have to be written. Logical choices are to write mass balances for $Zn(+II)_T$ and Ac_T. The resulting mass balances on soluble species are:

$$Zn(+II)_T = [Zn^{2+}] + [ZnOH^+] + [Zn(OH)_2^0] + [Zn(OH)_3^-] + [Zn(OH)_4^{2-}] + [ZnAc^+] + [Zn(Ac)_2^0] + [Zn(Ac)_3^-]$$

$$Ac_T = [Ac^-] + [HAc] + [ZnAc^+] + 2[Zn(Ac)_2^0] + 3[Zn(Ac)_3^-]$$

Make sure that you understand where the mass balance equations come from and can explain the coefficient of each term in the mass balances.

The fifth step is to write the electroneutrality constraint. The charge balance for this system is:

$$[H^+] + 2[Zn^{2+}] + [ZnOH^+] + [ZnAc^+] = [OH^-] + [Zn(OH)_3^-] + 2[Zn(OH)_4^{2-}] + [Zn(Ac)_3^-]$$

Again, make sure that you can justify the stoichiometric coefficient of each term in the charge balance equation.

The sixth step is to list other constraints in the system. In solving this system, we shall assume that $Zn(Ac)_2(s)$ dissolves completely and that its activity is zero. In addition, we shall assume that $Zn(OH)_2(s)$ does not form at all. (You can verify this assumption in the pH range of 2-6.5 after completing Chapter 19.) Finally, the activity of pure water will be assumed to be unity. As a final assumption, we shall assume in this case study that activities and concentrations are interchangeable.

After eliminating the ignored species and translating the equilibria into mathematical constraints, the final mathematical form of the system is as follows:

Unknowns:
$[Zn(Ac)_2(s)]$, $[Zn^{2+}]$, $[Ac^-]$, $[H_2O]$, $[H^+]$, $[OH^-]$, $[ZnOH^+]$, $[Zn(OH)_2^0]$, $[Zn(OH)_3^-]$, $[Zn(OH)_4^{2-}]$, $[HAc]$, $[ZnAc^+]$, $[Zn(Ac)_2^0]$, and $[Zn(Ac)_3^-]$

Equilibria:
$$K_W = [H^+][OH^-]/[H_2O]$$
$$K_1 = [ZnOH^+]/([Zn^{2+}][OH^-])$$
$$K_2 = [Zn(OH)_2^0]/([Zn^{2+}][OH^-]^2)$$
$$K_3 = [Zn(OH)_3^-]/([Zn^{2+}][OH^-]^3)$$
$$K_4 = [Zn(OH)_4^{2-}]/([Zn^{2+}][OH^-]^4)$$
$$K_6 = [H^+][Ac^-]/[HAc]$$
$$K_7 = [ZnAc^+]/([Zn^{2+}][Ac^-])$$
$$K_8 = [Zn(Ac)_2^0]/([Zn^{2+}][Ac^-]^2)$$
$$K_9 = [Zn(Ac)_3^-]/([Zn^{2+}][Ac^-]^3)$$

Mass balances:
$$Zn(+II)_T = [Zn^{2+}] + [ZnOH^+] + [Zn(OH)_2^0] + [Zn(OH)_3^-] + [Zn(OH)_4^{2-}] + [ZnAc^+] + [Zn(Ac)_2^0] + [Zn(Ac)_3^-]$$

$$Ac_T = [Ac^-] + [HAc] + [ZnAc^+] + 2[Zn(Ac)_2^0] + 3[Zn(Ac)_3^-]$$

Charge balance:
$$[H^+] + 2[Zn^{2+}] + [ZnOH^+] + [ZnAc^+] = [OH^-] + [Zn(OH)_3^-] + 2[Zn(OH)_4^{2-}] + [Zn(Ac)_3^-]$$

Other:
activity of $H_2O = 1$ (standard state)
$[Zn(Ac)_2(s)] = 0$ (complete dissolution)
activities and concentrations are interchangeable

The mathematical system consists of 14 unknowns and 14 equations. The 14 equations include 9 equilibria, 2 mass balances, 1 charge balance, and 2 other constraints. You can verify that the equations are linearly independent.

SUMMARY OF KEY IDEAS

- To analyze chemical equilibrium systems, define the system, identify the species of interest, and write the constraints on species concentrations

- To enumerate species, identify the starting materials, identify the initial hydrolysis products, enumerate subsequent hydrolysis products, and enumerate other reaction products

- Always include water as a starting material

- An H usually is available for transfer to water if it is bonded to O or N in the starting material

- Always include H_2O, H^+, and OH^- in the species list

- Na^+, K^+, and F^- are nonhydrolyzing and do not react appreciably with water

- Use equilibria to determine whether species on the species list react to form other species

- When generating a species list, include all species for which valid thermodynamic data are available

- Always include the self-ionization of water as an equilibrium

- Make sure the equilibria in the equilibrium list are linearly independent

- Translate the equilibrium expressions into mathematical equations (as described in Chapter 4)

- Remember that each species concentration in a mass balance equation is multiplied by a stoichiometric coefficient equal to the number of equivalents of the total mass per mole of the species

- Remember that each species concentration in the charge balance equation is multiplied by a stoichiometric coefficient equal to the number of equivalents per mole of the species (charges per ion)

PROBLEMS

6.1 Write a complete mathematical model for a closed system consisting of water only at equilibrium. Be sure to include starting material(s), a species list, equilibria (if any), mass balances (if any), a charge balance equation, and other constraints. Make sure you have n independent equations for n unknowns. Assuming the activity and concentration are interchangeable, how do $[H^+]$ and $[OH^-]$ compare at equilibrium?

6.2 Given your answer to Problem 6.1 and that $K_W = 10^{-14}$ at 25°C, can you guess why pH 7 is called *neutral* pH? A fluid with $[H^+]$ or $[OH^-]$ greater than 1 M is no longer called water. What is the approximate pH range of water? (Your answer is approximate because you have ignored the differences between activity and concentration in this problem.)

6.3 Write a complete mathematical model for a closed system at equilibrium consisting of NaCl(s) added to water. See Problem 6.1 for a description of what to include. Assume NaCl(s) dissolves completely. *Without calculating species concentrations*, how do $[H^+]$ and $[OH^-]$ in this system compare to $[H^+]$ and $[OH^-]$ in Problem 6.1? (Again, ignore the differences between activity and concentration.)

6.4 Write a complete mathematical model for a closed system at equilibrium consisting of HCl added to water. See Problem 6.1 for a description of what to include. If you assume that $[Cl^-]$ is about equal to the total amount of HCl added (a reasonable assumption if the amount of HCl added is small), how does the pH of this system compare to the pH in Problem 6.1? (You do not need to obtain a quantitative answer. You should be able to decide which system has the largest $[H^+]$ without actually calculating equilibrium concentrations.) Does your answer make sense?

6.5 Write the equation for the mass balance on total soluble zinc for the species list in Example 6.3.

6.6 Write the charge balance equation (electroneutrality constraint) for the system in Example 6.2.

6.7 How would the mathematical model of the NH_4Cl/H_2O system differ if one-half the starting ammonium chloride is replaced with NH_4OH?

6.8 In preparing a cake recipe, a baker adds some baking soda ($NaHCO_3$) to water. Generate a species list diagram (such as Figure 6.2) for this system. You will need to use at least some of the equilibria in Example 6.2.

6.9 Write a complete mathematical model for the system in Problem 6.8. Assume the system is closed (is this reasonable?) and $NaHCO_3$ dissolves completely. Again, you will need to use at least some of the equilibria in Example 6.2.

6.10 Would the pH be larger or smaller if the baker had added baking powder (primarily $Na_2CO_3{}^\dagger$) rather than baking soda? Assume the number of moles per liter of Na_2CO_3 added here was the same as the number of moles per liter of $NaHCO_3$ added in Problem 6.8, the system is closed, and Na_2CO_3 dissolves completely. You should be able to decide which system has the largest $[H^+]$ without actually calculating equilibrium concentrations. (Hint: you can modify the model in Problem 6.9 *very slightly* and use it to represent the system in this problem.)

6.11 Show how equations 6.1 through 6.6 can be used to model the systems described in Problems 6.1, 6.3, 6.4, 6.8, or 6.9.

---

† Commercial baking soda is mainly $NaHCO_3 \rightarrow HCO_3^- + Na^+$. Applying Le Chatelier's Principle to the equilibria in Example 6.2, you can see that baking soda requires the addition of acid (often vinegar or soured dairy products) to produce $CO_2(g)$ and leaven baked items. Commercial baking powder contains Na_2CO_3 (or sometimes a mixture of Na_2CO_3 and $NaHCO_3$) and an acid, usually tartaric acid ($C_4O_6H_6$) or cream of tartar (a potassium salt of tartaric acid). Thus, baking powder does not require an additional acid to produce $CO_2(g)$ since carbon dioxide is produced when baking powder mixes with water. As a result, commercial baking powder often contains starch to absorb moisture and prevent $CO_2(g)$ formation in the baking powder container.

CHAPTER 7

Algebraic Solutions to Chemical Equilibrium Problems

7.1 INTRODUCTION

After working through Chapter 6, you should feel confident that you can translate a chemical equilibrium system into a mathematical system consisting of n equations in n unknowns. The unknowns are the equilibrium concentrations of the n species. The n equations come from equilibria, mass balances, electroneutrality, and other constraints.

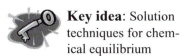 **Key idea**: Solution techniques for chemical equilibrium systems must be capable of solving n equations in n unknowns

In this chapter, algebraic solution methods for solving a system of n equations in n unknowns will be discussed. First, you will investigate the nature and linearity of the mathematical equations to be solved. Second, a solution method based on conversion of n equations in n unknowns into one equation in one unknown will be introduced. Third, you will use information about the chemical system to make approximations and simplify the mathematical equation(s) to be solved. A shared example will be used to demonstrate both solution techniques.

7.2 BACKGROUND ON ALGEBRAIC SOLUTIONS

7.2.1 Nature of the mathematical system

Before discussing solution strategies, it is necessary to examine the nature of the equations. You may be familiar with techniques to solve n *linear* equations in n unknowns (e.g., $5x + 4y = 13$ and $x - y = -1$). Recall that a system of equations is linear in the unknowns if each term in each equation is a constant or an unknown multiplied by a constant. Linear systems are quite easy to solve. Unfortunately, chemical equilibrium systems do not give rise to linear systems of equations.

Thoughtful Pause

What equations make the mathematical form of chemical equilibrium systems nonlinear?

The equilibrium expressions are *nonlinear* in the unknowns since they are of the form $K = [C]^c[D]^d/([A]^a[B]^b)$ (for the equilibrium $aA + bB = cC + dD$). Thus, you must use *nonlinear* solution techniques to calculate species concentrations in chemical equilibrium systems.

Key idea: Chemical equilibrium systems give rise to nonlinear systems of equations because equilibrium expressions contain products of species concentrations

7.2.2 Chapter example

To illustrate the nonlinear solution techniques presented in this chapter, you will calculate the species concentrations at equilibrium in a 1:100 dilution of vinegar. Vinegar can be modeled as a 0.7 M acetic acid solution. Recall from Chapter 2 that this notation means that the total concentration of acetic acid and related species is $0.7/100 = 7 \times 10^{-3}$ M. In other words, the acetic acid concentration is 7×10^{-3} M before any hydrolysis reactions are allowed to occur.

To set up the problem, follow the steps in Chapter 6. You may wish to try this on your own before reading further. Assume in this example that the solution is dilute and activities and concentrations are interchangeable.

Step 1: Define the system

The system is closed with no solids and $0.7 \text{ M}/100 = 7 \times 10^{-3}$ M acetic acid.

Step 2: Generate a species list

Acetic acid (CH_3COOH, abbreviated here HAc, where Ac stands for the fragment CH_3COO) hydrolyzes in water to form acetate (CH_3COO^-, abbreviated Ac^-). Thus, the species list is H_2O, H^+, OH^-, HAc, and Ac^-.

Step 3: Define the constraints

Equilibrium constraints:
$$H_2O = H^+ + OH^- \qquad K_W$$
$$HAc = Ac^- + H^+ \qquad K$$
Mass balance constraint:
$$\text{Total acetate} = Ac_T = 7 \times 10^{-3} \text{ M} = [HAc] + [Ac^-]$$
Electroneutrality constraint:
$$[H^+] = [Ac^-] + [OH^-]$$
Other constraints:
Activity of water = 1
All species concentrations ≥ 0

The mathematical system is:

Unknowns: $[H_2O]$, $[H^+]$, $[OH^-]$, $[HAc]$, and $[Ac^-]$
Equations:
$$K_W = [H^+][OH^-]/[H_2O] \qquad \text{eq. 7.1}$$
$$K = [Ac^-][H^+]/[HAc]$$

$$Ac_T = 7 \times 10^{-3} \text{ M} = [HAc] + [Ac^-]$$
$$[H^+] = [Ac^-] + [OH^-]$$
$$\text{Activity of } H_2O = 1 \qquad\qquad \text{eq. 7.2}$$
$$\text{All species concentrations} \geq 0$$

7.3 METHOD OF SUBSTITUTION

7.3.1 Introduction

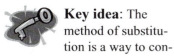

Key idea: The method of substitution is a way to convert a system of n equations in n unknowns (i.e., n species concentrations) into one equation in one unknown (i.e., one species concentration - often [H$^+$])

The method of substitution is a commonly used approach to solve both linear and nonlinear systems. The goal of the method of substitution is to convert a system of n equations in n unknowns into one equation in one unknown. In the language of equilibrium chemistry, we seek to convert a system of n equations in n species concentrations into one equation in one species concentration.

In general, the first step in the method of substitution is to write equations expressing the concentration of each variable as a function of *one* variable. For the simple example in Section 7.2.1 ($5x + 4y = 13$ and $x - y = -1$), you might rewrite the second equation to express y in terms of x: $y = x + 1$. The second step in the method of substitution is to insert the expressions from the first step into the as yet unused equation to create one equation in one unknown. In the simple example, insert $y = x + 1$ into $5x + 4y = 13$ to get $5x + 4(x + 1) = 13$. Now solve the one equation in one unknown (here, $x = 1$) and back-substitute to find the values of the other variables (here, $y = x + 1 = 2$).

The key to applying the method of substitution to equilibrium systems is to write equations expressing the concentration of *each* species as a function of the concentration of *one* species. Which species should be used as the common currency? In most cases, any species will do. Usually, one selects a meaningful species that is expected to be found at a reasonable concentration. If the equilibrium concentration of the selected species is very small, rounding errors may crop up in the calculations. For H$^+$ transfer (acid/base) reactions, it is logical to express the concentration of each species in the system as a function of H$^+$. Later in the text, you will use the method of substitution to express species concentrations in terms of species other than H$^+$.

7.3.2 Method of substitution procedure

Key idea: Use equilibria and mass balance equations to write each species concentration in terms of one species concentration

To apply the method of substitution, use the following five-step approach. First, simplify the system by setting the concentrations of all pure liquids and solids equal to unity and all solids that dissolve completely equal to zero. In almost all cases, this means setting the activity of water equal to one. In this way, you reduce the number of unknowns and simplify the K_W expression.

Second, express each species concentration in terms of one species concentration. This step is the key to the method of substitution. You

Key idea: Substitute the equations for each species concentration as a function of one species concentration into the charge balance equation to derive one equation in one unknown

Key idea: In engineering calculations, manipulate symbols first, then substitute known values for constants (in other words, when you plug and chug, chug first, then plug)

***Example 7.1*: Method of Substitution**

What are the species concentrations at equilibrium for a closed system consisting of 1×10^{-4} M NH_4Cl in water?

Solution:
From Chapter 6, the mathematical statement of the system is:
Unknowns:
    $[NH_4Cl(s)]$, $[H_2O]$, $[NH_4^+]$,
    $[Cl^-]$, $[H^+]$, $[OH^-]$, $[NH_3]$,
    and $[HCl]$
Equilibria:
    $K_1 = [NH_3][H^+]/[NH_4^+]$
    $K_3 = [H^+][Cl^-]/[HCl]$
    $K_W = [H^+][OH^-]/[H_2O]$
Mass balances:
    $N_T = [NH_4^+] + [NH_3]$
    $Cl_T = [Cl^-] + [HCl]$
Charge balance:
    $[NH_4^+] + [H^+] =$
        $[Cl^-] + [OH^-]$
Other:
    Activity of $H_2O = 1$
    $[NH_4Cl(s)] = 0$ (complete
        dissolution)

Following the steps for the method of substitution:

usually will find the equations for writing every species concentration in terms of one species concentration through the equilibria and mass balances.

Third, substitute the equations derived in the previous step into the unused equation. If the equilibria and mass balance equations are used to express each species concentration in terms of one species, then the charge balance equation will remain unused. Thus, substitute into the charge balance equation to end up with one equation in one unknown.

Fourth, insert known equilibrium constants and total masses and solve the one equation for the one unknown. Notice that to this point, you have been writing equations with no numbers. *It is always a good idea in engineering calculations to manipulate symbols first, then substitute known values for constants*. Thus, you derive one equation in one unknown first and *then* substitute in the equilibrium constants and total masses. There are many ways to solve the resulting equation. Several solution methods are outlined in Appendix A.

Finally, back-substitute into the previously developed equations to calculate the equilibrium concentration of each species. This five-step solution process is summarized in Appendix C (Section C.3.1). An example using the diluted vinegar system is shown below. Another illustration of the method is shown in Example 7.1.

Step 1: Simplify the system by setting the concentrations of all pure liquids and solids equal to unity and all solids that dissolve completely equal to zero

In the example, you can use the fact that the activity of water is unity (eq. 7.2) to modify the K_W expression (eq. 7.1). The system becomes:

Unknowns: $[H^+]$, $[OH^-]$, $[HAc]$, and $[Ac^-]$
Equations:
    $K_W = [H^+][OH^-]$ eq. 7.3
    $K = [Ac^-][H^+]/[HAc]$ eq. 7.4
    $Ac_T = 7 \times 10^{-3}$ M $= [HAc] + [Ac^-]$ eq. 7.5
    $[H^+] = [Ac^-] + [OH^-]$ eq. 7.6
    All species concentrations ≥ 0

Step 2: Express each species concentration in terms of one species concentration

In the example, let us express each species concentration in terms of $[H^+]$. This is accomplished most easily by using the equilibria and the mass balance equation:

Step 1: Simplify the system
The activity of $H_2O = 1$ and $[NH_4Cl(s)] = 0$ (complete dissolution). Thus:

Unknowns:
  $[NH_4^+]$, $[Cl^-]$, $[H^+]$, $[OH^-]$, $[NH_3]$, and $[HCl]$
Equations:
  $K_1 = [NH_3][H^+]/[NH_4^+]$
  $K_3 = [H^+][Cl^-]/[HCl]$
  $K_W = [H^+][OH^-]$
  $N_T = [NH_4^+] + [NH_3]$
  $Cl_T = [Cl^-] + [HCl]$
  $[NH_4^+] + [H^+] =$
    $[Cl^-] + [OH^-]$

Step 2: Express each species concentration in terms of $[H^+]$

  $[OH^-] = K_W/[H^+]$
  $[NH_4^+] = [NH_3][H^+]/K_1$
  $[HCl] = [Cl^-][H^+]/K_3$
  $[NH_4^+] = N_T - [NH_3]$
  $[HCl] = Cl_T - [Cl^-]$
So: $N_T - [NH_3] = [NH_3][H^+]/K_1$
    $Cl_T - [Cl^-] = [Cl^-][H^+]/K_3$
Or: $[NH_3] = N_TK_1/(K_1 + [H^+])$
    $[NH_4^+] = N_T[H^+]/(K_1 + [H^+])$
    $[Cl^-] = Cl_TK_3/(K_3 + [H^+])$
    $[HCl] = Cl_T[H^+]/(K_3 + [H^+])$

Step 3: Substitute into the charge balance equation

$N_T[H^+]/(K_1 + [H^+]) + [H^+] =$
  $Cl_TK_3/(K_3 + [H^+]) + K_W/[H^+]$

Step 4: Insert numbers and solve

With $N_T = Cl_T = 1\times10^{-4}$ M, $K_1 = 10^{-9.3}$, $K_3 = 10^{+3}$, and $K_W = 10^{-14}$, $[H^+] = 2.45\times10^{-7}$ M or pH 6.61 (assuming that activity and concentration are nearly equal)

Step 5: Back-calculate other species concentrations

$[OH^-] = K_W/[H^+]$ (from eq. 7.3)
$[HAc] = [Ac^-][H^+]/K$ (from eq. 7.4)
Also: $[HAc] = Ac_T - [Ac^-]$ (from eq. 7.5)
So: $Ac_T - [Ac^-] = [Ac^-][H^+]/K$
Or: $[Ac^-] = Ac_TK/(K + [H^+])$
And: $[HAc] = Ac_T - [Ac^-] = Ac_T[H^+]/(K + [H^+])$

Summarizing:

$[OH^-] = K_W/[H^+]$ eq. 7.7
$[Ac^-] = Ac_TK/(K + [H^+])$ eq. 7.8
$[HAc] = Ac_T[H^+]/(K + [H^+])$ eq. 7.9

Step 3: Substitute the equations derived in Step 2 into the unused equation in Step 1

After Step 2, one equation will remain used. In the example, the unused equation is the electroneutrality constraint (charge balance equation). Substituting eqs 7.7 through 7.9 into eq. 7.6:

$$[H^+] = K_W/[H^+] + Ac_TK/(K + [H^+])   \quad \text{eq. 7.10}$$

This gives one equation in the one unknown, $[H^+]$.

Step 4: Insert known equilibrium constants and total masses and solve

There are many ways to solve for $[H^+]$ in eq. 7.10. For example, one could rewrite eq. 7.10 as:

$$[H^+] - K_W/[H^+] - Ac_TK/(K + [H^+]) = 0$$

Or, clearing fractions:

$$[H^+]^3 + K[H^+]^2 - (Ac_TK + K_W)[H^+] - KK_W = 0$$

Using the methods described in Appendix A and for $Ac_T = 7\times10^{-3}$ M, $K = 10^{-4.7}$, and $K_W = 10^{-14}$, you can show that $[H^+] = 3.64\times10^{-4}$ M or pH 3.44 (if activity and concentration are nearly equal).

Step 5: Back-substitute to calculate the equilibrium concentration of each species

From eqs 7.7-7.9:

$[H^+] = 2.45 \times 10^{-7}$ **M**

$[OH^-] = K_W/[H^+] =$ **4.08×10^{-8}**
M

$[NH_3] = N_T K_1/(K_1 + [H^+]) =$
   2.04×10^{-7} M

$[NH_4^+] = N_T[H^+]/(K_1 + [H^+]) =$
   9.98×10^{-5} M

$[Cl^-] = Cl_T K_3/(K_3 + [H^+]) =$
   1.00×10^{-4} M

$[HCl] = Cl_T[H^+]/(K_3 + [H^+]) =$
   2.45×10^{-14} M

$[H^+] = 3.64 \times 10^{-4}$ M

$[OH^-] = K_W/[H^+] = 2.75 \times 10^{-11}$ M

$[Ac^-] = Ac_T K/(K + [H^+]) = 3.64 \times 10^{-4}$ M

$[HAc] = Ac_T[H^+]/(K + [H^+]) = 6.64 \times 10^{-3}$ M

You may wish to verify that these values satisfy equations 7.3-7.6.

A final note about the method of substitution. Please think of this method as a *process* rather than as a way to generate equations to be memorized. Although eq. 7.10 is the general equation for $[H^+]$ in systems with one acid that dissociates to form one H^+ (such as acetic acid), *do not* bother to memorize it. While memorization of the end equations is unnecessary, it *is* important to practice the process so you can derive a single equation in one unknown (e.g., eq. 7.10) quickly.

7.4 METHOD OF APPROXIMATION

7.4.1 Introduction

The method of substitution described in Section 7.3 works well for aqueous systems. The substitution approach will be used many times in this text. However, the method has some drawbacks. First, the method of substitution is tedious. Its use may lead to errors due to the number of manipulations required in larger systems. Second, as a brute force method, it does not increase your appreciation for the underlying chemistry.

The chances for errors can be reduced by making the system simpler. How can the system be made simpler? The system can be made simpler by reducing the number of unknowns. One way to reduce the number of unknowns is to make approximations. In making approximations, you disregard species that you believe to be smaller in concentration than other species in the same equation.

 **Key idea**: To make approximations, disregard species that you think are smaller in concentration than other species in the same equation

7.4.2 Where are approximations applied?

A key point in making approximations is *selecting the appropriate equation (or equations) to which the approximations are applied*. The nature of the equilibrium, mass balance, electroneutrality, and other constraints dictate where approximations can be used.

Thoughtful Pause

To which equations in the mathematical model of equilibrium can approximations be applied? (See Section 7.2.2 for an example of the mathematical form of a chemical equilibrium system.)

Apply approximations only to the equations where the species concentrations are added. This point is very important. It does not make sense to ignore a species in an equilibrium expression (e.g., $K = [C]^c[D]^d/[A]^a[B]^b$).

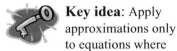

Key idea: Apply approximations only to equations where the species concentrations are added (i.e., only to mass balance and charge balance equations)

Even small concentrations must be included in the equilibrium expressions since the concentrations are multiplied by one another. However, you *can* apply approximations to the mass balance and charge balance equations, where the species concentrations are additive. For additive equations (e.g., $Ac_T = [HAc] + [Ac^-]$ or $[H^+] = [Ac^-] + [OH^-]$), you often can ignore species at low concentrations relative to the other species without introducing significant error.

Approximations typically are applied to either mass balance equations or the charge balance equation or both. For simple systems (especially where H^+ transfer is the main process), it makes sense to apply approximations to the charge balance equation. Why? You often can guess about the relative importance of $[H^+]$ and $[OH^-]$. Also, $[H^+]$ and $[OH^-]$ are both in the charge balance equation. Recall from Section 1.3 that $[H^+]$ is large under acidic conditions (low pH) and $[H^+]$ is small under basic conditions (high pH). From the self-ionization of water ($K_w = [H^+][OH^-]$, if the activity of water is unity), you know that when $[H^+]$ is large (low pH, acidic conditions), $[OH^-]$ will be relatively small, and when $[H^+]$ is small (high pH, basic conditions), $[OH^-]$ will be relatively large. This information is summarized in Table 7.1.

Table 7.1: Relative Values of pH, $[H^+]$, and $[OH^-]$ Under Acidic, Neutral pH, and Basic Conditions

Condition	pH	$[H^+]$	$[OH^-]$
Acidic	low	high	low
Neutral pH	near 7	moderate	moderate
Basic	high	low	high

Note: The exact pH, $[H^+]$, and $[OH^-]$ at neutral pH depend on the temperature, pressure, and concentrations of ions in solution

Thus, a good first approximation is to guess whether the system is acidic or basic.

Thoughtful Pause

How can you use the guess of acidity or basicity to reduce the number of unknowns in the system?

If the system is thought to be acidic, then ignore $[OH^-]$ relative to $[H^+]$ in any equation where they are additive. Typically, you ignore $[OH^-]$ relative to $[H^+]$ in the charge balance equation. If the system is thought to be basic, then you can ignore $[H^+]$ relative to $[OH^-]$ in the charge balance equation.

Key idea: By estimating the equilibrium pH, you usually can ignore either $[H^+]$ or $[OH^-]$ in equations where they are additive (i.e., the charge balance equation)

 Key idea: After calculating species concentrations, ***check approximations*** and the original equations; if the assumption proves incorrect, make a different assumption/approximation and iterate

---

***Example 7.2:* Method of Approximation**

Determine the equilibrium concentrations when 7×10^{-3} M of sodium acetate (NaAc, the sodium salt of acetic acid) is added to water. Use the method of approximation.

Solution:
The system is identical to the chapter example, with the added species $[Na^+]$ and $[NaAc(s)]$ and an additional mass balance on sodium.

Following the steps for the method of approximation:

Step 1: Simplify the system
The activity of $H_2O = 1$ and $[NaAc(s)] = 0$ (complete dissolution, since it is a sodium salt).

Approximations also can be applied to mass balance equations. A discussion of approximations in mass balance equations will be postponed until Chapter 11 because it is often difficult to guess which species should be ignored in a mass balance equation. In the acetic acid example, it is not clear which of $[HAc]$ or $[Ac^-]$ is larger at equilibrium. After you examine the nature of acid/base equilibria in Chapter 11, it will become easier to make approximations in the mass balances equations.

7.4.3 Checking approximations
It is essential to *check the validity of every approximation* made in chemical equilibrium calculations. If you guess that the system is basic (and therefore $[OH^-] \gg [H^+]$), you must check the calculated equilibrium values to confirm that $[OH^-]$ is much greater than $[H^+]$. How large should $[OH^-]$ be relative to $[H^+]$? The answer to this question depends on how much error you is willing to accept. Error analysis in approximations will be discussed in Section 7.4.5. In addition to checking the assumptions, you should check to see how well the original equilibrium and mass balance equations are satisfied.

Students approaching equilibrium calculations for the first time often are hesitant about making approximations. You may worry that you will make the wrong approximation. Do not be afraid to make approximations - *as long as you check them*. An incorrect assumption always will be revealed *if checked*. For example, if you assume the solution is acidic, but the calculated equilibrium $[H^+]$ is much *smaller* than the calculated equilibrium $[OH^-]$, then the assumption is wrong. The calculation should be repeated with another assumption (in most cases, with the *opposite* assumption - here, that the system is basic and $[H^+]$ can be ignored relative to $[OH^-]$). Making an incorrect assumption costs a little time and effort but will not result in a permanent error as long as the approximation is checked and revised if found to be inappropriate. (An exception is illustrated in the footnote at the end of Section 7.4.4.)

7.4.4 Method of approximation procedure
The substitution method can now be revised to include approximations. To apply approximations, use the following six-step approach. First, simplify the system by setting the concentrations of all pure liquids and solids equal to unity and all solids that dissolve completely equal to zero. Second, make approximations in the additive equations (mass balance and/or charge balance equations). A common approach is to assume acidic conditions and ignore $[OH^-]$ relative to $[H^+]$ in the charge balance equation or assume basic conditions and ignore $[H^+]$ relative to $[OH^-]$ in the charge balance equation. Third, manipulate the equations (as modified by the approximations) to end up with one equation in one unknown. Fourth, insert known equilibrium constants and total masses and solve the one equation for the one unknown. Fifth, back-substitute into the previously developed

Thus:

Unknowns:

$[Na^+]$, $[Ac^-]$, $[HAc]$, $[H^+]$, and $[OH^-]$

Equations:

$K = [Ac^-][H^+]/[HAc]$
$K_W = [H^+][OH^-]$
$Ac_T = [HAc] + [Ac^-]$
$Na_T = [Na^+]$
$[Na^+] + [H^+] =$
$\qquad [Ac^-] + [OH^-]$

Step 2: Make approximations in the additive equations

Looking at the charge balance, it seems logical that $[Na^+] >$ $[Ac^-]$ (since the salt dissociates completely to Na^+, but some Ac^- is converted to HAc). Thus, $[OH^-] > [H^+]$ and you might guess that the solution is basic.

Step 3: Write one equation in one unknown

Substituting into the charge balance equation:

$Na_T = Ac_T K/(K + [H^+]) +$
$\qquad K_W/[H^+]$

Or: $Na_T[H^+]^2 + [K(Na_T - Ac_T) -$
$\qquad K_W][H^+] - KK_W = 0$

Step 4: Insert numbers and solve

With $Na_T = Ac_T = 7 \times 10^{-4}$ M, K $= 10^{-4.7}$, and $K_W = 10^{-14}$, then $[H^+] = 5.34 \times 10^{-9}$ M or pH 8.27 (assuming activity and concentration are nearly equal)

Step 5: Back-calculate other species concentrations

$[H^+] = 5.34 \times 10^{-9}$ M (pH 8.27)

equations to calculate the equilibrium concentration of each species. Finally, check the assumption and original equations. If the assumption is shown to be invalid (or the original equations are not satisfied to the degree required), make the opposite assumption and iterate. This solution process is summarized in Appendix C, Section C.3.2. Examples using the acetic acid and sodium acetate systems are shown below and in Example 7.2, respectively.

Step 1: Simplify the system by setting the concentrations of all pure liquids and solids equal to unity and all solids that dissolve completely equal to zero

In the example, use the fact that the activity of water is unity to modify the K_W expression. The system becomes (equations from Section 7.2.2 are repeated here for convenience):

Unknowns: $[H^+]$, $[OH^-]$, $[HAc]$, and $[Ac^-]$
Equations:

$K_W = [H^+][OH^-]$	eq. 7.3
$K = [Ac^-][H^+]/[HAc]$	eq. 7.4
$Ac_T = 7 \times 10^{-3}$ M $= [HAc] + [Ac^-]$	eq. 7.5
$[H^+] = [Ac^-] + [OH^-]$	eq. 7.6

All species concentrations ≥ 0

Step 2: Make approximations in the additive equations

Since acetic acid is being added to water, it makes sense that the resulting solution may be acidic. Thus, assume as a starting point that $[H^+] \gg [OH^-]$. If $[H^+] \gg [OH^-]$ and $[H^+]$ $= [Ac^-] + [OH^-]$ (charge balance, eq. 7.6), then:

$$[H^+] \approx [Ac^-] \qquad\qquad \text{eq. 7.11}$$

Step 3: Write one equation in one unknown

Insert eq. 7.11 into the K equilibrium (eq. 7.4) and mass balance equation (eq. 7.5):

$K = [Ac^-][H^+]/[HAc] = [H^+]^2/[HAc]$
$\qquad$ (from eqs. 7.4 and 7.11)
Also: $[HAc] = Ac_T - [Ac^-] = Ac_T - [H^+]$
$\qquad$ (from eqs. 7.5 and 7.11)
So: $K = [H^+]^2/(Ac_T - [H^+])$
Or: $[H^+]^2 + K[H^+] - KAc_T = 0$

$[OH^-] = K_W/[H^+] =$
    1.87×10^{-6} M
$[Ac^-] = Ac_T K/(K + [H^+]) =$
    7.00×10^{-3} M
$[HAc] = Ac_T[H^+]/(K + [H^+]) =$
    1.87×10^{-6} M
$[Na^+] = 7.00 \times 10^{-3}$ M

Step 6: Check assumptions and original equations

$[OH^-]$ is about 350 times $[H^+]$. Thus, there is expected to be little error introduced from the assumption.

To summarize:

$[H^+] = 5.34 \times 10^{-9}$ M
(pH 8.27)
$[OH^-] = 1.87 \times 10^{-6}$ M
$[Ac^-] = 7.00 \times 10^{-3}$ M
$[HAc] = 1.87 \times 10^{-6}$ M
$[Na^+] = 7.00 \times 10^{-3}$ M

Solving:[†] $[H^+] = \dfrac{-K \pm \sqrt{K^2 + 4KAc_T}}{2}$

Step 4: Insert known equilibrium constants and total masses and solve

For $Ac_T = 7 \times 10^{-3}$ M and $K = 10^{-4.7}$, you can show that $[H^+] = 3.64 \times 10^{-4}$ M or -3.84×10^{-4} M. Ignoring the negative root (since $[H^+]$ and all other species concentrations must be nonnegative): $[H^+] = 3.64 \times 10^{-4}$ M or pH 3.44 (assuming activity and concentration are nearly equal).

Step 5: Back-substitute to calculate the equilibrium concentration of each species

Thus:
$[H^+] = 3.64 \times 10^{-4}$ M
$[OH^-] = K_W/[H^+] = 2.75 \times 10^{-11}$ M   (from eq. 7.3)
$[Ac^-] \approx [H^+] = 3.64 \times 10^{-4}$ M   (from eq. 7.11)
$[HAc] = Ac_T - [H^+] = 6.64 \times 10^{-3}$ M

Step 6: Check the assumptions and original equations and iterate if necessary

The assumption is reasonable: $[H^+]$ ($= 3.64 \times 10^{-4}$ M) is in fact much greater than $[OH^-]$ ($= 2.75 \times 10^{-11}$ M). Checking the original equations, you will find:

$$[Ac^-][H^+]/[HAc] = (3.64 \times 10^{-4} \text{ M})^2/6.64 \times 10^{-3} \text{ M}$$
$$= 2.00 \times 10^{-5} \text{ M} = 10^{-4.7} \approx K \text{ - ok}$$
$$[HAc] + [Ac^-] = 6.64 \times 10^{-3} \text{ M} + 3.64 \times 10^{-4}$$
$$= 7.00 \times 10^{-3} \text{ M} \approx Ac_T \text{ - ok}$$

Thus, the assumption and original equations are satisfied very well and no iteration is necessary.

[†] Recall that for the equation $ax^2 + bx + c = 0$, $x = \dfrac{-b \pm \sqrt{b^2 - 4ac}}{2a}$.

In this case, the assumption that the system is acidic was shown to be valid. What if you had made the wrong assumption initially? It is hard to make the wrong assumption in this simple system. For example, it makes no sense to assume that $[OH^-] \gg [H^+]$ since the charge balance equation ($[H^+] = [Ac^-] + [OH^-]$) shows that $[H^+]$ must be equal to or greater than $[OH^-]$ (because $[Ac^-]$ cannot be negative).

To demonstrate what happens when an incorrect assumption is made, rework the problem with $Ac_T = 1 \times 10^{-10}$ M. If you assume the solution is acidic, then $[Ac^-] \approx [H^+]$ and eq. 7.11 is regenerated. With $Ac_T = 1 \times 10^{-10}$ M and $K = 10^{-4.7}$, you can calculate that $[H^+] \approx 1 \times 10^{-10}$ M (pH 10) and $[OH^-] \approx 1 \times 10^{-4}$ M. Thus, the assumption that $[H^+] \gg [OH^-]$ is incorrect when $Ac_T = 1 \times 10^{-10}$ M. As stated earlier, you cannot assume that $[OH^-] \gg [H^+]$ in this case because the charge balance shows that $[H^+]$ cannot be smaller than $[OH^-]$.

The only other approximation available with the charge balance equation is that $[OH^-] \gg [Ac^-]$. With this assumption: $[H^+] \approx [OH^-] = 1 \times 10^{-7}$ M (pH 7.0), $[Ac^-] \approx 1 \times 10^{-10}$ M, and $[HAc] \approx 5 \times 10^{-13}$ M. In this case, the new assumption (i.e., $[OH^-] \gg [Ac^-]$) is valid. The solution makes sense: the pH will not be much different than 7 if a *very* small amount of the acid is added to water ($Ac_T = 1 \times 10^{-10}$ M is equivalent to about 10 drops of vinegar in the swimming pool used in the 2000 Summer Olympic Games in Sydney, Australia).

This example points out that incorrect approximations will be disclosed *if the approximations are checked*. Changing the assumptions, recalculating, and rechecking the results will lead eventually to the correct equilibrium concentrations.[†]

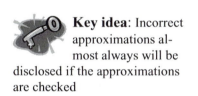

Key idea: Incorrect approximations almost always will be disclosed if the approximations are checked

7.4.5 Error analysis in approximations

The approximation made in the acetic acid problem with $Ac_T = 7 \times 10^{-3}$ M was that $[H^+]$ is much greater than $[OH^-]$. In checking the assumption, you discovered that $[H^+]$ was over 7 orders of magnitude greater than the concentration of OH^-. In this case, it is obvious that the assumption was valid. What if $[H^+]$ was only 100 times greater than $[OH^-]$? Or only 10 times greater? In such cases, a significant error may be introduced if the assumption is accepted.

[†] There are exceptions to this statement. Try reworking Example 7.1 by using the reasonable approximations that $[Cl^-] = Cl_T$ and $[NH_4^+] = N_T$. You will calculate an equilibrium pH of 7.0 and all assumptions will check out. The exact pH is 6.61. Why the difference? As you shall see in Chapter 8, the charge balance is not very sensitive in this case. A new kind of equation will be introduced in Chapter 8 to account for this situation.

As a general rule of thumb, you can usually ignore a species concentration in an additive equation if it is at least 100 times smaller than other concentrations in the additive equation.[†] However, the magnitude of the error depends on the nature of the chemical system. For simple acids such as acetic acid, the equilibrium pH depends on two parameters: the total amount of acid added (C) and K. The error caused by an approximation also depends on C and K. The error in the equilibrium pH at different values of the total acid added and K is shown in Figure 7.1.

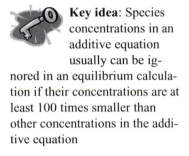

Key idea: Species concentrations in an additive equation usually can be ignored in an equilibrium calculation if their concentrations are at least 100 times smaller than other concentrations in the additive equation

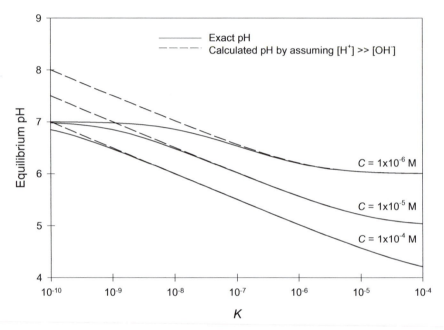

Figure 7.1: Exact pH and pH Predicted from the Approximation [H⁺] >> [OH⁻]

Thoughtful Pause

From Figure 7.1, what values of C and K result in the greatest error between the exact pH and pH estimated through approximation? Does this make sense?

[†] The species concentration may be ignored in terms of its contribution to the linear equation (i.e., in terms of its contribution to the charge balance or mass balance equations). The species may still play an important role in the chemical system even at low concentration. For example, the species may interact with other species or exert toxicity. You ignore the species only in magnitude relative to other species strictly for the purposes of equilibrium calculations.

Note that the error is largest for very small C and very small K. This makes sense. The addition of a very small amount of acid (or addition of an acid that does not dissociate very much to "produce" H^+) will result in a pH close to that of water without the acid. In other words, if C is very small and/or K is very small, you would expect the pH to be near 7. For such values of C and K, the assumption that $[H^+] \gg [OH^-]$ is incorrect and a larger error in the equilibrium pH will result. Remember that an error in the equilibrium pH may translate into sizable errors in the equilibrium concentrations of other species.

At first glance, the information in Figure 7.1 is discouraging. From Figure 7.1, it appears that significant errors in the calculated equilibrium pH may result if you make the incorrect assumption. How do you know an error occurs if you evaluate the system only with an approximation and never calculate the exact pH? *You know an error will occur if you check the assumption.* In this case, check the assumption by comparing the calculated values of $[H^+]$ and $[OH^-]$. If you assume that $[H^+] \gg [OH^-]$ but the calculated ratio of $[H^+]$ to $[OH^-]$ is small, then the assumption should be questioned.

For the example of an acid in water, the error in the equilibrium pH is plotted as a function of the calculated ratio of $[H^+]$ to $[OH^-]$ in Figure 7.2. In the calculation, it was assumed that $[H^+] \gg [OH^-]$. Note that the error is significant only when $[H^+]$ is less than about 100 times greater than the OH^- concentration. Imagine conducting an equilibrium calculation for an acid similar to acetic acid and assuming that $[H^+] \gg [OH^-]$. If you found that $[H^+]$ was only 10 times larger than $[OH^-]$, then the assumption was inappropriate and significant errors in the equilibrium pH may arise (see Figure 7.2). The system should be reanalyzed with a different assumption.

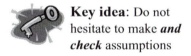

Key idea: Do not hesitate to make *and* *check* assumptions

Figure 7.2 only applies to systems with acids that hydrolyze to produce one H^+. However, the discussion in this section demonstrates that if you check the assumptions, you will know if errors are creeping into the analysis. The bottom line: do not be afraid to make approximations, *as long as you check them.*

7.5 SUMMARY

In this chapter, algebraic methods for the solution of chemical equilibrium problems were presented. Two methods were discussed. In the method of substitution, each species concentration is expressed in terms of the concentration of one species. For acid/base chemistry problems, it is useful to express every species concentration in the species list as functions of $[H^+]$. This usually is accomplished by using the equilibria and mass balance equations. The expressions can be substituted into the charge balance equation to derive one equation in one unknown. The nonlinear equation can be solved numerically by the techniques outlined in Appendix A.

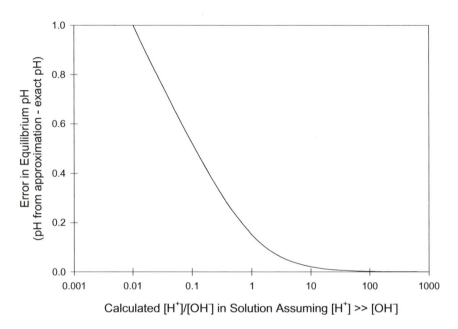

Figure 7.2: Error in the Equilibrium pH as a Function of $[H^+]/[OH^-]^{\dagger}$
(calculation assumes $[H^+] \gg [OH^-]$)

In the method of approximation, one or more species concentrations are ignored in the additive equations (mass and charge balance equations). This approach generally makes the resulting one equation in one unknown easier to solve. Be sure to check the assumption and original equations, changing the assumption and iterating if the assumption proves to be incorrect.

7.6 PART II CASE STUDY: HAVE YOU HAD YOUR ZINC TODAY?

The Part II case study concerns the speciation of zinc in the mouth and stomach upon consumption of a zinc acetate tablet. Recall from Section 5.8 that 20 mg zinc as Zn were dissolved in 100 mL of water and consumed. We are interested in the zinc species formed in the transition from pH 6.5-8.5 (saliva) to pH 1-3 (gastric juices). As a result of your efforts in this chapter, you should be able to determine the zinc species at fixed pH. For the present, the speciation of zinc will be determined at pH 6.5 and 3. Other pH values will be explored in Chapter 8.

---

† Figure 7.2 was created by using the data in Figure 7.1. Thus, K was varied from 10^{-10} to 10^{-4} and C was varied from 10^{-6} to 10^{-4} M. Over this range of K and C values, the relationship between $pH_{approx} - pH_{exact}$ and $[H^+]/[OH^-]$ is essentially independent of K and C. Caution should be exercised when using the line in Figure 7.2 outside of the indicated ranges of K and C.

The chemical system was developed in Section 6.8. It will be revised slightly and restated here. The total soluble zinc concentration, $Zn(+II)_T$, is $(0.02 \text{ g Zn})/(65.38 \text{ g Zn/mol Zn}) = 3.06 \times 10^{-4}$ mole in 100 mL or 3.06×10^{-3} M. Since zinc acetate is $Zn(Ac)_2(s)$, the total soluble acetate, Ac_T, is $2Zn(+II)_T$ or 6.12×10^{-3} M. As stated in Section 6.8, we assume that $Zn(Ac)_2(s)$ dissolves completely and we ignore the formation of $Zn(OH)_2(s)$. In addition, the activity of water is assumed to be unity, and activities and concentrations will be used interchangeably.

In this portion of the case study, the pH is fixed. Fixing $[H^+]$ adds an equation. Since unknown species are required to fix the pH, the usual approach in systems of fixed pH is to ignore the charge balance equation. Using the equilibrium constants from Appendix D, we have:

Unknowns:
  $[Zn^{2+}]$, $[Ac^-]$, $[H^+]$, $[OH^-]$, $[ZnOH^+]$, $[Zn(OH)_2^0]$, $[Zn(OH)_3^-]$, $[Zn(OH)_4^{2-}]$, $[HAc]$, $[ZnAc^+]$, $[Zn(Ac)_2^0]$, and $[Zn(Ac)_3^-]$

Equilibria:
  $K_W = [H^+][OH^-]/[H_2O] = 10^{-14}$
  $K_1 = [ZnOH^+]/([Zn^{2+}][OH^-]) = 10^{+5}$
  $K_2 = [Zn(OH)_2^0]/([Zn^{2+}][OH^-]^2) \, 10^{+10.2}$
  $K_3 = [Zn(OH)_3^-]/([Zn^{2+}][OH^-]^3) = 10^{+13.9}$
  $K_4 = [Zn(OH)_4^{2-}]/([Zn^{2+}][OH^-]^4) = 10^{+15.5}$
  $K_6 = [Ac^-][H^+]/[HAc] = 10^{-4.7}$
  $K_7 = [ZnAc^+]/([Zn^{2+}][Ac^-]) = 10^{+1.57}$
  $K_8 = [Zn(Ac)_2^0]1/([Zn^{2+}][Ac^-]^2) \approx 10^{+1.1}$
  $K_9 = [Zn(Ac)_3^-]/([Zn^{2+}][Ac^-]^3) \approx 10^{+1.57}$

Mass balances:
$$Zn(+II)_T = [Zn^{2+}] + [ZnOH^+] + [Zn(OH)_2^0] + [Zn(OH)_3^-]$$
$$+ [Zn(OH)_4^{2-}] + [ZnAc^+] + [Zn(Ac)_2^0]$$
$$+ [Zn(Ac)_3^-]$$
$$= 3.06 \times 10^{-3} \text{ M}$$

$$Ac_T = [Ac^-] + [HAc] + [ZnAc^+] + 2[Zn(Ac)_2^0] + 3[Zn(Ac)_3^-]$$
$$= 6.12 \times 10^{-3} \text{ M}$$

Assumptions:
  Activity of $H_2O = 1$ (standard state)
  Activity of $Zn(Ac)_2(s) = 0$ (complete dissolution)
  $Zn(OH)_2(s)$ formation ignored
  pH fixed and charge balance ignored
  Activities and concentrations interchangeable

The system could be solved exactly without approximation. However, the algebra is much simpler if we assume that some species concentrations are negligible. Of course, any assumptions must be checked and shown to be valid. Logical assumptions are in the number of Zn-OH species and number of Zn-Ac species formed. The pH is reasonably low. Thus, $[OH^-]$ is small, and it is reasonable to start with the assumption that the formation of higher Zn-OH species (e.g., $Zn(OH)_2^0$, $Zn(OH)_3^-$, and $Zn(OH)_4^{2-}$) can be ignored. In addition, Ac_T is fairly small, so the formation of $[Zn(Ac)_2^0]$ and $[Zn(Ac)_3^-]$ will be ignored as a starting point. The system becomes:

Unknowns:
   $[Zn^{2+}]$, $[Ac^-]$, $[H^+]$, $[OH^-]$, $[ZnOH^+]$, $[HAc]$, and $[ZnAc^+]$

Equilibria:
   $$K_W = [H^+][OH^-]/[H_2O] = 10^{-14}$$
   $$K_1 = [ZnOH^+]/([Zn^{2+}][OH^-]) = 10^{+5}$$
   $$K_6 = [Ac^-][H^+]/[HAc] = 10^{-4.7}$$
   $$K_7 = [ZnAc^+]/([Zn^{2+}][Ac^-]) = 10^{+1.57}$$

Mass balances:
   $$Zn(+II)_T \approx [Zn^{2+}] + [ZnOH^+] + [ZnAc^+] = 3.06\times10^{-3}\,M$$
   $$Ac_T \approx [Ac^-] + [HAc] + [ZnAc^+] = 6.12\times10^{-3}\,M$$

There are many ways to proceed. As discussed in this chapter, a useful approach is to express the concentration of each species in terms of a small number of species. Since $[H^+]$ is fixed, it would be valuable to write $[OH^-]$ as a function of $[H^+]$: $[OH^-] = K_W/[H^+]$. The species concentrations $[ZnOH^+]$, $[ZnAc^+]$, and $[HAc]$ also can be written in terms of $[Zn^{2+}]$ and $[Ac^-]$:

   $$[ZnOH^+] = K_1[Zn^{2+}][OH^-] = K_1K_W[Zn^{2+}]/[H^+]$$
   $$[ZnAc^+] = K_7[Zn^{2+}][Ac^-]$$
   $$[HAc] = [Ac^-][H^+]/K_6$$

Thus, the mass balances become:

$$Zn(+II)_T = [Zn^{2+}](1 + K_1K_W/[H^+] + K_7[Ac^-])$$
$$= 3.06\times10^{-3}\,M \qquad\qquad \text{eq. 7.12}$$
$$Ac_T = [Ac^-](1 + [H^+]/K_6 + K_7[Zn^{2+}]) = 6.12\times10^{-3}\,M \qquad \text{eq. 7.13}$$

Equations 7.12 and 7.13 can be solved as two equations in two unknowns. Another approach is to solve for $[Ac^-]$ in eq. 7.13 and substitute the new expression for $[Ac^-]$ into eq. 7.12. After substituting and rearranging:

$$a[Zn^{2+}]^2 + b[Zn^{2+}] + c = 0, \text{ where:} \qquad\qquad \text{eq. 7.14}$$

$$a = K_7(1 + K_1 K_W/[H^+])$$
$$b = (1 + [H^+]/K_6)(1 + K_1 K_W/[H^+]) + K_7(Ac_T - Zn(+II)_T)$$
$$c = -Zn(+II)_T(1 + [H^+]/K_6)$$

Solving eq. 7.14 with the quadratic equation at pH 6.5: $[Zn^{2+}] = 10^{-2.60}$ M. You can back-substitute and solve for $[Ac^-]$ and the other zinc-containing species:

$$[Ac^-] = Ac_T/(1 + [H^+]/K_6 + K_7[Zn^{2+}]) = 10^{-2.26} \text{ M}$$
$$[ZnOH^+] = K_1 K_W[Zn^{2+}]/[H^+] = 10^{-5.10} \text{ M}$$
$$[ZnAc^+] = K_7[Zn^{2+}][Ac^-] = 10^{-3.26} \text{ M}$$

Checking the approximations, the other zinc-containing species are all below 10^{-6} M and can be ignored (at least relative to Zn^{2+} and $ZnAc^+$). Thus, at pH 6.5, about 83% of the zinc is Zn^{2+}, with about 17% as $ZnAc^+$. About 90% of the total acetate is in the form of Ac^- at pH 6.5, with most the remainder as $ZnAc^+$.

You may wish to verify that at pH 3, the approximations are even more valid. At pH 3, over 99.5% of the zinc is in the form Zn^{2+}. About 98% of the total acetate is found as HAc at pH 3.

SUMMARY OF KEY IDEAS

- Solution techniques for chemical equilibrium systems must be capable of solving n equations in n unknowns

- Chemical equilibrium systems give rise to nonlinear systems of equations because equilibrium expressions contain products of species concentrations

- The method of substitution is a way to convert a system of n equations in n unknowns (i.e., n species concentrations) into one equation in one unknown (i.e., one species concentration, often $[H^+]$)

- Use equilibria and mass balance equations to write each species concentration in terms of one species concentration

- Substitute the equations for each species concentration as a function of one species concentration into the charge balance equation to derive one equation in one unknown

- In engineering calculations, manipulate symbols first, then substitute known values for constants (in other words, when you plug and chug, chug first, then plug)

- To make approximations, disregard species that you think are smaller in concentration than other species in the same equation

- Apply approximations only to equations where the species concentrations are added (i.e., only to mass balance and charge balance equations)

- By estimating the equilibrium pH, you usually can ignore either $[H^+]$ or $[OH^-]$ in equations where they are additive (i.e., the charge balance equation)

- After calculating species concentrations, ***check approximations*** and the original equations; if the assumption proves incorrect, make a different assumption/approximation and iterate

- Incorrect approximations almost always will be disclosed if the approximations are checked

- Species concentrations in an additive equation usually can be ignored in an equilibrium calculation if their concentrations are at least 100 times smaller than other concentrations in the additive equation

- Do not hesitate to make ***and check*** assumptions

PROBLEMS

7.1 Professor Whoops thinks it is possible to transform the equations representing a chemical system at equilibrium into a linear set of equations. Professor Whoops performs a log transformation of the equilibria; for example: $\log K = \log[Ac^-] + \log[H^+] - \log[HAc]$. This is now a linear equation in three unknowns: $x = \log[Ac^-]$, $y = \log[H^+]$, and $z = \log[HAc]$. Will Professor Whoops's idea generate a linear system of equations in the new unknowns? Why or why not?

7.2 At 30°C, $K_W = 10^{-13.83}$. Determine the equilibrium pH of pure water (containing only H_2O, H^+, and OH^-) at 30°C.

7.3 The mineral acids (e.g., HCl) significantly reduce the pH when added to water. For a 1×10^{-3} M HCl solution (i.e., the addition of 1×10^{-3} moles of HCl per liter of water), find the pH and [HCl] at equilibrium. Solve this problem by guessing whether the solution will be acidic or basic. Is it true that HCl dissociates nearly completely in this system? Use the equilibrium: $HCl = H^+ + Cl^-$, $K = 10^{+3}$

7.4 From your answer to Problem 7.3, it appears that HCl dissociates nearly completely. Rework Problem 7.3 using the assumption that HCl dissociates completely rather than the assumption about the relative concentrations of H^+ and OH^-.

7.5 Resolve the acetic acid problem in this chapter (with $Ac_T = 7\times10^{-4}$ M) by expressing each species concentration in terms of $[Ac^-]$ rather than in terms of $[H^+]$. Solve the system in three ways: (*a*) using no assumptions, (*b*) assuming $[Ac^-] \gg [HAc]$, and (*c*) assuming $[HAc] \gg [Ac^-]$.

7.6 Find the equilibrium pH for a 1×10^{-4} M solution of NaOH. State and check all assumptions.

7.7 One drop (0.05 mL) of 21% by weight ammonium hydroxide (NH_4OH) is added to 1 liter of water. What is the equilibrium pH if the system is closed? Assume ammonium hydroxide dissociates completely. The density of 21% by weight NH_4OH is 0.957 kg/L.

7.8 Cyanide (CN^-) can bind with an enzyme to cause catastrophic interference with oxygen utilization leading to death. How far below the pK would one have to acidify water so that the cyanide concentration was less than 5% of $[CN^-] + [HCN]$? In other words, find the value of pK − pH so that the cyanide concentration is less than 5% of $[CN^-] + [HCN]$.

7.9 The final equations in one unknown for the acetic acid system in the chapter example are repeated below:

> Exact solution: $[H^+]^3 + K[H^+]^2 - (Ac_T K + K_w)[H^+] - KK_w = 0$
> Assumption that $[H^+] \gg [OH^-]$: $[H^+]^2 + K[H^+] - KAc_T = 0$

Compare these two equations and discuss whether their differences are consistent with the assumption that $[H^+] \gg [OH^-]$ in the second equation. Hint: Multiply the assumption equation through by $[H^+]$.

7.10 Verify that a C M HCl solution has an equilibrium pH of about $-\log C$. For example, the pH of a 10^{-2} M HCl solution is about $-\log(10^{-2}) = 2$. Below what values of C does this relationship fail?

7.11 An industry has a waste stream consisting only of the organic chemical propanoic acid (informally, propionic acid). Propanoic acid, CH_3CH_2COOH, undergoes chemistry similar to acetic acid. Its conjugate base, $CH_3CH_2COO^-$, is called propanoate.

  A. When a 10^{-3} M propanoic acid solution is prepared, the equilibrium pH is 3.97. Find the equilibrium constant for the dissociation of propanoic acid to propanoate.

  B. The waste stream (consisting only of propanoic acid and propanoate in water) is at pH 4.7. Using your answer from part A, find the total propanoic acid concentration (= $[CH_3CH_2COOH]$ + $[CH_3CH_2COO^-]$).

7.12 Orange juice can be modeled crudely as a mixture of 8 g/L of citric acid (formally, 2-hydroxy-1,2,3-propanetricarboxylic acid; $C_6H_8O_7$) and 1.6 g/L potassium ion. (Note: concentrations are **g/L** not **mg/L**). Find the equilibrium pH of orange juice. Citric acid can be written as H_3C. Although the chemistry of citric acid is more complex, for this problem consider only the equilibrium: $H_3C = H_2C^- + H^+$; $K = 10^{-3.14}$.

7.13 The other important cations in orange juice are Na^+ (50 mg/L), Mg^{2+} (100 mg/L), and Ca^{2+} (100 mg/L). The other important acid is malic acid (formally, hydroxybutanedioic acid; $C_4H_4O_4$), present in orange juice at about 2 g/L. The pertinent equilibrium for maleic acid is: $H_2M = HM^- + H^+$; $K = 10^{-3.4}$. What is the revised equilibrium pH of orange juice, including all the cations listed and both citric and malic acid? Assume that none of the cations hydrolyze at reasonably low pH.

CHAPTER 8

Graphical Solutions to Chemical Equilibrium Problems

8.1 INTRODUCTION

Key idea: Graphical solutions help to (1) visualize species concentrations, (2) show how species concentrations change with a change in the master variable, and (3) identify easily which species contribute significantly and which can be ignored in the additive equations

Chapter 7 provided some basic techniques for calculating species concentrations at equilibrium. In theory, you now have all the tools you need to determine equilibrium concentrations. Through equilibria, mass balances, a charge balance, and other thermodynamic constraints, a system of n equations in n unknowns can be generated. The system can be solved by the algebraic methods in Chapter 7.

Graphical solutions to chemical systems at equilibrium were popularized by the Swedish geochemist Lars Gunnar Sillén (1916-1970). Sillén's wonderful review of graphical techniques begins with the same question you may have: "Why should one bother to represent equilibria graphically?" (Sillén, 1959). One reason for learning another solution technique is that algebraic methods become very tedious as the number of species (and, hence, the number of unknowns) increases. Computer solutions (Chapter 9) are one way to ease the computational burden. Graphical solutions to aquatic equilibrium systems, the subject of this chapter, are useful to help visualize species concentrations.

A second reason for mastering graphical solution methods is as follows. Although algebraic techniques give the species concentrations at equilibrium, they do not tell you how species concentrations change with a change in the master variable (see Section 1.3 for a review of the master variable concept). For example, you found in section 7.3.2 that the equilibrium pH of a hundredfold dilution of vinegar was about 3.4, and the equilibrium concentrations of acetic acid and acetate were about 6.6×10^{-3} and 3.7×10^{-4} M, respectively. What are the acetic acid and acetate concentrations if the pH is raised to 7? To answer this question with algebraic methods requires significant additional work. Thus, it is desirable to develop a solution method that readily yields information on how species concentrations change with a change in the master variable.

Finally, as you discovered in Chapter 7, algebraic methods become much easier if you can make assumptions about which species dominate the equations and which species can be ignored. However, it is not always easy to guess which species can be ignored. You will find in this chapter that graphical methods allow you to identify easily which species contribute significantly and which can be ignored in the additive equations (i.e., mass balance and charge balance equations). Sillén (1959) summarized the last two reasons for employing graphical methods succinctly when he wrote: "A graph may tell us at one glance how a complicated equilibrium is shifted on varying conditions, which species predominate, and which are negligible."

In this chapter, one type of graphical solution technique, the log concentration diagram, will be developed. In general, graphical methods are used for two purposes. First, graphs will be created to show how the concentration of each species varies with the master variable. These graphs will give you the concentration of each species at *every* value of the master variable. Second, graphical methods can be used to determine equilibrium concentrations of each species. Equilibrium occurs at *one* value of the master variable.

Key idea: Graphical solutions can show the concentration of each species with each value of the master variable and the equilibrium concentrations of each species

8.2 LOG CONCENTRATION AND p*C*-pH DIAGRAMS

8.2.1 Introduction

As stated in Section 8.1, we are interested in developing graphs to show *quantitatively* how species concentrations vary with a master variable. The graphs to be developed will be similar to most other plots in science and engineering; namely, the independent variable (in this case, the master variable) will be plotted on the abscissa (or *x*-axis) and the dependent variables (in this case, species concentrations) will be plotted on the ordinate (or *y*-axis).[†] Typically, we plot the log of the species concentrations, since the concentrations vary over many orders of magnitude. A plot of the log (base 10) of the species concentrations against the master variable is called a *log concentration diagram*. An example of the axes for a log concentration diagram is shown in Figure 8.1.

log concentration diagram: a plot of the log of the species concentrations against the master variable (usually pH)

[†] For those readers interested in the etymology of scientific terms, the word *abscissa* comes from the Latin *linea absicca* (*linea* = line and *ab-* + *scindere* = to cut off), literally a line that cuts off the distance from the origin. The word *ordinate* comes from the Latin *linea ordinate applicata*, literally a line applied in an orderly manner.

Log concentration diagrams are sometimes confusing at first because the ordinate (*y*-axis) typically is numbered "down" from zero. You are probably used to *y*-axes being numbered "up" from zero. Do not be alarmed. In the log concentration diagram, species concentrations *increase* as one moves "northward" up the ordinate, in accordance with your common experiences with plotting data. Remember that the ordinate is a log scale: one unit increase in a log concentration diagram represents a *tenfold* increase in the species concentration. Also note that the choice of zero for the maximum value of the log of the species concentration ($\log[C] = 0$ or $[C] = 1$ M) is arbitrary. Species concentrations can be greater than 1 M, and if so, the ordinate must be rescaled.

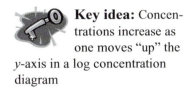

Key idea: Concentrations increase as one moves "up" the *y*-axis in a log concentration diagram

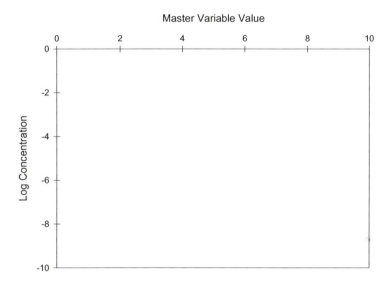

Figure 8.1: Axes for a General Log Concentration Diagram

8.2.2 A taste of acid-base chemistry and p*C*-pH diagrams

Log concentration diagrams are used in particular with equilibrium descriptions of acid-base chemistry. Although acid-base chemistry will be presented in detail in Chapter 11, a taste of acid-base chemistry is presented here to provide common examples of log concentration diagrams.

Of all possible equilibria, a special name is given to equilibria of the form of the reaction in eq. 8.1:

$$\text{acid} = \text{base} + n\text{H}^+ \qquad K \qquad\qquad \text{eq. 8.1}$$

acid dissociation equilibrium: a reaction of the form: acid = base + nH$^+$

This is called an *acid dissociation equilibrium*. One of the reactants is an acid, and the products include a corresponding (or conjugate) base and one or more H$^+$. If $n = 1$, we say that the acid is a *monoprotic acid*. Acetic acid, introduced in Chapter 7, is a monoprotic acid: $CH_3COOH = CH_3COO^- + H^+$

or $HAc = Ac^- + H^+$, where $Ac = CH_3COO$. The equilibrium constants for equilibria with monoprotic acids written in the form of eq. 8.1 are given a special name: K_a. Thus:

$$monoprotic\ acid = conjugate\ base + H^+ \qquad K = K_a$$

There are many types of log concentration diagrams which differ only in the choice of the master variable.

Thoughtful Pause

What is the logical master variable for acid/base chemistry?

---

pC-pH diagram: a log concentration diagram with pH on the abscissa and $pC = p$(concentration) $= -\log$(concentration) values on the ordinate

With acid/base reactions, the logical master variable is pH. Thus, it makes sense to plot pH values on the abscissa and log concentration values on the ordinate. A common approach is to plot p(concentration) $= -\log$(concentration) values on the ordinate, rather than log(concentration) values. The resulting plots are called *pC-pH diagrams*.

An example of a pC-pH diagram is shown in Figure 8.2. Again, the interpretation of the pC-pH diagram is straightforward. Species concentrations *increase* as you move up the ordinate. Also, pH *increases* as you move to the right, again in line with common scientific plots. In other words, more acidic conditions are on the left, and more basic conditions are on the right in pC-pH diagrams. Note, however, that this means the H^+ activity *decreases* as one moves to the right. Also, pC-pH diagrams are log-log plots: a one-unit decrease in pC or pH represents a tenfold increase in the species concentration or H^+ activity.

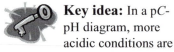 **Key idea:** In a pC-pH diagram, more acidic conditions are on the left (so the H^+ activity decreases to the right)

8.2.3 Plotting species concentrations on pC-pH diagrams

You can draw lines representing the species concentrations as functions of pH on a pC-pH diagram. We know that aqueous systems always contain H_2O, H^+, and OH^-.

Thoughtful Pause

What does the line representing $[H^+]$ look like on a pC-pH diagram?

To determine how the line representing any species appears on a pC-pH diagram, do what you have done many times in your science and engineering courses: write an expression for the dependent variable as a function of the independent variable. In this case, write an expression for pC as a function of pH. Writing $-\log[H^+]$ as a function of pH, you can show that

$p[H^+] \approx pH$.[†] In other words, the line representing $[H^+]$ on a pC-pH diagram is a straight line with a slope of 1 ($p[H^+] = m pH + b$, where m = slope = 1 and b = intercept = 0). How about OH^-? Recall that all aqueous systems must satisfy the equilibrium constraint given by the self-ionization of water (K_W = $[H^+][OH^-]$ = 10^{-14} at 25°C and with the activity of water as unity). From this equilibrium, you can verify that $pK_W = pH + pOH$ or:

$$pOH = -pH + pK_W$$

Thus, the line representing $[OH^-]$ on a pC-pH diagram is a straight line with a slope of -1 and an intercept of pK_W (intercept = 14 if $K_W = 10^{-14}$), assuming dilute conditions.[††]

Key idea: On a pC-pH diagram, the line representing $[H^+]$ slopes downward to the right with a slope of 1 and the line representing $[OH^-]$ slopes upward to the left with a slope of -1

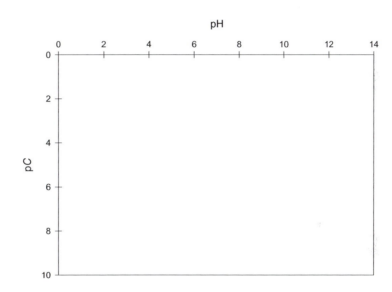

Figure 8.2: Axes for a General pC-pH Diagram

It is customary to refrain from plotting the H_2O line since the activity of water usually is constant, very large relative to other species, and independent of pH.

[†] This expression assumes that the solution is dilute enough so that activity and concentration are equal.

[††] At first glance, the signs of the slopes of the $[H^+]$ and $[OH^-]$ lines may seem confusing. You are used to a positive slope, meaning that the y values increase with increasing x values and the line slopes *upward* to the right. In the pC-pH diagram, a positive slope (as with the $[H^+]$ line) also indicates that the y values, p(concentration) values, increase with increasing pH. This means that the species concentration *decreases* with *increasing* pH. A positive slope trends *downward* to the right.

The H^+ and OH^- concentration lines must always have slopes of 1 and −1, respectively, because of the definition of pH and the self-ionization of water (as long as activity and concentration are equal). Thus, a pC-pH diagram for a dilute aqueous system will always have the H^+ and OH^- lines as shown in Figure 8.3 (if dilute at 25°C and 1 atm). As with any well-constructed scientific plot, you must label the axes and lines. The concentration units in pC-pH diagrams are assumed to be mol/L, unless otherwise indicated.

Key idea: Always label the axes and concentration lines in pC-pH diagrams

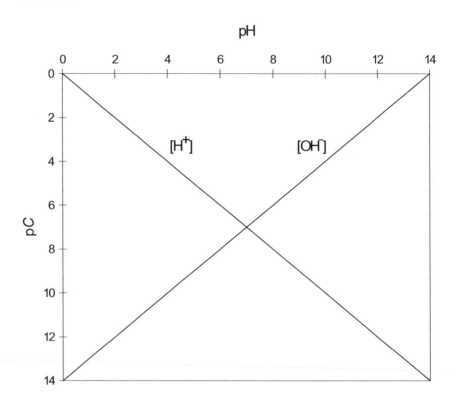

Figure 8.3: General pC-pH Diagram with H^+ and OH^- Lines

Key idea: Recall that concentrations on a pC-pH diagram are logarithmic and species concentrations are equal at the pH where the lines cross

Before using pC-pH diagrams to solve equilibrium problems, it is worthwhile to become very familiar with the interpretation of lines on a pC-pH diagram. Three points are important. First, you read off concentrations on a pC-pH diagram as you do with any other plot. Read off the concentration corresponding to the point where an imaginary line perpendicular to the x-axis passes through the pH of interest and intersects the line representing the species concentration. Thus, in Figure 8.3, pOH at pH 8 is 6 or $[OH^-] = 10^{-6}$ M at pH 8. Second, as always, species concentrations are equal at the pH where their lines cross on a pC-pH diagram. From Figure 8.3, you can see that $[OH^-] = [H^+] = 10^{-7}$ M at pH 7. Third, do not forget

Example 8.1: **Interpreting p***C***-pH Diagrams I**

For the pC-pH diagram below, find (1) the [Ac$^-$] at pH 8, (2) the relative concentrations of OH$^-$ and HAc at pH 6, and (3) the pH where [HAc] and [OH$^-$] are equal and their concentrations at this pH.

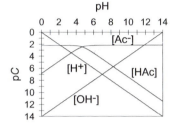

Solution:

(1) At pH 8, the Ac$^-$ concentration is a little less than 10^{-2} M, because the Ac$^-$ line is a little below pC = 2 at pH 8.

(2) At pH 6, [HAc] is significantly larger than [OH$^-$] since the HAc line is several log units above the OH$^-$ line. At pH 6, [HAc] $\approx 10^{-3.5}$ M and [OH$^-$] = 10^{-8} M, so [HAc] is about $10^{4.5}$ $\approx$ 30,000 times larger than the concentration of OH$^-$.

(3) [HAc] and [OH$^-$] are equal where their lines cross. This occurs at about pH 8.3, where their concentrations are about $10^{-5.7}$ M.

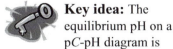

Key idea: The equilibrium pH on a pC-pH diagram is the pH where the charge balance is satisfied

that the y-axis is logarithmic. Two species with pC one unit apart differ in concentration by *tenfold*. For example, the concentration of [OH$^-$] at pH 7.5 (= $10^{-6.5}$ M) is 10 times greater than the concentration of [H$^+$] at pH 7.5 (= $10^{-7.5}$ M). For more practice, see Example 8.1. Another example of interpreting pC-pH diagrams and finding the equilibrium pH value is presented in Section 8.2.4.

8.2.4 Solving a simple equilibrium system with a pC-pH diagram

The pC-pH diagram would be useful if it could only be used to show species concentrations as a function of pH, as shown in Section 8.2.3. However, pC-pH diagrams are much more powerful. *They also can be used to determine the equilibrium pH and equilibrium concentrations of all plotted species.* The use of pC-pH diagrams for solving equilibrium systems is demonstrated in this section with a very simple example: pure water. A formal process for finding the equilibrium position from a pC-pH diagram will be developed in more detail in Section 8.3.2.

A system of pure water has three species (H$_2$O, H$^+$, and OH$^-$), one equilibrium ($K_W = $ [H$^+$][OH$^-$]/[H$_2$O]), no mass balances, a charge balance ([H$^+$] = [OH$^-$]), and the constraint that the activity of water is unity. Using the self-ionization of water, plot the concentrations of H$^+$ and OH$^-$ (see Section 8.2.3). The resulting pC-pH diagram was shown in Figure 8.3. To get this far, you have used the equilibrium constraint (with the activity of water set equal to unity).

Thoughtful Pause
What other information is available to determine the equilibrium position of the system?

You have not yet used the charge balance. The charge balance is:

$$[H^+] = [OH^-]$$

Where in Figure 8.3 are the concentrations of H$^+$ and OH$^-$ equal? As with any ordinate (y-axis) values, the concentrations are equal where the lines intersect. This occurs at pH 7. What are the concentrations of each species at pH 7? To determine species concentrations at pH 7, read the values off the graph as you have done many times in the past with x-y graphs. Locate the equilibrium pH (here, pH 7) on the x-axis. Draw a dotted line at pH 7 parallel to the y-axis (Figure 8.4). The species concentrations are given by the y-axis values at the points where your first dotted line intersects each species line. In this case, both [H$^+$] and [OH$^-$] are 10^{-7} M at pH 7 (Figure 8.4). The key point here is that *the equilibrium position of the system is the one point where the charge balance is satisfied.*

Now, admittedly, determining the equilibrium pH and concentrations of all species in pure water is not a difficult task. However, using graphical

Example 8.2: **Interpreting pC-pH Diagrams II**

What is [OH$^-$] at pH 3? Is pH 3 the equilibrium pH of a system of pure water?

Solution: From Figure 8.4, [OH$^-$] = 10^{-11} M (pOH = 11) at pH 3. This is not the equilibrium pH since the charge balance is not satisfied at pH 3. At pH 3, [OH$^-$] = 10^{-11} M and [H$^+$] = 10^{-3} M, so [H$^+$] ≠ [OH$^-$] and the charges do not balance.

methods, you were able to calculate the equilibrium pH and species concentrations at equilibrium without doing any algebra. You also can find the species concentrations at nonequilibrium pH values. For example, you can find the concentrations of [H$^+$] and [OH$^-$] at pH 9. The algebra involved in finding the species concentrations at pH 9 is not difficult, but the pC-pH diagram immediately shows that [H$^+$] = 10^{-9} M and [OH$^-$] = 10^{-5} M at pH 9, as seen in Figure 8.5. It is clear from Figure 8.5 that [H$^+$] << [OH$^-$] at pH 9 (see also Example 8.2).

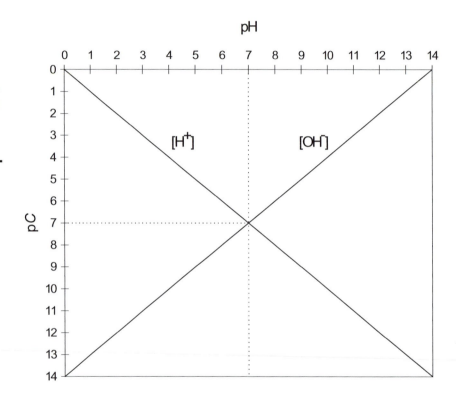

Figure 8.4: Equilibrium pH and Equilibrium Species Concentrations for Pure Water (25°C, 1 atm, dilute)

8.3 USING pC-pH DIAGRAMS WITH MORE COMPLEX SYSTEMS

8.3.1 Superpositioning

superpositioning: the ability to superimpose concentration lines from numerous species on one pC-pH diagram

Graphical methods would not be very useful if they were applicable only to pure water. The power of pC-pH diagrams comes from *superpositioning*. Superpositioning here refers to the ability to superimpose lines representing the concentrations of chemical species in a chemical system on a single log concentration diagram. What does this mean to you? It means that you can use pC-pH diagrams to solve more complex aqueous systems. A system with 100 species could, in theory, be solved with a pC-pH diagram containing 100 lines. In practice, pC-pH diagrams usually are used with small chemical systems consisting of less than about 10 species or so.

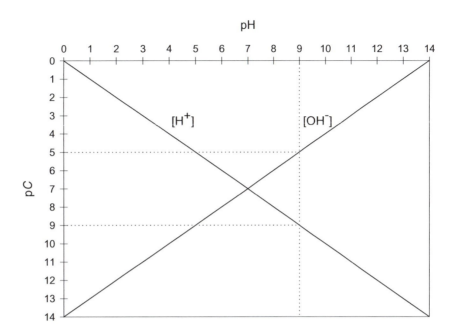

Figure 8.5: Species Concentrations at pH 9 for Pure Water
(25°C, 1 atm, dilute)

8.3.2 A systematic approach to using pC-pH diagrams

The simple problem of pure water from Section 8.2 gives us clues about how to solve general equilibrium problems by using pC-pH diagrams. Start by setting up the equilibrium problem as described in Chapter 6. Next, follow the first two steps of the method of substitution presented in Section 7.3. In the first step, simplify the system by setting the concentrations of all pure liquids and solids equal to unity and all solids that dissolve completely equal to zero. In the second step, express each species concentration in terms of one species concentration, namely, [H$^+$].

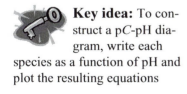

Key idea: To construct a pC-pH diagram, write each species as a function of pH and plot the resulting equations

Consider the example used in Chapter 7 - a 1:100 dilution of vinegar (i.e., total acetate = Ac_T = [HAc] + [Ac$^-$] = 7×10^{-3} M). After simplifying the system and expressing each species concentration in terms of [H$^+$] in Chapter 7, you derived eqs. 7.8 and 7.9, repeated below for convenience as eqs. 8.1 and 8.2:

$$[Ac^-] = Ac_T K/(K + [H^+]) \qquad \text{where } K = 10^{-4.7} \qquad \text{eq. 8.1}$$
$$[HAc] = Ac_T [H^+]/(K + [H^+]) \qquad\qquad\qquad\qquad \text{eq. 8.2}$$

In addition to eqs. 8.1 and 8.2, you know how to express $[H^+]$ and $[OH^-]$ in terms of $[H^+]$ ($[H^+] = [H^+]$ and $[OH^-] = K_W/[H^+]$, where $K_W = 10^{-14}$). It is a simple matter to calculate values of $[Ac^-]$ and $[HAc]$ at any pH (i.e., at any $[H^+]$, assuming concentration and activity are interchangeable) by using eqs. 8.1 and 8.2. For example, you might set up a spreadsheet with pH values in column A (e.g., 0.0, 0.2, 0.4, ..., 13.6, 13.8, 14.0), $[H^+]$ in column B (using the formula: $[H^+] = 10^{-pH}$, assuming a dilute solution), $[OH^-]$ in column C (using $[OH^-] = K_W/[H^+]$), $[Ac^-]$ in column D (using eq. 8.1), and $[HAc]$ in column E (using eq. 8.2). A screen shot of an example spreadsheet from Excel (with the formulas shown in row 8) is illustrated in Figure 8.6. The resulting pC-pH diagram would look like Figure 8.7.

	A	B	C	D	E
1	pH	[H⁺]	[OH⁻]	[Ac⁻]	[HAc]
2	0	1	1E-14	1.39666E-07	0.000318182
3	0.2	0.630957344	1.58489E-14	2.21352E-07	0.000200759
4	0.4	0.398107171	2.51189E-14	3.50813E-07	0.00012667
5	0.6	0.251188643	3.98107E-14	5.55986E-07	7.99236E-05
6	0.8	0.158489319	6.30957E-14	8.81137E-07	5.04284E-05
7	1	0.1	1E-13	1.39641E-06	3.18182E-05
8	1.2	=10^(-1*A8)	=1e-14/b8	=7e-3*10^-4.7/(10^-4.7+b8)	=7e-3*b8/(10^-4.7+b8)

Figure 8.6: Example Spreadsheet for 7×10^{-3} M Acetic Acid System

Notice in Figure 8.7 that the $[H^+]$ and $[OH^-]$ are identical to the previous figures in this chapter. To the H^+ and OH^- lines, superimpose the lines representing the concentrations of $[HAc]$ and $[Ac^-]$. Although it is not obvious at first glance,[†] $[HAc] + [Ac^-] = 0.007$ M $= 10^{-2.15}$ M at all pH values since mass is conserved.

8.3.3 Finding the equilibrium pH

Figure 8.7 shows the lines representing the concentrations of all species in the system (except H_2O) at all pH values. However, you also can use pC-pH diagrams to find the equilibrium state of the system. So far, you have used a mass balance and an equilibrium to derive eqs. 8.1 and 8.2 and draw Figure 8.7. To revisit the question posed in the *Thoughtful Pause* of Section 8.2.4: what other information is available to determine the equilibrium position of the system? You have not yet used the charge balance equation.

[†] It takes a little practice to *add* concentrations on pC-pH diagrams. Since the y-axis is logarithmic, you cannot just add pHAc + pAc to get $-\log([HAc] + [Ac^-])$. Recall that: $\log(a) + \log(b) = \log(ab)$, ***not*** $\log(a + b)$.

Figure 8.7 represents the system at all pH values but *there is only one pH on Figure 8.7 at which the charge balance is satisfied.* In other words, there is only one pH where $[H^+] = [Ac^-] + [OH^-]$. In still other words, the equilibrium pH of the system is the pH in Figure 8.7 where $[H^+] = [Ac^-] + [OH^-]$.

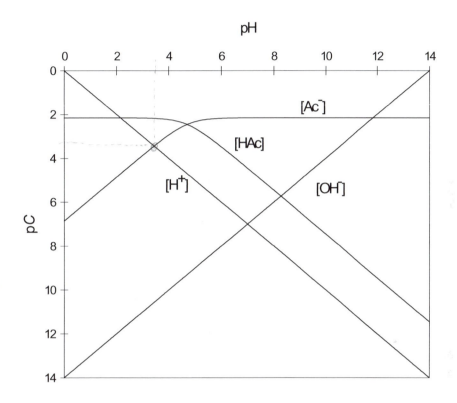

Figure 8.7: pC-pH Diagram for 7×10^{-3} M Acetic Acid

Key idea: To find the equilibrium conditions with a pC-pH diagram, find the pH where the charge balance is satisfied

Key idea: To find the pH where the charge balance is satisfied, assume that one or more species concentrations in the charge balance are negligible and check your assumptions

How can you use Figure 8.7 to find the pH where $[H^+] = [Ac^-] + [OH^-]$? It is a little difficult to mentally track three species concentrations at the same time. It is much easier to compare the concentrations of one cation and one anion. To simplify the charge balance to one cation and one anion, assume that one of the anion concentrations is small relative to the other anion. As an example, start by *assuming* that $[Ac^-] \ll [OH^-]$. Thus, the charge balance becomes $[H^+] \approx [OH^-]$.

Thoughtful Pause

Where in Figure 8.7 are the concentrations of H^+ and OH^- equal?

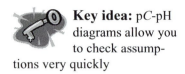

Key idea: pC-pH diagrams allow you to check assumptions very quickly

As in Section 8.2, $[H^+] \approx [OH^-]$ when the $[H^+]$ and $[OH^-]$ lines intersect. As shown by the dotted line in Figure 8.8, this occurs at about pH 7.0. Now you must, as always, check your assumptions: is $[Ac^-] \ll [OH^-]$ at pH 7? One of the strengths of the graphical method is that you can check your assumptions graphically and quickly. From Figure 8.8, the acetate concentration at pH 7.0 is about equal to $Ac_T = 10^{-2.15}$ M. Thus, $[Ac^-]$ is almost five orders of magnitude larger than $[OH^-]$ at pH 7.0. Clearly, the assumption is wrong.

These results might lead you to make the opposite assumption: $[OH^-] \ll [Ac^-]$ or, from the charge balance: $[H^+] \approx [Ac^-]$. At what pH is $[H^+] \approx [Ac^-]$ in Figure 8.7? Again, $[H^+]$ is about equal to $[Ac^-]$ at the pH where the lines representing the H^+ and acetate concentrations intersect. As shown by the dashed line in Figure 8.8, this occurs at about pH 3.4. You can read off the species concentrations at equilibrium (i.e., at pH about 3.4) from Figure 8.8: $[HAc] \approx Ac_T = 10^{-2.15}$ M, $[H^+] \approx [Ac^-] \approx 10^{-3.4}$ M, and $[OH^-] \approx 10^{-10.6}$ M.

Example 8.3: **Finding the Equilibrium pH with a pC-pH Diagram**

Use a pC-pH diagram to find the equilibrium pH of 1 L of pure water to which 1×10^{-3} mole of ammonia have been added.

Solution:
From Example 7.1 [which involved adding $NH_4Cl(s)$ to water], you developed the following system:

Unknowns:
  $[H_2O]$, $[NH_4^+]$, $[H^+]$, $[OH^-]$, and $[NH_3]$
Equilibria:
  $K = [NH_3][H^+]/[NH_4^+]$
    $= 10^{-9.3}$
  $K_W = [H^+][OH^-]/[H_2O]$
Mass balances:
  $N_T = [NH_4^+] + [NH_3]$
Charge balance:
  $[NH_4^+] + [H^+] = [OH^-]$
Other:
  Activity of $H_2O = 1$

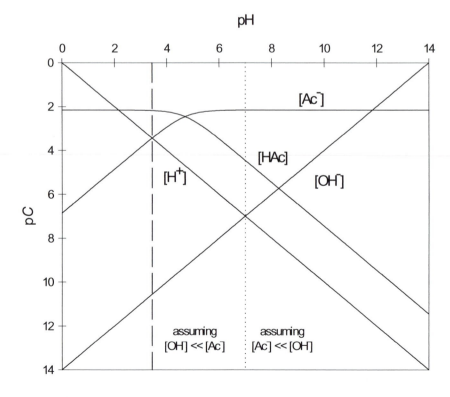

Figure 8.8: Determination of the Equilibrium pH for 0.007 M Acetic Acid

After simplifying and expressing each species concentration as a function of $[H^+]$ (see Example 7.1):

$$[H^+] = [H^+]$$
$$[OH^-] = K_W/[H^+]$$
$$[NH_3] = N_T K/(K + [H^+])$$
$$[NH_4^+] = N_T[H^+]/(K + [H^+])$$

Plotting the species on a pC-pH diagram:

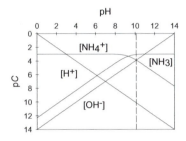

Equilibrium occurs where the charge balance is satisfied: $[NH_4^+] + [H^+] = [OH^-]$. Ignoring $[NH_4^+]$ relative to $[H^+]$ yields pH 7, but a glance at the pC-pH diagram reveals that $[NH_4^+]$ is not negligible compared to $[H^+]$ at pH 7. Ignoring $[H^+]$ relative to $[NH_4^+]$ yields pH 10.1 (dashed line) and the assumption is valid.

The equilibrium pH is about 10.1.

Of course, you need to check this assumption. It is clear from Figure 8.8 that $[OH^-]$ is very small compared to $[Ac^-]$ at pH 3.4; in fact, $[OH^-]$ is seven orders of magnitude smaller than $[Ac^-]$ at pH 3.4.

To summarize the graphical method (see also Appendix C, Section C.4.1):

Step 1: Set up the pC-pH diagram

Set up the chemical system and prepare a pC-pH diagram with lines representing the H^+ and OH^- concentrations

Step 2: Simplify

Simplify the system by setting the concentrations of all pure liquids and solids equal to unity and all solids that dissolve completely equal to zero

Step 3: Express concentrations as functions of $[H^+]$

Express each species concentration in terms of $[H^+]$ by using equilibria and mass balance(s)

Step 4: Plot

Plot the functions derived in Step 3 on the pC-pH diagram

Step 5: Find the equilibrium pH

To find the equilibria pH (and equilibrium concentrations of all plotted species), find the pH where the charge balance is satisfied

Step 6: Check assumptions

If assumptions were made in Step 5 (i.e., if ions were assumed to be of negligible concentration), check all assumptions and iterate if the original assumptions are invalid

Another example of finding the equilibrium pH of a monoprotic acid solution with a pC-pH diagram is shown in Example 8.3.

8.4 SPECIAL SHORTCUTS FOR MONOPROTIC ACIDS

8.4.1 Introduction

As promised in Section 8.1, the log concentration diagram provides information about all species concentrations at all pH values and allows for rapid identification of the equilibrium concentrations. Using spreadsheets, you can draw pC-pH diagrams fairly rapidly. For acids, shortcuts allow for the creation of approximate pC-pH diagrams in a few seconds. The shortcuts will be developed here for monoprotic acids. The shortcuts will be extended to other acids in Chapter 11.

8.4.2 Factors determining the equilibrium pH with monoprotic acids

Equations 8.1 and 8.2 show the dependency of acetate and acetic acid concentrations on pH. You can write these equations for a general monoprotic acid, HA. Assume HA dissociates to A^- as follows: HA = A^- + H^+, with $K = K_a$. In general:

$$[A^-] = A_T K_a/(K_a + [H^+]) \qquad \text{where } A_T = [HA] + [A^-] \qquad \text{eq. 8.3}$$
$$[HA] = A_T[H^+]/(K_a + [H^+]) \qquad\qquad\qquad\qquad \text{eq. 8.4}$$

The concentrations of A^- and HA clearly depend on pH. To generalize a pC-pH diagram for any monoprotic acid, you must look at factors other than pH that influence the acid and conjugate base concentrations.

Thoughtful Pause
What factors determine the A^- and HA concentrations at any pH?

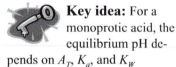 **Key idea:** For a monoprotic acid, the equilibrium pH depends on A_T, K_a, and K_W

From eqs. 8.3 and 8.4, it appears that A_T and K_a will influence the [HA] and [A^-] concentrations at any pH. What factors determine the equilibrium pH? Recall that you must apply a charge balance to the pC-pH diagram to determine the equilibrium pH. The charge balance equation for a monoprotic acid is: $[H^+] = [A^-] + [OH^-]$. Since $[OH^-]$ depends on K_W, it makes sense that *the equilibrium pH in general will depend on A_T, K_a, and K_W*. (If A_T and K_a are sufficiently large, the solution will be acidic, and the appropriate approximation is $[H^+] \approx [A^-]$. In this common case, K_W does not affect the equilibrium pH very much.)

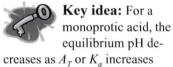 **Key idea:** For a monoprotic acid, the equilibrium pH decreases as A_T or K_a increases (unless A_T or K_a is very small and the solution pH is near neutral)

In a *qualitative* sense, how do A_T and K_a affect the equilibrium pH? It makes sense that as you add more acid (increase A_T), the equilibrium pH should decrease (the solution becomes more acidic). Also, as K_a increases, the acid dissociation equilibrium favors H^+ liberation. Thus, as K_a increases, the solution should become more acidic and the equilibrium pH should decrease. These observations are illustrated in Figure 8.9. Comparing Figures 8.9A, B, and C, which differ only in A_T, you can see that the equilibrium pH (indicated by the dotted line) decreases as A_T increases (from C to B to A). Similarly, comparing Figures 8.9D, E, and F, which differ only in K_a, you can see that the equilibrium pH decreases as K_a increases (from F to E to D).

Animation: To run an animation showing how the species concentrations and the equilibrium pH vary continuously with A_T and K_a, see the example in Appendix E, Section E.2

The effects of A_T and K_a can be shown more dramatically by allowing A_T and K_a to vary continuously. You can run an animation to show how the species concentrations and the equilibrium pH vary continuously with A_T and K_a using the example in Appendix E (Section E.2).

8.4.3 The system point and equilibrium pH

This chapter started with the premise that graphical solutions were much easier than algebraic solutions. So far, it does not appear that graphical solutions are all that simple. To determine the equilibrium pH, you must draw the $[H^+]$ and $[OH^-]$ lines and then plot the $[HA]$ and $[A^-]$ lines using eqs. 8.3 and 8.4.

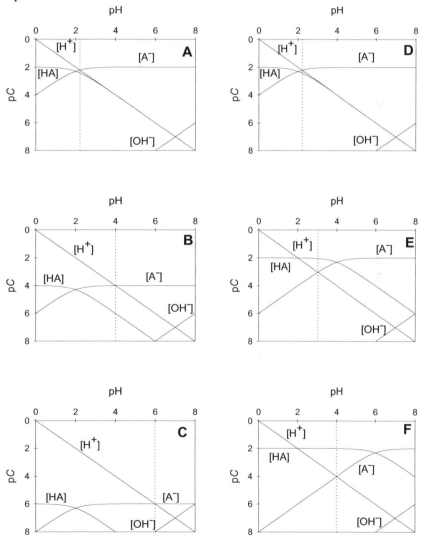

Figure 8.9: Effect of A_T and K_a on the Equilibrium pH of a Solution Containing a Monoprotic Acid - see Table 8.1 for conditions)

system point: the point on the pC-pH diagram where pH = pK_a and pC = pA_T

The simplicity of the graphical method comes from a clever shortcut for solving systems involving acids. To use the shortcut, remember that the equilibrium concentrations depend on A_T and K_a (and, to a lesser extent, on K_W). To follow the shortcut, first locate the system point. The *system point* is the point on the pC-pH diagram where pH = pK_a and pC = pA_T. The system point is located in Figure 8.10A for a monoprotic acid with pK_a = 5 and pA_T = 3.

Table 8.1: Conditions for the pC-pH Diagrams in Figure 8.9

Diagram	A_T (M)	K_a
Fig. 8.9A	10^{-2}	10^{-2}
Fig. 8.9B	10^{-4}	10^{-2}
Fig. 8.9C	10^{-6}	10^{-2}
Fig. 8.9D	10^{-2}	10^{-2}
Fig. 8.9E	10^{-2}	10^{-4}
Fig. 8.9F	10^{-2}	10^{-6}

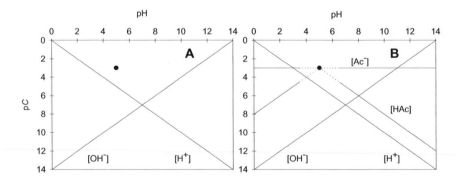

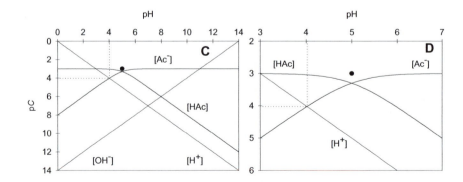

Figure 8.10: Steps Followed in Applying the Shortcut Method
(pK_a = 5 and pA_T = 3; parts A, B, C, and D discussed in the text)

Why is the system point important? From the pC-pH diagrams in Figure 8.10, note that slopes of the [HA] and [A$^-$] lines change at about the system point. In fact, [HA] is about equal to A_T when the pH is at least about 1.5 units *smaller* than the pK_a. At pH values *above* the pK_a, the concentration of [HA] decreases by tenfold for each unit increase in pH (i.e., dpHA/dpH = 1). Similarly, [A$^-$] is about equal to A_T when the pH is at least about 1.5 log units *larger* than the pK_a. At pH values *below* the pK_a, the concentration of [A$^-$] decreases by tenfold for each unit decrease in pH (i.e., dpA/dpH = −1). These observations can be confirmed with eqs. 8.3 and 8.4 (see Problem 8.5).

Key idea: Acid and conjugate base concentration lines away from the system point should intersect the system point if extended

From this analysis, the second step in the shortcut (after locating the system point) is to sketch the [HA] and [A$^-$] lines to within about 1.5 pH units on either side of the pK_a. When sketching the lines, *make sure that the lines, if extended, would go through the system point.* Lines for [HA] and [A$^-$] to within about 1.5 pH units of the pK_a are sketched in Figure 8.10B. Note again that the lines, if extended (dotted lines in Figure 8.10B), would intercept the system point.

For the third step in the shortcut, sketch the lines for [HA] and [A$^-$] near the system point.

Thoughtful Pause

How do the [HA] and [A$^-$] lines look near the system point?

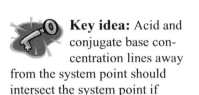

Key idea: Monoprotic acid and conjugate base concentration lines intersect about 0.3 log units below the system point (at pH = pK_a and about 0.3 log units below pA_T)

From eqs. 8.3 and 8.4, at pH = pK_a, you can see that [HA] = [A$^-$]. From the mass balance, [HA] = [A$^-$] = ½A_T. Thus: pHA = pA = −log(½) − log(A_T) ≈ pA_T + 0.3. Therefore, the [HA] and [A$^-$] lines intersect at pH = pK_a and about 0.3 log units "below" pA_T. In other words, the [HA] and [A$^-$] lines intersect about 0.3 log units below the system point. In a crude pC-pH diagram, you can curve the [HA] and [A$^-$] lines by eye so that they intersect about 0.3 log units below the system point (see Figure 8.10C, shown in detail in Figure 8.10D).

Finally, once again find the equilibrium pH by noting where the charge balance is satisfied; that is, where [H$^+$] = [A$^-$] + [OH$^-$]. The equilibrium pH is indicated in Figures 8.10C and 8.10D.

A summary of the shortcut approximation to the graphical method for monoprotic acids follows (see also Appendix C, Section C.4.2):

Example 8.4: Sketching pC-pH Diagrams

Sketch a pC-pH diagram and find the equilibrium pH of a 10^{-4} M HCN solution

Solution:
The pK_a of HCN is 9.2. Thus, the system point is at pC = total cyanide = 4 and pH = pK_a = 9.2. The pC-pH diagram for the system is sketched below:

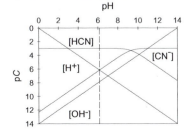

Step 1: Locate the system point

Prepare a pC-pH diagram with lines representing the H$^+$ and OH$^-$ concentrations and locate the system point (where pH = pK_a and pC = pA_T).

Step 2: Draw species concentration lines away from the system point

The equilibrium pH occurs at the pH where the charge balance is satisfied. The species list is H_2O, H^+, OH^-, CN^-, and HCN. Thus, the charge balance is:

$$[H^+] = [OH^-] + [CN^-]$$

This occurs at about pH 6.1 to 6.2.

Draw the [HA] and [A^-] lines at least 1.5 pH units away from the system point by noting that (1) *below* the pK_a, pHA ≈ pA_T and pA increases by 1 log unit for every unit decrease in pH below the pK_a, and (2) *above* the pK_a, pA ≈ pA_T and pHA increases by 1 log unit for every unit increase in pH above the pK_a. Make sure that the [HA] and [A^-] lines, if extended, would both go through the system point.

Step 3: Curve the lines

Curve the [HA] and [A^-] lines so that they intersect about 0.3 log units below the system point.

Step 4: Find the equilibrium pH

To find the equilibria pH (and equilibrium concentrations of all plotted species), find the pH where the charge balance is satisfied. Check any assumptions made in the charge balance and iterate if necessary.

Key idea: The graphical shortcut allows for very rapid estimation of the equilibrium pH of monoprotic acid systems

Animation: To run an animation showing how to draw a pC-pH diagram for a monoprotic acid, see the example in Appendix E, Section E.3

With this shortcut, a pC-pH diagram for a monoprotic acid can be sketched in literally a few seconds. Armed with only pencil and paper, you can use this approach to estimate the equilibrium pH of a monoprotic acid system to within a few tenths of a log unit in a minute or two. Another illustration is provided in Example 8.4. The graphical shortcut will be extended to other acid systems in Chapter 11. You can run an animation to show how to create a pC-pH diagram for a monoprotic acid by using the example in Appendix E (Section E.3).

8.5 WHEN GRAPHICAL METHODS FAIL: THE PROTON CONDITION

8.5.1 Ambiguous results from the graphical method

Armed with the graphical method and its handy shortcut, you should feel confident in solving for the equilibrium composition of monoprotic acid systems. To test the general applicability of the graphical method, consider a simple modification of the acetic acid problem solved in Section 8.3.2. We shall attempt to find the equilibrium pH of a 0.007 M solution of sodium acetate (abbreviated NaAc). Assume, as usual, that sodium acetate dissociates completely to Na^+ and Ac^-.

Thoughtful Pause

How will the calculation of the equilibrium pH of a NaAc solution differ from the calculation of the equilibrium pH of an HAc solution by the graphical approach?

Using the shortcut approach, you have the same acid system at the same total acetate concentration. Thus, you have the same A_T and K_a and therefore the same system point. You could sketch the pC-pH diagram in the same way as Figure 8.7 (except that, for completeness, you might add a horizontal line at $pC = 10^{-2.15}$ to represent the concentration of sodium ion). The pC-pH diagram for NaA with the line representing the sodium ion concentration is given in Figure 8.11.

In the last step of the shortcut, we finally see a big difference between the HAc and NaAc systems: they have different charge balances. For the NaA system, the charge balance is:

$$[Na^+] + [H^+] = [Ac^-] + [OH^-]$$

Where is this satisfied? You know that the Na^+ concentration is relatively large ($[Na^+] = 10^{-2.15}$ M at all pH), and thus you might guess that $[Na^+] \gg [H^+]$. So the charge balance becomes:

$$[Na^+] \approx [Ac^-] + [OH^-]$$

If you assume that $[Ac^-]$ is negligible compared to $[OH^-]$, you would look for the pH where $[Na^+] \approx [OH^-]$. This happens at pH $\approx$ 11.2. At this pH, $[Ac^-]$ is clearly not negligible compared to $[OH^-]$. Thus, make the opposite assumption: $[OH^-]$ is negligible compared to $[Ac^-]$, and thus $[Na^+] \approx [Ac^-]$.

Thoughtful Pause

At what pH in Figure 8.11 is $[Na^+] \approx [Ac^-]$ with $[OH^-]$ negligible compared to $[Ac^-]$?

A big problem looms. The pC-pH diagram in Figure 8.11 shows that the charge balance apparently is satisfied at *every* pH between about pH 6 and pH 10. However, thermodynamics tell you that *only one* equilibrium pH can exist.

8.5.2 The proton condition

What is the problem here and how can we resolve it? In fact, *there is only one pH where the charge balance is satisfied exactly*. The problem is that the graphical method in its present form is simply not sensitive enough to point to the one equilibrium pH.

You can get around this problem by modifying the charge balance slightly. Recall that the charge balance is: $[Na^+] + [H^+] = [Ac^-] + [OH^-]$. It was noted above that $[Na^+] = Ac_T = 10^{-2.15}$ M. Thus:

$$Ac_T + [H^+] = [Ac^-] + [OH^-] \qquad \text{eq. 8.5}$$

Key idea: pC-pH diagrams sometimes *appear* to point to more than one equilibrium pH value because of poor resolution

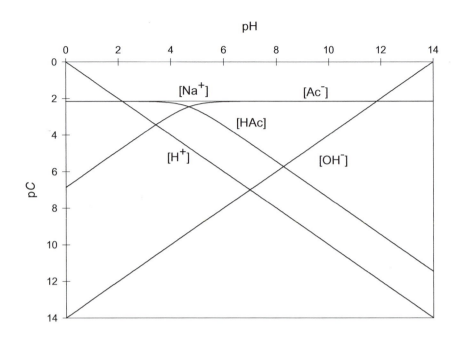

Figure 8.11: pC-pH Diagram for the NaAc/H$_2$O System
(pAc_T = 2.15)

Key idea: You can eliminate species of large constant concentration in the charge balance (e.g., Na$^+$, K$^+$, and Cl$^-$) by using mass balances

proton condition: a mass balance on H$^+$

Example 8.5: **Writing the Proton Condition from Mass and Charge Balances**

Using the mass and charge balances, write the proton condition for a solution formed by adding sodium carbonate (Na$_2$CO$_3$) to water.

Solution:
The starting material Na$_2$CO$_3$ will dissociate initially to Na$^+$ and CO$_3^{2-}$ (carbonate). Carbon-

From the mass balance: Ac_T = [HAc] + [Ac$^-$]. Substituting into eq. 8.5 and simplifying:

$$[HAc] + [H^+] = [OH^-] \qquad \text{eq. 8.6}$$

Is eq. 8.6 more valuable in finding the equilibrium pH than the original charge balance? Find where eq. 8.6 is satisfied on Figure 8.11. You can see that eq. 8.6 is satisfied at about pH 8.2, where [HAc] ≈ [OH$^-$] ≈ 10$^{-5.8}$ M and [H$^+$] (≈ 10$^{-8.2}$ M) is negligible.

This new equation (eq. 8.6) is called the ***proton condition***. Another example of using mass and charge balances to derive the proton condition is given in Example 8.5. As we have seen, the proton condition is very useful. It is a good way to eliminate species of large, constant concentration. The proton condition is a linear combination of the charge balance and the mass balance. Therefore, the set of charge balance/mass balance/proton condition is *not* linearly independent (see Section 6.4). Thus, you can use only two of the charge balance, mass balance, and proton condition. In general, use the proton condition as a replacement for the charge balance. Thus, you can use the proton condition to find the equilibrium pH with graphical methods, and we can use the mass balance and proton condition together in the algebraic approach.

ate reacts with water to form HCO_3^- and H_2CO_3. The charge balance is:

$$[H^+] + [Na^+] =$$
$$[HCO_3^-] + 2[CO_3^{2-}] +$$
$$[OH^-]$$

If enough sodium carbonate solution is initially C M, then the mass balances are:

$$[Na^+] = 2C$$
$$[H_2CO_3] + [HCO_3^-] +$$
$$[CO_3^{2-}] = C$$

Substituting the mass balances into the charge balance equation and simplifying, the proton condition becomes:

$$[H^+] + 2[H_2CO_3] +$$
$$[HCO_3^-] = [OH^-]$$

---

zero level of protons: the number of protons in each starting material

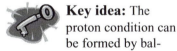

Key idea: The proton condition can be formed by balancing protons; that is, by equating the number of excess protons over starting materials with deficient protons over starting materials

8.5.3 The proton condition as a proton balance

The proton condition was developed in Section 8.5.2 as an alternative to the charge balance equation. So far, deriving the proton condition seems a bit hit or miss: combine the mass balance and charge balance equations until you eliminate species such as Na^+ that are present at high concentrations at all pH values.

The proton condition has a much deeper significance than a seemingly random combination of the mass balance and charge balance equations. *The proton condition is really a mass balance on protons.*[†] To do a mass balance on protons, you must define a *zero level of protons*. With a zero level, you can balance species contributing to a excess of H^+ over the zero level and a deficiency of protons relative to the zero level.

The choice of the zero level is arbitrary. However, it is convenient to use the starting materials as the zero level of protons. Recall from Section 6.3.1 that the starting materials are water and the chemical species added to water. Water is always a starting material. For the addition of NaAc to water, the starting materials are H_2O and NaAc. (Remember that NaAc dissociates completely to Na^+ and Ac^-. Thus, you could list the starting materials as H_2O, Na^+, and Ac^-.) To do a proton balance (i.e., to write the proton condition), you simply identify (1) the species in your species list with *more* protons than their source in the starting materials and (2) the species in your species list with *fewer* protons than their source in the starting materials. Multiply each species by the number of excess or deficient protons relative to the starting materials and equate the concentrations of excess and deficient protons. Note that in this approach, you do not include any species with the *same* number of protons as its starting material.

An example will make this clearer. For the NaAc problem, the starting materials and species list are given below:

Starting materials: H_2O, NaAc ($\rightarrow Na^+$ and Ac^-)
Species list: H_2O, H^+, OH^-, Na^+, Ac^-, and HAc

[†] The word *proton* here is used as a synonym for H^+ (both representing the hydrogen atom with an electron removed).

Example 8.6: **Writing the Proton Condition from the Species List and Starting Materials**

Using the species list and starting materials, write the proton condition for a solution formed by adding sodium carbonate (Na_2CO_3) to water.

Solution:

The starting materials are H_2O and Na_2CO_3 (which dissociates initially to Na^+ and CO_3^{2-}). The species list contains H_2O, H^+, OH^-, Na^+, H_2CO_3, HCO_3^-, and CO_3^{2-}. The number of excess or deficient protons for each species relative to its starting material is as follows (species, starting material, number of excess/deficient protons):

Species	Source	Excess
H_2O	H_2O	0
H^+	H_2O	+1
OH^-	H_2O	−1
Na^+	Na^+	0
H_2CO_3	CO_3^{2-}	+2
HCO_3^-	CO_3^{2-}	+1
CO_3^{2-}	CO_3^{2-}	0

Setting the number of excess protons equal to the number of deficient protons:

$$[H^+] + 2[H_2CO_3] + [HCO_3^-] = [OH^-]$$

As required, this is the same proton condition as in Example 8.5.

What is the source of each species in the species list? Always assume that H_2O, H^+, and OH^- come from water. It is reasonable to write that Na^+ comes from Na^+ and both Ac^- and HAc come from Ac^-. Now H_2O has the same number of protons as its source (namely, water itself), whereas H^+ and OH^- have, respectively, one more proton and one less proton than water.[†] How about Na^+, Ac^-, and HAc? Sodium ion has the same number of protons as its source (namely, Na^+). Acetic acid has one more proton than its source (Ac^-), and acetate has the same number of protons as its source (Ac^-). This information is summarized in Table 8.2.

Table 8.2: Summary of Proton Balance Information for the NaAc/H_2O System

Species	Source	Number of excess or deficient protons relative to the source
H_2O	H_2O	0
H^+	H_2O	1 excess
OH^-	H_2O	1 deficient
Na^+	Na^+	0
Ac^-	Ac^-	0
HAc	Ac^-	1 excess

Now you can balance protons. For each of the species with protons in excess of their source, determine the concentration of excess protons. The concentration of excess protons is given by: (number of excess protons)(species concentration). In a similar fashion, determine the concentration of deficient protons. Since protons are neither created nor destroyed, you can equate the concentration of excess and deficient protons. In the example:

$$(1 \text{ excess proton})[H^+] + (1 \text{ excess proton})[HAc] = (1 \text{ deficient proton})[OH^-]$$

Or: $[H^+] + [HAc] = [OH^-]$

[†] It may seem a little strange to view H^+ as having one more proton than water. Recall from Section 1.4.3 that the symbol H^+ represents several chemical species, including H_3O^+. If you think of H^+ as H_3O^+, it is much easier to accept H^+ as having one more proton than water.

Writing proton conditions takes a little practice. There are a several aspects of developing proton conditions that you should keep in mind. First, remember that H^+ always is considered to have one excess proton, and OH^- is considered to have one deficient proton. Second, it is easier to look for excess and deficient protons if you allow salts to dissociate completely before doing a proton balance (as with the NaAc example).

One more example: the HAc/H_2O system. Try developing the proton condition for the HAc/H_2O system on your own before you read further.

> Starting materials: H_2O, HAc
> Species list: H_2O, H^+, OH^-, Ac^-, and HAc

The proton condition is:

> (1 excess proton)$[H^+]$ =
> > (1 deficient proton)$[Ac^-]$ + (1 deficient proton)$[OH^-]$

> Or: $[H^+] = [Ac^-] + [OH^-]$

In this case, the proton condition is identical to the charge balance equation. At least, the charge balance and the proton condition for the HAc/H_2O system are *nearly* identical.

Thoughtful Pause

What are the differences between the charge balance and the proton condition for the HAc/H_2O system?

In the charge balance, the coefficients in front of the species concentrations are the number of charges per ion (or number of equivalents per mole, where charge is the currency of equivalence). In the proton balance, the coefficients in front of the species concentrations are the number of excess/deficient protons per mole (or number of equivalents per mole, where protons relative to the starting material is the currency of equivalence). This is, admittedly, a subtle difference, but it points to the importance of the units of the coefficients in such expressions. The proton condition for the Na_2CO_3/H_2O system (examined in Example 8.5) is derived by using the excess/deficient proton concept in Example 8.6. Note that this approach allows proton conditions to be written almost by inspection.

8.5.4 Uses of the proton condition

The proton condition can be used to solve equilibrium systems by either algebraic or graphical methods. What role should the proton condition play in the set of equations to be solved? As stated in Section 8.5.2, the mass balances, charge balance, and proton condition are not linearly independent.

You can generate any equation from the other two (see Problem 8.9). As a result, you can use only two of the three equations in your list of equations. *Usually, it is easiest simply to replace the charge balance equation with the proton condition.*

When should you use the charge balance equation and when should you use the proton condition? The simple answer is that the proton condition will *always* give an acceptable solution.[†] The proton condition is very useful for systems with salts as starting materials. When the starting material does not include a salt, the proton condition and the charge balance are identical. In general, use the proton condition to solve chemical equilibrium systems. Although the proton condition may seem awkward at first, you will find that, after some practice, you can write the proton condition as easily as you can write the charge balance equation.

8.6 CONCLUSIONS

Graphical solutions to equilibrium systems are used to show how species concentrations change with a change in the master variable and to determine the equilibrium pH and species concentrations. Graphical solutions also allow you to identify easily which species contribute significantly and which can be ignored in the additive equations. The most common graphical tool is the pC-pH diagram, a type of log concentration diagram where -1 times the log of each species concentration is plotted as a function of pH.

To make a pC-pH diagram, write each species concentration in terms of $[H^+]$. This process requires the use of the equilibrium expressions and mass balances equations. With monoprotic acids (and, as you shall see in Chapter 11, with other acids as well), a shortcut procedure can be employed to draw species concentration lines on a pC-pH diagram. In the shortcut method, the system point is identified. The system point occurs where $pC = p([HA] + [A^-])$ and $pH = pK_a$. The pHA and pA lines are drawn away from the system point and allowed to intersect 0.3 log units below the system point.

[†] In fact, the charge balance also always gives the correct equilibrium pH. However, the typical resolution of a pC-pH diagram is not sufficient to use the charge balance to find the equilibrium pH if the charge balance includes ions with relatively large, constant concentrations.

To find the equilibrium pH with a pC-pH diagram, simply find the pH where the charge balance is satisfied. You may wish to use and check assumptions about ignoring certain species relative to other species to find the equilibrium pH. The equilibrium pH also is found at the pH where the proton condition is satisfied. The proton condition comes from a mass balance on protons and balances protons in excess and deficient to the starting materials. The proton condition works better than the charge balance equation because the proton condition does not contain unreactive ions which limit the resolution of graphical methods for determining equilibrium concentrations. With a little practice, the proton condition can be written as readily as the charge balance equation.

8.7 PART II CASE STUDY: HAVE YOU HAD YOUR ZINC TODAY?

Recall that the Part II case study was probing the speciation of zinc as a zinc acetate was swallowed. The Part II case study can be completed by drawing a pC-pH diagram for the $Zn(Ac)_2(s)/H_2O$ system. Most of the work in solving this system was done in Section 7.6. The Zn^{2+} concentration can be determined by solving the equation:

$$a[Zn^{2+}]^2 + b[Zn^{2+}] + c = 0, \text{ where:}$$

$$a = K_7(1 + K_1 K_W/[H^+])$$
$$b = (1 + [H^+]/K_6)(1 + K_1 K_W/[H^+]) + K_7(Ac_T - Zn(+II)_T)$$
$$c = -Zn(+II)_T (1 + [H^+]/K_6)$$

The equilibrium constants and total masses are:

$$K_W = [H^+][OH^-] = 10^{-14}$$
$$K_1 = [ZnOH^+]/([Zn^{2+}][OH^-]) = 10^{+5}$$
$$K_6 = [Ac^-][H^+]/[HAc] = 10^{-4.7}$$
$$K_7 = [ZnAc^+]/([Zn^{2+}][Ac^-]) = 10^{+1.57}$$
$$Zn(+II)_T \approx [Zn^{2+}] + [ZnOH^+] + [ZnAc^+] = 3.06 \times 10^{-3} \text{ M}$$
$$Ac_T \approx [Ac^-] + [HAc] + [ZnAc^+] = 6.12 \times 10^{-3} \text{ M}$$

The approximations were shown to be valid in the pH range of 2 to 6.5 in Section 7.8. The quadratic equation can be solved for $[Zn^{2+}]$ at any pH. After calculating $[Zn^{2+}]$, you can determine the concentrations of acetate ion and the other zinc-containing species by back-substitution as follows:

$$[Ac^-] = Ac_T/(1 + [H^+]/K_6 + K_7[Zn^{2+}])$$
$$[ZnOH^+] = K_1 K_W[Zn^{2+}]/[H^+]$$
$$[ZnAc^+] = K_7[Zn^{2+}][Ac^-]$$

The calculations are performed most easily with a spreadsheet. The resulting pC-pH diagram for the zinc-containing species is shown in Figure 8.12. Note that $[Zn^{2+}]$ is the predominant zinc-containing species at low pH. As the pH approaches the pK_a of acetic acid (4.7), the $[ZnAc^+]$ becomes a greater fraction of the total zinc. Figure 8.12 also can be used to verify the assumptions that $Zn(Ac)_2^0$ and $Zn(Ac)_3^-$ can be ignored relative to $ZnAc^+$ and also that $Zn(OH)_2^0$, $Zn(OH)_3^-$, and $Zn(Ac)_4^{2-}$ can be ignored relative to $ZnOH^+$.

What is the significance of this equilibrium calculation? The significance lies in the relative absorption of the zinc-containing species. If the most highly absorbed species is Zn^{2+}, then pH would make little difference in the ability of the body to absorb zinc from the zinc acetate tablets. However, if any other zinc-containing species dominate zinc absorption, then pH may play a major role in the uptake of zinc. Note that the concentrations of all zinc species except Zn^{2+} are larger in the low pH of the gastric juices than in the near neutral pH of saliva.

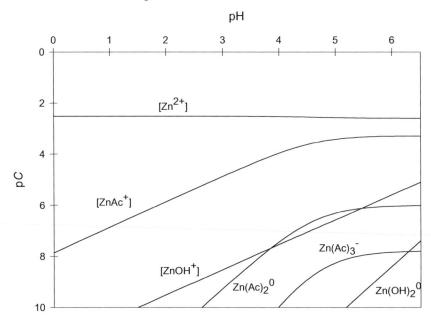

Figure 8.12: pC-pH Diagram for the Part II Case Study

SUMMARY OF KEY IDEAS

- Graphical solutions help to (1) visualize species concentrations, (2) show how species concentrations change with a change in the master variable, and (3) identify easily which species contribute significantly and which can be ignored in the additive equations

- Graphical solutions can show the concentration of each species with each value of the master variable and the equilibrium concentrations of each species

- Concentrations increase as one moves "up" the y-axis in a log concentration diagram

- In a pC-pH diagram, more acidic conditions are on the left (so the H$^+$ activity decreases to the right)

- On a pC-pH diagram, the line representing [H$^+$] slopes downward to the right with a slope of 1 and the line representing [OH$^-$] slopes upward to the left with a slope of -1

- Always label the axes and concentration lines in pC-pH diagrams

- Recall that concentrations on a pC-pH diagram are logarithmic, and species concentrations are equal at the pH where the lines cross

- The equilibrium pH on a pC-pH diagram is the pH where the charge balance is satisfied

- To construct a pC-pH diagram, write each species as a function of pH and plot the resulting equations

- To find the equilibrium conditions with a pC-pH diagram, find the pH where the charge balance is satisfied

- To find the pH where the charge balance is satisfied, assume that one or more species concentrations in the charge balance are negligible and check your assumptions

- pC-pH diagrams allow you to check assumptions very quickly

- For a monoprotic acid, the equilibrium pH depends on A_T, K_a, and K_W

- For a monoprotic acid, the equilibrium pH generally decreases as A_T or K_a increase

- Acid and conjugate base concentration lines away from the system point should intersect the system point if extended

- Monoprotic acid and conjugate base concentration lines intersect about 0.3 log units below the system point (at pH = pK_a and about 0.3 log units below pA_T)

- The graphical shortcut allows for very rapid estimation of the equilibrium pH of monoprotic acid systems

- pC-pH diagrams sometimes *appear* to point to more than one equilibrium pH value because of poor resolution

- You can eliminate species of large constant concentration in the charge balance (e.g., Na^+, K^+, and Cl^-) by using mass balances

- The proton condition can be formed by balancing protons; that is, by equating the number of excess protons over starting materials with deficient protons over starting materials

PROBLEMS

8.1 Find the equilibrium pH of a 10^{-3} M NaCN solution by using graphical methods. Try to find the equilibrium pH with the charge balance and separately with the proton condition.

8.2 An engineer is designing a chlorine contact chamber to disinfect poultry- processing wastes. (Note: these wastes are often contaminated with the pathogenic bacteria *Salmonella*). She can design the system to use chlorine gas (equivalent to using HOCl) or liquid bleach (equivalent to using sodium hypochlorite, NaOCl) as the chlorine source. The total chlorine concentration to be added is 7.1 mg/L as Cl_2 (= 10^{-4} M chlorine). The species HOCl is a much stronger disinfectant than the species OCl^-. The pK_a for HOCl is 7.54.

A. If chlorine chemistry controls the pH, which system provides more disinfection? (Hint: this problem requires you to calculate equilibrium concentrations of HOCl and OCl^- in 10^{-4} M HOCl and 10^{-4} M NaOCl systems.)

B. In a complex waste stream such as this, is the assumption that chlorine chemistry controls the pH reasonable? Why or why not?

8.3 Solve Problem 7.7 using graphical methods.

8.4 In Problem 7.11, you verified that a C M HCl solution has an equilibrium pH of about $-\log C$ (as long as C was greater than some value). Reverify this relationship (and find the minimum value of C) by using a graphical approach. (Hint: print out a general pC-pH diagram, like Figure 8.3, on a transparency sheet. Print generic [HA] and $[A^-]$ lines on another sheet. Use the sheets to show the relationship in question.)

8.5 For a monoprotic acid, verify that:

A. At pH values above the pK_a, the concentration of [HA] decreases by tenfold for each unit increase in pH (i.e., $dpHA/dpH = 1$).

B. At pH values below the pK_a, the concentration of [A⁻] decreases by tenfold for each unit decrease in pH (i.e., $dpA/dpH = -1$).

8.6 It is possible to write general approximations for a monoprotic acid, HA.

A. Show graphically that the equilibrium pH for a solution obtained by adding C moles/L of HA to water is about $(pK_a + pC)/2$. What assumptions did you make?

B. Show that the equilibrium pH for a solution obtained by adding C moles/L of NaA to water is about $(14 + pK_a - pC)/2$. What assumptions did you make?

8.7 Why does the proton condition eliminate the readily dissociable ions such as Na⁺, K⁺, and Cl⁻ from the charge balance? Why is the proton condition for adding HA to acid identical to the charge balance equation?

8.8 Write the proton condition when the following salts are added to water:

    A. NH_4Cl
    B. Na_2S
    C. KH_2PO_4
    D. $(NH_4)_2CO_3$

See Appendix D for ideas on the pertinent equilibria to generate species lists.

8.9 Show that the proton condition for a NH_4Cl solution is a linear combination of the mass balance equations and the charge balance equation.

CHAPTER 9

Computer Solutions to Chemical Equilibrium Problems

9.1 INTRODUCTION

Many equilibrium systems of interest in the environment are sufficiently complicated to render algebraic and graphical solution methods very tedious. For systems with many species, computer methods can be used to calculate equilibrium concentrations quickly. As with any computer solution methods, it is imperative that you understand both the chemical system itself and how to interpret the computer output. In this chapter, you will become familiar with several computer solution methods. The purpose of this chapter is to increase your comfort level with one or more computer solution techniques so that you can solve increasingly complex systems.

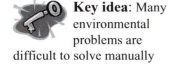 **Key idea**: Many environmental problems are difficult to solve manually

In Section 9.2, a problem is introduced that will be solved by all computer methods presented in the chapter. The chapter equilibrium problem is simple and familiar to you. It will allow you to try out the software on well-trodden ground. Section 9.3 introduces you to spreadsheet solutions. You will find that spreadsheets are ideal for solving equilibrium problems. Using a consistent format for your spreadsheets will prove beneficial when tackling large systems.

Equilibrium calculation software is discussed in general in Section 9.4. Two software packages are presented in more detail. In Section 9.5, a program called *Nanoql* is introduced. *Nanoql* was developed specifically for use with this text and follows the species/equilibrium/mass balance/charge balance approach you have used so far. In Section 9.6, you will be introduced to MINEQL. MINEQL is a very popular equilibrium chemistry software package. It uses an approach called the Tableau method, which is common to many equilibrium calculation programs.

9.2 CHAPTER PROBLEM

To illustrate the solution of equilibrium problems by computer, a sample problem will be set up and solved in several ways in this chapter. The chapter problem is to determine the species concentrations in a system consisting of 1×10^{-3} M acetic acid (HAc) in water. The system is:

Species list: H_2O, H^+, OH^-, HAc, and Ac^- (Ac^- = acetate)

Equilibria:
$$H_2O = H^+ + OH^- \quad K_W$$
$$HAc = Ac^- + H^+ \quad K$$

Or:

$$K_W = [H^+][OH^-]/[H_2O]$$
$$K = [Ac^-][H^+]/[HAc]$$

Mass balance:
$$Ac_T = 1 \times 10^{-3} \text{ M} = [HAc] + [Ac^-]$$

Charge balance:
$$[H^+] = [Ac^-] + [OH^-]$$

Other:
Activity of $H_2O = 1$
All species concentrations ≥ 0

In this system, any two species (other than water) could be used as components. Useful components are H^+ and Ac^-.

9.3 SPREADSHEET SOLUTIONS

9.3.1 Introduction

Spreadsheets are invaluable for the solution of small- to medium-sized chemical equilibrium problems. The use of spreadsheets for finding roots of equations was discussed in Chapter 7 and can be found in Appendix A. In this chapter, spreadsheets will be used to set up and solve equilibrium systems in a more general fashion.

The general approach to using spreadsheets for the solution of equilibrium problems is as follows. Equilibrium constants and added total masses are entered. In addition, the *equations* for the mass balances (sum of species concentrations weighted by their mass balance stoichiometric coefficients) and electroneutrality equation (charge balance or proton condition) are entered. Components are entered into the spreadsheet with guesses for their concentrations. Species are entered into the spreadsheet with equations for their concentrations as functions of the components, other species concentrations, and equilibrium constants. If the species concentration is fixed (e.g., H_2O), then its fixed concentration is entered rather than an equation.

Values of the component concentrations are calculated by the spreadsheet software to meet the mass balance and electroneutrality

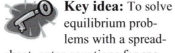
Key idea: To solve equilibrium problems with a spreadsheet, enter equations for species (based on equilibria) and vary the $-\log(\text{conc.})$ values of components until the mass and charge balances are satisfied

constraints. In most equilibrium calculations by spreadsheet, it is useful to have the spreadsheet program calculate pC values for each component [i.e., $-\log$(concentration) values] rather than have the software calculate concentrations for each component. Calculating pC values minimizes round-off and underflow errors when component concentrations are very small.

You have several constraining functions on the system: one for each mass balance plus electroneutrality. In a general optimization routine (such as found in Excel's *Solver*), the constraining functions are divided between the objective function and constraints. To meet the objective function, one cell in the spreadsheet (called the *target cell*) is set equal to a fixed value. To meet the constraints, other cells are set equal to other values. You can use the target cell to satisfy one mass balance or the electroneutrality condition and the constraints to account for the other mass balances and/or electroneutrality condition.

File available:
The spreadsheet for the chapter example may be found on the CD - file location
/Chapter 9/Acetic acid.xls

9.3.2 Setting up the spreadsheet

Spreadsheet solutions are much easier if you take the time to set up your spreadsheet carefully. Divide the spreadsheet into six regions for components, species, mass balances, electroneutrality, equilibria, and notes. A nice approach is shown in Figure 9.1. The spreadsheet shown in Figure 9.1 may be found on the CD at **/Chapter 9/Acetic acid.xls**. Note the six regions in the spreadsheet. For each of the two components, guesses of their pC values are entered. In the example, pC guesses are entered in cells D2, D3, E2, and E3. (The spreadsheet program eventually will vary your initial guesses. Entering the values twice allows you to keep a record of your initial guesses.) The formulas in cells C2 and C3 calculate the concentrations from cells D2 and D3. The formula in cell C2 is **=−1*LOG(D2)** and the formula in cell C3 is **=−1*LOG(D3)**.

Equations for the species concentrations are found in cells C7 through C11. The equations express the concentration of each species as a function of the components, other species concentrations, and equilibrium constants. (For illustrative purposes, the formulas in cells C7 through C11 are shown in cells D7 through D11 in Figure 9.1.) For example, cell D9 contains the formula **=I3/C3** to show that $[OH^-] = K_W/[H^+]$. For water, the fixed concentration of 1 is entered in cell C7.

The equation for the total acetate mass balance, expressing the total mass as the sum of the weighted species concentrations, is in cell D15. Cell D15 contains the formula **=C10+C11** to indicate that $Ac_T = [HAc] + [Ac^-]$. The numerical value of the added total mass is placed in cell C15.

Either the charge balance or proton condition can be used to account for electroneutrality. The charge balance formula is entered in cell C19. It is useful to enter the electroneutrality constraint as a function that is equal to zero. For example, the charge balance is entered as the sum of the eq/L of positive charges – the sum of the eq/L of negative charges (see formula in cell D19 in Figure 9.1).

The equilibrium information is written in columns H and I. It is recommended that you write out the chemical expressions to help you remember which equilibrium is found in each row (see column H in Figure 9.1). If you wish, pK or log K values could be entered into another column and K values calculated in column I.

Finally, write a few notes to yourself about how you set up the problem. It is useful to note what was fitted (component concentrations or pC values of the components) and to identify both the objective function and constraints.

Key idea: Write out the equilibrium expressions and write notes to help you remember the details of the chemical system and calculation

	A	B	C	D	E	F	G	H	I	J
1	Components		Conc.	pC	Guesses		Equilibria		K	
2		Ac⁻	1.00E-04	4	4			HAc = Ac⁻ + H⁺	2.00E-05	
3		H⁺	1.00E-07	7	7			H₂O = H⁺ + OH⁻	1.00E-14	
4										
5										
6	Species		Conc.				Notes			
7		H₂O	1.00E+00	=1				Problem:	1e-3 M HAc	
8		H⁺	1.00E-07	=C3				Fitting:	pAC, pH	
9		OH⁻	1.00E-07	=I3/C3				Obj. function:	Fit Ac_T	
10		Ac⁻	1.00E-04	=C2				Constraints:	Charge balance = 0	
11		HAc	5.01E-07	=C2*C3/I3						
12										
13										
14	Mass Balances		Added	Calculated						
15		Ac_T	1.00E-03	1.01E-04	=C10+C11					
16										
17										
18	Electroneutrality		Calculated							
19		Charge bal.	-1.00E-04	=C8-C9-C10						
20		Proton cond.								
21										

Figure 9.1: Spreadsheet for the Chapter Problem

9.3.3 Solving systems with spreadsheets

The dialog box for *Solver* is shown in Figure 9.2. Note that the total mass is being set to 0.001 M and the calculated excess charge is being set equal to 0. After you press the *Solve* button, the optimum values of the component concentrations are calculated by varying their pC values. The species concentrations in column C are updated automatically. The final concentrations are shown in Figure 9.3. You can verify that the solution is correct.

Whenever a generalized optimization routine is used, it is up to the user to decide if the results are accurate enough. To judge accuracy, examine

Key idea: Check the optimization results to see if sufficient resolution has been obtained by examining the predicted total masses and charge balance

the calculated mass balances and calculated electroneutrality (cells D15 and C19, respectively, in Figure 9.3). If the difference between the calculated mass balance and the added total mass is too large for your purposes (or if the calculated charge balance is very different from zero), the calculation should be repeated. The calculation also must be repeated if *Solver* does not converge. The accuracy of the results can be changed and convergence problems addressed in two ways. First, the initial component pC guesses can be altered. Second, the Precision setting (accessed through the Option menu on the *Solver* dialog box) can be adjusted. The results in Figure 9.3 were obtained with a precision setting of 1×10^{-10}.

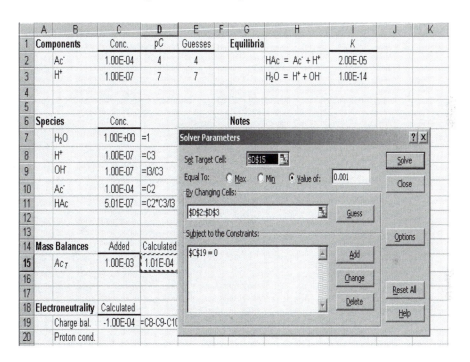

Figure 9.2: *Solver* **Dialog Box for the Chapter Problem**

This approach can be extended to larger systems with ease. A system with 41 species is solved by using the spreadsheet method in Chapter 23. The template spreadsheet on the accompanying CD can be modified to your purposes. For example, columns can be added to automatically calculate the species concentrations in mass units, calculate pC values for each species, or enter the equilibrium constants as pK values.

9.4 EQUILIBRIUM CALCULATION SOFTWARE

9.4.1 General approaches
Chemical equilibrium systems generally are solved by using one of two approaches. In the first approach, the Gibbs free energy of the system is

minimized with mass balances as constraints. An example of minimizing Gibbs free energy to determine the equilibrium composition of a system was given in Section 3.8.4. The minimization of free energy approach is very common in chemical engineering, where temperatures and pressures often are very different than standard state conditions.

In environmental engineering and science applications, a different approach to solving equilibrium systems usually is used. This approach is the basis of Part II of this text: find the species activities that satisfy the mathematical system created from equilibrium expressions, mass balances, a charge balance (if applicable), and other constraints (such as species with fixed activities). As discussed in Chapter 4, the mathematical system is nonlinear in the species activities because the equilibrium expressions are nonlinear in the species activities.

	A	B	C	D	E	F	G	H	I	J
2		Ac⁻	1.32E-04	3.880648	4			HAc = Ac⁻ + H⁺	2.00E-05	
3		H⁺	1.32E-04	3.880647	7			H₂O = H⁺ + OH⁻	1.00E-14	
4										
5										
6	Species		Conc.				Notes			
7		H₂O	1.00E+00	=1				Problem:	1e-3 M HAc	
8		H⁺	1.32E-04	=C3				Fitting:	pAC, pH	
9		OH⁻	7.60E-11	=I3/C3				Obj. function:	Fit Ac_T	
10		Ac⁻	1.32E-04	=C2				Constraints:	Charge balance = 0	
11		HAc	8.68E-04	=C2*C3/I3						
12										
13										
14	Mass Balances		Added	Calculated						
15		Ac_T	1.00E-03	1.00E-03	=C9+C10					
16										
17										
18	Electroneutrality		Calculated							
19		Charge bal.	-1.82E-11	=C7-C8-C9						
20		Proton cond.								
21										

Figure 9.3: Final Results from the Spreadsheet Solution

9.4.2 Equilibrium calculation programs

A number of software packages have been developed to set up and solve chemical equilibrium systems. Most of the programs are based on *Mineql*. Many commercial packages exist, and a short list of available software is provided in Table 9.1. For a more extensive review of available software, see Butler (1998).

9.5 *Nanoql*

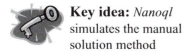

Key idea: *Nanoql* simulates the manual solution method

9.5.1 Introduction

Nanoql was designed to solve small- to moderate-sized aquatic equilibrium systems. The name acknowledges the legacy of MINEQL and the focus on smaller systems. The motivation behind the development of *Nanoql* was to produce an equilibrium calculation software package that simulates the manual solution method. Thus, users are encouraged to select species (rather than components) and enter equilibria and mass balance information. Based on reasonable components, *Nanoql* will provide guesses for the pertinent equilibria and mass balances. As with the other software packages in Table 9.1, *Nanoql* will provide you with many forms of output, including tables of concentrations; pC-pH diagrams; titration curves (Chapter 12); and variation in species concentrations with equilibria constants, total masses, temperature, and ionic strength (see Chapter 21).

The best way to learn *Nanoql* is to simply try it out. Install the software from the CD as directed in Appendix F. Many *Nanoql* examples will be given in this text. Additional hints for using *Nanoql* to solve specific types of chemical equilibrium problems are given in Appendix F.

File available: The *Nanoql* chemical system for the chapter example may be found on the CD - file location **/Chapter 9/Acetic acid.nqs**

9.5.2 Using *Nanoql* to Solve the Chapter Problem

To get your feet wet with *Nanoql*, run the program and open the chapter problem file. The file is on the CD at **/Chapter 9/Acetic acid.nqs**. To calculate equilibrium concentrations, simply press the Σ button (or select *Solve* under the *Calculate* menu). *Nanoql* will prompt you for initial guesses for each species. To avoid the tedium of entering each value separately, select *Helper*. *Nanoql* now will let you guess the pH (guesses can be very crude) and will assign all species to some user-entered fraction of the total concentration of one of their components[†]. Press *Ok* and *Nanoql* will calculate the species concentrations.

If you want to draw a pC-pH diagram, click on the *Calculate* menu item and select *Vary*, then *Species Concentration*. Select (i.e., click on) the species H^+. At the next step, enter the varying range and step size (try 0 to 14 first with a step size of 0.1 by entering **p0**, **p14**, and **p0.1**). Again, you will be prompted for initial concentration guesses at the first pH (pH 0 in this case). You can save the tabular output or plot the results. In plotting, first select the species to plot. Then set your own plot characteristics or choose pC-pH from the predetermined plot type to load preconfigured values.

[†] Numbers in *Nanoql* can be entered several different ways to save you time. Enter $10^{-3.5} = 3.16 \times 10^{-4}$ as **3.16e−4** or **p3.5** or **l−3.5**. Here, **p** means that the number that follows is -1 times $\log_{10}$ of the value to be entered and **l** (lowercase letter ell) means that the number that follows is $\log_{10}$ of the value to be entered. Always press the Enter key after typing in a number.

Table 9.1: Chemical Equilibrium Software Packages Used in Environmental Engineering and Science Applications

CHESS
    Developed by: Jan van der Lee
    System set up: Tableau method
    Application: Geochemical
    Commercial version: **chess.ensmp.fr/index.html**

MINEQL
    Developed by: F.M.M. Morel
    System set up: Tableau method
    Application: General
    Commercial version: MINEQL+ by Environmental Research Software
        (**www.mineql.com**)
    Free version: DOS ver 3.01 is free at **www.mineql.com**

MINTEQA2
    Developed by: U.S. EPA Center for Exposure Assessment Modeling
    System set up: Tableau method
    Application: General
    Commercial version: Scientific Software Group
        (**www.scisoftware.com**)
    Free version: U.S. EPA at **www.epa.gov/ceampubl/minteq.htm**

Nanoql
    Developed by: J.N. Jensen
    System set up: As per Part II of this text
    Application: General
    Commercial version: Included with this text

PHREEQC
    Developed by: U.S. Geological Survey (USGS)
    System set up: Tableau method
    Application: Geochemistry
    Commercial version: PHREEQE at Scientific Software Group
        (**www.scisoftware.com**)
    Free version: **wwwbrr.cr.usgs.gov/projects/GWC_coupled/phreeqc/**
        (Windows version: **www.geo.vu.nl/users/posv/phreeqc.html**)

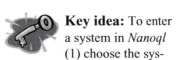

Key idea: To enter a system in *Nanoql* (1) choose the system characteristics (including whether to use the electroneutrality condition), (2) choose the species (either individually or by components), (3) enter the equilibria, (4) enter the mass balances, and (5) enter User Defined Functions, if desired

You should also practice entering chemical systems in *Nanoql*. Open a new system (click on the *New* button on the right side of the toolbar, or select *New* under the *File* menu). You will be prompted to open a Nanoql Possible Species file (.nps file). This is a database of species characteristics and thermodynamic data. Choose the default file (although you can build your own .nps files later if you wish).

You will now be led through five steps for entering a new system. In Step 1, choose the system characteristics. For the chapter example, choose to use the electroneutrality condition. Press *Next*. In Step 2, choose the species. You can do this by clicking on the species individually in the top table or clicking on the components in the top list box to select all species in a given component. In either way, add the species HAc and Ac⁻. The species type can be changed, but for now leave the new species as dissolved species.

In Steps 3 and 4, equilibria and mass balances are entered. You can click on species in the top table to add the information manually or click on *Helper* to let *Nanoql* assist you with the process. Be sure to enter the equilibrium constant and total mass. The fifth step is optional. You can enter ***User Defined Functions***: linear combinations of species concentrations. If you wish, you could enter the charge balance to see if it really is zero at equilibrium. Now save the system and calculate equilibrium concentrations as described in the beginning of this section.

User Defined Function: in *Nanoql*, a linear combination of species concentrations that can be entered and tracked by the user

9.6 THE TABLEAU METHOD AND OTHER EQUILIBRIUM CALCULATION PROGRAMS

9.6.1 Importance of components

One of the weaknesses of *Nanoql* is that the software must perform a large number of calculations to solve a system of n nonlinear equations in n unknowns (i.e., the n species concentrations). With modern personal computers, the calculation time is still very fast with small- to medium-sized systems. For very large systems (with dozens or hundreds of species), it may be necessary to recast the mathematical system to reduce the number of unknowns. This is accomplished through the use of components.

Recall from Chapter 3 that components are the minimum set of building blocks from which all species can be formed. Most equilibrium calculation software is based on the concept of components to make the calculations more efficient. You have a lot of freedom in choosing the components to describe a system. In fact, the components do not have to be real chemical species (see Morel and Hering, 1993). However, there are two common rules used to form components. First, it is common practice to use the set of fully dissociated ions as components. For example, the species $CdCl^+$ can dissociate into Cd^{2+} and Cl^-. Neither Cd^{2+} nor Cl^- can dissociate further. Thus, Cd^{2+} and Cl^- would be used as components of the system. On the other hand, H_2CO_3 can dissociate into HCO_3^- and H^+. Bicarbonate (HCO_3^-) can dissociate further into CO_3^{2-} and H^+. Thus, the appropriate components would be CO_3^{2-} and H^+. From CO_3^{2-} and H^+, you can "make" bicarbonate (from the equilibrium $CO_3^{2-} + H^+ = HCO_3^-$) and H_2CO_3 (from the equilibrium $CO_3^{2-} + 2H^+ = H_2CO_3$). Second, it is common practice to include H_2O and H^+ as components in all aqueous chemical systems.

9.6.2 Expressing equilibrium information with components

One way to determine how to make up each species from your set of components is to combine equilibrium expressions and write the species concentration as a function of component concentrations. The stoichiometric coefficients for the components are given by the exponents in this expression. Thus, from the equilibrium $CO_3^{2-} + 2H^+ = H_2CO_3$ it is easy to see that: $[H_2CO_3] = K[CO_3^{2-}]^1[H^+]^2$. Therefore, we say that H_2CO_3 is made up of one part of the component CO_3^{2-} and two parts of the component H^+.

Thoughtful Pause

How would you express the species OH^- with components H_2O and H^+?

To "make" OH^- from the components H_2O and H^+, manipulate the equilibrium for the self-ionization of water ($H_2O = H^+ + OH^-$):

$$[OH^-] = K_W[H_2O]^1[H^+]^{-1}$$

Thus, OH^- is made up of $+1$ part of the component H_2O and -1 part of the component H^+.

If you have selected the components carefully, you should be able to write the concentration of each species as a linear combination of the component concentrations (assuming here that activities and concentrations are interchangeable). For example, consider the simplest aqueous system: pure water, with species H_2O, H^+, and OH^-. The logical components are H_2O and H^+. Making up each species from components (with the pertinent equilibria and equilibrium expressions in parentheses):

H_2O: 1 part component H_2O and 0 parts component H^+
             ($H_2O = H_2O$ or $[H_2O] = [H_2O]^1$)

H^+:   0 parts component H_2O and 1 part component H^+
             ($H^+ = H^+$ or $[H^+] = [H^+]^1$)

OH^-:   1 part component H_2O and -1 part component H^+
             ($H_2O = H^+ + OH^-$ or $[OH^-] = K_W[H_2O]^1[H^+]^{-1}$)

You can write this information compactly in a tabular form, with components as columns and species as rows:

		Components	
		H_2O	H^+
Species	H_2O	1	0
	H^+	0	1
	OH^-	1	-1

Adding in the equilibrium constants for the equilibria used to express the species as components:

		Components		
		H_2O	H^+	$\log K$
Species	H_2O	1	0	0 ($K = 1$)
	H^+	0	1	0 ($K = 1$)
	OH^-	1	-1	-14 ($K = 10^{-14}$)

The entries in the table help to recreate the information in the system equilibria. The columns with component headings give the stoichiometric coefficients for each species. Consider the component columns as an n by m matrix, where each element, X_{ij}, gives the stoichiometric coefficient for component j in species i (for n species and m components). The log of the equilibrium constant is listed in the rightmost column. For the equilibrium $H_2O - H^+ = OH^-$:

$$\log K_W = -14 = \log[OH^-] - \log[H_2O] + \log[H^+]$$
$$\text{Or: } \log[OH^-] = \log[H_2O] - \log[H^+] + \log K_W$$

In matrix form:

$$\begin{pmatrix} \log[H_2O] \\ \log[H^+] \\ \log[OH^-] \end{pmatrix} = \begin{pmatrix} 1 & 0 \\ 0 & 1 \\ 1 & -1 \end{pmatrix} \begin{pmatrix} \log[H_2O] \\ \log[H^+] \end{pmatrix} + \begin{pmatrix} 0 \\ 0 \\ \log K_W \end{pmatrix}$$

In general:

$$\mathbf{S} = \mathbf{XC} + \mathbf{K}, \text{ where} \qquad\qquad \text{eq. 9.1}$$

$\mathbf{S} = n \times 1$ matrix of log of the species concentrations
  $\mathbf{S}(i) = $ log of the concentration of species i
$\mathbf{X} = n \times m$ matrix of component coefficients
  $\mathbf{X}(i,j) = $ coefficient for making up species i from component j
$\mathbf{C} = m \times 1$ matrix of log of the component concentrations
  $\mathbf{C}(j) = $ log of the concentration of component j

$\mathbf{K} = n \times 1$ matrix of log of the equilibrium constants for making up species from components

$\mathbf{K}(i) = $ log of the equilibrium constant for making up species i

As stated in Section 9.6.1, one of the advantages of using components is that it is possible to solve for the m component concentrations rather than the n species concentrations. Thus, it is necessary to rearrange eq. 9.1 to solve for $\mathbf{C}$ as follows:

$$\mathbf{C} = (\mathbf{X^T X})^{-1} (\mathbf{X^T}(\mathbf{S} - \mathbf{K}))$$

where $\mathbf{X^T}$ is the transpose of $\mathbf{X}$ (i.e., the matrix formed when the rows and columns of X are interchanged) and $\mathbf{A}^{-1}$ is the inverse of $\mathbf{A}$ (i.e., $\mathbf{A}^{-1}\mathbf{A} = $ the identity matrix = a matrix with 1s in the main diagonal and 0s elsewhere).

In the pure water example, you can verify that $\mathbf{X} = \begin{pmatrix} 1 & 0 \\ 0 & 1 \\ 1 & -1 \end{pmatrix}$, $\mathbf{X^T} = \begin{pmatrix} 1 & 0 & 1 \\ 0 & 1 & -1 \end{pmatrix}$, $\mathbf{X^T X} = \begin{pmatrix} 1 & 0 & 1 \\ 0 & 0 & -1 \\ 1 & -1 & 2 \end{pmatrix}$, and $(\mathbf{X^T X})^{-1} = \begin{pmatrix} 1 & 1 & 0 \\ 1 & -1 & -1 \\ 0 & -1 & 0 \end{pmatrix}$.

Tableau method: a common way to set up chemical systems by expressing species as functions of components

This approach is part of a general method called the ***Tableau method***. In the tableau, it is customary to leave table entries blank when the stoichiometric coefficients or log K values are equal to zero. Thus, the tableau accounting for the equilibrium information in pure water is:

		Components		
		H_2O	H^+	log K
Species	H_2O	1		
	H^+		1	
	OH^-	1	-1	-14

Now it is time to come up with a set of components for the chapter problem. Recall that the species list is H_2O, H^+, OH^-, HAc, and Ac^-.

Thoughtful Pause
What components can be used to describe the chapter problem?

There are many different components that can be used to describe the system in the chapter problem. Using the approach from Section 9.6.1 (i.e., use the most dissociated forms and always include H_2O and H^+ as components), suitable components are H_2O, H^+, and Ac^-. The species can be formed from the components as follows:

H_2O: 1 part component H_2O, 0 part component H^+, and 0 parts component Ac^-
$$(H_2O = H_2O; \log K = 0 \text{ or } [H_2O] = [H_2O]^1)$$

H^+: 0 parts component H_2O, 1 part component H^+, and 0 parts component Ac^-
$$(H^+ = H^+; \log K = 0 \text{ or } [H^+] = [H^+]^1)$$

OH^-: 1 part component H_2O, -1 part component H^+, and 0 parts component Ac^-
$$(H_2O = H^+ + OH^-; \log K_W = -14 \text{ or }$$
$$[OH^-] = K_W[H_2O]^1[H^+]^{-1})$$

HAc: 0 parts component H_2O, 1 part component H^+, and 1 part component Ac^-
$$(HAc = Ac^- + H^+; \log K = -4.7 \text{ or }$$
$$[HAc] = (1/K)[Ac^-]^1[H^+]^1)$$

Ac^-: 0 parts component H_2O, 0 parts component H^+, and 1 part component Ac^-
$$(Ac^- = Ac^-; \log K = 0 \text{ or } [Ac^-] = [Ac^-]^1)$$

The partial tableau is then:

		Components			
		H_2O	H^+	Ac^-	$\log K$
Species	H_2O	1			
	H^+		1		
	OH^-	1	-1		-14
	HAc		1	1	4.7
	Ac^-			1	

9.6.3 Expressing mass balance information with components

The bottom half of the tableau will include information on the mass balances. As always, the mass balance constraints are determined by the starting materials (often called the ***recipe*** in the Tableau method). It is necessary to express each starting material as a function of the components. One approach for doing this is to write a mass balance for the elements in each starting material as a linear function of the components by determining the necessary stoichiometric coefficients for each component.

recipe: a way in the Tableau method of expressing the starting materials as functions of the components

An example will make this process clearer. The starting materials for the chapter example are HAc and H_2O. You need to find the stoichiometric coefficients a, b, c, d, e, and f so that the elements balance in the following expressions:

$$HAc = (H_2O)_a(H^+)_b(Ac^-)_c$$
$$H_2O = (H_2O)_d(H^+)_e(Ac^-)_f$$

The unique solution is: $a = e = f = 0$ and $b = c = d = 1$. Incorporating this information into the tableau, the complete tableau for the chapter example becomes:

		Components			
		H_2O	H^+	Ac^-	$\log K$
Species	H_2O	1			
	H^+		1		
	OH^-	1	-1		-14
	HAc		1	1	4.7
	Ac^-			1	
Recipe	HAc		1	1	
	H_2O	1			

The mass balance information in the tableau is interpreted as follows. If you look down each component column, then the stoichiometric coefficients show the mass balances for each component. Thus:

$$TOTH_2O = [H_2O] + [OH^-] + [Ac^-]$$
$$TOTH = [H^+] - [OH^-] + [HAc]$$
$$TOTAc = [HAc] + [Ac^-]$$

(Total component concentrations typically are expressed with the symbol TOT in the Tableau method.) The rows in the Recipe section yield the expressions for the total mass of each component:

$$TOTH_2O = [H_2O] \text{ added} \approx 55.5 \text{ M}$$
$$TOTH = [HAc] \text{ added} = Ac_T = 1 \times 10^{-3} \text{ M}$$
$$TOTAc = [HAc] \text{ added} = Ac_T = 1 \times 10^{-3} \text{ M}$$

Combining:

$$TOTH_2O = [H_2O] + [OH^-] + [Ac^-] \approx 55.5 \text{ M}$$
$$TOTH = [H^+] - [OH^-] + [HAc] = 1 \times 10^{-3} \text{ M}$$
$$TOTAc = [HAc] + [Ac^-] = 1 \times 10^{-3} \text{ M}$$

A final comment on the expression of mass balance information in the tableau: be aware that the stoichiometric coefficients may not always be positive. An example of negative stoichiometric coefficients in a tableau mass balance statement is given in the *Thoughtful Pause* below.

Thoughtful Pause

How would you write the elemental composition of NaOH as a function of the components Na^+, H_2O, and H^+?

For example, for NaOH you would write: $NaOH = (Na^+)_1(H_2O)_1(H^+)_{-1}$.

9.6.4 Expressing the electroneutrality equation with components

The electroneutrality equation can be generated by a linear combination of rows of the tableau. From the mass balances on the components:

$$TOTH = [H^+] - [OH^-] + [HAc] = TOTAc = [HAc] + [Ac^-]$$
$$\text{Or: } [H^+] = [OH^-] + [Ac^-]$$

You can generate the electroneutrality equation more easily if only one component is charged. Since H^+ is usually included as a component, the electroneutrality equation will be given by $TOTH$ if H^+ is the only charged component. For example, you can rework the chapter example with components H_2O, H^+, and HAc. The resulting tableau is shown in Figure 9.4.

		Components			
		H_2O	H^+	HAc	log K
Species	H_2O	1			
	H^+		1		
	OH^-	1	-1		-14
	HAc		1		
	Ac^-		-1	1	-4.7
Recipe	HAc			1	
	H_2O	1			

Figure 9.4: Final Tableau for the Chapter Example

Note that the equilibrium information (Species portion of the tableau) yields $TOTH = [H^+] - [OH^-] - [Ac^-]$ and the mass balance information (Recipe portion of the tableau) yields $TOTH = 0$. Combining:

$$[H^+] - [OH^-] - [Ac^-] = 0$$
$$\text{Or: } [H^+] = [OH^-] + [Ac^-]$$

9.6.5 Solving chemical systems with the Tableau method

The tableau summarizes the equations necessary to solve the chemical system. The equations usually are solved by computer rather than by hand. An example of manual calculation with a tableau is given here. Computer solutions are illustrated in Section 9.6.7.

The tableau for the chapter example with components H_2O, H^+, and Ac^- is repeated below for convenience. From the component columns of the tableau (see also Section 9.6.4):

		Components			
		H_2O	H^+	Ac^-	$\log K$
Species	H_2O	1			
	H^+		1		
	OH^-	1	-1		-14
	HAc		1	1	4.7
	Ac^-		1		
Recipe	HAc		1	1	
	H_2O	1			

$$TOTH_2O = [H_2O] + [OH^-] + [Ac^-] \approx 55.5 \text{ M}$$
$$TOTH = [H^+] - [OH^-] + [HAc] = 1\times10^{-3} \text{ M}$$
$$TOTAc = [HAc] + [Ac^-] = 1\times10^{-3} \text{ M}$$

From the species rows of the tableau (ignoring rows with $\log K = 0$):

$$[OH^-] = K_w[H_2O]^1[H^+]^{-1} = K_w[H_2O]/[H^+]$$
$$[HAc] = (1/K)[Ac^-]^1[H^+]^1 = (1/K)[Ac^-][H^+]$$

Because of the large fixed concentration of water, the $TOTH_2O$ equation becomes meaningless. Ignoring the $TOTH_2O$ equation and recognizing that $\{H_2O\} = 1$ in the $[OH^-]$ expression:

$$TOTH = [H^+] - [OH^-] + [HAc] = 1\times10^{-3} \text{ M}$$
$$TOTAc = [HAc] + [Ac^-] = 1\times10^{-3} \text{ M}$$
$$[OH^-] = K_w/[H^+]$$
$$[HAc] = (1/K)[Ac^-][H^+]$$

Substituting for $[OH^-]$ and $[HAc]$ in $TOTH$ and $TOTAc$ yields:

$$TOTH = [H^+] - K_w/[H^+] + (1/K)[Ac^-][H^+] = 1\times10^{-3} \text{ M} \qquad \text{eq. 9.1}$$
$$TOTAc = (1/K)[Ac^-][H^+] + [Ac^-] = 1\times10^{-3} \text{ M} \qquad \text{eq. 9.2}$$

Equations 9.1 and 9.2 represent two nonlinear equations in two unknowns, namely [Ac^-] and [H^+].

As noted above, the $TOTH_2O$ equation usually adds little information since the concentration of water is large and constant. Thus, water is sometimes merely assumed to be a component and not listed explicitly in the tableau as a component, species, or starting material. For example, a simplified tableau for the chapter problem is shown in Figure 9.5.

		Components		
		H^+	Ac^-	log K
Species	H^+	1		
	OH^-	-1		-14
	HAc	1	1	4.7
	Ac^-		1	
Recipe	HAc	1	1	

Figure 9.5: Simplified Tableau for the Chapter Example

9.6.6 Summary of the Tableau method

The steps involved in making a tableau can be summarized as follows (also listed in Appendix C, Section C.5):

Step 1: Make a list of species and equilibria

Step 2: Select the components

The usual approach is to include H_2O, H^+, and the most dissociated forms of each species family as components.

Step 3: Draw the initial tableau

Make one column for each component and one column labeled log K. Make rows for each species and rows at the bottom of the tableau for each starting material.

Step 4: Enter the component stoichiometric coefficients from equilibria

Rearrange the equilibria to express each species as a product of the components raised to a power. For example, the self-ionization of water can be rearranged to: [OH^-] = $K_w[H_2O]^1[H^+]^{-1}$. Enter the coefficients for each component on the row for that species.

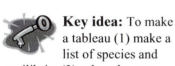 **Key idea:** To make a tableau (1) make a list of species and equilibria, (2) select the components, (3) draw the initial tableau, (4) enter the component stoichiometric coefficients from equilibria, and (5) enter the component stoichiometric coefficients from mass balances

Step 5: Enter the component stoichiometric coefficients from mass balances

Express the elemental composition of each starting material as a linear combination of the components. For example: $NaOH = (Na^+)_1(H_2O)_1(H^+)_{-1}$. Enter the coefficients for each component on the row for that starting material.

9.6.7 Using software based on the Tableau method

Software packages based on the Tableau method require the user to select species and enter the tableau information. Tableau information generally is entered as a table, similar to Figure 9.5. After entering the tableau information, the system is solved by the software for the component concentrations. Species concentrations are back-calculated by the software and can be examined in tabular output or plotted.

The true power and beauty of the Tableau method is shown more clearly in larger examples. The main advantages of the Tableau method are computational efficiency and expandability to larger systems. For example, the final equations in the Tableau method (eqs. 9.1 and 9.2) usually are solved numerically. The system has been simplified to two unknowns, namely, the concentrations of the components $[Ac^-]$ and $[H^+]$ (assuming that the component H_2O has activity = 1). The method employed by *Nanoql* yields a system of four equations in four unknowns, namely, the concentrations of the species $[H^+]$, $[OH^-]$, $[Ac^-]$, and $[HAc]$ (assuming that the component H_2O has activity = 1). Since the number of computational steps in nonlinear solution methods generally increases dramatically with the number of unknowns, the Tableau method offers great computational efficiency.

The main disadvantage with the Tableau method is that the user must preprocess the system information to generate the tableau. The process of making a tableau will become more familiar to you with practice. In addition, the choice of components sometimes can influence the results of the calculation.

9.7 SUMMARY

Computer software can speed up equilibrium calculations significantly and allow for the solution of more complex systems than with manual calculations. Equilibrium calculations lend themselves to solution by spreadsheets. Orderly arrangement of the spreadsheets will aid in keeping track of calculated results.

The computer program *Nanoql* was written to simulate the solution process taught in Part II of this book. The main unit in *Nanoql* is the species. Although less computationally efficient than other software packages, *Nanoql* requires little preprocessing of the chemical system before entering the system into the software.

The Tableau method was developed to be computationally efficient. The main unit in Tableau-based programs is the component. MINEQL is an example of an equilibrium calculation software package that uses the Tableau method. It requires some preprocessing of the chemical system by the user to write it in terms of components and not species.

9.8 PART II CASE STUDY: HAVE YOU HAD YOUR ZINC TODAY?

File available:
The *Nanoql* chemical system for the Part II case study may be found on the CD - file location **/Chapter 9/Case study.nqs**

Recall that the Part II case study concerns the speciation of zinc in the mouth and stomach upon consumption of a zinc acetate tablet. The system was set up in Section 7.6. If you use computer software, it is fairly straightforward to calculate the concentrations of all species at any pH of interest. As an example, the *Nanoql* file for the Part II case study is included on the CD at **/Chapter 9/Case study.nqs**. You should explore the file and try to regenerate the species concentrations at pH 3 and 6.5 from Section 7.6 and the pC-pH diagram from Section 8.6. Note that the chemical system is incomplete in that the species responsible for fixing the pH are not listed. In this case, H^+ should be changed in *Nanoql* to a fixed species and the electroneutrality equation should be ignored (since not all species are included). Generating a tableau for the case study system may be difficult now but will be within your abilities by the conclusion of Part IV of this text.

SUMMARY OF KEY IDEAS

- Many environmental problems are difficult to solve manually

- To solve equilibrium problems with a spreadsheet, enter equations for species (based on equilibria) and vary the $-\log$(conc.) values of components until the mass and charge balances are satisfied

- Write out the equilibrium expressions and write notes to help you remember the details of the chemical system and calculation

- Check the optimization results to see if sufficient resolution has been obtained by examining the predicted total masses and charge balance

- *Nanoql* simulates the manual solution method

- To enter a system in *Nanoql* (1) choose the system characteristics (including whether to use the electroneutrality condition), (2) choose the species (either individually or by components), (3) enter the equilibria, (4) enter the mass balances, and (5) enter User Defined Functions, if desired

- To make a tableau (1) make a list of species and equilibria, (2) select the components, (3) draw the initial tableau, (4) enter the component stoichiometric coefficients from equilibria, and (5) enter the component stoichiometric coefficients from mass balances

PROBLEMS

9.1 Use a spreadsheet to find the equilibrium pH of a 1×10^{-3} M sodium acetate solution. Assume that sodium acetate dissociates completely to Na^+ and Ac^-. In solving the problem, set up a spreadsheet similar to Figure 9.1.

9.2 Enter the chapter example into *Nanoql* and solve for the equilibrium pH. Make a pC-pH diagram inside *Nanoql*. Show that the only place on the pC-pH diagram where charge is balanced is at the equilibrium pH.

9.3 Use *Nanoql* to find the equilibrium pH of a 1×10^{-3} M sodium acetate solution and plot its pC-pH diagram. Hint: You can take your file created in Problem 9.2, add the species Na^+, make it a fixed species, and set its concentration equal to 1×10^{-3} M.

9.4 Explain the meanings of each row and column in the tableaux in Figures 9.4 and 9.5.

9.5 Enter the chapter example in a computer program that employs the Tableau method and solve for the equilibrium pH. Make a pC-pH diagram.

9.6 Write the tableau for the sodium acetate system in Problem 9.1. Use a computer program that employs the Tableau method to solve for the equilibrium pH. Draw a pC-pH diagram.

9.7 Using the computer method of your choice, find the equilibrium pH for the following systems:

A. 1×10^{-2} M NH_3

B. 1×10^{-2} M NH_4Cl

C. 1×10^{-3} M H_2CO_3

Acid-Base Equilibria in Homogenous Aqueous Systems

... the number of gram atoms of hydrogen ions per liter ...
can thus be placed equal to 10^{-P}. For the number P,
I have chosen the name "hydrogen ion exponent" and
the written expression P_H.
Sören P.L. Sörenson

The general principles of chemical equilibrium can
similarly be used in discussion of a weak base ... and also
of salts formed by weak acids and weak bases.
Linus Pauling

CHAPTER 10

Getting Started with Acid-Base Equilibria in Homogeneous Aqueous Systems

10.1 INTRODUCTION

At this point, you have nearly all the tools you need to solve chemical systems at equilibrium. You know how to set up chemical systems for solution: listing species, writing pertinent equilibria, devising mass balances, and writing the electroneutrality condition. You know how to solve systems: making approximations to simplify the mathematics (and subsequently checking the assumptions), converting systems to one equation in one unknown, or grinding it out by using algebra or computer software.

Given that you now know nearly everything you need to know, you may wonder why there are another dozen or so chapters in this text. The answer is simple: there is a lot more to aquatic chemistry than being able to solve for species concentrations at equilibrium. The richness of the field comes from the variety of reaction types and the myriad applications in natural and engineered systems.

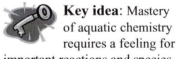 **Key idea**: Mastery of aquatic chemistry requires a feeling for important reactions and species in addition to fluency in equilibrium calculations

Another reason for the rest of the text is that you need more than algebra to understand chemical processes in the environment. To be sure, performing equilibrium calculations is a necessary part of your mastery of aquatic chemistry. However, *fluency in aquatic chemistry also requires a feeling for important reactions and species*. You need to be able to estimate the values of the master variables and the effects of chemical change on speciation. We begin this odyssey in Part III of this text, where you will explore the first type of chemical equilibrium: acid-base equilibria in homogeneous systems.

10.2 HOMOGENEOUS SYSTEMS

homogeneous system: a chemical system consisting of only one phase

10.2.1 What are homogeneous systems?

A **homogeneous system** (from the Greek *homo-* same + *-genes* kind) consists of only one phase. In this text, *homogeneous system* will refer to an aqueous chemical system with no gas or solid phases. In addition, we

193

are excluding for the moment other liquid phases, such as pure benzene or oils or gasoline. To put it another way, we are considering systems in which all the species (except one) are dissolved aqueous species that carry the suffix (aq). The one exception is water itself. Water is the only pure liquid included in the homogeneous systems in this text.[†]

The definition of a homogeneous system seems fairly straightforward. Unfortunately, several classes of species carry names that are somewhat ambiguous. For example, what are we to make of the dissolved gases? Are they dissolved species? Gas-phase species? Are you supposed to include or exclude them in homogeneous systems? ***Dissolved gases*** are dissolved aqueous species that may equilibrate appreciably with a gas phase. For example, the species $O_2(aq)$ (also called dissolved oxygen, or DO) is an appropriate species for inclusion in a homogeneous system since it is an aqueous species. At sufficiently high concentrations (or appropriate temperature and pressure conditions), the dissolved gases can form bubbles. At this point, you no longer have a homogeneous system and must use the tools of Chapter 18 to calculate species concentrations at equilibrium.

Similarly, ***dissolved solids*** are considered to be dissolved aqueous species. Under the proper conditions, dissolved solids can precipitate, forming one or more solid phases. Again, precipitation creates a nonhomogeneous system. The tools presented in Chapter 19 will be needed to calculate species concentrations.

10.2.2 How realistic are homogeneous models?

In Parts III and IV of this text, systems will be *assumed* to be homogeneous. By assuming that the system is homogeneous, we clearly are placing large restrictions on the applicability of the model.

Thoughtful Pause

Can you think of a natural or engineered system that is truly homogeneous?

In fact, it is difficult to conjure up a system with any environmental relevance that is truly homogeneous. For example, the top layers of lakes and rivers may be in equilibrium with the atmosphere. To describe these waters, it is necessary to devise a nonhomogeneous system consisting of aqueous and gas phases. Similarly, the bottom layers of a lake may be in equilibrium with the solids in the sediments, necessitating a nonhomogeneous system. Even the middle layers may contain suspended solids in chemical equilibrium with the water. Groundwaters usually are in equilibrium with the surrounding soil. Even water flowing in pipes is not necessarily homogeneous. Important equilibria may occur between water

dissolved gases: dissolved aqueous species that may equilibrate with a gas phase

dissolved solids: dissolved aqueous species that may equilibrate with a solid phase

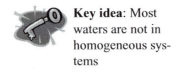

Key idea: Most waters are not in homogeneous systems

[†] Any single pure liquid could be considered. For example, much of the text could be repeated by using, say, methanol instead of water.

and the walls of the pipe. Turbulent flow may even induce cavitation and thus a small gas phase. Reasonable descriptions of each of these waters may require nonhomogeneous systems.

So when can you assume that a system is homogeneous? We usually use the homogeneous assumption for three reasons. First, the system of interest may be *nearly homogeneous*. For example, waters at middepth in a deep lake may be sufficiently isolated from the atmosphere and sediments. If suspended solids do not dissolve appreciably, then a homogeneous model may be reasonable.

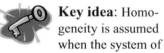

Key idea: Homogeneity is assumed when the system of interest is: nearly homogeneous, not significantly affected by exchange with other phases, or treated on purpose as an ideal case

Second, the system may not be homogeneous at all, but the equilibria with other phases may not affect the processes of interest. A good example is dissolved nitrogen, $N_2(aq)$. Dissolved nitrogen is nearly inert (see Section 6.3.6) and is usually ignored in equilibrium calculations. If you were interested in the conversion of OCl^- to $HOCl$ in an open beaker, you would most likely ignore the exchange of chlorine between the atmosphere and water. How do you know if nonhomogeneity affects the processes of interest? In short, you do not know unless you perform the equilibrium calculations twice, first assuming homogeneous conditions and then repeating the calculation in the presence of other phases.

Third, we sometimes impose homogeneous conditions as an idealized case. For example, suppose you wish to understand the relationships between apparently related compounds such as carbonic acid, bicarbonate, and carbonate. It might make sense to study the chemistry first as an isolated homogeneous system without influence from gas or solid phases. After you complete the analysis of the idealized case, more realistic conditions could be introduced.

10.3 TYPES OF REACTIONS IN HOMOGENEOUS SYSTEMS

Parts III and IV of this text are concerned with reactions that occur in homogeneous systems. The focus is on two chemical species: H^+ (the proton) and e^- (the electron). Recall from Section 1.3 that these two species control a great deal of aquatic chemistry. Thus, pH ($= -\log_{10}\{H^+\}$) and pe ($= -\log_{10}\{e^-\}$) are logical choices for master variables.

Protons and electrons can undergo two types of reactions. First, the proton or electron can be *transferred* from one chemical species to another. Proton transfer reactions are called *acid-base reactions*. Thus, proton-transfer reactions are the subject of Part III. Electron-transfer reactions are called *oxidation-reduction* (or *redox*) reactions. Redox reactions are the focus of Chapter 16 in Part IV.

Key idea: Protons and electrons can be transferred or shared between chemical species

Second, the proton or electron can be *shared* between chemical species. The sharing of a proton results in a special kind of low-energy bond called a *hydrogen bond*. Hydrogen bond formation allows for very rapid transfer of protons. As a result, acid-base reactions usually are very fast in water. Electron-sharing reactions are called *complexation reactions*. Complexation reactions are the subject of Chapter 15 in Part IV. The names for the proton- or electron-transfer reactions and proton- or electron-sharing reactions are summarized in Table 10.1.

**Table 10.1: Summary of Proton or Electron Transfer
or Sharing Reactions**

Species	Transfer Reactions	Sharing Reactions
Protons (H^+)	acid-base reactions	hydrogen bond formation
Electrons (e^-)	redox reactions	complexation reactions

10.4 THE WONDERFUL WORLD OF ACIDS AND BASES

10.4.1 Equilibrium calculations with acids and bases

In Chapter 11, you will learn how to perform and interpret equilibrium calculations with systems containing acids and bases. You have been exposed to equilibrium calculations with simple acids and bases in Chapters 7 and 8. Chapter 11 will provide much greater detail concerning systems of acids and bases.

Before you leap into the details of Chapter 11, a qualitative view of the effects of chemical addition on equilibrium pH is in order. This exercise will help develop your intuition with acids and bases in water. Recall that pure water has an equilibrium pH at 25°C of 7.0. You saw in Chapter 7 that when some compounds were added to water, the equilibrium pH was less than 7. Other compounds produced an equilibrium pH greater than 7 when added to water.

Thoughtful Pause

How can you predict whether the equilibrium pH will be greater than or less than 7?

The key to estimating whether the equilibrium pH will be greater than, less than, or about equal to 7 is in the charge balance. Recall that for pure water, the charge balance is:

$$[H^+] = [OH^-] \qquad \text{eq. 10.1}$$

Now add an uncharged compound to water and do the thought experiment described in Section 6.3.2; namely, look for its hydrolysis products. Say that the chemical addition produces only one monovalent (i.e., singly charged) anion as the added material reacts with water. Assume for the moment that this initial hydrolysis product does not react further with water or anything else.

Thoughtful Pause

What is the new electroneutrality equation?

The new electroneutrality equation will be:

$$[H^+]_{new} = [OH^-]_{new} + [A^-]_{new} \qquad \text{eq. 10.2}$$

where A^- is the anionic hydrolysis product. Now, how does the hydrogen ion concentration in eq. 10.2 compare with the hydrogen ion concentration in eq 10.1? If the terms in eqs. 10.1 and 10.2 were *unconstrained* variables, then we have no idea how $[H^+]_{new}$ compares to $[H^+]$. However, $[H^+]$ and $[OH^-]$ are constrained by the self-ionization of water: $[H^+][OH^-] = [H^+]_{new}[OH^-]_{new} = 10^{-14}$ at 25°C (assuming the concentrations and activities are interchangeable).

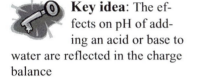

Key idea: The effects on pH of adding an acid or base to water are reflected in the charge balance

Thoughtful Pause

Given that the hydrogen and hydroxide ion concentrations are constrained, how do $[H^+]_{new}$ and $[H^+]$ compare?

Even if $[A^-]_{new}$ is small (say, 10^{-7} M), the hydrogen ion concentration must increase and the hydroxide ion concentration must decrease when the hydrolyzing compound is added. In other words, $[H^+]_{new} > [H^+]$. This point is illustrated in Figure 10.1. Note that the electroneutrality condition is satisfied for both conditions in Figure 10.1. When the compound is added, $[H^+]$ increases and $[OH^-]$ decreases to maintain the charge balance. If $[H^+]_{new} > [H^+]$, then the pH after addition must be *smaller* than the pH before addition (assuming that activity and concentration are interchangeable in both systems). Thus, adding a compound that produces anions but not cations will cause the pH to decrease.

This little exercise raises two questions. First, how can an added compound produce anions but not cations? After all, if a salt such as NaCl(s) is added to water, both a cation (Na^+) and an anion (Cl^-) are produced. In fact, an electron balance tells you that it is impossible to add a neutral compound that dissociates to form an anion but not a cation. However, there are compounds that hydrolyze to produce anions and only the cation H^+. Thus, no *new* cations are produced compared to pure water, and the situation in Figure 10.1 is valid. Compounds that hydrolyze to form only anions and H^+ are part of the class of species called *acids*.[†] In Chapter 11, you will learn much more about the properties of acids.

[†] The many definitions of acids will be explored in Chapter 11. The concept of an acid is much more general than just compounds hydrolyzing in water for form anions and H^+. As a simple example, ammonium (NH_4^+) is an acid. It dissociates to form ammonia (NH_3) and H^+.

The second question raised by Figure 10.1 is as follows: how do we know that $[A^-]_{new} = 10^{-7}$ M? In general, we do not know $[A^-]_{new}$ without doing a complete equilibrium calculation. The value of $[A^-]_{new} = 10^{-7}$ M was used for illustration. Again, Chapter 11 will provide the quantitative tools to allow for the calculation of the concentration of each species at equilibrium.

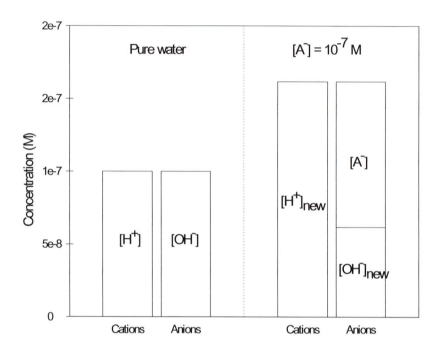

Figure 10.1: Charge Balances with and Without Chemical Addition

10.4.2 Mixtures of acids and bases

In nearly all aquatic systems, there are mixtures of acids and bases. In fact, as you shall see in Chapter 11, water itself is both an acid and a base. Therefore, the addition of an acid or base to water means that you have a mixture of acids and bases. Thus, to determine the master variable pH in a real water, it is necessary to be able to handle mixtures of acids and bases. Another important example of acid-base mixtures in engineered systems is the neutralization of acidic (or basic) wastes.

A simple example will demonstrate how you can treat mixtures of acids and bases qualitatively. Suppose you take the solution described in Section 10.4.1 and add NaOH to it. Before NaOH is added, the charge balance is:

$$[H^+] = [OH^-] + [A^-] \qquad \text{eq. 10.3}$$

(Equation 10.3 is just eq. 10.2 with the subscript "new" removed for clarity.) How will the charge balance change after adding NaOH? Sodium hydroxide is expected to dissociate completely to Na^+ and OH^-. The new charge balance is:

$$[Na^+] + [H^+] = [OH^-] + [A^-] \qquad \text{eq. 10.4}$$

Note that eq. 10.4 is just eq. 10.3 with an additional term, the concentration of sodium ion, on the lefthand side. There are only three ways to balance eq. 10.4 after adding the additional term on the lefthand side: (1) $[H^+]$ could decrease, (2) $[OH^-]$ could increase, or (3) $[A^-]$ could increase. Now, possibilities (1) and (2) are really the same since $[H^+][OH^-] = 10^{-14}$. Thus, we really have only two possibilities: either $[H^+]$ decreases and $[OH^-]$ increases or $[A^-]$ increases. Now $[A^-]$ cannot increase without limit. It is constrained by both equilibrium constraints and by at least one mass balance. For certain values of $[Na^+]$ (i.e., for certain doses of NaOH), the concentration of A^- is very constrained. At these NaOH doses, the pH increases very strongly since $[OH^-]$ must increase to maintain the charge balance. At other values of $[Na^+]$, the concentration of A^- is not very constrained. At these NaOH doses, the pH increases very little, since $[A^-]$ can increase within its constraints to maintain the charge balance.

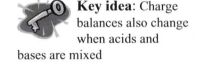

Key idea: Charge balances also change when acids and bases are mixed

In Chapter 12, you will learn to translate these qualitative arguments into quantitative tools as solution techniques for mixtures of acids and bases are developed. You will find that mixtures can be analyzed by a small extension of the principles learned in Chapter 11. Why devote an entire chapter to a small extension of Chapter 11? The analysis of mixtures of acids and bases is a good example of the importance of developing intuition and approximations for aquatic systems. In many cases, you can sketch the dependence of the equilibrium pH on the added acid or base concentration without doing any calculations at all. You will develop this "feel" for acid-base mixtures in Chapter 12.

10.4.3 Responses to acidic or basic inputs

The qualitative analysis of acid-base mixtures in the previous section can be generalized as follows: aquatic systems vary greatly in their response to acid or base additions depending on their chemical composition. As stated in Section 10.4.2, the tools for quantifying that response will be developed in Chapter 12. In Chapter 13, water quality parameters that influence the response to acid or base addition will be elucidated. In particular, you will develop a set of related parameters that describe how the system composition influences the response to acidic or basic inputs. One important parameter is *alkalinity*. The meaning and uses of alkalinity are presented in Chapter 13.

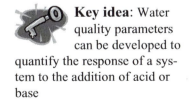

Key idea: Water quality parameters can be developed to quantify the response of a system to the addition of acid or base

10.5 PART III ROAD MAP

Part III of this book discusses the applications of equilibrium calculation techniques to systems containing acids and bases. In addition, an intuitive feel for the behavior of acids and bases will be cultivated in Part III. By the conclusion of this part of the text, you should know what acids and bases are and have a qualitative and quantitative understanding of their behavior in water.

In Chapter 11, the history of the concepts of acids and bases is reviewed, culminating in their definitions. Calculation techniques are developed for systems of individual acids and bases. Dimensionless parameters are developed to describe the relative concentrations of acids and bases in the same chemical family. By the conclusion of Chapter 11, you should be very comfortable with answering quantitative questions about individual acids and bases in aquatic systems.

Chapter 12 provides solution techniques for mixtures of acids and bases. This allows for the analysis of much more complicated and realistic problems in aquatic chemistry. Graphical analysis techniques play an important role in summarizing how the equilibrium pH changes with acid or base addition.

In Chapter 12, you will find that the change in equilibrium pH depends on the system. In other words, each system responds differently to acid or base addition. In Chapter 13, the tools for quantifying the response of a system to acid or base addition will be developed.

10.6 SUMMARY

In this chapter, some qualitative ideas were developed to aid in your understanding of acids and bases in homogeneous systems. A homogeneous system in this text is an aqueous chemical system with no gas or solid phases. Many systems of environmental interest are not homogeneous. However, we use homogeneous models when the system of interest is (1) nearly homogeneous, (2) not significantly affected by exchange with other phases, and/or (3) treated on purpose as an ideal case.

The species H^+ (proton) and e^- (electron) are related to the master variables pH and pe, respectively. Both protons and electrons can undergo two types of reactions: transfer and exchange. Part III deals with proton-transfer reactions. The details of equilibrium calculations with proton-transfer reactions are considered in Chapters 11 through 13. The charge balance was used in this chapter as a qualitative tool in assessing the effects on pH of acid addition. A more quantitative discussion is the focus of the remainder of this part of the text.

10.7 PART III CASE STUDY: ACID RAIN

The application of scientific principles to analyze and solve pollution problems is relatively recent. As with all scientific ventures, the analysis of pollution problems has matured. In the third quarter of the twentieth century, it was common to segregate pollution by the type of media impacted. Air, water, and solid waste pollution were studied separately.

Only in the last 30 years or so has the strong relationship among air, water, and solid waste pollution been recognized. The focus now is on intermedia transport of pollutants (e.g., transport from air to water or water to air) and multimedia effects.

A prime mover in the development of a multimedia focus is *acid rain*. It is now well-established that certain air pollutants from natural and anthropogenic sources can dissolve in rain, snow, and fog to produce highly acidic conditions. For example, sulfur dioxide from the combustion of coal and other fossil fuels can react with water in the troposphere to produce sulfuric acid, $H_2SO_4(aq)$. Similarly, oxides of nitrogen [$NO(g)$ and $NO_2(g)$], produced primarily from transportation sources, can be oxidized and can combine with atmospheric water vapor to form nitric acid, $HNO_3(aq)$. As a final example, hydrochloric acid, HCl, is produced in the gas phase from the combustion of chlorine-containing polymers (such as polyvinyl chloride, PVC) and liberated from volcanos. Thus, fairly acidic rain can be produced.

In this case study, the impact of acidic inputs on the pH of rainfall will be examined. In addition, the effects of acid rain on other waters will be explored. In Chapter 11, the pH of acid rain will be determined. The impact of acid rain will be elucidated in Chapter 12, where the mixture of acid rain and a surface water will be analyzed. In Chapter 13, new water quality parameters will be used to simplify the calculations of the impacts of acid rain on surface water pH.

SUMMARY OF KEY IDEAS

- Mastery of aquatic chemistry requires a feeling for important reactions and species, as well as fluency in equilibrium calculations

- Most waters are not in homogeneous systems

- Homogeneity is assumed when the system of interest is: nearly homogeneous, not significantly affected by exchange with other phases, or treated on purpose as an ideal case

- Protons and electrons can be transferred or shared between chemical species

- The effects on pH of adding an acid or base to water are reflected in the charge balance

- Charge balances also change when acids and bases are mixed

- Water quality parameters can be developed to quantify the response of a system to the addition of acid or base

CHAPTER 11

Acids and Bases

11.1 INTRODUCTION

As stated in section 1.3, pH is a master variable in aquatic systems. Why the emphasis on pH? We choose pH as a master variable of interest because H^+ has such a strong influence on aquatic chemistry and toxicity endpoints and because pH is easily measured. In this chapter, you shall begin to explore the equilibria in which protons, H^+, are transferred between chemical species. These equilibria constitute equilibrium acid-base chemistry.

The chapter begins by examining the formal definitions of the words *acid* and *base*. The definitions of acid and base have changed over time. This history will be followed to arrive at definitions that are both consistent and useful. Acids and bases will be defined by equilibria, as well as words. The thermodynamic concepts of acid and base strength will be quantified by the equilibrium constants of the defining equilibria.

Next, the discussion of acids and bases will be extended to polyprotic species - those acids and bases that donate or accept more than one proton. The solution techniques from Part II will be applied to polyprotic acid and base systems. In particular, the shortcut graphical method for monoprotic acids developed in Chapter 8 will be broadened and applied to systems containing polyprotic acids.

Finally, compact equations called *alpha values* will be developed. Alpha values show concisely the effects of pH on the percentage contributions of acidic and basic forms to the total mass.

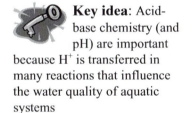 **Key idea**: Acid-base chemistry (and pH) are important because H^+ is transferred in many reactions that influence the water quality of aquatic systems

11.2 DEFINITIONS OF ACIDS AND BASES

11.2.1 The early history of acids and bases

You will notice that the title of section 11.2 refers to more than one definition of acids and bases. The concept of acids and bases has evolved over the last 200 years or so[†]. In several cases, an old definition of acid or base had to be replaced when exceptions were found.

[†] The progression of acid-base concepts presented here was influenced by material at the ChemTeam web site (**dbhs.wvusd.k12.ca.us/ChemTeamIndex.htm**).

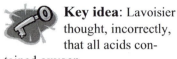

Key idea: Lavoisier thought, incorrectly, that all acids contained oxygen

In the early days of modern chemistry (mid-eighteenth century), acids were thought to be acids because they contained some common substance. The substance made liquids containing acids taste sour, hence the name acid (from the Latin *acere* sour). Antoine Laurent Lavoisier (1743-1794) concluded that the common substance was oxygen. His commitment to this idea is shown in the name he selected for oxygen - *principe oxygine*, meaning acidifying principle.

Thoughtful Pause

Can you think of a substance that is categorized as an acid today but that does not fit Lavoisier's definition of an acid?

Lavoisier was wrong, of course. Many acids do not contain oxygen, including common acids (e.g., HCl), less common acids (HCN and HSCN), and some compounds that may not look like acids at first glance (e.g., NH_4^+). One of the troublesome compounds of the day that did not fit Lavoisier's definition was chlorine. Chlorine was originally named oxymuriatic acid (after muriatic acid, HCl, named from the Latin *muria* brine), with the prefix *oxy-* because all acids were supposed to contain oxygen. Humphry Davy (1778-1829) showed that molecular chlorine did not contain oxygen.[†] He suggested, based on acids such as HCl and H_2S, that perhaps hydrogen was the common substance among acids. To explain why some hydrogen-containing substances were not acids, Justus von Liebig (1803-1873) proposed that an acid was a hydrogen-containing substance in which the hydrogen could be replaced by a metal. Thus, HCl is an acid, according to this definition, because the hydrogen can be replaced by, say, sodium to form NaCl. On the other hand, methane (CH_4) is not an acid because CH_3Na does not exist in nature.

All was well for 50 years. Then, in his Ph.D. dissertation of 1884, Svante August Arrhenius (1859-1927) shocked his dissertation committee with the idea that molecules could break apart in a solvent (such as water) to form charged species called *ions*. Arrhenius received the Nobel Prize in Chemistry in 1903 for this work, later called the ionic theory. Arrhenius extended his ionic theory to acids and bases. The Arrhenius definitions of acids and bases are as follows:

Arrhenius acid: a substance that produces H^+ in aqueous solution

Arrhenius base: a substance that produces OH^- in aqueous solution

> *Arrhenius acids* are substances that produce H^+ in aqueous solution, and *Arrhenius bases* are substances that produce OH^- in aqueous solution

---

[†] To see why molecular chlorine was considered an acid, recall that chlorine undergoes the following equilibrium in water: $Cl_2 + H_2O = HOCl + H^+ + Cl^-$. HOCl is hypochlorous acid.

11.2.2 Brønsted-Lowry definitions

Arrhenius had taken a giant leap forward in linking H^+ to acids. He also provided a definition for a base that was consistent with the observations that bases neutralize acids. However, the Arrhenius definitions did not describe the behavior of some common bases, such as ammonia and sodium carbonate. The next step in the evolution of acids and bases was provided independently and nearly simultaneously by the Danish chemist Johannes Nicolaus Brønsted (1879-1947) and the English chemist Thomas Martin Lowry (1874-1936). Brønsted postulated that acids donated a proton (H^+) to the solvent and bases accepted a proton from the solvent. Lowry used the following formalization:

$$HA + H_2O = A^- + H_3O^+$$

Together, the Brønsted-Lowry definitions are as follows:

> ***Brønsted-Lowry acids*** are substances that *donate* H^+ to water[†] (to produce H_3O^+), and ***Brønsted-Lowry bases*** are substances that *accept* a proton from water (to leave OH^-)

These concepts of acids and bases as, respectively, *proton donors* and *proton acceptors* are very powerful. The definitions link acids and bases very tightly. They are consistent with Arrhenius's idea that bases are somehow connected with OH^-.

Brønsted-Lowry acid: a substance that donates H^+ (to water)

Brønsted-Lowry base: a substance that accepts a proton (from water)

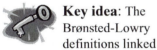 **Key idea**: The Brønsted-Lowry definitions linked acids and bases by relating them both to proton transfer

---

Thoughtful Pause

How are the Brønsted-Lowry definitions consistent with Arrhenius' ideas about bases?

With the Brønsted-Lowry definitions, you can see that the OH^- comes from *water*, not necessarily from the base itself.

We shall pause in our romp through history here, although the definition of acids and bases will undergo one more major change in Chapter 14. For the majority of this text, the Brønsted-Lowry definitions will be used: acids are proton donors and bases are proton acceptors. In other words, *acid-base chemistry (in the Brønsted-Lowry sense) means proton transfer chemistry.*

 Key idea: Acid-base equilibria are proton-transfer equilibria

---

[†] The focus here is on water because this is a text on aquatic chemistry. However, one of the strengths of the Brønsted-Lowry definitions is that they can be applied to proton-donating and proton-accepting solvents other than water (e.g., ethanol).

11.2.3 Acid dissociation reactions and base association reactions

Following the approach of Brønsted and Lowry, you also can define acids and bases by chemical equilibria. (This material was presented briefly in Section 8.2.2 but will be developed in more detail here.) For acids, you can write:

$$H_nA^a + H_2O = H_{n-1}A^{a-1} + H_3O^+ \qquad \text{eq. 11.1}$$

A **monoprotic acid** is acid that donates one proton: $n = 1$ in eq. 11.1. If $n > 1$, the acid is called a **polyprotic acid**. For example, diprotic acids and triprotic acids can donate two and three protons, respectively.

For an uncharged, monoprotic acid:

$$HA + H_2O = A^- + H_3O^+ \qquad \text{eq. 11.2}$$

Equations 11.1 and 11.2 are called **acid dissociation reactions**. Any substance for which you can write an equilibrium in the form of eq. 11.1 or 11.2 is called an acid (formally, a Brønsted-Lowry acid). The equilibrium constant for the equilibria in eqs. 11.1 and 11.2 is given a special name. The equilibrium constants are called **acid dissociation constants** and are given the special symbol K_a. Remember: K_a is *just an equilibrium constant*. It is the equilibrium constant for equilibria of the form of eq. 11.1 or 11.2.

You can develop a similar system for bases. For bases, you can write:

$$B^- + H_2O = HB + OH^- \qquad \text{eq. 11.3}$$

Equation 11.3 is called a **base association reaction**. Any substance for which you can write an equilibrium in the form of eq. 11.3 is called a base (again, formally, a Brønsted-Lowry base). The equilibrium constant for the equilibrium in eq. 11.3 is given a special name. The equilibrium constants are called **base association constants** and are abbreviated K_b. Again, K_b is *just an equilibrium constant*. It is the equilibrium constant for equilibria of the form of eq. 11.3.

11.2.4 Ampholytes

Some species can both donate a proton to water and accept a proton from water. Such compounds are both acids and bases. They are given special names and are called amphoteric species (from the Greek *ampho* both) or **ampholytes** (short for *amphoteric electrolytes*).

Key idea: Brønsted-Lowry acids and bases also can be defined by acid dissociation reactions and base association reactions

monoprotic acid: an acid that donates one proton (to water)

polyprotic acid: an acid that donates more than one proton (to water)

acid dissociation reaction: a reaction of the form of eq. 11.1 or 11.2

acid dissociation constant (K_a): the equilibrium constant for an acid dissociation reaction

base association reaction: a reaction of the form of eq. 11.3

base association constant (K_b): the equilibrium constant for a base association reaction

ampholyte: a species that can both donate and accept a proton

---

Thoughtful Pause

Can you name the most common ampholyte in aqueous systems?

Key idea: Water is amphoteric (acts as both an acid and a base)

The most common ampholyte in aquatic systems is water itself. As eq. 11.4 shows, water can both *donate* a proton to water (and hence is an acid) and *accept* a proton from water (and thus is a base):

$$H_2O + H_2O = OH^- + H_3O^+ \qquad \text{eq. 11.4}$$

Example 11.1: **Identification of Acids, Bases, and Ampholytes**

Identify the following as an acid, base, or ampholyte, using the Brønsted-Lowry definitions: OH^-, HCN, NH_3, NH_4^+, and HCO_3^-

Solution:
You can write the following equilibria, analogous to eqs. 11.1 and 11.3:

$$OH^- + H_2O = H_2O + OH^-$$
$$HCN + H_2O = CN^- + H_3O^+$$
$$NH_3 + H_2O = NH_4^+ + OH^-$$
$$NH_4^+ + H_2O =$$
$$NH_3 + H_3O^+$$
$$HCO_3^- + H_2O =$$
$$H_2CO_3 + OH^-$$
$$HCO_3^- + H_2O =$$
$$CO_3^{2-} + H_3O^+$$

Thus, you would classify the species as follows: **OH^- is a base, HCN is an acid, NH_3 is a base, NH_4^+ is an acid, and HCO_3^- is amphoteric**

Other common ampholytes include species with intermediate degrees of protonation (i.e., species that are neither completely protonated nor completely deprotonated). For example, phosphoric acid (H_3PO_4) is a triprotic acid. It dissociates in water to form three other species: $H_2PO_4^-$, HPO_4^{2-}, and PO_4^{3-} (phosphate).

Thoughtful Pause

Are H_3PO_4, $H_2PO_4^-$, HPO_4^{2-}, and PO_4^{3-} acids, bases, or ampholytes?

Phosphoric acid is called an acid (can only donate protons to water) and phosphate a base (can only accept protons from water). $H_2PO_4^-$ and HPO_4^{2-} are ampholytes. For example, $H_2PO_4^-$ can donate a proton to water to form HPO_4^{2-} or accept a proton from water to form H_3PO_4. For more practice in identifying acids, bases, and ampholytes, see Example 11.1.

The example of phosphoric acid shows ampholytes formed from partially protonated species that are part of a polyprotic acid family. Sometimes, ampholytes can be formed when more than one acidic groups are found on the same molecule. For example, ethanedioic acid (commonly, oxalic acid: HOOC-COOH) and 1,2-benzenedicarboxylic acid (commonly, *o*-phthalic acid) each have two acidic functional groups. The partially protonated species (e.g., the monoanionic species HOOC-COO$^-$) are ampholytes. A good example of complex systems exhibiting ampholytic behavior are the macromolecules that contribute color to natural waters, called humic and fulvic acids.

Another type of ampholyte has both acidic and basic functional groups on the same backbone. Examples include the amino acids (see Problem 11.10) and some types of polymers.

11.3 ACID AND BASE STRENGTH

11.3.1 Introduction

How should you compare the relative strengths of acids and bases? Acids donate protons to water, so it makes sense to develop an acid strength scale that measures the tendency of a compound to donate a proton to water. There are several ways to construct such a scale. The most common approach is to compare the *thermodynamic tendency* of an acid to donate protons to water.

Thoughtful Pause

What is an appropriate quantitative measure of the thermodynamic tendency of an acid to donate protons to water?

---

acid strength: thermodynamic tendency of an acid to donate a proton to water

base strength: thermodynamic tendency of a base to accept a proton from water

 Key idea: Acid and base strength are thermodynamic properties

Example 11.2: **Acid and Base Strength**

Rank the following acids by their acid strength: H_2S (hydrogen sulfide; $pK_a = 7.1$), HNO_3 (nitric acid; $K_a = 10^{+3}$), and H_2O_2 (hydrogen peroxide; $pK_a = 11.7$). Rank the following bases by their base strength: NH_3 (ammonia; $K_b = 10^{-4.7}$), CN^- (cyanide; $pK_b = 4.8$), and CO_3^{2-} (carbonate; $pK_b = 3.7$).

Solution:
To compare, express the acid strength for all acids (or bases) in terms of either K_a (K_b for bases) or pK_a values (pK_b for bases). The choice is arbitrary, but pK_a and pK_b values are more convenient. Thus:

A reasonable quantitative measure is K_a (or pK_a). Thus, **acid strength** is defined as the thermodynamic tendency of an acid to donate a proton to water. Acid strength is quantified by K_a or pK_a. Similarly, you can define **base strength** as the thermodynamic tendency of a base to accept a proton from water and quantify base strength by K_b or pK_b.

11.3.2 Properties of acid and base strength

The concepts of acid and base strength have a number of important properties. First, since acid and base strength are quantified by an equilibrium constant, acid and base strength are thermodynamic properties. This means that acid and base strength (as with all thermodynamic properties) may change as a function of temperature, pressure, and the concentrations of other species.

Second, acid and base strength are defined for *one-proton transfer equilibria* (eqs. 11.2 and 11.3). For example, in considering the acid strength of carbonic acid (H_2CO_3), use the equilibrium constant for the transfer of *one* proton (i.e., K for $H_2CO_3 + H_2O = HCO_3^- + H_3O^+$), **not** the equilibrium constant for the transfer of two protons (i.e., not K for $H_2CO_3 + 2H_2O = CO_3^{2-} + 2H_3O^+$).

Third, one acid is said to be stronger than another acid if it dissociates more completely to donate a proton to water. Thus, from eqs. 11.1 and 11.2, a stronger acid has a *larger* K_a. This means that a stronger acid has a *smaller* pK_a. Thus, HOCl (hypochlorous acid), with a pK_a of 7.54, is a stronger acid than ammonium ($pK_a = 9.3$, both values at 25°C). Very strong acids have very small (and even negative) pK_a values: the pK_a of HCl is about -3. Similarly, stronger bases have larger K_b values (and thus smaller pK_b values). Other examples of ranking acids and bases by their strengths are given in Example 11.2.

11.3.3 Acid strength of H^+

What is the acid strength of H^+? To determine its acid strength, write an equilibrium similar to eq. 11.1 with H^+ as a reactant:

$$H^+ + H_2O = H_3O^+ \quad K_a = ?$$ eq. 11.5

What is K_a for H^+? Formally, the standard Gibbs free energy of reaction for this equilibrium is defined to be zero, and thus $K_a = 1$. Why was the

H_2S $pK_a = 7.1$
HNO_3 $pK_a = -3$
H_2O_2 $pK_a = 11.7$

NH_3 $pK_b = 4.7$
CN^- $pK_b = 4.8$
CO_3^{2-} $pK_b = 3.7$

The acid strength order is given by **HNO_3 (a strong acid)** >> **H_2S > H_2O_2**. The base strength order is given by **CO_3^{2-} > NH_3 ≈ CN^-**.

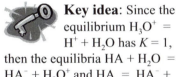

Key idea: Since the equilibrium $H_3O^+ = H^+ + H_2O$ has $K = 1$, then the equilibria $HA + H_2O = HA^- + H_3O^+$ and $HA = HA^- + H^+$ have the same equilibrium constant

conjugate base: the base formed when an acid dissociates

conjugate acid: the acid formed when a base accepts a proton

standard Gibbs free energy of reaction defined to be zero? Recall from Section 1.4.3 that H^+ is really an abbreviation for $H^+(H_2O)_n$. Thus, it makes sense that the free energy required to convert H^+ to H_3O^+ will be zero.

There are several important ramifications of the fact that $K_a = 1$ for the equilibrium in eq. 11.5. First, it means that H^+ is a strong acid with $pK_a = 0$. (You can show that the pK_a of H_3O^+ also is zero: see Problem 11.2.) Second, it means that you can save some ink when you write the K_a expressions in eqs. 11.1 and 11.2. For example, adding eq. 11.2 and the reverse of eq. 11.5 (see Chapter 4 for a review of the rules of adding and reversing equilibria) results in:

$$
\begin{array}{ll}
HA + H_2O = HA^- + H_3O^+ & K_1 = K_a \\
+ H_3O^+ = H^+ + H_2O & K_2 = 1/1 = 1 \\
\hline
HA = A^- + H^+ & K = K_1K_2 = K_a
\end{array}
$$

This little exercise shows that the equilibrium $HA + H_2O = A^- + H_3O^+$ and the equilibrium $HA = A^- + H^+$ have the *same equilibrium constant*. Thus, you can stop writing acid dissociation reactions in the form of eqs. 11.1 and 11.2 and write them as eqs. 11.6 and 11.7:

$$H_nA^a = H_{n-1}A^{a-1} + H^+ \qquad \text{eq. 11.6}$$

$$HA = A^- + H^+ \qquad \text{eq. 11.7}$$

The equilibria in eqs. 11.1 and 11.6 have the same equilibrium constant, as do the equilibria in eqs. 11.2 and 11.7. This text will use equilibria in the form of eqs. 11.6 and 11.7 in most instances and use the equilibrium forms in eqs. 11.1 and 11.2 only when the transfer of protons to water needs to be emphasized.

11.3.4 Conjugate bases and acids

When an acid donates a proton to water, it produces H_3O^+ and another species. The other species is called the ***conjugate base*** of the acid. Thus, from Example 11.1, you can see that CN^- is the conjugate base of HCN and ammonia is the conjugate base of ammonium ion. Similarly, when a base accepts a proton from water, it produces OH^- and the ***conjugate acid*** of the starting base.

Thoughtful Pause

How are the acid strength of an acid and the base strength of its conjugate base related?

You can write the acid dissociation equilibrium and K_b equilibrium for the conjugate base of an uncharged, monoprotic acid as follows:

$$HA = A^- + H^+ \qquad\qquad K_a$$

$$A^- + H_2O = HA + OH^- \qquad K_b \text{ for conjugate base}$$

Adding:

$$H_2O = H^+ + OH^- \qquad\qquad K = K_W = K_a K_b$$

To summarize in words: the product of the K_a of an acid and the K_b of its conjugate base is K_W. An equivalent statement is this: *the sum of the pK_a of an acid and the pK_b of its conjugate base is pK_W*. At 25°C, where $pK_W = 14$:

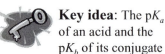

Key idea: The pK_a of an acid and the pK_b of its conjugate base sum to pK_W ($\approx$ 14 at 25°C)

$$pK_a + pK_b \text{ (of the conjugate base)} = 14 \qquad\qquad \text{eq. 11.8}$$

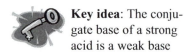

Key idea: The conjugate base of a strong acid is a weak base

Equation 11.8 is very useful. It gives you a lot of information from the knowledge of just one equilibrium constant. If you know the pK_a of an acid, then you can calculate easily the pK_b of its conjugate base (from $pK_b = 14 - pK_a$). Equation 11.8 implies a truism in aquatic chemistry - namely, *the conjugate base of a strong acid is a weak base* (and, conversely, the conjugate base of a weak acid is a strong base). For example, HCl is a strong acid. Its pK_a is about -3.

***Example 11.3:* Conjugate Acids and Bases**

The K_a for phosphoric acid (H_3PO_4) is $10^{-2.1}$. What can you say about the acid strength of phosphoric acid? What can you say about the base and acid strengths of $H_2PO_4^-$?

Solution:
The pK_a of phosphoric acid (= 2.1) is fairly small, so **phosphoric acid is a relatively strong acid**. The conjugate base of H_3PO_4 is $H_2PO_4^-$ (since $H_3PO_4 = H_2PO_4^- + H^+$). Thus, **$H_2PO_4^-$ is a relatively weak base**, with $pK_b = 14 - 2.1 = 11.9$. With the information provided, **you can make no comment about the acid strength of $H_2PO_4^-$**. Its acid strength is quantified by the equilibrium constant of $H_2PO_4^- = HPO_4^{2-} + H^+$.

Thoughtful Pause
What can you say about the base strength of chloride (Cl^-)?

Chloride must be a very weak base since it is the conjugate base of a very strong acid. In fact, pK_b for chloride is 14 minus the pK_a for HCl = $14 - (-3) = 17$. The implication is that chloride will *not* accept a proton appreciably from water to form HCl. As your common experiences tell you, dumping table salt (NaCl) into a glass of water does not lead to the formation of appreciable quantities of hydrochloric acid. The relationship between conjugate acids and bases is explored further in Example 11.3.

11.3.5 Limits to acid and base strength
Are there limits to the strengths of acids and bases? For all practical purposes, you can put limits on the strengths of acids and bases in aqueous systems. Fluids with pH < 0 or pH > 14 generally are considered to be nonaqueous (that is, the solvent is no longer water). Very strong acids (i.e., acids with $pK_a < 0$) are essentially completely deprotonated (i.e., donate at least one proton completely to water) at all pH values in water. Thus, you should not expect to see appreciable concentrations of the most protonated forms of strong acids such as HCl or H_2SO_4 in water. Only the conjugate bases of these acids (Cl^- and HSO_4^-, along with its conjugate base, SO_4^{2-})

are expected to be found in water at reasonable concentrations. In the same vein, you should not expect to see appreciable concentrations of strong bases such as NH_2^- or S^{2-} in water. Only their conjugate acids will be present (e.g., NH_3 and its conjugate acid NH_4^+, and HS^- along with H_2S).[†]

11.3.6 A brief review of equilibrium calculations with monoprotic acids and bases

You have been exposed to equilibrium calculations with monoprotic acids and bases in Chapters 7 through 9. In this section, the key points in the equilibrium calculations will be reviewed. Recall that to perform algebraic equilibrium calculations with monoprotic acids, you should write the species list, equilibria (one acid dissociation equilibrium for each acid plus the self-ionization of water), mass balance equation(s) (one mass balance equation for each acid family), and the charge balance or proton condition. As a shortcut for the graphical method, draw the lines for the H^+ and OH^- concentrations, identify the system point (the point where $pC = p$(total acid) $= pA_T$ and $pH = pK_a$), and draw lines for the acid and conjugate base. The line for the acid (HA) concentration is about equal to A_T at $pH \ll pK_a$, and the line for the conjugate base (A^-) concentration is about equal to A_T at $pH \gg pK_a$. The [HA] and [A^-] lines cross at $pH = pK_a$ and $pC \approx pA_T + 0.3$ (i.e., 0.3 log units below the system point). The equilibrium position of the system is the single point on the pC-pH diagram where the charge balance or proton condition is satisfied. The proton condition is more useful in systems having a large concentration of a species with constant concentration, such as when the salt of an acid (e.g., NaA) is added to water.

A typical pC-pH diagram for a monoprotic acid is shown in Figure 11.1. The charge balance (and proton condition) for the addition of HA to water is: $[H^+] = [A^-] + [OH^-]$. This is satisfied near point A in Figure 11.1, where $[H^+] \approx [A^-]$ and $[OH^-]$ is small enough to be ignored. The proton condition for the addition of NaA to water is: $[H^+] + [HA] = [OH^-]$. This is satisfied near point B in Figure 11.1, where $[HA] \approx [OH^-]$ and $[H^+]$ is small enough to be ignored.

Several results from the algebraic and graphical solutions are important to remember. First, the HA and A^- concentrations depend on A_T, pH, and pK_a. Specifically, from eqs. 8.3 and 8.4:

$$[A^-] = A_T K_a / (K_a + [H^+]) \text{ and}$$
$$[HA] = A_T [H^+] / (K_a + [H^+])$$

[†] Acids stronger than about the strength of 100% sulfuric acid are called *superacids*. An example of a superacid is fluorosulfonic acid, HSO_3F. Fluorosulfonic acid is about 40 times stronger than 100% sulfuric acid and, of course, completely deprotonated in water. The acid strength of superacids cannot be measured by their pK_a values. In fact, the concept of a pK_a is meaningless, since all superacids are completely deprotonated in water. Instead, their acid strengths are quantified by *acidity functions*. Acidity functions are based on equilibria where the superacids are allowed to protonate bases other than water.

Thus, the concentration of HA is greater than $[A^-]$ at $pH < pK_a$. Conversely, the concentration of A^- is greater than $[HA]$ at $pH > pK_a$. Second, at $pH = pK_a$, the HA and A^- concentrations are equal. You may wish to verify these conclusions for the monoprotic acid in Figure 11.1.

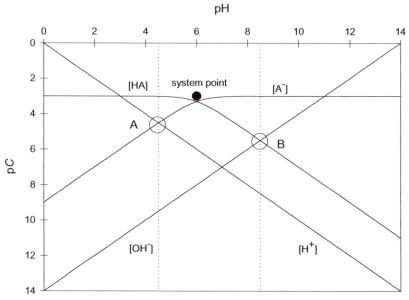

Figure 11.1: pC-pH Diagram for a Monoprotic Acid with pK_a = 6 and pA_T = 3

[dotted lines indicate the approximate equilibrium pH for an HA solution (left line, point A) and a NaA solution (right line, point B)]

11.4 POLYPROTIC ACIDS

11.4.1 Definitions

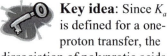

Key idea: Since K_a is defined for a one-proton transfer, the dissociation of polyprotic acids *usually* is written as a stepwise process

As stated in Section 11.2.3, a polyprotic acid can donate more than one proton to water. Common examples of diprotic acids (which can donate two protons to water) in the aquatic environment include H_2CO_3, H_2S, and H_2SO_4. The donation of protons to water from polyprotic acids usually is written as a stepwise process.[†] For a general diprotic acid, H_2A:

$$H_2A = HA^- + H^+ \quad K_{a,1}$$
$$HA^- = A^{2-} + H^+ \quad K_{a,2}$$

[†] It is not *required* to write polyprotic acid dissociation as a stepwise process. Usually, the equilibria are written as one-proton transfers because K_a is defined for one-proton transfers. You could write any consistent set of equilibria to express the relationships among H_2A, HA^-, and A^{2-}; for example, $H_2A = A^{2-} + 2H^+$ and $HA^- = A^{2-} + H^+$.

Because of the stepwise nature of proton transfer, $K_{a,1}$ is always greater than $K_{a,2}$ and thus $pK_{a,1} < pK_{a,2}$.

For diprotic acids, the amphoteric form (HA⁻ in the equilibria above) is given the name of the most basic form with the prefix *bi-*. Examples include the carbonate family H_2CO_3 (carbonic acid), HCO_3^- (bicarbonate), and CO_3^{2-} (carbonate), as well as the sulfide family H_2S (hydrogen sulfide), HS^- (bisulfide), and S^{2-} (sulfide).

11.4.2 Solving equilibrium systems with polyprotic acids

You can use the algebraic, graphical, and computer methods discussed in Part II to solve equilibrium systems containing polyprotic acids. The example of a 10^{-3} M H_2CO_3 solution (closed system) will be solved by each method in this section. We must pause for a moment and introduce one new symbol. Recall from Section 1.4.3 that H^+, $(H_2O)H^+$, $(H_2O)_2H^+$, and so on are almost indistinguishable. Thus, the symbol $[H^+]$ is used to represent the sum $[H^+] + [(H_2O)H^+] + [(H_2O)_2H^+] +$ other similar species. Similarly, the species $H_2CO_3(aq)$ and $CO_2(aq)$ are nearly indistinguishable, and it is common to represent the sum of their concentrations by one symbol: $[H_2CO_3^*] = [H_2CO_3(aq)] + [CO_2(aq)]$. The system is as follows (if activity and concentration are interchangeable):

Unknowns: $[H_2O]$, $[H^+]$, $[OH^-]$, $[H_2CO_3^*]$, $[HCO_3^-]$, and $[CO_3^{2-}]$
Starting materials: H_2O and $H_2CO_3^*$
Equilibria:

$$H_2O \ = \ H^+ + OH^- \qquad K_W = 10^{-14}$$
$$H_2CO_3^* \ = \ HCO_3^- + H^+ \qquad K_{a,1} = 10^{-6.3}$$
$$HCO_3^- \ = \ CO_3^{2-} + H^+ \qquad K_{a,2} = 10^{-10.3}$$

Equations:
Equilibria:
$K_W = [H^+][OH^-]/[H_2O]$
$K_{a,1} = [HCO_3^-][H^+]/[H_2CO_3^*]$
$K_{a,2} = [CO_3^{2-}][H^+]/[HCO_3^-]$
Mass balance:
$C_T = 1\times10^{-3}$ M $= [H_2CO_3^*] + [HCO_3^-] + [CO_3^{2-}]$
Charge balance:
$[H^+] = [HCO_3^-] + 2[CO_3^{2-}] + [OH^-]$
(or you can use the proton condition, also: $[H^+] = [HCO_3^-] + 2[CO_3^{2-}] + [OH^-]$)
Other:
Activity of $H_2O = 1$
All species concentrations > 0

(Remember that any two of the mass balance, charge balance, and proton condition can be combined to generate the third equation.) This system can be solved by the algebraic, graphical, and computer methods. Examples of each solution technique will be presented.

Example 11.4: **Salt of a Diprotic Acid**

Following the example in the text with the $H_2CO_3^*$ system, find the equilibrium concentrations for a 1×10^{-3} M sodium bicarbonate solution.

Solution:
If $NaHCO_3$ ($\rightarrow Na^+ + HCO_3^-$) was added to water instead of carbonic acid, the following changes would be made to the in the text
(1) Add Na^+ as a species
(2) Change starting materials to $NaHCO_3 \rightarrow Na^+ + HCO_3^-$ and H_2O
(3) Add a new mass balance equation: $Na_T = 10^{-3}$ M $= [Na^+]$
(4) Include Na^+ as a cation in the charge balance
(5) Change the proton condition to: $[H^+] + [H_2CO_3^*] = [CO_3^{2-}] + [OH^-]$.

Algebraic solution:

The system has the same equilibria as in the text, so eqs. 11.9-11.11 are still valid. Since $[Na^+]$ is present, it is easier to use the proton condition than the charge balance. Plugging eqs. 11.9-11.11 into the charge balance yields a variation of eq. 11.12:

$$[H^+] + C_T[H^+]^2/\Delta =$$
$$C_T K_{a,1} K_{a,2}/\Delta + K_W/[H^+] \quad (*)$$

where $\Delta = [H^+]^2 + K_{a,1}[H^+] + K_{a,1}K_{a,2}$.

Solving:
$$[H^+] = 5.28 \times 10^{-9} \text{ M}$$
$$(\text{pH } 8.28)$$
$$[OH^-] = 1.89 \times 10^{-6} \text{ M}$$
$$[H_2CO_3^*] = 1.03 \times 10^{-5} \text{ M}$$
$$[HCO_3^-] = 9.80 \times 10^{-5} \text{ M}$$
$$[CO_3^{2-}] = 9.30 \times 10^{-6} \text{ M}$$

Approximation method:

You expect the pH to be closer to neutral pH than in the case of $H_2CO_3^*$ addition. Thus, Δ is about equal to $K_{a,1}[H^+]$ and you might try ignoring $[OH^-]$ relative to $[CO_3^{2-}]$ in the proton condition. Thus, (*) becomes:

$$[H^+] + C_T[H^+]/K_{a,1} = C_T K_{a,2}/[H^+]$$

Or: $[H^+]^2(1 + C_T/K_{a,1}) = C_T K_{a,2}$

Since $C_T/K_{a,1} \gg 1$: $[H^+] = [(C_T K_{a,2})/(C_T/K_{a,1})]^{1/2} = (K_{a,1}K_{a,2})^{1/2} = 10^{-16.2/2} = 10^{-8.3}$ or **pH 8.3**.

Checking the assumptions, you can show that at pH 8.26, $\Delta = 2.6 \times 10^{-15} \approx K_{a,1}[H^+] = 2.5 \times 10^{-15}$. Also: $[OH^-] = 10^{-5.7}$ M and $[CO_3^{2-}] \approx C_T K_{a,2}/[H^+] = 10^{-5}$ M. The assumption that $[OH^-]$ can be ignored relative to $[CO_3^{2-}]$ is of questionable validity.

In the brute force algebraic method, express each species concentration in terms of the concentration of one species (usually $[H^+]$). From the equilibria and mass balances (after a little algebra, which you should confirm):

$$[OH^-] = K_W/[H^+]$$
$$[H_2CO_3^*] = C_T[H^+]^2/([H^+]^2 + K_{a,1}[H^+] + K_{a,1}K_{a,2}) \qquad \text{eq. 11.9}$$
$$[HCO_3^-] = C_T K_{a,1}[H^+]/([H^+]^2 + K_{a,1}[H^+] + K_{a,1}K_{a,2}) \qquad \text{eq. 11.10}$$
$$[CO_3^{2-}] = C_T K_{a,1}K_{a,2}/([H^+]^2 + K_{a,1}[H^+] + K_{a,1}K_{a,2}) \qquad \text{eq. 11.11}$$

Substituting into the charge balance (or proton condition) yields:

$$[H^+] = \frac{C_T K_{a,1}[H^+] + 2C_T K_{a,1} K_{a,2}}{[H^+]^2 + K_{a,1}[H^+] + K_{a,1}K_{a,2}} + \frac{K_W}{[H^+]} \qquad \text{eq. 11.12}$$

Solving (see Appendix A for hints on solving this equation):

$$[H^+] = 2.22 \times 10^{-5} \text{ M (pH 4.65)}$$
$$[OH^-] = 4.51 \times 10^{-10} \text{ M}$$
$$[H_2CO_3^*] = 9.78 \times 10^{-4} \text{ M}$$
$$[HCO_3^-] = 2.21 \times 10^{-5} \text{ M}$$
$$[CO_3^{2-}] = 4.99 \times 10^{-11} \text{ M}$$

Using the method of approximation, you might guess that the solution is mildly acidic since a relatively weak acid (pK_a of $H_2CO_3^*$ is 6.3) is being added to water. Thus, the righthand side of the charge balance should be dominated by the weaker bases.

Thoughtful Pause

Why is the charge balance dominated by the weaker bases?

The weaker bases generally are higher in concentration because the stronger bases will be more protonated at an acidic pH. Therefore, from the charge balance: $[H^+] \approx [HCO_3^-]$ or, using eq. 11.9:

$$[H^+] \approx C_T K_{a,1}[H^+]/([H^+]^2 + K_{a,1}[H^+] + K_{a,1}K_{a,2})$$

The denominator of the expression for $[HCO_3^-]$ (eq. 11.10) is about equal to $[H^+]^2$ under acidic conditions.[†] Thus: $[H^+] \approx C_T K_{a,1}/[H^+]$ or $[H^+]^2 \approx C_T K_{a,1}$ or pH $\approx (pC_T + pK_{a,1})/2 = (3 + 6.3)/2 = 4.65$.

[†] Why is the denominator in eq. 11.9 about equal to $[H^+]^2$ under acidic conditions? The term $[H^+]^2$ is larger than $K_{a,1}[H^+] = 10^{-6.3}[H^+]$ if the pH is less than 6.3. Also, $[H^+]^2$ is larger than $K_{a,1}K_{a,2} = 10^{-16.6}$ if the pH is less than 16.6/2 = 8.3. Thus, $[H^+]^2$ is the largest term in the denominator if the pH is less than 6.3.

Graphical method:
Figure 11.2 is still valid. We seek the point where the proton condition is satisfied, i.e., where

$$[H^+] + [H_2CO_3^*] = [CO_3^{2-}] + [OH^-]$$

Near neutral pH, you expect $[H^+]$ and $[OH^-]$ to be small, so the proton condition becomes $[H_2CO_3^*] \approx [CO_3^{2-}]$. From Figure 11.2, the $[H_2CO_3^*]$ and $[CO_3^{2-}]$ lines cross at about **pH 8.2-8.3**.

Computer method:
In *Nanoql*, add the species Na^+ and fix its concentration at 10^{-3} M. (Alternatively, add the species Na^+ and the mass balance $Na_T = 10^{-3}$ M $= [Na^+]$).

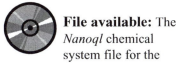

File available: The *Nanoql* chemical system file for the $H_2CO_3^*$ system may be found on the CD - file location **/Chapter 11/carbonic acid.nqs**

You may wish to check that the assumptions are valid. The assumptions were:

In the charge balance: $[HCO_3^-] \gg 2[CO_3^{2-}]$ and $[HCO_3^-] \gg [OH^-]$
In the denominator: $[H^+]^2 \gg K_{a,1}[H^+] + K_{a,1}K_{a,2}$

This example points out the value of chemical intuition in selecting the approximation. The wise assumption of mildly acidic conditions here converted a fourth-order polynomial equation in $[H^+]$ to the arithmetic of $pH \approx (3 + 6.3)/2 = 4.65$. Working problems like this is one of the best ways to strengthen your chemical intuition and save you work in the long run.

To use the graphical method, merely plot the expressions in eqs. 11.9-11.11 and find where the charge balance (or proton condition) is satisfied. A pC-pH diagram for the carbonic acid system is shown in Figure 11.2. The equilibrium pH is again approximated from the charge balance at the pH where $[H^+] \approx [HCO_3^-]$ (about where the lines for $[H^+]$ and $[HCO_3^-]$ cross). This occurs at about pH 4.6-4.7 (dotted line in Figure 11.2). The approximations in the charge balance ($[HCO_3^-] \gg 2[CO_3^{2-}]$ and $[HCO_3^-] \gg [OH^-]$) can be checked easily in the pC-pH diagram. A shortcut to the graphical method with polyprotic acids will be discussed in Section 11.4.3.

Of course, polyprotic acid systems can be solved with computer methods. The *Nanoql* chemical system file for the $H_2CO_3^*$ system may be found on the CD at **/Chapter 11/carbonic acid.nqs**. Try running the file and calculating equilibrium concentrations. A related chemical system is solved in Example 11.4 by algebraic, approximation, graphical, and computer methods.

11.4.3 A shortcut to the graphical method for polyprotic acids

You can develop a shortcut method for graphical solutions to polyprotic acid systems similar to the shortcut method developed for monoprotic acids in Section 8.4. For a diprotic acid, consider three pH ranges: $pH < pK_{a,1}$, $pK_{a,1} < pH < pK_{a,2}$, and $pH > pK_{a,2}$. In the lowest pH range, the denominator in eqs. 11.9-11.11 is about equal to $[H^+]^2$. Thus:

$$[H_2CO_3^*] = C_T[H^+]^2/([H^+]^2 + K_{a,1}[H^+] + K_{a,1}K_{a,2}) \approx C_T$$

$$[HCO_3^-] = C_TK_{a,1}[H^+]/([H^+]^2 + K_{a,1}[H^+] + K_{a,1}K_{a,2}) \approx C_TK_{a,1}/[H^+]$$

$$[CO_3^{2-}] = C_TK_{a,1}K_{a,2}/([H^+]^2 + K_{a,1}[H^+] + K_{a,1}K_{a,2}) \approx C_TK_{a,1}K_{a,2}/[H^+]^2$$

Therefore, where $pH < pK_{a,1}$, you expect:

1. The $[H_2CO_3^*]$ line should have a slope of zero and be about equal to C_T.

2. The $[HCO_3^-]$ line should be downward sloping to the left with a slope equal to -1 (parallel to the $[OH^-]$ line, where $[OH^-] = K_W/[H^+]$).

3. The $[CO_3^{2-}]$ line should be downward sloping to the left with a slope equal to -2. Also note that $[CO_3^{2-}]$ is much smaller than either $[H_2CO_3^*]$ or $[HCO_3^-]$ since $[H^+] \gg K_{a,2}$ (pH < p$K_{a,2}$ = 10.3) in this pH range.

Be sure to work with the above equations until these conclusions make sense to you.

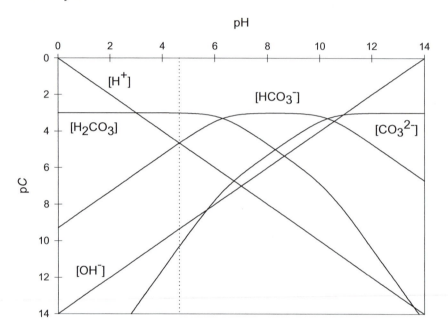

Figure 11.2: pC-pH Diagram for the Addition of
10^{-3} M $H_2CO_3^*$ to Water
(dotted line represents the equilibrium pH - where the charge balance and proton condition are satisfied)

In the middle pH range, the denominator in eqs. 11.9-11.11 is about equal to $K_{a,1}[H^+]$. Thus:

$$[H_2CO_3^*] = C_T[H^+]^2/([H^+]^2 + K_{a,1}[H^+] + K_{a,1}K_{a,2}) \approx C_T[H^+]/K_{a,1}$$

$$[HCO_3^-] = C_T K_{a,1}[H^+]/([H^+]^2 + K_{a,1}[H^+] + K_{a,1}K_{a,2}) \approx C_T$$

$$[CO_3^{2-}] = C_T K_{a,1}K_{a,2}/([H^+]^2 + K_{a,1}[H^+] + K_{a,1}K_{a,2}) \approx C_T K_{a,2}/[H^+]$$

Therefore, where $pK_{a,1} < pH < pK_{a,2}$ you expect:

1. The $[H_2CO_3*]$ line should be downward sloping to the right with a slope equal to +1 (parallel to the $[H^+]$ line).

2. The $[HCO_3^-]$ line should have a slope of zero and be about equal to C_T.

3. The $[CO_3^{2-}]$ line should be downward sloping to the left with a slope equal to -1.

Finally, in the highest pH range ($pH < pK_{a,2}$), the denominator in eqs. 11.9-11.11 is about equal to the product of the K_a values ($= K_{a,1}K_{a,2}$). Thus:

$$[H_2CO_3*] = C_T[H^+]^2/([H^+]^2 + K_{a,1}[H^+] + K_{a,1}K_{a,2}) \approx C_T[H^+]^2/(K_{a,1}K_{a,2})$$

$$[HCO_3^-] = C_TK_{a,1}[H^+]/([H^+]^2 + K_{a,1}[H^+] + K_{a,1}K_{a,2}) \approx C_T[H^+]/K_{a,2}$$

$$[CO_3^{2-}] = C_TK_{a,1}K_{a,2}/([H^+]^2 + K_{a,1}[H^+] + K_{a,1}K_{a,2}) \approx C_T$$

Therefore, where $pH > pK_{a,2}$ you expect:

1. The $[H_2CO_3*]$ line should be downward sloping to the right with slope = +2. Also note that $[H_2CO_3*]$ is much smaller than either $[HCO_3^-]$ or $[CO_3^{2-}]$ since $[H^+] << K_{a,2}$ ($pH > pK_{a,2}$ = 10.3) in this pH range.

2. The $[HCO_3^-]$ line should be downward sloping to the right with a slope equal to +1 (parallel to the $[H^+]$ line).

3. The $[CO_3^{2-}]$ line should have a slope of zero and be about equal to C_T.

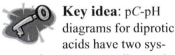

Key idea: pC-pH diagrams for diprotic acids have two system points (corresponding to the two pK_a values), with the slope of the pC lines changing at the system points

The graphical shortcut method takes advantage of the two system points for a diprotic acid. The first system point occurs at $pH = pK_{a,1}$ and $pC = pC_T$. The second system point occurs at $pH = pK_{a,2}$ and $pC = pC_T$.

The behavior of the species in the diprotic acid system with respect to pH can be summarized. The p(concentration) of the most protonated form of the diprotic acid (H_2CO_3* in the example) is about equal to C_T at pH values less than $pK_{a,1}$ (i.e., to the left of the first system point), increases in a 1:1 ratio with increasing pH at pH values between the two pK_a values (i.e., between the system points), and increases in a 2:1 ratio with increasing pH at pH values greater than $pK_{a,2}$ values (i.e., to the right of the second system point).

Similarly, the p(concentration) of the next most protonated form of the diprotic acid (HCO_3^- in the example) decreases in a 1:1 ratio with pH at increasing pH values less than $pK_{a,1}$ (i.e., to the left of the first system point), is about equal to C_T at pH values between the two pK_a values (i.e., between the system points), and increases in a 1:1 ratio with increasing pH at pH values greater than $pK_{a,2}$ (i.e., to the right of the second system point).

Finally, the p(concentration) of the least protonated form of the diprotic acid (CO_3^{2-} in the example) decreases in a 2:1 ratio with pH at increasing pH values less than $pK_{a,1}$ (i.e., to the left of the first system point), decreases in a 1:1 ratio with increasing pH at pH values between the two pK_a values (i.e., between the system points), and is about equal to C_T at pH values greater than $pK_{a,2}$ (i.e., to the right of the second system point). As with monoprotic acids, it is quite easy to sketch the pC-pH diagram for a polyprotic acid.

The extension of the shortcut method of Section 8.4 to n-protic acids is fairly straightforward. Consider an n-protic acid, H_nA, which dissociates in water to form n other species: $H_{n-1}A^-$, $H_{n-2}A^{2-}$, ..., A^{n-}. Each species is of the form H_iA. The sum of the concentrations of the n species containing the fragment -A is A_T. To sketch the pC-pH diagram (see also Appendix C, Section C.4.3), use the following approach:

Step 1: Locate system points

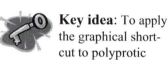

Key idea: To apply the graphical short-cut to polyprotic acids, locate system points (at $pC = pA_T$ and $pH = pK_{a,i}$), draw species lines (making sure that species lines intersect below the system points), and find the equilibrium pH where the proton condition is satisfied

Prepare a pC-pH diagram with lines representing the H^+ and OH^- concentrations. Locate the n system points. These occur where $pH = pK_{a,1}$ and $pC = pA_T$, $pH = pK_{a,2}$ and $pC = pA_T$, ..., and $pH = pK_{a,n}$ and $pC = pA_T$.

Step 2: Draw species lines

Draw the lines for the p(concentration) of each species at least 1.5 pH units away from each system point. The p(concentration) line of species H_iA has the following slopes: $i - n$ at pH values less than the first system point, $i - n + j$ between the jth and $(j + 1)^{th}$ system point, and $-i$ at pH values greater than the final (n^{th}) system point. (Recall that a negative slope means that the pC decreases with increasing pH and thus the line is upward sloping to the right.) Make sure that the lines, if extended, would go through the system points.

Step 3: Make species lines intersect below the system points

Curve the lines so that they intersect about 0.3 log units below the system points.

Step 4: Find the equilibria pH

To find the equilibria pH (and equilibrium concentrations of all plotted species), find the pH where the charge balance or proton condition is satisfied. Check any assumptions made in the charge balance and iterate if necessary.

With this shortcut, a pC-pH diagram for any acid can be sketched quickly and the equilibrium pH estimated with little fuss.

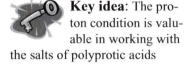 **Key idea**: The proton condition is valuable in working with the salts of polyprotic acids

11.4.4 The power of the proton condition

The real value of the proton condition shows itself with polyprotic acids. Once you have solved the system formed by adding H_nA to water, you can easily calculate the equilibrium conditions for adding various salts to water (generalized as Na_mH_pA, where $m + p = n$). For example, after putting in the work on the H_2CO_3* system, you can find the equilibrium pH for systems where $NaHCO_3$ (sodium bicarbonate) and Na_2CO_3 (sodium carbonate) are added to water. The three systems (adding H_2CO_3*, $NaHCO_3$, or Na_2CO_3 to water) have the same set of equilibria and differ only in their mass balance on the sodium ion and thus differ in their proton condition.

Example 11.4 demonstrated the ease in extending the H_2CO_3* system to $NaHCO_3$ solutions. You also can apply the previous work to Na_2CO_3 solutions. Begin by writing the proton condition for a Na_2CO_3 solution (starting materials are $Na_2CO_3 \rightarrow 2Na^+ + CO_3^{2-}$ and H_2O):

$$[H^+] + [HCO_3^-] + 2[H_2CO_3*] = [OH^-]$$

(Note the factor of 2 in front of the carbonic acid concentration, because H_2CO_3* has *two* more protons than its source in the starting materials, namely CO_3^{2-}.) You might expect the pH to be high.

Thoughtful Pause

Why should the pH be high for a sodium carbonate solution?

The righthand side of the proton condition contains only $[OH^-]$. Thus, $[OH^-]$ must be much larger than $[H^+]$.[†] At high pH, $[H^+]$ and $[H_2CO_3*]$ are expected to be small and the proton condition becomes: $[HCO_3^-] \approx [OH^-]$. From Figure 11.2, this occurs at about pH 10.5-10.6 (exact solution: pH 10.56).

[†] The same conclusion can be reached from the charge balance: $[H^+] + [Na^+] = [HCO_3^-] + 2[CO_3^{2-}] + [OH^-]$. Since $[Na^+] = 2 \times 10^{-3}$ M is large, the species concentrations on the righthand side also must be large and $[OH^-]$ is expected to be much larger than $[H^+]$.

11.5 DISTRIBUTION FUNCTIONS (ALPHA VALUES)

11.5.1 Introduction

In developing analytical solutions for monoprotic acids (Section 8.4) and polyprotic acids (Section 11.4), you combined the equilibria and mass balances to express the concentration of acids, bases, and ampholytes in terms of the total concentration multiplied by a function of $[H^+]$ and K_a values. For example, eqs. 8.3 and 8.4 stated that, for a monoprotic acid: $[A^-] = A_T K_a/(K_a + [H^+])$ and $[HA] = A_T[H^+]/(K_a + [H^+])$, where $A_T = [HA] + [A^-]$ (see also Section 11.3.6). Note that these expression are of the form:

$$[A^-] = A_T f([H^+], K_a)$$
$$[HA] = A_T g([H^+], K_a)$$

distribution functions (ionization fractions, alpha values): functions of pH and the K_a values that describe the fractional contribution of each species to the total concentration in its mass balance

where f and g are the functions $K_a/(K_a + [H^+])$ and $[H^+]/(K_a + [H^+])$, respectively.

The function f and g are of the same form for any monoprotic acid. They are called ***distribution functions*** (or ***ionization fractions***). The functions have been given many names over the years. Since f and g usually are given the symbols α_1 and α_0, respectively, they shall be referred to as ***alpha values*** in this text.

The alpha values for a monoprotic acid have a number of useful interpretations and characteristics. First, the alpha values represent the *fraction of the total concentration found as either [HA] or [A$^-$]*. The alpha values are truly fractions and vary in magnitude between zero and one. Second, the alpha values encapsulate all the information on how [HA] and [A$^-$] change with pH and K_a. In fact, alpha values separate the dependence of species concentrations on pH and K_a from the dependency on A_T. Third, the alpha values sum to one. You can show with a little algebra that $\alpha_0 + \alpha_1 = 1$. It is perhaps more important to see that α_0 plus α_1 **must** equal one. Since the alpha values represent the fraction of the total concentration that is [HA] or [A$^-$], the sum of the alpha values must be unity. Fourth, the alpha values are easily interconvertible. Note that:

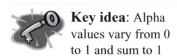

Key idea: Alpha values vary from 0 to 1 and sum to 1

$$\alpha_1 = (K_a/[H^+])\alpha_0 \text{ and } \alpha_0 = ([H^+]/K_a)\alpha_1$$

For monoprotic acids, these interconversions are pretty obvious. They are more useful with polyprotic acids because they save time in the calculations.

11.5.2 Alpha values of diprotic acids

The alpha values for diprotic acids can be generalized from eqs. 11.9-11.11 as follows:

$$[H_2A] = \alpha_0 A_T, \text{ where } \alpha_0 = [H^+]^2/([H^+]^2 + K_{a,1}[H^+] + K_{a,1}K_{a,2})$$
$$[HA^-] = \alpha_1 A_T, \text{ where } \alpha_1 = K_{a,1}[H^+]/([H^+]^2 + K_{a,1}[H^+] + K_{a,1}K_{a,2})$$
$$[A^{2-}] = \alpha_2 A_T, \text{ where } \alpha_2 = K_{a,1}K_{a,2}/([H^+]^2 + K_{a,1}[H^+] + K_{a,1}K_{a,2})$$

Note that there are *three* alpha values to accommodate the three species in the family. Also, note that the alpha values represent different expressions for monoprotic acids and diprotic acids, although they have the same interpretation as fractions of the total concentration. In other words, α_0 for a monoprotic acid and α_0 for a diprotic acid are calculated with different formulas, but both are equal to the fraction of total acid made up by the *most protonated* form of the acid.

You also can see the patterns in the nomenclature: the symbol α_0 always is used for the most protonated form of the acid (H_2A for diprotic acids and HA for monoprotic acids). You can show (both by logic and algebra) that the alpha values for diprotic acids sum to one. In addition, the alpha values are interconvertible. Comparing α_0 and α_1:

$$\alpha_1 = (K_{a,1}/[H^+])\alpha_0 \text{ and } \alpha_0 = ([H^+]/K_{a,1})\alpha_1$$

Comparing α_1 and α_2:

$$\alpha_2 = (K_{a,2}/[H^+])\alpha_1 \text{ and } \alpha_1 = ([H^+]/K_{a,2})\alpha_2$$

Finally, comparing α_0 and α_2:

$$\alpha_2 = (K_{a,1}K_{a,2}/[H^+]^2)\alpha_0 \text{ and } \alpha_0 = \{[H^+]^2/(K_{a,1}K_{a,2})\}\alpha_2$$

11.5.3 Alpha values for *n*-protic acids

You can extrapolate the formulas above to alpha values for *n*-protic acids. For *n*-protic acids, the ith alpha value (denoted α_i, where $i = 0$ to n) relates the concentration of the species $H_{n-i}A^{i-}$ to the total concentration: $[H_{n-i}A^{i-}] = \alpha_i A_T$. You can derive the following relationships:

Key idea: Alpha values can be extended to *n*-protic acids

$$\alpha_i = \frac{[H^+]^{n-i} \prod_{k=1}^{i} K_{a,k}}{\sum_{j=0}^{n} \left([H^+]^{j-i} \prod_{k=1}^{j} K_{a,k} \right)}$$

$$\sum_{i=0}^{n} \alpha_i = 1$$

$$\alpha_i = \left([H^+]^{j-i} \prod_{k=1}^{j} K_{a,k} \right) \alpha_j, i < j$$

You should verify these relationships for $n = 1$ (monoprotic acids) and $n = 2$ (diprotic acids).

11.6 SUMMARY

Although the definitions of acid and base have changed over time, the Brønsted-Lowry definitions are very useful: acids donate protons to water and bases accept protons from water. Acids and bases can be defined by equilibria as well. For example, the acid dissociation reaction for an uncharged monoprotic acid is $HA + H_2O = A^- + H_3O^+$ (thermodynamically equivalent to $HA = A^- + H^+$). The equilibrium constant for equilibria of this form are called *acid dissociation constants* and denoted by the symbol K_a. Similarly, bases can be defined by equilibria of the form $A^- + H_2O = HA + OH^-$, with $K = K_b$, the *base association constant*.

Acid strength is measured by K_a or pK_a values and base strength by K_b or pK_b values. Thus, acid and base strengths are thermodynamic properties and change with thermodynamic conditions.

Acids and bases can donate or accept more than one proton. Acids that donate more than one proton to water are called *polyprotic acids*. We usually think of polyprotic acids as donating protons stepwise in a series of one-proton reactions each with its own K_a value. In this chapter, the solution techniques from Part II were applied to polyprotic acid and base systems. A shortcut graphical method for polyprotic acids and their salts was developed. In the shortcut, a diprotic acid was found to have two system points, corresponding to the two pK_a values.

Compact equations were developed to describe how the fraction of the total concentration contributed by each species depends on pH. These equations are called *distribution functions* or *alpha values*. Alpha values can be written for *n*-protic acids.

11.7 PART III CASE STUDY: ACID RAIN

After completing this chapter, you now have a good understanding of how to perform equilibrium calculations with acids and bases. In this first portion of the Part III case study, you will apply your newly minted skills to the analysis of acid rain.

You collected a sample of potentially acidic rainwater and submitted it to an analytical laboratory. Being familiar with the chemistry of acid rain (as reviewed briefly in Section 10.7), you asked the lab to measure the concentrations of common ions in the sample. You received the following results: $[SO_4^{2-}] = 0.5$ mg/L as S, $[NO_3^-] = 0.6$ mg/L as N, and $[Cl^-] = 0.7$ mg/L as Cl. The lab also reported that the total inorganic carbon concentration ($= [H_2CO_3^*] + [HCO_3^-] + [CO_3^{2-}]$) is 0.1 mg/L as C. Upon reading this list, you realized that you forgot to ask the lab to measure pH. The sample has been discarded, but you trust the analytical results (including the fact that no cations other than H^+ apparently are in the sample in appreciable concentrations). You are a little confused about why the lab measured inorganic carbon but did not report the concentrations of HCO_3^- or CO_3^{2-}.

Being too embarrassed to tell your boss that you forgot to have pH measured, you are determined to calculate the pH of the rainwater. You know that H_2SO_4, HNO_3, and HCl are very strong acids. Thus, it makes sense that their conjugate bases (SO_4^{2-}/HSO_4^-, NO_3^-, and Cl^-) are very weak bases and likely will not accept protons from water. Therefore, it is a good approximation that SO_4^{2-}, NO_3^-, and Cl^- are the only forms of S(+VI), N(+V), and Cl(−I) in the sample. You develop a mathematical model of the system:

Unknowns: $[H_2O]$, $[H^+]$, $[OH^-]$, $[H_2CO_3^*]$, $[HCO_3^-]$, and $[CO_3^{2-}]$
Starting materials: H_2O and some form of inorganic carbon
Equilibria:

$$H_2O = H^+ + OH^- \qquad\qquad K_W = 10^{-14}$$
$$H_2CO_3^* = HCO_3^- + H^+ \qquad K_{a,1} = 10^{-6.3}$$
$$HCO_3^- = CO_3^{2-} + H^+ \qquad K_{a,2} = 10^{-10.3}$$

Equations:
Equilibria:

$$K_W = [H^+][OH^-]/[H_2O]$$
$$K_{a,1} = [HCO_3^-][H^+]/[H_2CO_3^*]$$
$$K_{a,2} = [CO_3^{2-}][H^+]/[HCO_3^-]$$

Mass balances:

$$C_T = [H_2CO_3^*] + [HCO_3^-] + [CO_3^{2-}] = 0.1 \text{ mg/L as C}$$
$$= (1\times10^{-4} \text{ g/L})/(12 \text{ g/mol}) = 8.3\times10^{-6} \text{ M}$$
$$S_T = [SO_4^{2-}] = .5 \text{ mg/L as S}$$
$$= (5\times10^{-4} \text{ g/L})/(32 \text{ g/mol}) = 1.6\times10^{-5} \text{ M}$$
$$N_T = [NO_3^-] = 0.6 \text{ mg/L as N}$$
$$= (6\times10^{-4} \text{ g/L})/(14 \text{ g/mol}) = 4.3\times10^{-5} \text{ M}$$
$$Cl_T = [Cl^-] = 0.7 \text{ mg/L as Cl}$$
$$= (7\times10^{-4} \text{ g/L})/(35.45 \text{ g/mol}) = 2.0\times10^{-5} \text{ M}$$

Charge balance:

$$[H^+] = [HCO_3^-] + 2[CO_3^{2-}] + [OH^-] + 2[SO_4^{2-}]$$
$$+ [NO_3^-] + [Cl^-]$$
$$\text{Or: } [H^+] = [HCO_3^-] + 2[CO_3^{2-}] + [OH^-] + 9.5\times10^{-5} \text{ eq/L}$$

Other:
Activity of $H_2O = 1$
All species concentrations > 0

You might guess that the pH is acidic. Thus, you can ignore $[CO_3^{2-}]$ and $[OH^-]$ in the charge balance. The charge balance becomes:

$$[H^+] = [HCO_3^-] + 9.5\times10^{-5} \text{ eq/L}$$

Based on your experience, you know that $[HCO_3^-] = \alpha_1 C_T$, where $\alpha_1 = K_{a,1}[H^+]/([H^+]^2 + K_{a,1}[H^+] + K_{a,1}K_{a,2})$. At low pH (well below pH 6.3), $\alpha_1 \approx K_{a,1}/[H^+]$. Thus:

$$[H^+] = C_T K_{a,1}/[H^+] + 9.5 \times 10^{-5} \text{ eq/L}$$

So:

$$[H^+]^2 - 9.5 \times 10^{-5}[H^+] - C_T K_{a,1} = 0 \qquad \text{eq. 11.13}$$

Solving: $[H^+] = 9.5 \times 10^{-5}$ M or pH 4.0. Now you realize why the lab failed to report bicarbonate and carbonate concentrations. At pH 4.0 and with 0.1 mg/L as C total inorganic carbon, the concentrations of bicarbonate and carbonate are very, very small. In other words, the inorganic carbon in the sample is primarily $H_2CO_3^*$.

Should you have a great deal of confidence in this calculated pH value? In a word, no. Note that the term $C_T K_{a,1}$ in eq. 11.13 is small compared to the other terms. Thus, $[H^+] = 9.5 \times 10^{-5}$ M $= 2[SO_4^{2-}] + [NO_3^-] + [Cl^-]$. Any slight error in the measurement of sulfate, nitrate, or chloride may have a significant effect on the calculated pH. Also, the presence of any unmeasured cations or anions may skew calculated pH as well. In the absence of a measured pH value, we will proceed in the case study with the calculated value of the acid rain pH as 4.0.

SUMMARY OF KEY IDEAS

- Acid-base chemistry (and pH) are important because H^+ is transferred in many reactions that influence the water quality of aquatic systems

- Lavoisier thought, incorrectly, that all acids contained oxygen

- The Brønsted-Lowry definitions linked acids and bases by relating them both to proton transfer

- Acid-base equilibria are proton-transfer equilibria

- Brønsted-Lowry acids and bases also can be defined by acid dissociation reactions and base association reactions

- Water is amphoteric (acts as both an acid and a base)

- Acid and base strength are thermodynamic properties

- Since the equilibrium $H_3O^+ = H^+ + H_2O$ has $K = 1$, then the equilibria: $HA + H_2O = HA^- + H_3O^+$ and $HA = HA^- + H^+$ have the same equilibrium constant

- The pK_a of an acid and the pK_b of its conjugate base sum to pK_W ($= 14$ at 25°C)

- The conjugate base of a strong acid is a weak base

- Since K_a is defined for a one-proton transfer, the dissociation of polyprotic acids *usually* is written as a stepwise process

- pC-pH diagrams for diprotic acids have two system points (corresponding to the two pK_a values), with the slope of the pC lines changing at the system points

- To apply the graphical shortcut to polyprotic acids, locate system points (at $pC = pA_T$ and $pH = pK_{a,i}$), draw species lines (making sure that species lines intersect below the system points), and find the equilibrium pH where the proton condition is satisfied

- The proton condition is valuable in working with the salts of polyprotic acids

- Alpha values vary from 0 to 1 and sum to 1

- Alpha values can be extended to *n*-protic acids

PROBLEMS

Note: Most of these problems can be solved by either algebraic, graphical, or computer methods. Try using more than one approach with some problems to get a feel for the advantages and disadvantages of the three approaches.

11.1 How many acids, bases, and ampholytes are there in the species formed when an *n*-protic acid is added to water? Recall that an *n*-protic acid can donate *n* protons to water.

11.2 What is the pK_a of H_3O^+? What is the pK_b of OH^-?

11.3 Show quantitatively that the conjugate base of a strong acid is a weak base.

11.4 If the pK_a of a monoprotic acid, HA, is 4.2, write the K_b equilibrium and calculate the K_b value for its conjugate base, A^-.

11.5 An industry is discharging an acetic acid waste stream. (Recall that acetic acid is CH_3COOH.) Its National Pollutant Discharge Elimination System (NPDES) permit requires pH > 4 and total organic carbon (TOC) < 250 mg/as C. What is the maximum strength of acetic acid waste (as TOC in mg/L) that can be discharged without violating either permit requirement? (Recall that TOC, in mg/L, measures the number of milligrams of *carbon* per liter.)

11.6 Besides baking soda ($NaHCO_3$) and baking powder (mainly Na_2CO_3), another leavening agent in cooking is ammonium carbonate, $(NH_4)_2CO_3$, also called baker's ammonia or hartshorn.[†]

   A. What is the proton condition for the addition of ammonium carbonate to water?

   B. Using graphical techniques, calculate the equilibrium pH and the equilibrium concentrations of all the dissolved species for a 10^{-3} M $(NH_4)_2CO_3$ solution.

11.7 Three manufacturers are producing carbonated bottled water. One company introduces carbonate by bubbling carbon dioxide into water until $C_T = 1 \times 10^{-4}$ M (equivalent to adding $H_2CO_3^*$ until $C_T = 1 \times 10^{-4}$ M). Another manufacturer adds sodium bicarbonate until $C_T = 1 \times 10^{-4}$ M, and the third company adds sodium carbonate until $C_T = 1 \times 10^{-4}$ M. Using one pC-pH diagram, estimate the pH of the bottled water from each of the three manufacturers.

11.8 Chlorendic acid [1,4,5,6,7,7-hexachlorobicyclo-(2.2.1)-hept-5-ene-2,3-dicarboxylic acid] is a highly chlorinated compound used as a flame retardant. It is a diprotic acid with $pK_{a,1} = 3.1$ and $pK_{a,2} = 4.6$. Chlorendic acid inhibits the growth of the alga *Chlorella* at pH 3.5, but not at pH 5.0 or pH 7.5 (Hendrix et al.,1983).

   A. Using graphical techniques, which form of chlorendic acid (neutral, mono-anionic, or di-anionic) is **not** responsible for the inhibitory effect?

   B. Using graphical techniques, calculate the distribution (i.e., the fraction) of chlorendic acid (in uncharged, mono-anionic and di-anionic forms) at pH 3.5, 5.0, and 7.5.

11.9 Professor Whoops attempts to make a 10^{-2} M NaH_2PO_4 solution. The good professor accidently grabs the Na_2HPO_4 bottle. What pH did Professor Whoops expect the solution to be? What was the actual pH of the solution as made? (One approach: sketch a pC-pH diagram to solve this problem. For phosphoric acid, $pK_{a,1} = 2.1$, $pK_{a,2} = 7.2$, and $pK_{a,3} = 11$.)

[†] Hartshorn originally was made by grinding the horns of harts or reindeer. Modern hartshorn is ammonium carbonate.

11.10 Amino acids have a monoprotic acid group and a monoprotic base in the same molecule. The acid group is a carboxylic acid, with the following acid dissociation reaction: $-COOH = -COO^- + H^+$. The basic group is an amine, with the following base association reaction: $-NH_2 + H^+ = -NH_3^+$. A fully protonated amino acid can be written as $HOOC-R-NH_3^+$, where R is an organic functional group. Typically, the pK_a values for the $-COOH$ group are about 2 to 3 and the pK_b values for the $-NH_2$ group are about 4 to 5.

A. In principle, an amino acid could take four forms, depending on the pH: $HOOC-R-NH_2$, $^-OOC-R-NH_2$, $HOOC-R-NH_3^+$, and $^-OOC-R-NH_3^+$. Explain why the species $HOOC-R-NH_2$ is *not* formed in significant quantities.

B. Find the pH range where each of the other three species dominates. The other three forms are $^-OOC-R-NH_2$, $HOOC-R-NH_3^+$, and overall neutral form $^-OOC-R-NH_3^+$. (The doubly-charged form, $^-OOC-R-NH_3^+$, is called a *zwitterion*, from the German *Zwitter* hermaphrodite[†]).

C. The isoelectric point (or isoelectric pH, abbreviated pI) is defined as the pH where the fraction of the acid group dissociated is equal to the fraction of basic group that is protonated. At pH = pI, the amino acid has no overall charge. Develop an expression for pI as a function of pK_a and pK_b. What is the pI of the amino acid alanine if the pK_a of the carboxylic acid group is 2.3 and the pK_a of the protonated amino group is 9.7?

11.11 It is possible to write general expressions for the alpha values of a monoprotic acid at any pH relative to the pK_a of the acid.

A. Verify that $\alpha_0 = 0.760$ and $\alpha_1 = 0.240$ at pH $= pK_a - 0.5$.

B. What are the values of α_0 and α_1 for any monoprotic acid at pH $= pK_a - 1$?

C. Show that at $\alpha_0 = 1/(1 + 10^{\pm\epsilon})$ and $\alpha_1 = 1/(1 + 10^{\mp\epsilon})$ at pH $= pK_a \pm \epsilon$.

11.12 Show that the equilibrium pH of a NaHA solution is about equal to the average of $pK_{a,1}$ and $pK_{a,2}$ if the concentrations of H^+ and OH^- can be ignored relative to other species. Here, NaHA is the sodium salt of HA^- and part of the diprotic acid system: $H_2A/HA^-/A^{2-}$. State any assumptions you made. Does this trick apply to Example 11.4?

[†] Hermaphrodite was the son of Hermes and Aphrodite. According to Greek myth (popularized in Ovid's *Metamorphosis*), Hermaphrodite was attracted to the nymph Salmacis. The two became one being, who inherited the worst characteristics of both males and females.

11.13 Biochemists often use the Henderson-Hasselbach equation to relate the equilibrium pH to the ratio of the acid and conjugate base concentrations. The Henderson-Hasselbach equation for a monoprotic acid is $pH = pK_a + \log([A^-]/[HA])$.

    A. Derive the Henderson-Hasselbach equation for a monoprotic acid.

    B. Derive the following version of the Henderson-Hasselbach equation for a diprotic acid $(H_2A/HA^-/A^{2-})$: $pH = \frac{1}{2}\{pK_{a,1} + pK_{a,2} + \log([A^{2-}]/[H_2A])\}$.

11.14 For a 1×10^{-3} M Na_2S solution ($pK_{a,1} = 7.1$ and $pK_{a,2} = 14$):

    A. Calculate the pH at equilibrium.

    B. At equilibrium, which is larger $[HS^-]$ or $[S^{2-}]$?

11.15 Find the total carbonate concentration ($= [H_2CO_3^*] + [HCO_3^-] + [CO_3^{2-}]$) of a pure Na_2CO_3 solution so that $[HCO_3^-] = [CO_3^{2-}]$ at equilibrium.

11.16 A truism in acid-base chemistry is that the pH of a solution consisting of equimolar concentrations of an acid and its conjugate base is about equal to the pK_a.

    A. Show that the pH of a solution consisting of C mole/L of the monoprotic acid HA and C mole/L of the conjugate base NaA is about equal to pK_a. What restrictions are necessary for the equilibrium pH to be about pK_a?

    B. For a diprotic acid, show that the pH of a solution consisting of C mole/L of H_2A and C mole/L of NaHA is about equal to $pK_{a,1}$.

CHAPTER 12

Titrations

12.1 INTRODUCTION

titration: the combining of acids and bases

Key idea: Many environmental problems involve titrations

You now have a good understanding of acids and bases and the determination of equilibrium concentrations in systems containing acids and bases. In this chapter, the acid-base concept will be taken one step further by investigating *mixtures* of acids and bases.

It may seem strange to devote an entire chapter to the mixing of acids and bases. However, as you shall see, many important environmental processes involve acid-base combinations. The act of combining acids and bases is called a ***titration***. The word *titration* comes from the French *titre*, meaning title or qualification, and refers to the use of titrations for the determination of the purity of gold or silver alloys. You may have been introduced to the concept of adding a base to an acid solution with a buret in the ubiquitous titration laboratory in high school and freshman chemistry courses. In an acid-base titration, small volumes of a strong acid (or base) are added to the solution of interest and monitor pH. Your experiences may have convinced you that mixing acids and bases is a "pure chemistry" concept best relegated to introductory chemistry courses.

In fact, many global and local problems in environmental engineering and science have at their heart the mingling of acids and bases in water. Three examples will show the wide applicability of combining acids and bases in the aquatic environment. The first example concerns the early development of the Earth's oceans. One prevailing theory purports that the oceans were created when the planet cooled sufficiently so that water vapor could condense into rain. The major anions in the early marine environment are thought to have come from the dissolution of volcanic gases into rainwater. The gases [CO_2, SO_2, and $HCl(g)$] combined with rainwater to form the corresponding acids [namely, H_2CO_3, H_2SO_4, and $HCl(aq)$]. (This chemistry also was discussed in the Part III case study.) The acidic rain titrated the basic materials in rock. The anions from the gases (e.g., Cl^-, SO_4^{2-}, HCO_3^-, and CO_3^{2-}) and the resulting dissolved cations (such as Ca^{2+}, Na^+, and K^+) became the mineral soup of the early oceans (Figure 12.1). In essence, the composition of the major chemical species in the early oceans came from the largest, most sustained titration in the 4 billion year history of the Earth[†].

[†] An early contributor to this theory was Lars Sillén, who gave us the p*C*-pH diagram discussed in Chapter 8. For a review, see Sillén (1967).

229

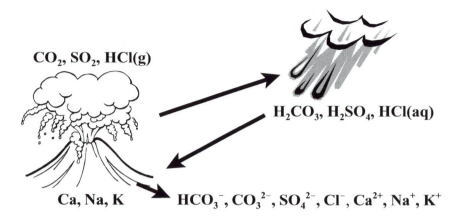

Figure 12.1: A Schematic View of the Formation of the Early Ocean by Titration of Minerals with Acids from Volcanic Gases

For the second example of acid-base mixtures, skip ahead a few billion years. In more modern times, anthropogenic contributions have increased the discharge of gases to the atmosphere, which dissolve to form acidic precipitation, acid rain or acid snow. The term *acid rain* was first coined by the Scottish chemist Robert Angus Smith (1817-1884) in 1856. Smith took note of the damage to buildings in London from acid rain produced primarily from the combustion of coal. The damage occurred as the acid rain titrated the bases (calcium and magnesium) in the building materials. This kind of defacement has been observed with revered structures throughout the world, from the U.S. Capitol Building and Westminster Abbey to the Parthenon and the Taj Mahal. Again, the equilibrium chemistry of acid rain is explored in the case study.

A third example of the intermingling of acids and bases comes from the design of treatment systems. The generation of acidic wastes is very common in industry. In fact, the strong inorganic acids (called mineral acids, e.g., HCl, HNO_3, and H_2SO_4) are among the most commonly reported chemicals released into the environment from industrial sources. In many cases, acidic waste streams are ***neutralized*** (i.e., brought to near-neutral pH) by the addition of a strong base before discharge to municipal wastewater treatment facilities[†]. The calculation of the dose of the base requires a quantitative understanding of acid-base interactions.

neutralization: to bring the pH to near-neutral pH by the addition of acids or bases

---

[†] Acidic wastes may come from liquid waste streams or gas scrubbers. In gas scrubbers, basic solutions are used to trap acidic components of off-gases. In either event, an acidic aqueous stream is generated and must be treated.

12.2 PRINCIPLES OF ACID-BASE TITRATIONS

12.2.1 Nomenclature

In an acid-base titration, an acid (or base) solution of known concentration is added to another solution of known volume. The solution added is called the ***titrant***. Usually we want to know the resulting pH when a specified volume, v, of the titrant is added to a known volume, V_0, of the original solution.

titrant: solution added during a titration

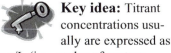

Key idea: Titrant concentrations usually are expressed as eq/L (i.e., moles of protons accepted or donated per liter)

The concentration of the titrant usually is expressed as equivalents per liter. Here, *equivalents refers to the number of moles of protons being exchanged*. For example, sodium hydroxide (NaOH) accepts one proton from water and thus has one equivalent per mole. If a 10^{-2} M NaOH solution is used as a titrant, its concentration would be written as $(10^{-2}$ mol/L$)(1$ eq/mol$) = 10^{-2}$ eq/L $= 10^{-2}$ N. A 0.2 M H_2SO_4 titrant would be expressed as $(0.2$ mol/L$)(2$ eq/mol$) = 0.4$ eq/L $= 0.4$ N since sulfuric acid donates two protons (at all but the lowest pH values in water).

In most cases, the titrant strength is assumed to be large, so the titrant volumes are small compared to the volume being titrated. Thus, dilution by the titrant usually is small (but is easily accounted for if desired).

12.2.2 Titrations and chemical equilibria

In the classic presentation of acid-base titrations, chemistry texts usually go through a standard set of combinations: titration of a strong acid with a strong base, titration of a weak monoprotic acid with a strong base, and so on. *It is not necessary to analyze all possible types of titrations if you approach the titration solution at any point in the titration as a chemical equilibrium problem.* The key to analyzing mixtures of acids and bases is to use the tools developed in Chapters 5-9.

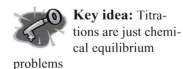

Key idea: Titrations are just chemical equilibrium problems

Imagine that you are titrating a solution containing a monoprotic acid (HA, with total concentration $= A_T$) with NaOH. What species would the system include? The system would contain H_2O, H^+, OH^-, HA, A^-, and Na^+ (since NaOH dissociates completely to Na^+ and OH^-).

Thoughtful Pause

Does the species list change during the titration?

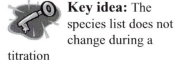

Key idea: The species list does not change during a titration

It is important to note that the species list does **not** change during the titration. The *concentrations* of the species at equilibrium change as NaOH is added. Adding NaOH changes $[Na^+]$ (since NaOH is assumed to dissociate completely). This will change the concentrations of all the other species.

You can account for the acid-base equilibria and mass balance on HA and A^- in several ways. One simple approach is to use alpha values. Thus:

$[HA] = \alpha_0 A_T$ and $[A^-] = \alpha_1 A_T$. As usual, plug these expressions into the charge balance or proton condition.

Thoughtful Pause

Is it better to use the charge balance or proton condition in this case?

Recall that the proton condition is extremely useful for eliminating the unreactive ions, such as Na^+, from consideration. In a titration, you *want* to include the sodium ion concentration since $[Na^+]$ is changing as you titrate. Thus, you are better off here with the charge balance. The charge balance is:

Key idea: Use the charge balance to solve chemical equilibrium problems involving titrations

$$[H^+] + [Na^+] = [A^-] + [OH^-] \qquad \text{eq. 12.1}$$

Recall that each term in the charge balance is in units of charges per liter or equivalents per liter, where equivalents means *the charge on the ion*. For titrations with base, replace $[Na^+]$ with a general symbol for the equivalents of base added (or, more specifically, the equivalents of cations added by titrating with a base that dissociated completely). The general symbol is C_B - the concentration of strong base cations. Why use the term *strong base*? We are assuming that the base in the titrant dissociates completely. This means that the titrant is a strong base (see Section 11.3).

C_B: the concentration of strong base cations in eq/L

Substituting the alpha values and C_B into eq. 12.1:

$$[H^+] + C_B = \alpha_1 A_T + [OH^-], \text{ or:}$$

$$C_B = \alpha_1 A_T + [OH^-] - [H^+] \qquad \text{eq. 12.2}$$

If the titrant volume (v) is expected to be large compared to the original HA volume (V_0), you can correct for dilution:

File available: The *Nanoql* chemical system file for the Na^+/acetic acid system may be found on the CD - file location **/Chapter 12/Acetic acid titration.nqs**

$$C_B\left(\frac{v}{v+V_0}\right) = \alpha_1 A_T\left(\frac{V_0}{v+V_0}\right) + [OH^-] + [H^+] \qquad \text{eq. 12.3}$$

Volume correction is needed where very accurate results are imperative or when v is greater than about 5% of V_0.

12.2.3 Titration curves

How do you use equations such as eq. 12.2 (or 12.3)? Recall from Section 12.2.1 that we want to find the pH for every value of C_B. For a monoprotic acid, this is a relatively easy task. The simple solution is to set up a spreadsheet with a column of pH (or $[H^+]$) values and a column with the formula $\alpha_1 A_T + [OH^-] - [H^+]$, or, more specifically $K_a A_T/([H^+] + K_a) + K_w/[H^+] - [H^+]$. The latter column will be the C_B values. The pair of

titration curve: a plot of pH against a measure of the amount of titrant added (usually C_B or C_A)

Animation: To run an animation showing how the equilibrium pH varies as C_B increases for the titration of a monoprotic acid, see the example in Appendix E, Section E.4

C_A: the concentration of strong acid anions in eq/L

Key idea: The same amount of strong base or strong acid added has a different effect on pH depending on how much base or acid has been added previously

---

Example 12.1: **Neutralization of an Acidic Waste**

The Sourpuss Vinegar Company produces spiced and herbed vinegar products for the domestic market. The factory discharges between 50,000 and 200,000 gallons per day (131 to 525 L/min) of a liquid waste stream that contains only acetic acid at a concentration of 80 mg/L as acetic acid. Size a metering pump to dose 50% sodium hydroxide at a rate to meet the effluent pH standard of 7.2.

Solution:
The waste stream is (0.08 g/L as CH_3COOH)/(60 g CH_3COOH per mol) = 1.33×10^{-3} M acetic acid. Performing a charge balance after NaOH addition yields eq. 12.2:

$$C_B = \alpha_1 Ac_T + [OH^-] - [H^+]$$

where $Ac_T = [HAc] + [Ac^-] = 1.33 \times 10^{-3}$ M

At pH 7.2, substitute:

columns gives pH-C_B pairs. An example spreadsheet for using this approach to calculate C_B from pH for the titration of 10^{-2} M acetic acid with strong base is shown in Figure 12.2. The C_B values are in column F. For clarity, the formulas used are shown in the column to the right of the calculated values.

	A	B	C	D	E	F	G
1	K_a	1.99526E-05					
2	A_T	0.01	M				
3							
4	pH	[H⁺]	[H⁺] formula	[A⁻]	[A⁻] formula	C_B	C_B formula
5	0	1	=10^(-1*A2)	2E-07	=$B$2*$B$1/($B$1+B5)	-0.9999998	=D5+1E-14/B5-B5
6	1	0.1	=10^(-1*A3)	1.99E-06	=$B$2*$B$1/($B$1+B6)	-0.09999801	=D6+1E-14/B6-B6
7	2	0.01	=10^(-1*A4)	1.99E-05	=$B$2*$B$1/($B$1+B7)	-0.00998009	=D7+1E-14/B7-B7
8	3	0.001	=10^(-1*A5)	0.000196	=$B$2*$B$1/($B$1+B8)	-0.00080438	=D8+1E-14/B8-B8

Figure 12.2: Example Spreadsheet for Back-Calculating C_B from pH for the Titration of 10^{-2} M Acetic Acid with Base

Alternatively, you could use a fitting routine such as *Solver* to find the equilibrium pH for each value of C_B. In this text, you are encouraged to use *Nanoql* to perform equilibrium calculations. With *Nanoql*, you can enter the acetic acid/acetate system and add Na^+ as a species. The *Nanoql* chemical system file for the Na^+/acetic acid system may be found on the CD at: **/Chapter 12/Acetic acid titration.nqs**. Now vary the concentration of Na^+ (using *Vary•Species Concentration...*) and plot pH against C_B (= $[Na^+]$).

As an example of the titration of a monoprotic acid with a strong base, consider the titration of 10^{-2} M acetic acid ($pK_a = 4.7$) with NaOH. You may wish to take a moment and use one of the approaches described in the previous paragraph to generate Figure 12.3. Figure 12.3 is called a ***titration curve***. A titration curve is a plot of pH versus C_B. Note that titration curves are log-linear plots: the y-axis (pH) is logarithmic, but the x-axis (C_B) is linear. To visualize how equilibrium pH varies with C_B for the titration of a monoprotic acid, run the animation in Appendix E, Section E.4. If you titrate with a strong acid (say, HCl), then the titration curve is a plot of pH versus C_A. Here, C_A is the concentration of anions associated with the strong acid: $[Cl^-]$ in the case of HCl as a titrant.

The titration curve contains a large amount of information. Even a cursory glance at Figure 12.3 tells you that the *same amount of NaOH has a different effect on pH depending on how much base has been added previously*. For example, adding the first 5×10^{-3} eq/L (= 5 meq/L) of base raises the pH from about 3.4 to about 4.7. The second 5 meq/L addition raises the pH to about 8.8 and the third addition causes the pH to increase to about 11.7. The pH increase for the first 0.9 meq/L added is only about one pH unit, whereas a 4.3 pH unit increase is observed for the addition of the next 0.2 meq/L aliquot.

$[H^+] = 10^{-7.2}$ M
$[OH^-] = 10^{-6.8}$ M, and:
$\alpha_1 = K_a/([H^+] + K_a)$
$= 10^{-4.7}/(10^{-7.2} + 10^{-4.7})$
$= 0.997$

Thus: $C_B = (0.997)(1.33 \times 10^{-3}$
M$) + 10^{-6.8}$ M $- 10^{-7.2}$ M $=$
1.33×10^{-3} M.

We must add enough NaOH to achieve a sodium ion concentration of 1.33×10^{-3} M. Fifty percent NaOH has a density of 1.53 kg/L. Thus, its concentration is (0.5 kg NaOH/kg solution)(1.53 kg solution/L solution) = 0.765 kg/L, or (0.765 kg/L)(1000 g/kg)/(40 g/mol) = 19.1 M.

The rate of acetic acid released is (131 L/min)(1.33×10^{-3} M) = 0.174 mole/min at the low flow rate and (525 L/min)(1.33×10^{-3} M) = 0.698 mole/min at the larger flow rate. The NaOH flow rate required is (0.174 mole/min)/(19.1 M) = 9.11 mL/min to (0.698 mole/min)/(19.1 M) = 36.5 mL/min.

Thus, the recommended flows are **9.11 to 36.5 mL/min (or 3.47 to 14.0 gpd)**.

The titration curve tells you how much base you must add to achieve a certain pH. This information also can be calculated directly from the charge balance. A use of the charge balance to determine the quantities of materials needed to neutralize an acid or base can be found in Example 12.1.

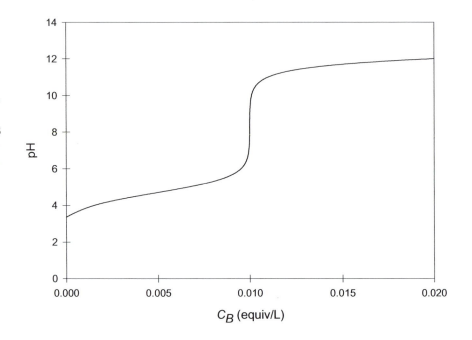

Figure 12.3: Titration Curve for the Titration of 10^{-2} M Acetic Acid with NaOH

12.3 EQUIVALENCE POINTS

12.3.1 Introduction

The approach discussed in Section 12.2 is all you need to do equilibrium calculations with acid and base mixtures. However, the brute force method discussed above is not very satisfactory. It does not give you much insight into the titration chemistry. For example, the grind-it-out approach does not help you understand why the pH is so constant at moderate and large base additions. It also does not tell you why the pH is so sensitive to the amount of base added at $C_B = 1 \times 10^{-3}$ N (see Figure 12.3). Why worry about such insights when you can just calculate the pH from eq. 12.2 or 12.3?

Key idea: Although titration curves are relatively easy to generate, an appreciation of the features of the titration curve requires chemical insight

f function: a normalized measure of the amount of base titrant added and equal to the eq/L of base added divided by the moles/L of initial material (units: eq/mol)

***Example 12.2:* Sensitivity of pH to C_B**

Based on Example 12.1, a pump with a nominal discharge of 3.5 gpd was selected. The actual discharge is 3.5 gpd ± 10%. What is the expected pH range of the treated waste at 50,000 gpd?

Solution:
The actual discharge is 3.15 to 3.85 gpd. Therefore, the actual C_B is (19.1 M)(3.15 gpd)/(50,000 gpd) = 1.20×10^{-3} M at 3.15 gpd. Similarly, C_B is 1.34×10^{-3} M and 1.47×10^{-3} M at 3.5 and 3.85 gpd, respectively. Inserting into the charge balance and solving for pH, **the actual pH values are 5.7, 9.0, and 10.1 at 3.15, 3.5, and 3.85 gpd, respectively**.

Note: to achieve a pH range of 7.15 to 7.25 (i.e., the target pH of 7.2±0.05), a discharge rate of between 3.4684 and 3.4707 gpd is required (3.4697 gpd for pH 7.2). This is a tolerance of about ±0.001 gpd, or about ±2.6 μL/min.

---

equivalence point: the point in a titration when the titrated solution has a charge balance identical with the proton condition of a solution formed using an acid (or base) or conjugate base (or conjugate acid) as a starting material

Without chemical intuition, we are doomed to make mistakes in environmental engineering and science. Take the case discussed in Example 12.1. A small error in the base addition rate may result in a significantly different pH than expected. This question is explored quantitatively in Example 12.2.

12.3.2 Equivalence points and the proton condition
Titration curves such as Figure 12.3 often are presented by normalizing the x-axis. This is accomplished by dividing the titrant concentration (in eq/L) by the concentration of the initial material being titrated (in moles/L). For the titration of acids with a base, this function is called *f*:

$$f = (\text{eq/L base added})/(\text{moles/L of initial material})$$
$$= C_B/A_T = \alpha_1 + ([OH^-] - [H^+])/A_T$$

The function *f* has units of eq/mol. You can redraw Figure 12.3 by using the *f* function, as shown in Figure 12.4.

Figure 12.4 allows for further interpretation of the titration curve. The slope of the titration curve is steepest at $f = 1$. The slope appears to be increasing at $f = 0$. At $f = 1.5$, the slope is quite small. It is also small at $f = 0.5$ and at the higher pH values ($f \geq 1.25$).

What is so special about $f = 0$ and $f = 1$? These points will be referred to, respectively, as the 0th ***equivalence point*** and the 1st equivalence point. The word *equivalence* here may (and should) lead you to ask (as always): equivalent to what? To answer this question, examine the charge balance at $f = 1$. At $f = 1$, $C_B = [Na^+] = A_T = [Ac^-] + [HAc]$. Thus, the charge balance ($[Na^+] + [H^+] = [Ac^-] + [OH^-]$) becomes: $[Ac^-] + [HAc] + [H^+] = [Ac^-] + [OH^-]$ or:

$$[HAc] + [H^+] = [OH^-] \qquad \text{eq. 12.4}$$

Thoughtful Pause
What solution has the same proton condition as eq. 12.4?

Equation 12.4 is the proton condition for a NaAc solution. In other words, a solution of acetic acid titrated to $f = 1$ is *identical to* (or *equivalent to*) an equimolar solution of sodium acetate. You *cannot tell the difference* between a 10^{-2} M acetic acid solution titrate to $f = 1$ and a 10^{-2} M sodium acetate solution (if the titrant does not dilute the solution being titrated appreciably).

What about $f = 0$? This point in the titration curve is even easier to interpret. It is obvious that an *untitrated* 10^{-2} M acetic acid solution is identical (equivalent) to a 10^{-2} M acetic acid solution[†]. Thus, the 0th and 1st equivalence points correspond to pure solutions of acetic acid and sodium acetate, respectively.

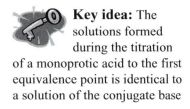

Key idea: The solutions formed during the titration of a monoprotic acid to the first equivalence point is identical to a solution of the conjugate base

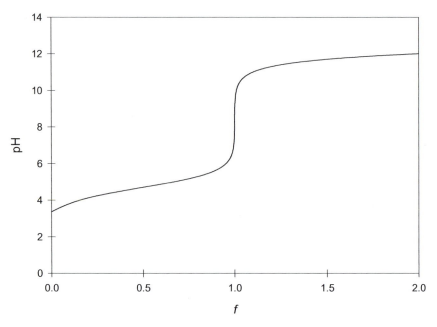

Figure 12.4: Titration Curve Using the *f* Function
(10^{-2} M acetic acid titrated with NaOH)

What about $f = \frac{1}{2}$? The point $f = \frac{1}{2}$ is called the *half-equivalence point* (or, formally, the first half-equivalence point). At $f = \frac{1}{2}$, $C_B = [Na^+] = \frac{1}{2}A_T = \frac{1}{2}([Ac^-] + [HAc])$. Thus, the charge balance (i.e., $[Na^+] + [H^+] = [Ac^-] + [OH^-]$) becomes:

$$\frac{1}{2}[Ac^-] + \frac{1}{2}[HAc] + [H^+] = [Ac^-] + [OH^-]$$

Or: $\frac{1}{2}[HAc] + [H^+] = \frac{1}{2}[Ac^-] + [OH^-]$

Now, *if [H$^+$] and [OH$^-$] are small* relative to the acetic acid and acetate concentrations, this becomes: $[HAc] \approx [Ac^-]$.

Thoughtful Pause
At what pH does $[HAc] = [Ac^-]$?

[†] Formally, at $f = 0$, $C_B = [Na^+] = 0$ and the charge balance (i.e., $[Na^+] + [H^+] = [Ac^-] + [OH^-]$) becomes $[H^+] = [Ac^-] + [OH^-]$. This is the charge balance and proton condition for an acetic acid solution.

An acid and its conjugate base are equal in concentration at pH = pK$_a$. (Recall that [HAc] = [Ac$^-$] means $\alpha_0 = \alpha_1$ or [H$^+$] = K_a or pH = pK$_a$.) Thus, f = ½ occurs at about pH = pK$_a$ as long as [H$^+$] and [OH$^-$] are small compared to the acid and conjugate base concentrations. You may wish to verify that pH $\approx$ pK$_a$ at f = ½ for the acetic acid titration curve in Figure 12.4.

Key idea: The pH of a monoprotic acid titrated to f = ½ is about equal to the pK$_a$ as long as [H$^+$] and [OH$^-$] are small compared to the acid and conjugate base concentrations

12.3.3 Shortcut method for titration curves of monoprotic acids

The information in Section 12.3.2 can be used to sketch a titration curve for any uncharged monoprotic acid (say, HA at a total concentration of C M). The pH values at f = 0, ½, and 1 can be estimated easily. The pH values at the 0th and 1st equivalence points are the equilibrium pH values of a C M HA and C M NaA solution, respectively. Recall that you can use *one* pC-pH diagram to estimated these pH values: simply apply the proton conditions for HA and NaA solutions. The proton condition for HA as the starting material is [H$^+$] = [Ac$^-$] + [OH$^-$], and the proton condition for NaA as the starting material is [H$^+$] + [HA] = [OH$^-$]. The pH at f = ½ is estimated as pK$_a$. Be sure to check graphically whether you can assume that [H$^+$] and [OH$^-$] can be ignored at this pH.

With three points estimated, the rest of the titration curve can be sketched by hand. The curve should be steepest at f = 1. The procedure can be summarized as follows (see also Appendix C, Section C.6.1):

Key idea: To sketch the titration curve for a monoprotic acid: (1) Prepare a pC-pH diagram for a C M HA solution; (2) Apply the proton conditions for an HA and NaA solution to the pC-pH diagram to find the pH at f = 0 and f = 1, respectively; (3) Locate the half equivalence point at about pH = pK$_a$ (use the pC-pH diagram to check if [H$^+$] and [OH$^-$] are much smaller than [HA] and [A$^-$]); and (4) Draw a smooth curve through the three points determined in the previous steps, making sure that the slope is steepest at f = 1

> **Step 1: pC-pH diagram**
> Prepare a pC-pH diagram for a C M HA solution
>
> **Step 2: Zeroth and first equivalence points**
> Apply the proton conditions for an HA and NaA solution to the pC-pH diagram to find the pH at f = 0 and f = 1, respectively.
>
> **Step 3: Half-equivalence point**
> The half equivalence point occurs at about pH = pK$_a$. Using the pC-pH diagram, check to make sure that [H$^+$] and [OH$^-$] are much smaller than [HA] and [A$^-$].
>
> **Step 4: Connecting the dots**
> Draw a smooth curve through the three points determined in the previous steps. Make sure the slope is steepest at f = 1.

An example will make the use of this procedure clearer. Hemoglobin (Hb) is an important protein in blood. It helps transport oxygen and serves a major role in regulating the pH and acid-base chemistry in red blood cells. Protonated hemoglobin (HHb$^+$) has a pK$_a$ of about 7.2. Since the oxygen-carrying capacity of Hb varies with pH, we wish to know how the pH in the red blood cell is affected by acid or base addition. To sketch the titration curve, first find the pH values at f = 0, 1, and ½. Blood contains about 15

***Example 12.3:* Titration of a Monoprotic Acid**

Both benzoic acid (formally, benzenecarboxylic acid, C_6H_5COOH) and the sodium salt of its conjugate base, sodium benzoate, are common preservatives in foods. A food producer wishes to transition from the manufacture of 10^{-2} M stock solutions of benzoic acid to 10^{-2} M stock solutions of sodium benzoate by adding NaOH. Can you estimate the pH of the mixture at the beginning, middle, and end of the transition? How does the pH of the stock solutions change during the transition?

Solution:
Let benzoic acid be represented as HA and benzoate as A^-, with $A_T = [HA] + [A^-] = 10^{-2}$ M. The pK_a of HA is 4.2.

A pC-pH diagram for the benzoic system is shown below:

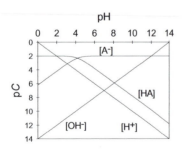

The beginning, middle, and end of the transition correspond to f = 0, ½, and 1. At $f = 0$, the mixture is equivalent to benzoic acid (proton condition: $[H^+]$ = $[A^-] + [OH^-]$), and thus the pH is 3.1. At $f = 1$, the mixture is equivalent to benzoate (proton

g of total hemoglobin per 100 mL or about 2.2×10^{-3} M since the average molecular weight of Hb is about 68,000 g/mole. Thus, total hemoglobin = $Hb_T = [HHb^+] + [Hb] = 2.2 \times 10^{-3}$ M. The pC-pH diagram for the HHb^+/Hb system is shown in Figure 12.5.

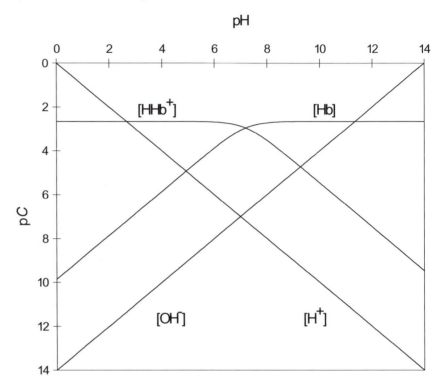

Figure 12.5: pC-pH Diagram for 2.2×10^{-3} M Hemoglobin

Imagine that a neutral salt, say $HHbCl$ ($\rightarrow HHb^+ + Cl^-$), is added to water at 2.2×10^{-3} M as a chemical model of a red blood cell and titrated with base. At $f = 0$, the untitrated solution is equivalent to a HHb^+ solution. It has the following proton condition: $[H^+] = [Hb] + [OH^-]$. Imposing this proton condition onto Figure 12.5, the equilibrium pH at $f = 0$ is about 4.9. At $f = 1$, the titrated solution is equivalent to a Hb solution (proton condition: $[H^+] + [HHb^+] = [OH^-]$), and so the pH is about 9.3. At $f = ½$, the pH is about equal to the $pK_a = 7.2$, since $[H^+]$ and $[OH^-]$ are negligible.

The pH values at $f = 0$, 1, and ½ are shown in Figure 12.6. You can sketch the titration curve by connecting the dots with a steeper slope at $f = 1$. For greater accuracy, you must develop the titration curve equation from the charge balance. At any point in the titration:

condition: $[HA] + [H^+] = [OH^-]$) and thus the pH is 8.1. At $f = \frac{1}{2}$, the pH is about equal to the pK_a if $[H^+]$ and $[OH^-]$ are small (confirmed in the pC-pH diagram), so: pH $\approx pK_a = 4.2$ at $f = \frac{1}{2}$.

At any point in the titration, the charge balance is:

$$[Na^+] + [H^+] = [A^-] + [OH^-]$$

Or:

$$\begin{aligned} C_B &= [A^-] + [OH^-] - [H^+] \\ &= \alpha_1 A_T + [OH^-] - [H^+] \end{aligned}$$

This titration curve is plotted below with the equivalence points and half-equivalence point highlighted.

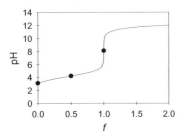

To summarize: **the pH is about 3.1, 4.2, and 8.1 at $f = 0$, $\frac{1}{2}$, and 1**.

$$[H^+] + [HHb^+] + C_B = [OH^-] + [Cl^-]$$

From mass balances on chloride and hemoglobin:

$$Hb_T = [Cl^-] = [HHb^+] + [Hb]$$

Thus, the charge balance becomes:

$$C_B = [OH^-] + [Hb] - [H^+]$$

This equation is plotted in Figure 12.6. Another illustration of the titration of a monoprotic acid is provided in Example 12.3.

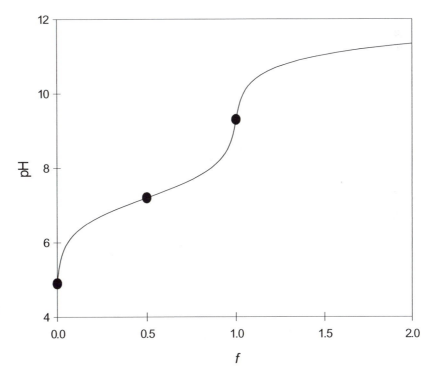

Figure 12.6: Titration Curve for 2.2×10^{-3} M Hemoglobin
(hemoglobin added as its chloride salt)

12.3.4 Titration of a base with strong acid

The discussion of the titration of an acid with base in Sections 12.3.2 and 12.3.3 can be repeated for titration of a base with acid. For a monoprotic base, NaA, we might titrate with a strong acid (such as HCl) that dissociates completely in water. A typical charge balance would be:

$$[Na^+] + [H^+] = [OH^-] + [A^-] + [Cl^-]$$

From mass balances:

$$Na_T = [Na^+] = A_T = [HA] + [A^-]$$

Substituting C_A for $[Cl^-]$ (see Section 12.2.3):

$$C_A = [HA] + [H^+] - [OH^-] = \alpha_0 A_T + [H^+] - [OH^-]$$

Once again, you can normalize the x-axis of the titration curve. For the titration of bases with an acid, the function is called g:

$$g = (\text{eq/L acid added})/(\text{moles/L of initial material})$$

As with f, the function g has units of eq/mol. As an example, you could titration a solution of ammonia with a strong acid (say, HCl). The resulting titration curve for a 10^{-3} M ammonia solution is shown in Figure 12.7.

Key idea: The titration of a monoprotic base is analogous to the titration of a monoprotic acid: the charge balance describes the titration curve

g function: a normalized measure of the amount of acid titrant added and equal to the eq/L of acid added divided by the moles/L of initial material (units: eq/mol)

Key idea: The solution formed by the titration of a monoprotic base to the first equivalence point is identical to a solution of the conjugate acid

Key idea: The titration of a monoprotic acid with base and the titration of the conjugate base of the acid with acid trace out the same titration curve

Key idea: For a monoprotic acid or base: f (titration of the acid) + g (titration of the conjugate base) = 1

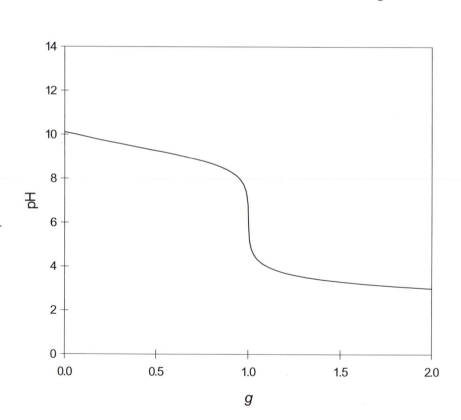

Figure 12.7: Titration Curve for the Titration of 1×10^{-3} M Ammonia with Strong Acid

***Example 12.4:* Titration of a Monoprotic Base**

A dye manufacturer uses aniline (benzenamine, $C_6H_5NH_2$) as a feedstock for synthesizing dyes. The various synthesis reactions require different pH values for optimum yield. The manufacturer wishes to know how the pH of a 10,000-gallon tank containing aniline at one-half its aqueous solubility changes as a function of the number of gallons of 40% (w/w) HCl added.

Solution:
At its heart, this is a titration problem. Let B = aniline, HB^+ = conjugate acid ($C_6H_5NH_3^+$), and $B_T = [B] + [HB^+]$.

$$B + H^+ = HB^+ \quad pK_b = 9.4$$

To find the pH values at the equivalence points, sketch the pC-pH diagram (see below). For aniline: $pK_a = 14 - pK_b = 4.6$. The aqueous solubility is 36.07 g/L = (36.07 g/L)/(93 g/mol) = 0.388 M, so B_T = (0.388 M)/2 = 0.194 M.

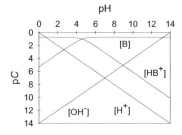

At $g = 0$ ($C_A = 0$), the mixture is equivalent to aniline (proton condition: $[H^+] + [HB^+] = [OH^-]$), and thus the pH is 8.95. At $g = 1$ ($C_A = 0.194$ eq/L), the mixture is equivalent to the conjugate acid (proton condition: $[H^+] = [B] + [OH^-]$), and thus the pH is 2.66. At $g = \frac{1}{2}$ ($C_A = $

The interpretation of the titration curve in Figure 12.7 is similar to the interpretation of titration curves with acid as the titrant. At $g = 0$ (the zeroth equivalence point), the solution is equivalent to a 10^{-3} M NH_3 solution (pH 10.1). At $g = 1$ (the first equivalence point), the solution is equivalent to a 10^{-3} M solution of the conjugate acid of ammonia, namely, ammonium (i.e., the solution is equivalent to, say, a 10^{-3} M NH_4Cl solution: equilibrium pH ≈ 6.1). At $g = \frac{1}{2}$, the concentration of base and conjugate acid are nearly equal ($[NH_3] \approx [NH_4^+]$) if the concentrations of H^+ and OH^- can be ignored. Thus, the pH at $g = \frac{1}{2}$ is about equal to the pK_a of the conjugate acid (NH_4^+, in this case, about pH 9.3). Another instance of the titration of a monoprotic base with strong acid is shown in Example 12.4.

The foregoing analysis shows the similar nature of the titration of an acid with base and the titration of the conjugate base of the acid with acid. For example, the titration of an acid, HA, with NaOH is related to the titration of NaA with HCl. You can show (see Problem 12.4) that $f + g = 1$ at any point in the titration.

An example with the titration of HOCl is shown in Figure 12.8. Using the *lower x*-axis, the titration curve (read left to right) shows the titration of HOCl with strong base. Using the *upper x*-axis, the titration curve (read right to left) shows the titration of the conjugate base with strong acid. Common salts of the conjugate base are NaOCl and $Ca(OCl)_2$. The titration curve in Figure 12.8 provides a great deal of information about the chemistry of HOCl and OCl^-.

Thoughtful Pause

Can you estimate the pK_a of HOCl from the titration curve in Figure 12.8?

At $f = g = \frac{1}{2}$, you expect the pH to be about equal to the pK_a if $[H^+]$ and $[OH^-]$ are negligible in the charge balance (here, they are small compared to $[OCl^-]$). Checking to see if $[H^+]$ and $[OH^-]$ are negligible, the pH is about 7.5 at $f = g = \frac{1}{2}$, so $[H^+] \approx 10^{-7.5}$ M, $[OH^-] \approx 10^{-6.5}$ M, and $[OCl^-] = \alpha_1 A_T \approx \frac{1}{2}(10^{-3}$ M$) = 5 \times 10^{-4}$ M. Thus, $[H^+]$ and $[OH^-]$ are negligible and the pK_a is about 7.5 (actual pK_a is 7.54).

12.3.5 Titration curves at extreme pH values

As discussed above, titration curves can be sketched over a reasonable pH range from a pC-pH diagram and knowledge of the pK_a of the acid in the system being titrated. The titration curves tend to level off at large positive or negative values of f and g. In other words, titration curves level off at extreme pH values. Why is this so? To answer this question, return once more to the charge balance. For the titration of a monoprotic acid (e.g., a C M HA solution) with a strong base, NaOH, the charge balance at any point in the titration is: $[Na^+] + [H^+] = [Ac^-] + [OH^-]$. At very high pH, the charge balance becomes: $[Na^+] \approx [OH^-]$ or $C_B \approx [OH^-]$ or $f \approx K_W/(C[H^+]) \approx$

0.097 eq/L), the pH is about equal to the pK_a if $[H^+]$ and $[OH^-]$ are small (confirmed in the pC-pH diagram), so: pH ≈ pK_a = 4.6 at g = ½.

The charge balance at any point in the titration curve is:
$[H^+] + [HB^+] = [Cl^-] + [OH^-]$
Or: $C_A = [H^+] + [HB^+] - [OH^-]$

The titration curve is shown below:

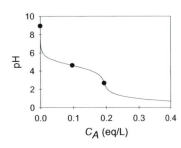

You must translate the units of C_A from eq/L to gallons of 40% HCl/10,000 gallons. The density of 40% HCl is 1.1977 g/cm³ = 1197.7 g/L. Thus, its concentration is [(1197.7 g/L)(0.40 g HCl/g 40% HCl)/(36.45 g HCl/mol)](1 eq/mol) = 13.14 N. So: 1 gal of 40% HCl per 10,000 gal represents a C_A of (13.14 N)(1 gal)/(10,000 gal) = $1.314×10^{-3}$ eq/L. To convert from C_A from eq/L to gallons of 40% HCl/10,000 gallons, divide $1.314×10^{-3}$, as shown below:

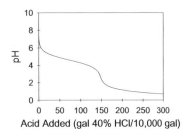

$(K_W/C)10^{pH}$. Rearranging: pH ≈ $\log(C/K_W) + \log(f)$. Go to your favorite spreadsheet program and plot the function: $y = \text{constant} + \log(x)$. You will find that this function flattens out at large values of x. A similar argument can be made to show that the titration of a base with an acid should show a fairly flat titration curve at large g values (see Example 12.5).

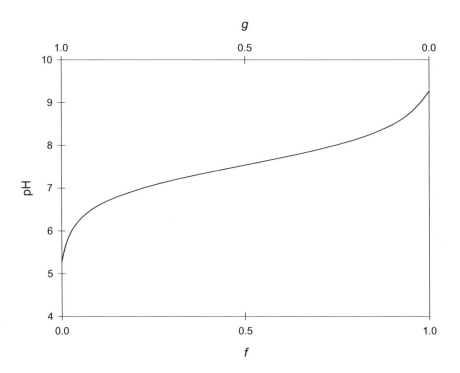

Figure 12.8: Titration Curve for the HOCl/OCl⁻ System
($[HOCl] + [OCl^-] = 1×10^{-3}$ M)

Return to the question, Why does the titration curve flatten out at extreme pH values? There are two factors contributing to this behavior. First, at extreme pH, nearly all the acid (or base) being titrated has been converted to its conjugate base (or conjugate acid) form. Thus, the pH is proportional to $\log(f)$ [or $-\log(g)$; see Example 12.5]. Second, the titration curve is a log-linear plot: a plot of $-\log\{H^+\}$ versus f (or g). These two factors combine to cause the flattening of the titration curve at extreme pH values.

One ramification of this flattening behavior is that features of the titration curve can be hidden. Consider the titration of a 10^{-3} M solution of phenol (pK_a = 9.9) with base. The titration curve is shown in Figure 12.9. Note that the titration curve is nearly featureless. The slope of the titration curve at $f = 1$ is flat, not steep.

***Example 12.5:* Titration of a Monoprotic Base at Low pH**

Why does the titration curve for the titration of a base flatten out at larger g values (low pH)?

Solution:
Consider the titration of a monoprotic base (e.g., a C M B solution, where: $B + H^+ = HB^+$) with a strong acid, say HCl. The charge balance at any point in the titration is: $[H^+] + [HB^+] = [Cl^-] + [OH^-]$. At very low pH, this becomes: $[H^+] \approx [Cl^-]$ or $C_A \approx [H^+]$ or $g \approx [H^+]/C \approx 10^{-pH}/C$. Rearranging: $pH \approx pC - \log(g)$. The function: $y = $ constant $- \log(x)$ flattens out at large values of x.

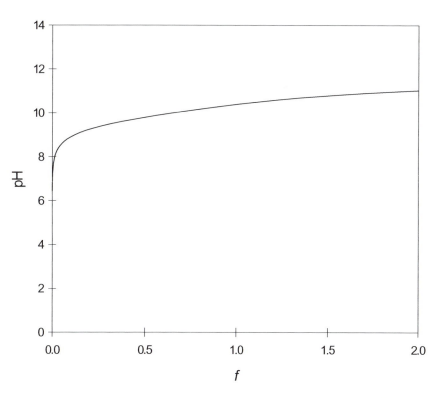

Figure 12.9: Titration Curve for the Titration of 1×10^{-3} M Phenol with Strong Base

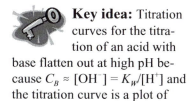 **Key idea:** Titration curves for the titration of an acid with base flatten out at high pH because $C_B \approx [OH^-] = K_W/[H^+]$ and the titration curve is a plot of pH versus C_B (or $f = C_B/C$)

 Key idea: Titration curves for the titration of a base with acid flatten out at low pH because $C_A \approx [H^+]$ and the titration curve is a plot of pH versus C_A (or $g = C_A/C$)

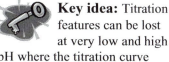 **Key idea:** Titration features can be lost at very low and high pH where the titration curve flattens out

---

Thoughtful Pause
Why is the titration curve for 10^{-3} M phenol featureless?

The phenol titration curve does not show the usual features because phenol is a very weak acid. By the point in the titration when $f = 1$, the pH is already high. By this point, $[Na^+]$ is already nearly equal to $[OH^-]$. (You may wish to confirm this point for yourself.) Thus, the flattening out of the titration curve at high pH is present in the titration curve for phenol even at lower f values.

12.4 TITRATION OF DIPROTIC ACIDS

Key idea: By using only one pC-pH diagram, you can calculate the pH at $f = 0$, 1, and 2 easily for a diprotic acid (H_2A) by imposing the proton conditions for H_2A, NaHA, and Na_2A systems

12.4.1 Introduction

The approach discussed in Section 12.3 also can be applied to diprotic acids. For the titration of a diprotic acid (say, a C M solution of H_2A), the titration curve once again is described by the charge balance equation:

$$[Na^+] + [H^+] = [HA^-] + 2[A^{2-}] + [OH^-]$$

At $f = 0$, 1, and 2, $C_B = [Na^+]$ is equal to 0, C ($= [H_2A] + [HA^-] + [A^{2-}]$), and $2C$ ($= 2[H_2A] + 2[HA^-] + 2[A^{2-}]$), respectively. You can show (Problem 12.5) that the titration solution is equivalent to H_2A, NaHA, and Na_2A solutions at $f = 0$, 1, and 2, respectively. The important equivalence points in the titration curve are summarized in Table 12.1.

Table 12.1: Equivalence Points for the Titration of a Diprotic Acid

Equiv. point	f	Proton condition	Equiv. to:
0th	0	$[H^+] = [HA^-] + 2[A^{2-}] + [OH^-]$	H_2A solution
1st	1	$[H_2A] + [H^+] = [A^{2-}] + [OH^-]$	NaHA solution
2nd	2	$2[H_2A] + [HA^-] + [H^+] = [OH^-]$	Na_2A solution

Thus, you can calculate the pH at $f = 0$, 1, and 2 easily by using *one* pC-pH diagram and imposing the proton conditions for H_2A, NaHA, and Na_2A systems.

The half-equivalence points are meaningful as well. At $f = \frac{1}{2}$, $C_B = [Na^+] = \frac{1}{2}C = \frac{1}{2}([H_2A] + [HA^-] + [A^{2-}])$, and the charge balance becomes:

$$0.5[H_2A] + [H^+] = 0.5[HA^-] + 1.5[A^{2-}] + [OH^-]$$

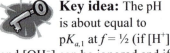

Key idea: The pH is about equal to $pK_{a,1}$ at $f = \frac{1}{2}$ (if $[H^+]$ and $[OH^-]$ can be ignored and if $[A^{2-}] \ll [HA^-]$) and the pH is about equal to $pK_{a,2}$ at $f = 3/2$ (if $[H^+]$ and $[OH^-]$ can be ignored and if $[H_2A] \ll [HA^-]$): check these assumptions carefully

If $[H^+]$ and $[OH^-]$ can be ignored and the pH is low enough that $[A^{2-}]$ is small compared to $[HA^-]$, then the charge balance becomes $[H_2A] \approx [HA^-]$, which is satisfied at about $pK_{a,1}$. Similarly, you can show that pH $\approx pK_{a,2}$ at $f = 3/2$ as long as $[H^+]$ and $[OH^-]$ can be ignored and the pH is high enough so that $[H_2A]$ is small compared to $[HA^-]$.

Be very careful about the use of these approximations at $f = \frac{1}{2}$ and $f = 3/2$ for diprotic acids. If the pK_a values of the acid are not near 7 (or the total concentration of the acid is small), then $[H^+]$ and $[OH^-]$ *cannot* be ignored. When they cannot be ignored, features in the titration curve are lost, as shown in the phenol example in Section 12.3.4. *Always check carefully the assumptions about ignoring species in titration curves.*

12.4.2 Shortcut method for the titration of diprotic acids

As a result of this analysis, a shortcut procedure for the titration curve of a diprotic acid titrated by strong base can be developed and is as follows (see also Appendix C, Section C.6.2):

Step 1: pC-pH diagram

Prepare a pC-pH diagram for a C M H$_2$A solution

Step 2: Zeroth, first, and second equivalence points

Apply the proton conditions for H$_2$A, NaHA, and Na$_2$A solutions to the pC-pH diagram to find the pH at $f = 0$, 1, and 2, respectively.

Step 3: Half-equivalence points

As a first guess, half-equivalence points are *estimated* to occur at about pH = p$K_{a,1}$ and pH = p$K_{a,2}$. Using the pC-pH diagram, check the assumptions: (1) at $f = \frac{1}{2}$, check that [H$^+$] and [OH$^-$] can be ignored and [A^{2-}] << [HA$^-$]; (2) at $f = 3/2$, check that [H$^+$] and [OH$^-$] can be ignored and [H$_2$A] << [HA$^-$].

Step 4: Connecting the dots

Draw a smooth curve through the three points determined in the previous steps. Make sure the slope is steepest at $f = 1$.

 Key idea: To sketch the titration curve for a diprotic acid (1) Prepare a pC-pH diagram for a C M H$_2$A solution; (2) Apply the proton conditions for H$_2$A, NaHA, and Na$_2$A solutions to the pC-pH diagram to find the pH at $f = 0$, 1, and 2, respectively; (3) Estimate the half-equivalence points at about pH = p$K_{a,1}$ (use the pC-pH diagram to check if [H$^+$] and [OH$^-$] can be ignored and [A^{2-}] << [HA$^-$]) and p$K_{a,2}$ (use the pC-pH diagram to check if [H$^+$] and [OH$^-$] can be ignored and [H$_2$A] << [HA$^-$]); and (4) Draw a smooth curve through the three points determined in the previous steps, making sure that the slope is steepest at $f = 1$.

As an example of the titration of a diprotic acid, consider the carbonate system (H$_2$CO$_3$*/HCO$_3^-$/CO$_3^{2-}$). The carbonate system is the dominant weak acid in natural waters. You will return to carbonate chemistry many times in this text. As you will calculate in Chapter 19, typical total carbonate concentrations in natural waters are about 2.5×10^{-3} M. The symbol C_T will be used to represent total carbonate:

$$C_T = [\text{H}_2\text{CO}_3\text{*}] + [\text{HCO}_3^-] + [\text{CO}_3^{2-}]$$

Recall that the pK_a values are about 6.3 and 10.3 for carbonic acid and bicarbonate, respectively. The pC-pH diagram is shown in Figure 12.10.

As an aside, you should develop a comfort level with determining the equivalence points in diprotic acid systems. A simple approach is to use a spreadsheet and a nonlinear solver to determine the pH at the equivalence points. A screen shot of the *Excel* spreadsheet used to find the pH of the equivalence points in Figure 12.10 is shown in Figure 12.11. The formulas are shown in column C of the spreadsheet. (The cell labeled "delta" is the denominator of the alpha value for a diprotic acid.) To find the pH of the equivalence points, use *Solver*. For the zeroth equivalence point, set f equal to 0 (in *Solver*, set target cell $B$14 equal to 0) by changing pH (in *Solver*, change cell $B$7).

Key idea: Use a spreadsheet and a nonlinear solver to find the pH values at the equivalence points

---

Example 12.6: **Titration of a Diprotic Acid**

The acid content of wines is expressed as *total acidity*. A total acidity of 0.6% means that the acids in the wine are approximately equivalent to 0.6 g tartaric acid per 100 mL (tartaric acid is 2,3-dihydroxybutanedioic acid, a diprotic acid). Typical red wines have total acidity values of about 0.6-0.7% and pH values of 3.3-3.7.

A Chardonnay has a total acidity of 0.58% and pH equal to 3.4. To what extent has the tartaric acid been titrated in the wine? Is potassium hydrogen tartarate (potassium bitartarate) a reasonable model for the acid-base chemistry of this wine?

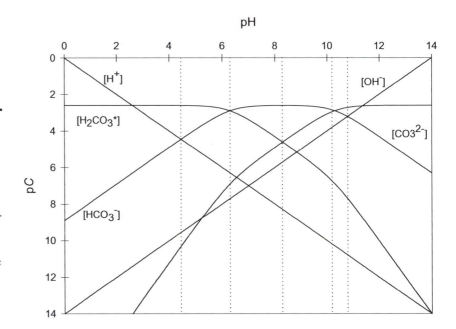

Figure 12.10: pC-pH Diagram for H$_2$CO$_3$* at $C_T = 2.5 \times 10^{-3}$ M
(dotted lines are equivalence points: from left to right, $f = 0, 0.5, 1.0, 1.5,$ and 2.0)

Solution:

Use the following abbreviations:
tartaric acid = H$_2$T
potassium hydrogen tartarate = KHT
total tartarate = [H$_2$T] + [HT$^-$] + [T^{2-}] = T_T.

The charge balance at any point in the titration (with, say, KOH) is:

$$[K^+] + [H^+] = [HT^-] + 2[T^{2-}] + [OH^-]$$

Or:
$$
\begin{aligned}
C_B &= [HT^-] + 2[T^{2-}] + [OH^-] \\
&\quad - [H^+] \\
&= (\alpha_1 + 2\alpha_2)T_T + [OH^-] \\
&\quad - [H^+] \qquad (*)
\end{aligned}
$$

	A	B	C
1	Symbol	Value	Formula
2			
3	C_T	2.50E-03	
4	p$K_{a,1}$	6.3	
5	p$K_{a,2}$	10.3	
6			
7	pH	10.78783959	
8	[H$^+$]	1.6299E-11	=10^(-1*B7)
9			
10	delta	3.3288E-17	=B8*B8+B8*10^(-1*B4)+10^(-1*B4)*10^(-1*B5)
11	[HA$^-$]	0.000613498	=B3*10^(-1*B4)*B8/B10
12	[A^{2-}]	0.001886482	=B3*10^(-1*B4)*10^(-1*B5)/B10
13			
14	f	2.00E+00	=(B11+2*B12+1E-14/B8-B8)/B3

Figure 12.11: Spreadsheet for Determining the pH of the Equivalence Points in Figure 12.10

For tartaric acid, $pK_{a,1} = 2.98$, $pK_{a,2} = 4.34$, and $T_T = (5.8$ g/L)/(150.09 g/mol) = 3.86×10^{-2} M.

Solving for C_B at pH 3.4: $C_B = 3.14 \times 10^{-2}$ eq/L. This corresponds to $f = (3.14 \times 10^{-2}$ eq/L)/(3.86×10^{-2} mol/L) = 0.81 eq/mol.

A solution of potassium hydrogen tartarate (KHT) would represent tartaric acid titrated to $f = 1$. From (*), the pH would be about 3.7 at $f = 1$.

Thus, the acid-base chemistry of this wine is similar to tartaric acid titrated to $f = 0.81$, not too different from a solution of potassium hydrogen tartarate ($f = 1$).

The titration curve comes from the charge balance: $[Na^+] + [H^+] = [HCO_3^-] + 2[CO_3^{2-}] + [OH^-]$, or:

$$C_B = [HCO_3^-] + 2[CO_3^{2-}] + [OH^-] - [H^+]$$
$$= (\alpha_1 + 2\alpha_2)C_T + [OH^-] - [H^+]$$

The titration curve is plotted in Figure 12.12 with the equivalence points highlighted. Note that the second half-equivalence point ($f = 3/2$ or $C_B = 1.5 \times 10^{-5}$ eq/L) does *not* occur at pH = $pK_{a,2} = 10.3$.

Thoughtful Pause

Why does the second half-equivalence point not occur at pH = $pK_{a,2}$?

The second half-equivalence point does not occur at pH = $pK_{a,2}$ because $[OH^-]$ is not negligible compared to $[HCO_3^-]$ or $[CO_3^{2-}]$ at this point in the titration. From the pC-pH diagram in Figure 12.10, $[OH^-]$ is larger than $[CO_3^{2-}]$ at the second half-equivalence point.

A similar procedure can be developed for the titration of the conjugate bases of a diprotic acid, for example, for starting with NaHA or Na₂A rather than H₂A. In such cases, you must include the Na⁺ formed by complete dissociation into the charge balance. Another example of the titration of a diprotic acid is shown in Example 12.6.

12.5 BUFFERS

12.5.1 Introduction

In many environmental engineering applications, acids or bases are added to water for purposes other than pH control. For example, in drinking water treatment, aluminum salts frequently are added to promote particle removal through *coagulation*. The chemistry of coagulation is pH-dependent. If you add a Al^{3+} salt such as alum, $Al_2(SO_4)_3$, the pH might drop since protons are released when Al^{3+} salts are added to water: $Al^{3+} + 3H_2O \rightarrow Al(OH)_3(s) + 3H^+$.

Thoughtful Pause

Will the pH of water change significantly after alum is added?

The answer to the *Thoughtful Pause* is: It depends. If the water behaves like the solution in Figure 12.8 at $f = g = 1$, then a small addition of acid (or base) *may* significantly change the pH. However, if the water behaves like the solution in Figure 12.8 at $f = g = \frac{1}{2}$, then a small addition of acid (or base) *may not* significantly change the pH.

buffering: the extent to which the pH changes when acids or bases are added

As this example illustrates, solutions vary in the change in pH they experience when combined with acids or bases. The extent to which the pH changes is called **buffering**. A water is said to be *well buffered* if its pH does not change significantly upon addition of acid or base. Conversely, a solution is *poorly buffered* if its pH changes substantially upon addition of acid or base.

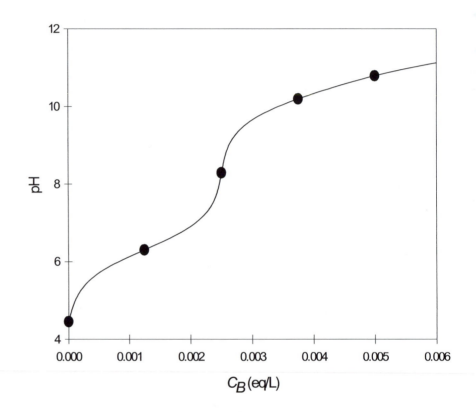

Figure 12.12: Titration Curve for the Titration of $H_2CO_3^*$ at $C_T = 2.5 \times 10^{-3}$ M with Strong Base
(filled circles are equivalence points)

12.5.2 Buffer intensity

It would be handy to develop a more quantitative measure of buffering. For inspiration, look at a titration curve, say Figure 12.8.

Thoughtful Pause

How does the pH change with respect to acid added (*f*) or base added (*g*) in Figure 12.8?

The change in pH is related to the slope: slope = dpH/df or dpH/dg. It would be perfectly reasonable to define your own measure of buffering: "[Insert your name here]'s buffering factor" = slope = dpH/df. Historically, a different measure of buffering was based on titration curves with C_A or C_B on the x-axis. The measure is called the **_buffer intensity_** (also called the buffer capacity or buffer index[†]). The buffer intensity is given the symbol β and is defined as $\beta = \partial C_B/\partial \text{pH} = -\partial C_A/\partial \text{pH}$. Partial derivatives are used because the buffer intensity is a thermodynamic property and thus dependent on temperature, pressure, and the concentration of other species. If the temperature, pressure, and the concentration of other species are about constant, then $\beta \approx dC_B/d\text{pH} = -dC_A/d\text{pH}$.

buffer intensity (β): a measure of the degree of buffering; $\beta = \partial C_B/\partial \text{pH} = -\partial C_A/\partial \text{pH}$

The buffer intensity has a number of important features. First, buffer intensity is *directly related to the degree of buffering*. If a solution requires only a small addition of base to make a large change in the pH, then you would say that the solution is poorly buffered. For such a solution, $\Delta C_B/\Delta \text{pH}$ is small so β is small. Similarly, a well-buffered solution has a large buffer intensity.

Second, the buffer intensity is related to the slope of the titration curve.

Thoughtful Pause
How is the buffer intensity related to the slope of the titration curve?

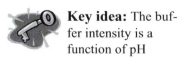 **Key idea:** The buffer intensity is equal to the reciprocal of the slope of the titration curve: a well-buffered solution has a large buffer intensity

The buffer intensity is equal to the *reciprocal* of the slope of the titration curve (or the negative of the reciprocal of the titration curve if the titration curve is plotted against C_A). Recall that the portions of the titration curve with a steep slope are poorly buffered. For such regions of the titration curve, the buffer intensity is small.

Third, the buffer intensity always is positive. Why? The pH always increases upon the addition of base (although the increase in pH is small at high pH), and so ΔC_B and ΔpH have the same sign. Similarly, the pH always *decreases* upon the addition of acid (albeit only slightly at low pH), so ΔC_A and ΔpH have *opposite* signs and $\beta \approx -\Delta C_A/\Delta \text{pH}$ is positive.

Key idea: The buffer intensity has units of eq/L per pH unit (or meq/L per pH unit)

Fourth, what are the units of buffer intensity? Both C_A and C_B have units of eq/L. The pH is usually expressed as pH units. Thus, β has units of eq/L per pH unit and is frequently written as meq/L per pH unit.

Key idea: The buffer intensity is a function of pH

Fifth, the buffer intensity changes at each point in the titration curve because the slope of the titration curve continually changes. Since the pH also changes at each point in the titration curve, *the buffer intensity is a function of pH*.

[†] The measure was developed by Van Slyke (1922), who named it the *buffer value*. In the aquatic chemistry literature, the names *buffer intensity* and *buffer capacity* are used most commonly.

12.5.3 Mathematical expressions for the buffer intensity

You can combine the titration curve equations (i.e., the charge balance) with the definition of β to develop mathematical expressions for the buffer intensity. For example, consider the titration of a monoprotic acid with a strong base. The titration curve equation is given by eq. 12.2: $C_B = \alpha_1 A_T + [OH^-] - [H^+]$. Thus:

$$\beta = dC_B/dpH = d(\alpha_1 A_T + [OH^-] - [H^+])/dpH$$

Each of these terms can be evaluated directly. Consider $d[H^+]/dpH$. Since $[H^+] \approx 10^{-pH}$, then:

$$d[H^+]/dpH \approx d(10^{-pH})/dpH$$

Recall that for any constant a: $d(a^u)/dx = \ln(a)a^u(du/dx)$. Here: $a = 10$, $u = -pH$, and $x = pH$, so:

$$d[H^+]/dpH \approx d(10^{-pH})/dpH = -\ln(10)[H^+] \approx -2.303[H^+]$$

Similarly: $d[OH^-]/dpH \approx K_w d(10^{pH})/dpH \approx +2.303 K_w/[H^+] = +2.303[OH^-]$. Finally:

$$\begin{aligned}
d(\alpha_1 A_T)/dpH &= A_T K_a d\{1/(K_a + [H^+])\}/dpH \\
&\approx 2.303 A_T K_a [H^+]/(K_a + [H^+])^2 \\
&= 2.303 A_T \alpha_0 \alpha_1 \\
&= 2.303 [HA][A^-]/A_T
\end{aligned}$$

Thus, putting the pieces together for a monoprotic acid:

$$\begin{aligned}
\beta &\approx 2.303([HA][A^-]/A_T + [OH^-] + [H^+]) & \text{eq. 12.5a} \\
&= 2.303(A_T \alpha_0 \alpha_1 + [OH^-] + [H^+]) & \text{eq. 12.5b} \\
&= 2.303\{A_T K_a [H^+]/(K_a + [H^+])^2 + K_w/[H^+] + [H^+]\} & \text{eq. 12.5c}
\end{aligned}$$

12.5.4 Regions of poor buffering

The buffer intensity for the titration of a 10^{-2} M acetic acid solution with NaOH is shown in Figure 12.13, along with its titration curve. The acetic acid solution is poorly buffered at the integral equivalence points. In other words, acetic acid solutions (0th equivalence point) and sodium acetate solutions (1st equivalence point) are poorly buffered. Why? You can determine why acetic acid and sodium acetate solutions are poorly buffered in two ways. First, use a mathematical approach. The contributions of each of the three terms in eqs. 12.5a, b, and c to the buffer intensity are broken out in Figure 12.14. Note that the minimum at the first equivalence point occurs when the first term is equal to the third term, or $[HA][A^-]/A_T \approx [H^+]$ (from eq. 12.5a). At pH $< pK_a$, $[A^-] \ll [HA]$ so $[HA] \approx A_T$. Thus,

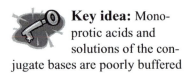

Key idea: Monoprotic acids and solutions of the conjugate bases are poorly buffered

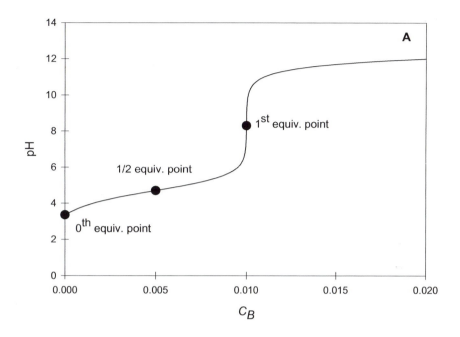

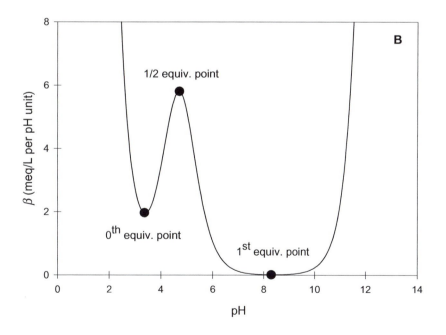

Figure 12.13: Titration Curve (A) and Buffer Intensity (B) for the Titration of 10^{-2} M Acetic Acid

$[A^-] \approx [H^+]$ at the buffer intensity minimum at low pH. Note that $[A^-] \approx [H^+]$ is the proton condition for an acetic acid solution! Thus, *a local minimum in buffer intensity is expected at the first equivalence point.* Similarly, a minimum in the buffer intensity at high pH occurs when the first and second terms are equal, that is, when $[HA] \approx [OH^-]$, which is the proton condition for a sodium acetate solution. Thus, *weak acid solutions and solutions of the conjugate bases of weak acids are poorly buffered.*

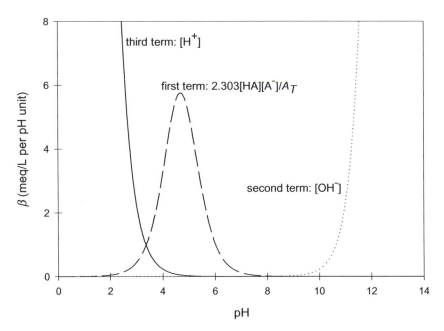

Figure 12.14: Contributions of Terms to the Buffer Intensity for the Titration of 10^{-2} M Acetic Acid

Second, you can use chemical intuition to determine why acetic acid and sodium acetate solutions are poorly buffered. Remember that the integral equivalence points are associated with certain points on the pC-pH diagram. The relationship between the pC-pH diagram and the titration curve for 10^{-2} M acetic acid is illustrated in Figure 12.15. The top part of the figure shows the pC-pH diagram for the system. The equilibrium pH values for the pure acetic acid system, pure sodium acetate system, and pH $\approx pK_a$ are indicated by circles. Note again that these points correspond to $f = 0$, 1, and ½, as shown in middle panel (the titration curve rotated 90°) and bottom panel of Figure 12.15 (the titration curve). What happens when strong acid or base is added to the system corresponding to acetic acid or sodium acetate? Note that the equilibrium pH is determined where two straight lines on the pC-pH diagram cross (H^+ and Ac^- for acetic acid and HA and OH^- for sodium acetate). For acetic acid, the charge balance after the addition of a little strong acid, say, HCl, becomes:

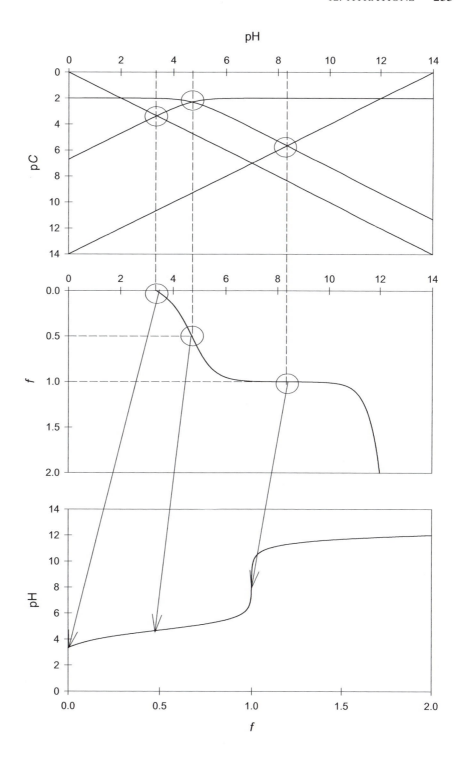

Figure 12.15: Relationship Between the pC-pH Diagram and Titration Curve for the 10^{-2} M Acetic Acid System

$$[H^+] = [Ac^-] + [Cl^-]$$

The charge balance after the addition of a little strong base, say NaOH, becomes:

$$[H^+] + [Na^+] = [Ac^-] \text{ or } [H^+] = [Ac^-] - [Na^+]$$

The addition of acid or base results in a new equilibrium pH that is determined by sliding up or down the Ac^- line. Since the slope of the line is 1, it does not take much acid or base to make a significant change in the equilibrium pH. A similar argument explains why the pH at $f = 1$ is so sensitive to the addition of strong acid or base.

12.5.5 Regions of good buffering

The titrated solution is very well buffered at extremely high and low pH values. Why? Mathematically, the $[H^+]$ or $[OH^-]$ terms dominate the buffer intensity at the extremes of pH. Chemical intuition tells you that at high pH, the titration curve flattens out as $[Na^+]$ approaches $[OH^-]$ (see Section 12.3.4). Conversely, low pH values would be obtained only with titration with acid, say HCl (see Figure 12.16). Under such conditions, the titration curve flattens out at very low pH as $[Cl^-]$ approaches $[H^+]$.

More interesting is the relatively high buffer intensity at the half-equivalence point. In other words, the solution appears to be fairly well buffered at pH near the pK_a of 4.7. Why? From eq. 12.5b, you can see that the first term dominates. Since $\alpha_1 = 1 - \alpha_0$, the buffer intensity is about $2.303 A_T \alpha_0 (1 - \alpha_0)$. The function $x(1 - x)$ has a maximum of 0.25 at $x = 0.5$ if x is limited to between 0 and 1. Thus, the buffer intensity has a local maximum at $\alpha_0 = 0.5$ or pH $\approx pK_a$. Thus, *weak acids show maximum buffering (other than at the pH extremes) at pH $\approx pK_a$*, and the maximum β is about $(2.303)(0.25)A_T = 0.576A_T$.

You also can use chemical intuition to justify why weak acid solutions are so well-buffered at a pH near the pK_a. At the pK_a, $[HA] = [A^-]$. Thus, the solution has the ability to accept added H^+ (because of the relatively high $[A^-]$) and the ability to accept OH^- (because of the relatively high $[HA]$). As a result, the solution is well buffered at pH near the pK_a. In the bottom panel of Figure 12.15, the slopes of the lines determining the pH (namely, HAc and Ac^-) are much less than one at pH $= pK_a$. Thus, from Section 12.5.4, a change in the charge balance from the addition of strong acid or base does not affect the pH significantly.

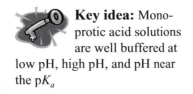 **Key idea:** Monoprotic acid solutions are well buffered at low pH, high pH, and pH near the pK_a

12.5.6 pH buffers

The discussion in Sections 12.5.4 and 12.5.5 can be used to create solutions that will buffer the pH well.

Key idea: The buffer intensity of a monoprotic solution at pH near the pK_a is about 0.576 multiplied by the total acid concentration

 Key idea: You can make a reasonable buffer by titrating a weak acid to near its pK_a or by combining equimolar concentrations of an acid and its conjugate base

---

Example 12.7: **pH Buffer**

You wish to maintain a 10 L treatment reactor at pH 4. The pH of the reactor contents cannot vary by more than 0.1 pH units if 1 mL of 1 N HCl or 1 N NaOH is added. Create a buffer for the reactor by using acetic acid.

Solution:
To maximize buffering, you need to titrate an acetic acid solution with strong base to pH 4. The only question is how strong the buffer should be. The addition of 1 mL of 1 N HCl or 1 N NaOH is equivalent to adding (1 eq/L)(0.001 L)/(10 L) = 1×10^{-4} eq/L of strong acid or base. Thus, you want the buffer intensity to be greater than 1×10^{-4} eq/L per 0.1 pH units = 1×10^{-3} eq/L per pH unit (or 1 meq/L per pH unit). From eq. 12.5b:

$$\beta \approx 2.303(A_T\alpha_0\alpha_1 + [OH^-] + [H^+])$$

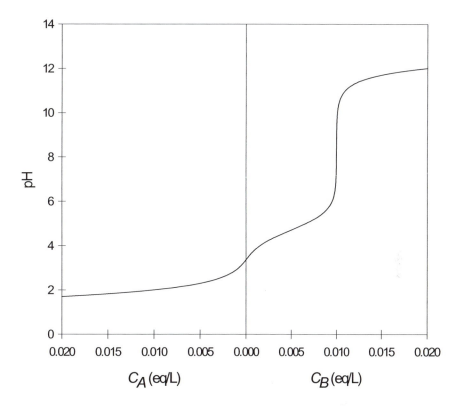

Figure 12.16: Titration Curve for the Titration of 10^{-2} M Acetic Acid with Strong Acid or Strong Base

---

Thoughtful Pause
How can you make a pH buffering solution?

You can maximize buffering at reasonable pH values by titration of a solution of a weak acid (or its conjugate base) to a pH about equal to the pK_a. Conversely, if you need to buffer the pH of something to pH 8, find a weak acid with $pK_a \approx 8$ and titrate it to pH 8. The buffer intensity will be about $0.576A_T$ (Section 12.5.5). This means that the buffer will be stronger if more weak acid is added initially.

Another common way to make a pH buffer is to combine approximately equimolar concentrations of an acid and its conjugate base. Recall from Problem 11.16 that the pH of a solution consisting of an acid and its conjugate base is about equal to the pK_a. You can tweak the solution a little, adding a little more of the acid or conjugate base, to form a good pH buffer near the pK_a.

At pH 4: $[H^+] = 1 \times 10^{-4}$ M, $[OH^-] = 1 \times 10^{-10}$ M, $\alpha_0 = 0.834$, and $\alpha_1 = 0.166$. Thus:

$$\beta = 1 \times 10^{-3} \text{ eq/L per pH unit}$$
$$= 2.303(0.139 A_T + 1 \times 10^{-4})$$

Solving: $A_T = 2.4 \times 10^{-3}$ M.

How much strong base do you need to add to create the buffer? You can calculate the base dose from the charge balance:

$$C_B = \alpha_1 A_T + [OH^-] - [H^+]$$
$$= 3.0 \times 10^{-4} \text{ N}$$

Thus, a suitable buffer would be 2.4×10^{-3} M acetic acid with 3.0×10^{-4} N strong base added.

A good example of this approach is the phosphate buffer. Phosphoric acid is a triprotic acid with pK_a values of 2.1, 7.2, and 11. The pK_a values are spaced far enough so that the only important species at neutral pH are $H_2PO_4^-$ and HPO_4^{2-}. You can make a good buffer at pH 7.0 by combining slightly more of the acidic form (e.g., KH_2PO_4) than the basic form (e.g., Na_2HPO_4). For example, a buffer used in the determination of the ultraviolet absorbance of natural waters is made by dissolving 4.08 g KH_2PO_4 and 2.84 g of Na_2HPO_4 in 1 L of water (APHA et al., 1998). What is the pH and buffer intensity of this solution? You can verify that the added concentrations are 0.03 M KH_2PO_4 and 0.02 M Na_2HPO_4. If the salts dissociate completely, then $[Na^+] = 0.04$ M, $[K^+] = 0.03$ M, and $P_T = 0.05$ M. A simplified charge balance near neutral pH is:

$$[Na^+] + [K^+] \approx [H_2PO_4^-] + 2[HPO_4^{2-}]$$

Near pH 7, the system can be approximated as a monoprotic acid with $[H_2PO_4^-] \approx P_T[H^+]/(K_a + [H^+])$ and $[HPO_4^{2-}] \approx P_T K_a/(K_a + [H^+])$, where $K_a = 10^{-7.2}$. Thus:

$$P_T[H^+]/(K_a + [H^+]) + 2P_T K_a/(K_a + [H^+]) \approx [Na^+] + [K^+] = 0.07 \text{ M}$$

Solving with $P_T = 0.05$ M, $[H^+] = 9.46 \times 10^{-8}$ M or pH 7.02. The buffer intensity (from eq. 12.5) is 27.3 meq/L per pH unit. Is this a strong buffer? The reciprocal of the buffer intensity tells us that the pH will change by only about 0.04 pH units upon the addition of 1 meq/L of strong acid or base. Another pH buffer is shown in Example 12.7.

12.6 INTERPRETATION OF TITRATION CURVES WITH COMPLEX MIXTURES

Key idea: Titrations can be used to characterize the acid-base chemistry of water samples

12.6.1 Titrations as a tool to characterize chemical systems

So far in this chapter, we have examined the titration curves of simple systems where the composition is *known*. You now should feel comfortable in *generating* a titration curve for many mixtures of acids and bases.

Titrations also can be used to probe and characterize systems where the composition is *unknown*. As a simple example, consider a waste stream from an industrial process. You measure the pH as 3.3 and need to determine how much base to add to meet a discharge permit of pH > 7.5. You could, of course, simply titrate a known volume of the waste with a base titrant of known strength and record the amount of titrant required to raise the pH to 7.5. However, by performing a titration and recording the pH at *each* addition of base, you gain much more information. For example, suppose your recorded titration curve looked like Data Set A or Data Set B in Figure 12.17.

Thoughtful Pause

If all you need to know is how much base to add to get to pH 7.5, why should you care whether the recorded data look like Data Set A or Data Set B in Figure 12.17?

As the two data sets show, buffer intensity is important, as well as the final pH. In Data Set A, the pH at pH 7.5 is very sensitive to the amount of base (or acid) added. Clearly, you would have to be very careful about monitoring pH if you wished to adjust the pH of this waste stream to pH 7.5. On the other hand, the pH at pH 7.5 is not very sensitive to the amount of base (or acid) added in Data Set B. You could be less careful about monitoring pH if you wished to adjust the pH of this waste stream to pH 7.5. Your knowledge of titration chemistry tells you that *the shape of the curve is important*, not just the amount of titrant to reach a specified pH.

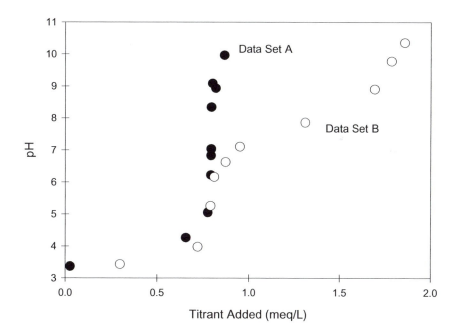

Figure 12.17: Examples of Titration Data

12.6.2 Loss of features in complex systems

The features of titration curves have been emphasized in this chapter. However, do not put too much faith in interpreting titration curves based solely on their shape. As a example, imagine that you titrated an unknown mixture and recorded the titration curve shown in Figure 12.18. You would probably agree that the titration data look like a fairly run-of-the-mill titration curve for the titration of a monoprotic acid with a strong base.

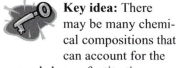 **Key idea:** There may be many chemical compositions that can account for the general shape of a titration curve

From the steep portion of the curve, you might be tempted to estimate that the monoprotic acid has a total concentration (A_T) of about 1.32×10^{-3} M. Thus, the first half-equivalence point is at about $C_B = 6.6 \times 10^{-4}$ eq/L, and the pK_a is therefore about 4.3. The line in Figure 12.18 represents the titration curve of a monoprotic acid with $A_T = 1.32 \times 10^{-3}$ M and $pK_a = 4.3$. The line seems to describe the data fairly well. Unfortunately, the original data are for the titration of a mixture of 10^{-4} M of a monoprotic acid with $pK_a = 4$ and 10^{-5} M of a monoprotic acid with $pK_a = 3.5$. Thus, there may be many chemical compositions that can account for the general shape of a titration curve.

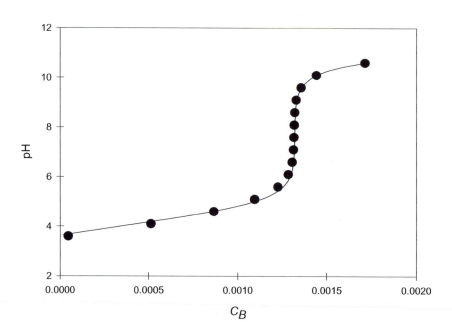

Figure 12.18: Synthetic Titration Curve Data of a Multicomponent System (line is the best-fit monoprotic acid curve)

One application in which titration curve features often are lost is in the titration of natural waters. The acid-base chemistry of natural waters is complex. Naturally occurring organics can accept and donate protons. However, the strength of the proton binding sites changes during the titration as the three-dimensional structure of the organics change. As a result, titration curves of natural waters often are fairly featureless. The rich detail of acid-base chemistry is smeared by the large number and changing nature of the organic acids and bases.

To illustrate this point, the titration curve is shown in Figure 12.19 for the titration of 1.85 g/L of natural organics (fulvic acid) isolated from the

Göta River (near Göteburg, Sweden; data from Plechanov et al., 1983). Note the relatively featureless titration curve. Plechanov and colleagues estimated that the carboxylic acid content of the natural material was 4.0 meq/g (or 7.4 meq/L for the sample titrated in Figure 12.19).

12.7 SUMMARY

An acid-base titration is a stepwise addition of acids or bases to a solution. Titrations describe a number of phenomena in the environment, from the creation of the early oceans to the neutralization of acidic waste streams. Each step in a titration is a separate equilibrium calculation. The equilibrium concentration of most of the species in the system change at each step because the charge balance changes as strong acid anions or strong base cations are added.

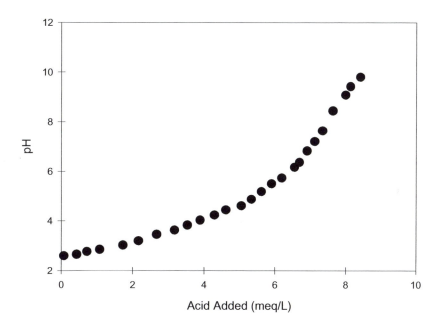

Figure 12.19: Titration Curve for a Fulvic Acid Solution
(see text for details)

Graphical representations of titrations, called *titration curves*, can be sketched easily by calculating the equilibrium pH at a few key points, called *equivalence points*. Integral equivalence points occur where the partially titrated solution is equivalent to (i.e., has the same proton condition as) a solution with simple starting materials. Half-equivalence points occur at the pK_a values if $[H^+]$ and $[OH^-]$ can be ignored. At extreme pH values, the titration curved tend to flatten out.

Titration curves give an indication of the degree of buffering of a water. Where the titration curve is flat, the solution is well buffered. Formally, the buffer intensity is the inverse of the slope of the titration curve. Very well buffered solutions (pH buffers) can be made by combining equimolar concentrations of an acid and its conjugate base. Care should be taken when interpreting the titration curves of complex mixtures, as several models (i.e., several hypothetical compositions) may show similar titration curves.

12.8 PART III CASE STUDY: ACID RAIN

In Chapter 11, you used acid-base chemistry to calculate the pH of rainwater contaminated with several anions. Suppose that a family in a rural area has filled a cistern with 15,000 L of river water for their water usage. The river water is at pH 7.5 and has a total carbonate concentration ($C_{T,river}$) of 1.2×10^{-3} M ($= 10^{-2.92}$ M). The cistern is recharged by rainfall from a 150 m^2 collection system on the roof of their house. The monotonous weather in the area produces 1 cm of rainfall each day. This means that the family can collect $(150\ m^2)(10^4\ cm^2/m^2)(1\ cm)(10^{-3}\ L/cm^3) = 1500$ L of rainwater each day. The family intends to use 1500 L of water each day, so that 15,000 L of water will remain in the cistern at the end of the day. How does the pH of the water in the cistern change over time?

To analyze this portion of the case study, assume the system is closed. In other words, ignore evaporation and exchange of carbon dioxide with the atmosphere. The addition of the rainwater to the river water is a titration. As with any titration problem, the chemical characteristics of both the titrant (rainwater) and water being titrated (river/cistern water) must be established. The water chemistry of the rainwater was elucidated in Section 11.7.

To analyze the river water chemistry, it is useful to ask whether the river water is just an H_2CO_3* solution or whether the river water already contains other strong base cations or strong acid anions. To answer this question, you can quickly estimate the pH of a 1.2×10^{-3} M H_2CO_3* solution. The charge balance is:

$$[H^+] = [HCO_3^-] + 2[CO_3^{2-}] + [OH^-] \qquad \text{eq. 12.6}$$

At lower pH, this simplifies to:

$$[H^+] \approx [HCO_3^-] = \alpha_1 C_{T,river} \approx K_{a,1} C_{T,river}/[H^+]$$

Thus: $[H^+]^2 \approx (10^{-6.3})(10^{-2.92}) = 10^{-10.22}$, so $[H^+] \approx 10^{-5.11}$ or about pH 5.1. Since the pH of the river water is much higher than 5.1, the river water must contain some strong base cations to maintain the charge balance in eq. 12.6. Use the symbol $C_{B,river}$ to represent the concentration of strong base cations in the river in eq/L. A charge balance for the river water is:

$$[H^+] + C_{B,river} = [HCO_3^-] + 2[CO_3^{2-}] + [OH^-]$$

Thus:

$$C_{B,river} = [HCO_3^-] + [CO_3^{2-}] + [OH^-] - [H^+] \qquad \text{eq. 12.7}$$

At pH 7.5 and $C_{T,river} = 1.2\times10^{-3}$ M, the righthand side of eq. 12.7 can be evaluated. You can show that $C_{B,river} = 1.13\times10^{-3}$ eq/L.

The rainwater and river water are adequately characterized. The rainwater has a pH of 4.0; a total carbonate concentration, $C_{T,rain}$, of 8.3×10^{-6} M; and a total strong acid anion concentration, $C_{A,rain}$, of 9.5×10^{-5} eq/L (see Section 11.7). The river water has a pH of 7.5; a total carbonate concentration, $C_{T,river}$, of 8.3×10^{-6} M; and a total strong base cation concentration, $C_{B,river}$, of 1.13×10^{-3} eq/L. When a volume v (= 1500 L) of rainwater is mixed with a volume V (= 15,000 L) of water in the cistern, the resulting concentrations are:

$$C_T = (vC_{T,rain} + VC_{T,cistern})/(v + V)$$
$$C_A = (vC_{A,rain} + VC_{A,cistern})/(v + V), \text{ and:}$$
$$C_B = (vC_{B,rain} + VC_{B,cistern})/(v + V)$$

where the subscript "cistern" represents the water in the cistern prior to the rainfall that day. (For day 1, the water in the cistern prior to the rainfall is just river water and $C_{T,cistern}$, $C_{A,cistern}$, and $C_{B,cistern}$ are equal to $C_{T,river}$, $C_{A,river}$, and $C_{B,river}$, respectively.) A charge balance on the water in the cistern after a rainfall is:

$$[H^+] + C_B = [HCO_3^-] + 2[CO_3^{2-}] + [OH^-] + C_A$$

With $[HCO_3^-] = \alpha_1 C_T$ and $[CO_3^{2-}] = \alpha_2 C_T$, the pH can be solved for from the charge balance after the rainfall each day. The results are plotted in Figure 12.20.

Note in Figure 12.20 that the pH of cistern water starts at pH 7.5 (since it contains only river water initially) and moves to pH 4.0 (as it approaches pure rainwater). Note the large daily changes in pH around day 25 as the river water has been titrated to about its first equivalence point. The pH plot is a little different from a normal titration curve because water is drained from the tank each day.

The titration analysis tools you learned about in this chapter were useful in determining the cistern pH. Another tool to analyze problems such as this one will be presented in Chapter 13.

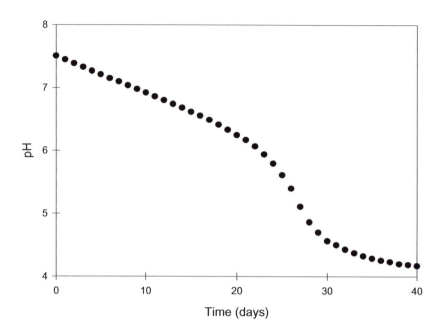

Figure 12.20: Trends in the Cistern Water pH Over Time

SUMMARY OF KEY IDEAS

- Many environmental problems involve titrations

- Titrant concentrations usually are expressed as eq/L (i.e., number of protons accepted or donated per liter)

- Titrations are just chemical equilibrium problems

- The species list does not change during a titration

- Use the charge balance to solve chemical equilibrium problems involving titrations

- The same amount of strong base or strong acid added has a different effect on pH depending on how much base or acid has been added previously

- Although titration curves are relatively easy to generate, an appreciation of the features of the titration curve requires chemical insight

- The solutions formed during the titration of a monoprotic acid to the first equivalence point is identical to a solution of the conjugate base

- The pH of a monoprotic acid titrated to $f = \frac{1}{2}$ is about equal to the pK_a as long as $[H^+]$ and $[OH^-]$ are small compared to the acid and conjugate base concentrations

- To sketch the titration curve for a monoprotic acid (1) Prepare a pC-pH diagram for a C M HA solution; (2) Apply the proton conditions for an HA and NaA solution to the pC-pH diagram to find the pH at $f = 0$ and $f = 1$, respectively; (3) Locate the half-equivalence point at about pH = pK_a (use the pC-pH diagram to check if $[H^+]$ and $[OH^-]$ are much smaller than $[HA]$ and $[A^-]$); and (4) Draw a smooth curve through the three points determined in the previous steps, making sure that the slope is steepest at $f = 1$

- The titration of a monoprotic base is analogous to the titration of a monoprotic acid: the charge balance describes the titration curve

- The solution formed by the titration of a monoprotic base to the first equivalence point is identical to a solution of the conjugate acid

- The titration of a monoprotic acid with base and the titration of the conjugate base of the acid with acid trace out the same titration curve

- For a monoprotic acid or base: f (titration of the acid) $+ g$ (titration of the conjugate base) $= 1$

- Titration curves for the titration of a base with acid flatten out at low pH because $C_A \approx [H^+]$ and the titration curve is a plot of pH versus C_A (or $g = C_A/C$)

- Titration features can be lost at very low and high pH where the titration curve flattens out

- By using only one pC-pH diagram, you can calculate the pH at $f = 0$, 1, and 2 easily for a diprotic acid (H_2A) by imposing the proton conditions for H_2A, NaHA, and Na_2A systems

- The pH is about equal to $pK_{a,1}$ at $f = \frac{1}{2}$ (if $[H^+]$ and $[OH^-]$ can be ignored and if $[A^{2-}] \ll [HA^-]$) and the pH is about equal to $pK_{a,2}$ at $f = 3/2$ (if $[H^+]$ and $[OH^-]$ can be ignored and if $[H_2A] \ll [HA^-]$): check these assumptions carefully

- To sketch the titration curve for a diprotic acid (1) Prepare a pC-pH diagram for a C M H_2A solution; (2) Apply the proton conditions for H_2A, NaHA, and Na_2A solutions to the pC-pH diagram to find the pH at $f = 0$, 1, and 2, respectively; (3) Estimate the half-equivalence points at about pH = $pK_{a,1}$ (use the pC-pH diagram to check if $[H^+]$ and $[OH^-]$

can be ignored and $[A^{2-}] \ll [HA^-]$) and $pK_{a,2}$ (use the pC-pH diagram to check if $[H^+]$ and $[OH^-]$ can be ignored and $[H_2A] \ll [HA^-]$); and (4) Draw a smooth curve through the three points determined in the previous steps, making sure that the slope is steepest at $f = 1$

- Use a spreadsheet and a nonlinear solver to find the pH values at the equivalence points

- The buffer intensity is equal to the reciprocal of the slope of the titration curve - a well-buffered solution has a large buffer intensity

- The buffer intensity has units of eq/L per pH unit (or meq/L per pH unit)

- The buffer intensity is a function of pH

- Monoprotic acids and solutions of the conjugate bases are poorly buffered

- Monoprotic acids solutions are well buffered at low pH, high pH, and pH near the pK_a

- The buffer intensity of a monoprotic solution at pH near the pK_a is about 0.576 multiplied by the total acid concentration

- You can make a reasonable buffer by titrating a weak acid to near its pK_a or by combining equimolar concentrations of an acid and its conjugate base

- Titrations can be used to characterize the acid-base chemistry of water samples

- There may be many chemical compositions that can account for the general shape of a titration curve

PROBLEMS

12.1 A monoprotic acid ($pK_a = 4.9$; total concentration $= 3.6 \times 10^{-3}$ M) is titration with strong base. Calculate and draw the titration curve.

   A. At what point in the titration curve is the solution buffered the poorest? (Do not calculate the buffer intensity. Just look at the shape of the titration curve.)

   B. At what value of C_B does $[HA] = [A^-]$?

   C. Calculate and plot the buffer intensity at each point on the titration curve.

D. How could your titration curve in part A be relabeled to generate the titration curve for the titration of a 3.6×10^{-3} M NaA solution?

12.2 Sketch the titration curves for the titration of 10^{-2} M Na_2CO_3 and 10^{-3} M Na_2CO_3 with strong acid. (Hint: use pC-pH diagrams to compute the pH values at $g = 0$, 1, and 2; that is, the zeroth, first, and second equivalence points.)

A. How do the pH values at the zeroth, first, and second equivalence points change with the total carbonate concentration?

B. Where are the regions of high buffer intensity? What species are responsible for the high buffer intensity in each of these regions?

12.3 Sketch the titration curve for 10^{-3} M H_3PO_4 when titrated with a strong base and indicate the significant points of the curve. Where are the regions of high buffer intensity? What species are responsible for the high buffer intensity in each of these regions? (Note: *sketch* means calculate the important points with the aid of a pC-pH diagram and draw the rest of the curve by hand.)

12.4 Consider two titrations: the titration of a monoprotic acid with a strong base and the titration of the salt of a monoprotic acid with a strong acid.

A. Using charge balances, show that the solutions have the same composition when $f + g = 1$.

B. For the titration of a diprotic acid (H_2A) and its salt (Na_2A), is it true that the solutions have the same composition when $f + g = 2$?

12.5 Show that the titration solution is equivalent to H_2A, NaHA, and Na_2A solutions at $f = 0$, 1, and 2, respectively, during the titration of a diprotic acid with a strong base.

12.6 When adding a strong base to an acidic waste stream, the industrial wastewater treatment technician noted that the pH did not change much when a little more base was added. At this point, he had added 2 gallons of 1 M NaOH to 750 gallons of waste. The pH was about 5.8. Estimate the pK_a and concentration of acid in the original waste (assuming it is monoprotic).

12.7 An industry discharges a waste stream that consists of a single monoprotic acid called HX. The titration curve for the titration of the waste stream with NaOH is given below. Unfortunately, the technician forgot to record the HX concentration in the waste stream, the NaOH concentration of the titrant, or amount of base added. (In the figure below, the tick marks are evenly spaced and the leftmost tick mark corresponds to zero base added.)

A. Estimate the pK_a of HX.

B. What is the total HX concentration $(= [HX] + [X^-])$ in the waste stream?

C. How many liters of 1 M NaOH are required to neutralize 100,000 L of the waste to pH 7?

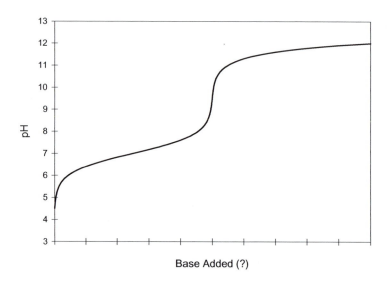

Base Added (?)

12.8 One liter of a monoprotic acid was titrated with 0.1 N NaOH. From the titration curve below, determine the pK_a of the acid and its total concentration. Explain your answer.

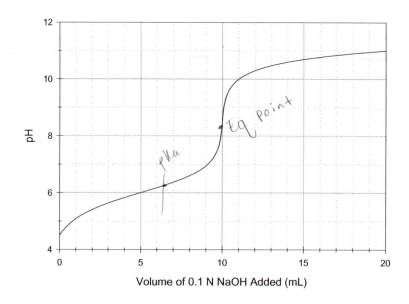

Volume of 0.1 N NaOH Added (mL)

CHAPTER 13

Alkalinity and Acidity

13.1 INTRODUCTION

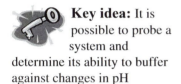 **Key idea:** It is possible to probe a system and determine its ability to buffer against changes in pH

Chapter 12 provided an overview of how to quantitatively analyze mixtures of acids and bases. In Chapter 12, you learned that titrations can be used to probe the chemistry of a system. In addition, you discovered that titration curves give important information about the buffering capacity of a water. (Remember that a water is well buffered if its pH does not change very much upon addition of an acid or base.) If we combine these two ideas, it should be possible to probe a system and determine its ability to buffer against changes in pH.

An important water quality parameter that comes from the probing of a water through titration and allows for the determination of the extent of buffering is *alkalinity*. In this chapter, the concept of alkalinity is developed. The properties and units of alkalinity will be presented. In addition, you will learn about a companion parameter called *acidity*. The practical applications of alkalinity and acidity to solve aquatic chemistry problems will be discussed.

13.2 ALKALINITY AND THE ACID NEUTRALIZING CAPACITY

13.2.1 Partially titrated systems

You know that the titration of a weak acid with a strong base generates a characteristic titration curve. The shape of the curve depends on the total acid concentration, the number of exchangeable protons on the acid, and the pK_a value(s). The weak acid partially neutralizes the basic titrant. In other words, the strong base from the titrant does not affect the solution pH as much as it would if the strong base was added to pure water. Similarly, the titration of a weak base with a strong acid generates a characteristic titration curve, the shape of which depends on the total base concentration, the number of exchangeable protons on the base, and the pK_b value(s). The weak base partially neutralizes the acidic titrant.

In this chapter, you shall explore *partially* titrated acids and bases. Imagine that a weak acid *already* has been combined with a strong base. In other words, the weak acid already has been titrated to some arbitrary point. You are at some unknown value of f that is greater than zero.

Thoughtful Pause

What happens if you now start to add strong acid to the system?

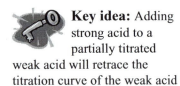 **Key idea:** Adding strong acid to a partially titrated weak acid will retrace the titration curve of the weak acid

To answer this question, recall from Section 12.3.4 that the titration of an acid with a strong base and the titration of the conjugate base of the acid with a strong acid trace out the same titration curve. Thus, adding strong acid to a partially titrated weak acid will retrace the titration curve of the weak acid with a strong base.

An example may make this important point clearer. Say that your lab partner secretly makes up a 10^{-2} M acetic acid solution and adds 3×10^{-3} N NaOH to it. She hands it to you, and you probe the system by titrating with a strong acid (say HCl). You would generate a titration curve that looks like Figure 13.1. It is clear that the original solution (at the beginning of the titration) is capable of partially neutralizing the acid added in the titration. To illustrate this point, the titration curve for the titration of pure water with strong acid is shown in Figure 13.1. Note that the pH of the unknown solution generally is larger. This indicates that the unknown solution is buffered to a greater extent that pure water.

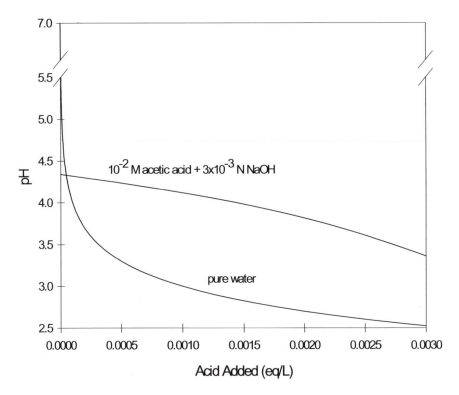

Figure 13.1: Titration Curves for a 10^{-2} M Acetic Acid + 3×10^{-3} N NaOH Solution and Pure Water

You could reorder the x-axis to obtain Figure 13.2. The titration curve for the titration of the acetic acid + base solution with strong acid (bottom x-axis in Figure 13.2; titration proceeds right to left) is the same as the titration curve for the titration of acetic acid with strong base (top x-axis in Figure 13.2; titration proceeds left to right). This makes sense: the starting solution (right side of Figure 13.2) is obtained by titrating acetic acid (left side of Figure 13.2) up to $f = 0.3$ (i.e., up to the addition of 3×10^{-3} N NaOH).

13.2.2 Base and acid neutralizing capacity

How much of the acidic titrant has been neutralized by the original solution? You can use any measure of the neutralizing ability that you want. By convention, the ***acid neutralizing capacity*** (***ANC***) of a water is defined as the equivalents per liter[†] of strong acid required to reach the zeroth equivalence point. Why pick the zeroth equivalence point? Recall from Chapter 12 that the buffer capacity is low at the zeroth equivalence point. The buffering at pH values less than the pH at $f = 0$ is not due to the original composition of the solution (see Section 12.3.5). Thus, it makes sense to stop at $f = 0$.

In a similar fashion, the ***base neutralizing capacity*** (***BNC***) of a water is defined to be the equivalents per liter of strong base required to reach the first equivalence point ($f = 1$ and $g = 0$). The ANC and BNC are very useful concepts, especially for systems partially titrated to between $f = 0$ and $f = 1$ (or, if starting with the conjugate base, partially titrated between $g = 0$ and $g = 1$). Recall from Chapter 12 that solutions are fairly well buffered between $f = 0$ and $f = 1$, with maximum buffering at $f = \frac{1}{2}$. Thus, *ANC and BNC tell you how much added acid or base could be added without significantly changing the pH of the system.*

The ANC is the amount of strong acid required to bring a solution to $f = 0$. For a monoprotic acid, ANC is the amount of acid required to make the solution equivalent to an HA solution. Thus, at the end of the titration, the solution is equivalent to an HA solution and its proton condition is:

$$[H^+] = [A^-] + [OH^-] \text{ or:}$$
$$\alpha_1 A_T + [OH^-] - [H^+] = 0$$

At the end of the titration, you are at $f = 0$. Obviously, it takes *no* acid to get from the end of the titration to $f = 0$. Thus, ANC = 0 at the end of the titration. At any other point in the titration, $\alpha_1 A_T + [OH^-] - [H^+] \neq 0$. Thus, you can think of *ANC as the amount of strong acid needed to change $\alpha_1 A_T + [OH^-] - [H^+]$ from its current value to zero.* Therefore, for a monoprotic acid:

[†] Formally, ANC has units of equivalents, not equivalents per liter. We will find it convenient to use the units of equivalents per liter in this text.

acid neutralizing capacity (***ANC***): equivalents per liter of strong acid required to reach the zeroth equivalence point of a solution ($f = 0$)

base neutralizing capacity (***BNC***): equivalents per liter of strong acid required to reach the first equivalence point of a solution ($g = 0$)

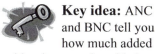

Key idea: ANC and BNC tell you how much added acid or base can be added without significantly changing the pH of the system

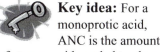

Key idea: For a monoprotic acid, ANC is the amount of strong acid needed to change $\alpha_1 A_T + [OH^-] - [H^+]$ from its current value to zero, so ANC = $\alpha_1 A_T + [OH^-] - [H^+]$

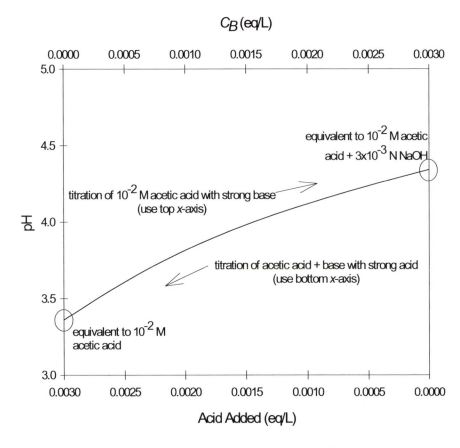

**Figure 13.2: Redrawn Titration Curves for a 10^{-2} M Acetic Acid +
3×10^{-3} N NaOH Solution**

$$\text{ANC} = \alpha_1 A_T + [\text{OH}^-] - [\text{H}^+] \qquad\qquad \text{eq. 13.1}$$

Similarly, BNC is the amount of strong base required to bring a solution to
$g = 0$. For a monoprotic base, BNC is defined by the proton condition for
the base (say, NaA):

$$\text{BNC} = [\text{HA}] + [\text{H}^+] - [\text{OH}^-] = \alpha_0 A_T + [\text{H}^+] - [\text{OH}^-] \qquad \text{eq. 13.2}$$

In this way, BNC for a monoprotic base is the amount of strong base
needed to change $\alpha_0 A_T + [\text{H}^+] - [\text{OH}^-]$ from its current value to zero.

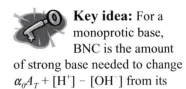 **Key idea:** For a
monoprotic base,
BNC is the amount
of strong base needed to change
$\alpha_0 A_T + [\text{H}^+] - [\text{OH}^-]$ from its
current value to zero, so BNC =
$\alpha_0 A_T + [\text{H}^+] - [\text{OH}^-]$

13.2.3 Alkalinity and acidity

In most natural waters, the main weak acids are HCO_3^- and H_2CO_3^* and the
main weak bases are HCO_3^- and CO_3^{2-}. By analogy with eq. 13.1 and
substituting the proton condition for the H_2CO_3 system:

$$\text{ANC} = [\text{HCO}_3^-] + 2[\text{CO}_3^{2-}] + [\text{OH}^-] - [\text{H}^+]$$
$$= (\alpha_1 + 2\alpha_2)C_T + [\text{OH}^-] - [\text{H}^+]$$

where $C_T = [\text{H}_2\text{CO}_3^*] + [\text{HCO}_3^-] + [\text{CO}_3^{2-}]$. The coefficient of 2 in front of α_2 stems from the fact that carbonate has two fewer protons than carbonic acid. Similarly, from the proton condition for Na_2CO_3, the BNC is given by:

$$\text{BNC} = 2[\text{H}_2\text{CO}_3^*] + [\text{HCO}_3^-] + [\text{H}^+] - [\text{OH}^-]$$
$$= (2\alpha_0 + \alpha_1)C_T + [\text{H}^+] - [\text{OH}^-]$$

*alkalinity (**Alk**)*: ANC in a water where the carbonic acid family is the dominant weak acid system and is used to define the ending point in the titration (i.e., titration to $f = 0$ for carbonic acid)

The carbonic acid system is the dominant weak acid system in most natural waters. Therefore, it makes sense to use $f = 0$ for the carbonic acid system as the ending point of the titration. For waters in which the carbonic acid system is used to define the ending point in the titration (i.e., for most natural waters), a special name is given to the ANC. The ANC is called ***alkalinity*** (often abbreviated Alk). Thus:

$$\text{Alk} = (\alpha_1 + 2\alpha_2)C_T + [\text{OH}^-] - [\text{H}^+] \qquad \text{eq. 13.3}$$

In carbonate-dominated waters, another special name is given to the BNC. It is called the ***acidity*** (often abbreviated Acy):

*acidity (**Acy**)*: BNC in a water where the bicarbonate and carbonate are the dominate weak bases and are used to define the ending point in the titration (titration to $g = 0$ for carbonate)

$$\text{Acy} = (2\alpha_0 + \alpha_1)C_T + [\text{H}^+] - [\text{OH}^-] \qquad \text{eq. 13.4}$$

You shall explore the characteristics of alkalinity and acidity further in Section 13.4. However, first the concept of alkalinity will be rederived in a different way to highlight another feature of the parameter.

13.3 ALKALINITY AND THE CHARGE BALANCE

So far in this text, you have performed charge balances on many aqueous systems where you *know* the starting materials. Can you develop a charge balance for a typical natural water where you *do not know* the starting materials? In short, no, you cannot write a complete charge balance without knowledge of the chemical species in the system. However, you can *guess* at the major ions in solution and write a fairly accurate charge balance.

Thoughtful Pause
What are the major cations and anions in natural waters?

From Section 12.1, you know that the major ions in the early oceans came from the weathering of rock caused by very, very acidic rainfall. Thus, you might guess the major ions from Figure 12.1.

The major ions in several water bodies from around the world are listed in Table 13.1. Although the cations and anions at the highest concentrations differ[†], the ions that dominate the charge balance usually are sodium, potassium, calcium, magnesium, sulfate, chloride, and bicarbonate ion. To this list, several ions usually are added that are important in aquatic chemistry, although typically much smaller in eq/L than the major ions. The minor ions are H^+, OH^-, and CO_3^{2-} (carbonate ion). Thus, a good guess at a charge balance would be:

Key idea: The major ions in most fresh waters are Na^+, K^+, Ca^{2+}, Mg^{2+}, SO_4^{2-}, Cl^-, and HCO_3^-

$$[H^+] + [Na^+] + [K^+] + 2[Ca^{2+}] + 2[Mg^{2+}] + \text{other cations} =$$
$$[OH^-] + 2[SO_4^{2-}] + [Cl^-] + [HCO_3^-] + 2[CO_3^{2-}] + \text{other anions}$$
$$\text{eq. 13.5}$$

Table 13.1: Major Ions in Selected Water Bodies[1]
(ion concentrations in meq/L)

	Great Lakes[2]	Lake Tahoe	Ganges River	Colorado River	Mississippi River	Dead Sea
$[Na^+]$	0.28	0.27	0.28	4.13	0.93	1519
$[K^+]$	0.03	0.04	0.06	0.13	0.08	193
$[Ca^{2+}]$	1.50	0.47	1.10	4.14	2.03	788
$[Mg^{2+}]$	0.59	0.21	0.40	1.98	0.93	3453
Sum of cations	2.39	0.98	1.84	10.37	3.97	5954
$[SO_4^{2-}]$	0.38	0.05	0.06	1.71	0.52	11
$[Cl^-]$	0.37	0.05	0.16	7.61	1.52	5859
$[HCO_3^-]$	1.65	0.66	1.70	2.21	2.03	4
Sum of anions	2.40	0.76	1.93	11.53	4.08	5874
TDS (mg/L)	176	64	149	694	280	309,040

Notes: 1. Source: United Nations Environment Programme's Global Environment Monitoring System Freshwater Quality Programme (UNEP GEMS/WATER)

2. Average values for the Laurentian Great Lakes (Superior, Michigan, Huron, Erie, and Ontario)

[†] For most temperate zone lakes, the order of predominance ions is $Ca^{2+} > Mg^{2+} \geq Na^+ > K^+$ and $HCO_3^- > SO_4^{2-} > Cl^-$. Examples in Table 13.1 include the Great Lakes, Lake Tahoe, the Ganges River, and the Mississippi River. For igneous and soft waters (e.g., Colorado River): $Ca^{2+} > Na^+ > Mg^{2+} > K^+$ and $Cl^- \geq SO_4^{2-} > HCO_3^-$. For saline lakes (e.g., the Dead Sea), Na^+ and Cl^- usually dominate, although the dominant anion can vary (Wetzel, 1983).

In eq. 13.5, the ion concentrations are in mol/L, and the coefficients in front of each concentration term (here, 1 or 2) have units of eq/mol = number of charges/ion. You can assign each ion in eq. 13.5 into one of three categories: (1) ions coming from the dissociation of strong acids or bases (e.g., SO_4^{2-} and Cl^- from strong acids and K^+, Ca^{2+}, and Mg^{2+} from strong bases), (2) ions coming from weak acids or bases (e.g., HCO_3^- and CO_3^{2-} from the weak acid H_2CO_3), and (3) H^+ and OH^-. As in Chapter 12, use the symbol C_A to represent the sum of the eq/L of strong acid anions and the symbol C_B to represent the sum of the eq/L of strong base cations. Equation 13.5 becomes:

$$[H^+] + C_B = [OH^-] + C_A + [HCO_3^-] + 2[CO_3^{2-}] \qquad \text{eq. 13.6}$$

Or:

$$C_B - C_A = [HCO_3^-] + 2[CO_3^{2-}] + [OH^-] - [H^+] \qquad \text{eq. 13.7}$$

Equations 13.6 and 13.7 include all "other" ions from strong bases and acids in C_B and C_A, respectively, and ignore "other" ions from weak acids and bases (but see Section 13.6). One more simplification: express the ions bicarbonate and carbonate in terms of their alpha values. Thus, the charge balance becomes:

$$C_B - C_A = (\alpha_1 + 2\alpha_2)C_T + [OH^-] - [H^+] \qquad \text{eq. 13.8}$$

In eq. 13.8, C_T again is the total dissolved carbonate concentration: $C_T = [H_2CO_3] + [HCO_3^-] + [CO_3^{2-}]$.

How does all this relate to alkalinity? Comparing eqs. 13.3 and 13.8:

alkalinity equation: Alk = C_B – C_A = $(\alpha_1 + 2\alpha_2)C_T$ + [OH⁻] – [H⁺]

$$\text{Alk} = C_B - C_A = (\alpha_1 + 2\alpha_2)C_T + [OH^-] - [H^+] \qquad \text{eq. 13.9}$$

Equation 13.9 is called the ***alkalinity equation*** . It is one of the few equations in this text that you should memorize (although you can derive it quickly from the charge balance).

13.4 CHARACTERISTICS OF ALKALINITY AND ACIDITY

13.4.1 What does alkalinity mean?

Remember that alkalinity is the acid neutralizing capacity for waters in which the acid-base chemistry is dominated by the carbonate family. Alkalinity tells you how much acid the water can accept to make it equivalent to a H_2CO_3 solution ($f = 0$). Sometimes, alkalinity is thought of as a measure of buffering: high alkalinity waters are thought to be well buffered. Although natural waters with high alkalinities usually *are* well buffered, you must be careful not to confuse alkalinity with buffer intensity.

Alkalinity is a *capacity*: it measures the total amount of acid a water can accept (to a specified equivalence point). Buffer intensity measures the *response* of a system to acid (or base) input.

As an example, consider a 1×10^{-3} N NaOH and 1×10^{-3} M NaHCO$_3$ solution. You can show (see Section 13.5.3) that both solutions have the same alkalinity. In other words, it takes the same amount of acid to make each solution equivalent to a H$_2$CO$_3$ solution containing the same C_T as the starting solution[†]. However, the pH values for the two solutions along the path from the starting solutions to $f = 0$ are *very* different (see Figure 13.3). The bicarbonate solution clearly provides better pH buffering near neutral pH.

Although we have emphasized that alkalinity is not a measure of the buffer intensity, *alkalinity still commonly is used as an indicator of buffering*. Why? Alkalinity is a great measure of buffering if the lower pH of interest is near $f = 0$ for the C_T in your system. As you shall see in Section 13.4.2, this is a pH of about 4.3-4.7 for most natural waters. Not coincidently, this lower pH of 4.3 to 4.7 corresponds to the pH at which aquatic biota begin to be adversely affected in natural water bodies.

If you are more interested in a different pH range, however, alkalinity may be misleading. Say, for example, that you have a water with pH 7.4, $C_T = 1.08\times10^{-3}$ M, and Alk = 1×10^{-3} eq/L (you should verify that these values are consistent with eq. 13.3), and you are concerned about the pH dropping below 6.8 as a result of an acidic input. Although the alkalinity is 1×10^{-3} eq/L, it takes only about 1.8×10^{-4} eq/L of strong acid (about 18% of the alkalinity) to reduce the pH to 6.8. In this case, alkalinity is not so valuable. *Use caution when accepting alkalinity as a indicator of buffering if the lower pH range of interest is different than 4.3-4.7.* Alternative types of alkalinity for different pH ranges of interest will be developed in Section 13.4.3. Another example of the relationship between alkalinity and the desired pH range is shown in Example 13.1.

13.4.2 Measuring alkalinity and acidity by titration

Alkalinity is the amount of strong acid required to make a water equivalent to an H$_2$CO$_3$ solution ($f = 0$ for H$_2$CO$_3$), and acidity is the amount of strong base required to make water equivalent to an Na$_2$CO$_3$ solution ($g = 0$ for Na$_2$CO$_3$). This means that alkalinity and acidity are intimately tied to the H$_2$CO$_3$ titration curve. It also means that titration with strong acid is an excellent way to measure alkalinity. Similarly, titration with strong base is used to measure acidity.

Alkalinity is measured by titrating the water sample to the zeroth equivalence point for H$_2$CO$_3$ *for that sample*.

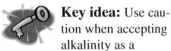
Key idea: Alkalinity commonly is used as an indicator of buffering, even though it is not a measure of buffer intensity

Key idea: Use caution when accepting alkalinity as a indicator of buffering if the lower pH range of interest is different than 4.3-4.

***Example 13.1:* Alkalinity and the pH Range of Interest**

A river water has $C_T = 7.8\times10^{-4}$ M and pH 7.9. If the trout population in the river is to be maintained, the river pH cannot drop below 7.5. If the trout are relocated, the river pH can drop to as low as 5.8. Explain the value of alkalinity in determining the vulnerability of the river pH.

[†] Note that the C_T values of the starting solutions are different: 1×10^{-3} N NaOH has $C_T = 0$, and 1×10^{-3} M NaHCO$_3$ has $C_T = 1\times10^{-3}$ M.

Solution:

First, calculate the alkalinity. At pH 7.9, $\alpha_1 = 0.971$ and $\alpha_2 = 0.00487$. From eq. 13.3, Alk = 7.66×10^{-4} eq/L. After an acidic input, the charge balance is:

$$[H^+] = (\alpha_1 + 2\alpha_2)C_T + [OH^-] + C_A$$

To get to pH 7.5 requires $C_A = 3 \times 10^{-5}$ eq/L, or 4% of the alkalinity. To get to pH 5.8 requires $C_A = 5.8 \times 10^{-4}$ eq/L, or 76% of the alkalinity.

Thus, alkalinity is a reasonable estimate of the acidic input required to reach pH 5.8 but greatly overestimates the acidic input required to get to pH 7.5.

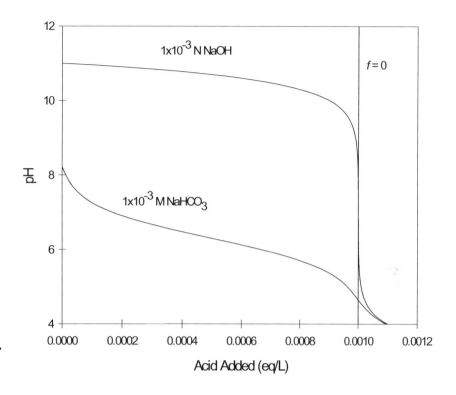

Figure 13.3: Comparison of the pH During Titration for 1×10^{-3} N NaOH and 1×10^{-3} M NaHCO$_3$ Solutions

---

Thoughtful Pause

How do you know what the zeroth equivalence point for H$_2$CO$_3$ is for a given water sample?

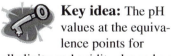

Key idea: The pH values at the equivalence points for alkalinity and acidity depend on the total carbonate concentration (C_T) of the sample

The pH of the H$_2$CO$_3$ equivalence point depends on C_T: waters with higher C_T have a lower pH at $f = 0$ (more weak acid present at $f = 0$). Similarly, waters with higher C_T have a higher pH at $g = 0$ (more weak base present at $g = 0$). This is shown graphically in Figures 13.4 and 13.5. In Figure 13.4, note that for natural waters ($C_T \approx 1$ mM), the pH at $f = 0$ is about 4.7 and the pH at $g = 0$ is about 10.5.

As shown in Figures 13.4 and 13.5, the exact pH at the $f = 0$ and $g = 0$ equivalence points varies with C_T. To be precise, you would have to measure C_T with every water sample, calculate the pH at the equivalence points, and titrate to those pH values. However, because the range of C_T values in most natural waters is small, we use one set of equivalence point

endpoint: a key point in a titration, usually selected to be an estimate of a true equivalence point in a titration

pH values for most waters. For routine analysis, alkalinity is determined by titration to pH 4.5. Since this pH value does not necessary correspond to exactly $f = 0$, we refer to it as an *endpoint*, rather than an equivalence point. In general, the endpoint pH depends on the alkalinity (see Problem 13.4).

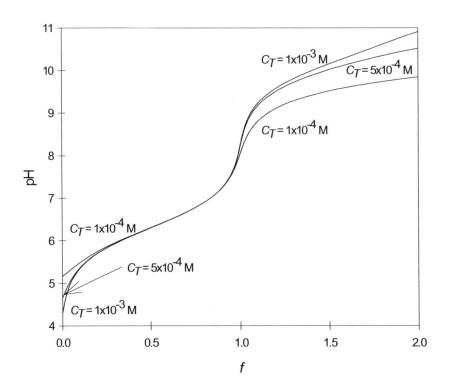

Figure 13.4: Carbonic Acid Titration Curve for Different C_T Values

methyl orange endpoint: a pH endpoint of about pH 4.5 to 3.2, formally used to estimate the alkalinity equivalence point ($f = 0$ for carbonic acid)

[acidic form (red)]

=

[basic form (orange)]

Endpoints generally are set in three ways: fixed pH (e.g., pH 4.5 as an indicator of $f = 0$), inflection points (where the *entire* titration curve is generated in the laboratory, and the strong acid or base required to reach an inflection point is recorded), and pH indicators. A pH indicator is an acid-base pair where the acid (or conjugate base) is one color in solution and the conjugate base (or acid) is another color or colorless in solution. Only a small amount of the indicator is added to the sample to avoid changing the sample pH. If the pK_a of the acid is near the pH of the endpoint, then a color change is observed at pH values near the endpoint. In the older literature, the endpoint approximating the $f = 0$ equivalence point is called the ***methyl orange endpoint*** because the pH indicator methyl orange (see the structure in the left margin) was used to visualize the endpoint. The acidic form of methyl orange has a pK_a of 3.8, with a transition in color (coming from higher pH) between pH 4.5 to 3.2 (Butler, 1998). The preferred pH indicator currently is bromcresol green (or the bromcresol green-methyl red mixed indicator).

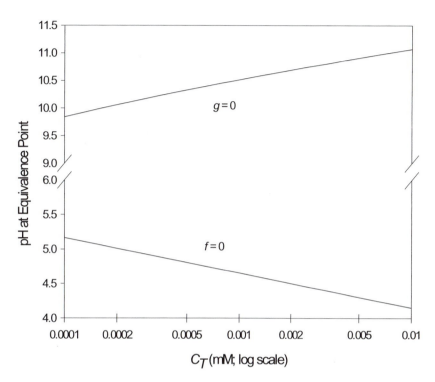

Figure 13.5: pH Values at the $f = 0$ and $g = 0$ Equivalence Points as a Function of C_T

13.4.3 Types of alkalinity

As discussed in Section 13.4.1, alkalinity is not always the best measure of buffering, especially if you are more interested in the ability to neutralize acids in the neutral pH range and are not interested in letting the pH drop to about 4.5. As a result, you can define several types of alkalinity, depending on the sample and pH range of interest (see also Sawyer and McCarty, 1978). Each type of alkalinity is related to the titration to a specific equivalence point (or, in practice, a specific endpoint). Each type will be discussed below and illustrated with example titration curves for three hypothetical samples. The samples all have $C_T = 1$ mM, where the equivalence points at $f = 0$, 1, and 2 are about pH 4.7, 8.2, and 10.5, respectively. The samples have different pH and alkalinity values: sample A has a pH value of 7.0 (between the zeroth and first equivalence points), sample B has a pH value of 9.5 (between the first and second equivalence points), and sample C has a pH value of 10.8 (greater than the second equivalence point).

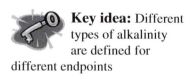 **Key idea:** Different types of alkalinity are defined for different endpoints

Alkalinity (also total alkalinity, or Alk). Alkalinity is defined as the amount of strong acid required to bring the sample to the $f = 0$ (H_2CO_3) equivalence point, an endpoint of about pH 4.5. The hypothetical titration curve for a sample at pH 7.0 is shown in Figure 13.6, along with the alkalinity. Alkalinity describes the total acid neutralizing capacity of a water, within the pH range of biological interest. It can be calculated from the proton condition for H_2CO_3: Alk $= (\alpha_1 + 2\alpha_2)C_T + [OH^-] - [H^+]$. Near neutral pH, the main contributor to alkalinity is HCO_3^-.

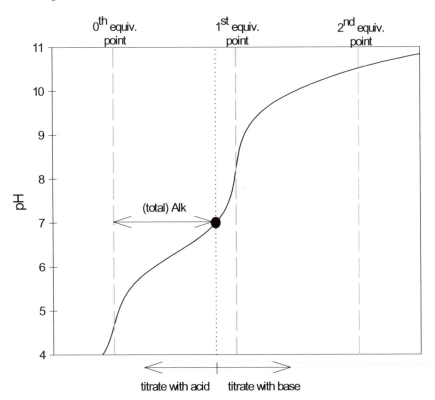

Figure 13.6: Example Alkalinity Titration for a Sample Initially at pH 7.0

phenolphthalein alkalinity: the amount of strong acid required to bring the sample to the $f = 1$ (NaHCO$_3$) equivalence point (an endpoint of about pH 8.2-8.3)

Phenolphthalein alkalinity (also *p*-alkalinity, or *p*-Alk). *Phenolphthalein alkalinity* is defined as the amount of strong acid required to bring the sample to the $f = 1$ (NaHCO$_3$) equivalence point, an endpoint of about pH 8.3. (Note from Figure 13.4 that the $f = 1$ equivalence point is fairly independent of C_T.) If the pH of the sample is less than 8.3, then the phenolphthalein alkalinity is less than zero, but usually reported as zero (i.e., no acid is needed to reach pH 8.3). The hypothetical titration curve for a sample at pH 9.5 is shown in Figure 13.7, along with the phenolphthalein alkalinity. Phenolphthalein alkalinity describes the acid neutralizing capacity of a water near neutral pH. Phenolphthalein[†] alkalinity can be calculated from the proton condition for $NaHCO_3$: p-Alk $= (\alpha_2 - \alpha_0)C_T + [OH^-] - [H^+]$. Except at very high pH, the main contributor to p-Alk is CO_3^{2-}.

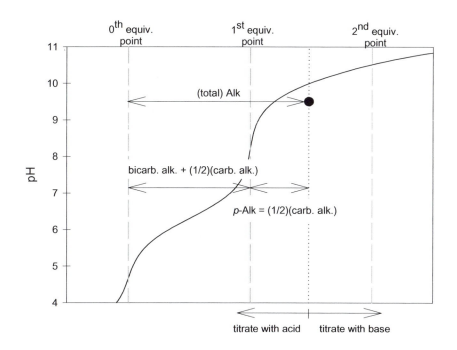

Figure 13.7: Example Alkalinity Titration for a Sample Initially at pH 9.5

Caustic alkalinity (also OH-Alk, or hydroxide alkalinity). *Caustic* (from the Greek *kaiein* to burn) **alkalinity** is defined as the amount of strong acid required to bring the sample to the $f = 2$ (Na_2CO_3) equivalence point, an endpoint of about pH 9.5-11. If the pH of the sample is less than

caustic alkalinity: the amount of strong acid required to bring the sample to the $f = 2$ (Na_2CO_3) equivalence point

---

[†] The name stems from the pH indicator phenolphthalein, which is commonly used to indicate the endpoint. However, the name "phenolphthalein alkalinity" is used regardless of the manner in which the endpoint is determined. For example, the endpoint could be determined by pH, inflection point, or another pH indicator (e.g., metacresol purple). The word *phenolphthalein* comes from *phenol* (i.e., hydroxy-benzene) + *phthal* (from naphtha) + *ein* (from -*ene*, meaning a double bond). The word *naphtha* comes from the Persian *neft*, perhaps related to the Greek *nephos*, meaning a cloud or mist.

the endpoint pH, then the caustic alkalinity is less than zero (but usually reported as zero). The hypothetical titration curve for a sample at pH 10.8 is shown in Figure 13.8, along with the caustic alkalinity. Caustic alkalinity can be calculated from the proton condition for Na_2CO_3: OH-Alk $= [OH^-] - (\alpha_1 + 2\alpha_0)C_T - [H^+]$. The main contributor to caustic alkalinity is OH^-.

Carbonate alkalinity. As long as the pH of the sample is less than about 10.5, the total alkalinity is primarily from HCO_3^- and CO_3^{2-}. The total alkalinity can be divided into two parts: *carbonate alkalinity* (the portion of the total alkalinity contributed by CO_3^{2-}) and bicarbonate alkalinity (the portion of the total alkalinity contributed by HCO_3^-). After a sample has been titrated to the phenolphthalein endpoint, almost all of the carbonate has been converted to bicarbonate (since pH $8.3 < pK_{a,2} = 10.3$), but almost none of the bicarbonate has been converted to carbonic acid (since pH $8.3 > pK_{a,1} = 6.3$). Thus, only one-half of the alkalinity from carbonate has been titrated at the pH 8.3 endpoint. This means that the carbonate alkalinity is about twice the *p*-Alk[†]. The relationship between the titration curve and the carbonate alkalinity is shown in Figures 13.7 and 13.8.

Bicarbonate alkalinity. The *bicarbonate alkalinity* is the difference between the total alkalinity and the carbonate alkalinity. The amount of strong acid required to move from the phenolphthalein endpoint to the pH 4.5 endpoint is the bicarbonate alkalinity plus the remaining one-half of the carbonate alkalinity. Thus:

$$
\begin{aligned}
\text{bicarbonate alkalinity} &= \text{total alkalinity} - \text{carbonate alkalinity} \\
&= \text{total alkalinity} - 2(p\text{-Alk}) + \text{OH-Alk} \\
&\approx \text{total alkalinity} - 2(p\text{-Alk})
\end{aligned}
$$

carbonate alkalinity: the portion of the total alkalinity contributed by CO_3^{2-}

bicarbonate alkalinity: the difference between the total alkalinity and the carbonate alkalinity

---

***Example 13.2:* Alkalinity Types from Titration Curves**

A 250 mL lake water sample was titrated with 0.1 N HCl, and the titration curve shown below was obtained. Find the alkalinity, *p*-Alk, OH-Alk, carbonate alkalinity, and bicarbonate alkalinity.

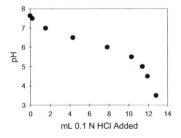

mL 0.1 N HCl Added

The relationship between the titration curve and the carbonate alkalinity is shown in Figures 13.7 and 13.8. Other illustrations of the different types of alkalinity are presented in Example 13.2.

13.4.4 Types of acidity

Different types of acidity also can be defined, depending on the endpoint. The three types of acidity are total acidity, CO_2 acidity, and mineral acidity.

Acidity (also total acidity, or Acy). Acidity is analogous to alkalinity. Acidity is defined as the amount of strong base required to bring the sample to the $g = 0$ (Na_2CO_3) equivalence point. Acidity describes the total base neutralizing capacity of a water, within the pH range of biological interest. It can be calculated from the proton condition for Na_2CO_3: Alk $= [H^+] - (2\alpha_0 + \alpha_1)C_T - [OH^-]$.

[†] Formally, you must correct for OH-Alk when calculating the carbonate and bicarbonate alkalinity. Thus: carbonate alkalinity $= 2(p\text{-Alk} - \text{OH-Alk})$, and bicarbonate alkalinity $= \text{total alkalinity} - 2(p\text{-Alk}) + \text{OH-Alk}$.

Solution:
The initial pH of the sample is well below the caustic alkalinity endpoint (pH > 9.5), so OH-Alk = 0. The initial pH also is below the p-Alk endpoint (about pH 8.3 - note that the inflection point around the $NaHCO_3$ equivalence point is not fully developed). Thus, p-Alk = 0. Since the carbonate alkalinity is 2(p-Alk), then the carbonate alkalinity = 0.

The total alkalinity is the acid added to reach about pH 4.5 or about 12 mL. The Alk is (12 mL)(1 N HCl)/(250 mL) = 4.8×10^{-3} eq/L or 240 mg/L as $CaCO_3$ (see Section 13.4.5 for alkalinity units). The bicarbonate alkalinity is equal to the total Alk since OH-Alk and carbonate alkalinity are zero.

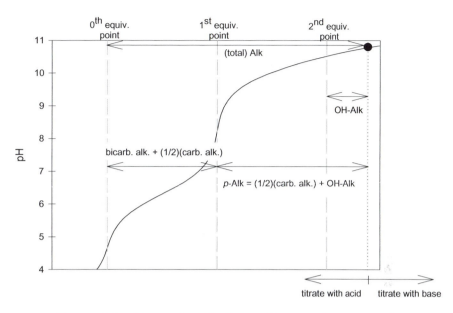

Figure 13.8: Example Alkalinity Titration for a Sample Initially at pH 10.8

To summarize:

Alk = 4.8×10^{-3} eq/L
     (240 mg/L as $CaCO_3$)
p-Alk = 0
OH-Alk = 0
carbonate alkalinity = 0
bicarbonate alkalinity =
     4.8×10^{-3} eq/L (240 mg/L
     as $CaCO_3$)

CO_2-Acy: the amount of strong acid required to bring the sample to the g = 1 ($NaHCO_3$) equivalence point

*mineral acidity (**H-Acy**)*: the amount of strong base required to bring the sample to the g = 2 (H_2CO_3) equivalence point

CO_2 acidity (also CO_2-Acy). *CO_2-Acy* is analogous to p-Alk. It is defined as the amount of strong base required to bring the sample to the g = 1 ($NaHCO_3$) equivalence point, an endpoint of about pH 8.3. CO_2 acidity describes the base neutralizing capacity of a water near neutral pH. It can be calculated from the proton condition for $NaHCO_3$: CO_2-Acy = $[H^+] + (\alpha_0 - \alpha_2)C_T - [OH^-]$.

Mineral acidity (also H-acidity, or H-Acy). *Mineral acidity* is analogous to caustic alkalinity. Mineral acidity is defined as the amount of strong base required to bring the sample to the g = 2 (i.e., H_2CO_3) equivalence point, an endpoint of about pH 4.5. For all but the most acidic waters, the mineral acidity is less than zero. The main contributor to H-Acy is H^+ (i.e., strong acids). Mineral acidity can be calculated from the proton condition for H_2CO_3: H-Acy = $[H^+] - (\alpha_1 + 2\alpha_2)C_T - [OH^-]$.

13.4.5 Units of alkalinity
One common set of units for alkalinity, from eqs. 13.3 and 13.9, is equivalents per liter (or meq/L). Clearly, from the alkalinity equation (eq. 3.9), alkalinity will be calculated in units of eq/L if C_B and C_A are in units of eq/L.

Thoughtful Pause

What are the units of alkalinity if calculated from $Alk = (\alpha_1 + 2\alpha_2)C_T + [OH^-] - [H^+]$?

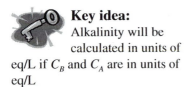

Key idea:
Alkalinity will be calculated in units of eq/L if C_B and C_A are in units of eq/L

Recall that eq. 13.9 was derived from a charge balance. Therefore, each term has units of eq/L = moles of charges/L. You could rewrite eqs. 13.3[†] and 13.9 as:

$$Alk = [(1 \text{ eq/mol})\alpha_1 + (2 \text{ eq/mol})\alpha_2]C_T + (1 \text{ eq/mol})[OH^-] - (1 \text{ eq/mol})[H^+]$$

The units of eq/mol on the coefficients usually are not written out but cannot be ignored. Thus, using either eq. 13.3 or 13.9, you can calculate the alkalinity in units of eq/L.

There is another common set of units for alkalinity based on the mass as nomenclature (see Section 2.4.2). *Alkalinity commonly is expressed in units of mg/L as $CaCO_3$.* Recall that this means we use the molecular weight of $CaCO_3$ for the molecular weight of alkalinity. Why use these units? As you shall see in Chapter 19, calcium carbonate (in the form of the mineral calcite) is a common source of alkalinity in natural waters. Also, calcium carbonate is a convenient choice for alkalinity units because the molecular weight of $CaCO_3$ is about 100 g/mol.

How do you convert units of eq/L to units of mg/L as $CaCO_3$? As calcium carbonate dissolves, it produces Ca^{2+}, which has 2 eq of charge/ion. Thus: 1 eq/L Alk = (1 eq/L)(100 g $CaCO_3$/mol $CaCO_3$)/(2 eq/mol $CaCO_3$) = 50 g/L Alk as $CaCO_3$. To summarize:

1 eq/L Alk = 50 g/L Alk as $CaCO_3$ = 50,000 mg/L Alk as $CaCO_3$
1 meq/L Alk = 50 mg/L Alk as $CaCO_3$

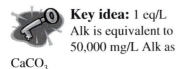

Key idea: 1 eq/L Alk is equivalent to 50,000 mg/L Alk as $CaCO_3$

As examples of units conversion, 2.1×10^{-3} eq/L Alk is equivalent to:

$$2.1 \times 10^{-3} \text{ eq/L} = (2.1 \times 10^{-3} \text{ eq/L})(50,000 \text{ mg/L as } CaCO_3 \text{ per eq/L})$$
$$= 105 \text{ mg/L as } CaCO_3$$

[†] It may not be immediately apparent that alkalinity will be in units of eq/L if calculated from eq. 13.3. Recall that eq. 13.3 was derived from a proton condition. Thus, the units of the coefficients in eq. 13.3 are "excess or deficient protons over the zero proton level". Since the zero proton level is set here for uncharged species (H_2CO_3 and H_2O), x excess protons and x deficient protons correspond to x eq of charge per mole. Thus, alkalinity will be in units of eq/L if calculated from eq. 13.3.

An alkalinity of 175 mg/L as $CaCO_3$ corresponds to:

$$175 \text{ mg/L as } CaCO_3 = (175 \text{ mg/L as } CaCO_3)/(50 \text{ mg/L as } CaCO_3 \text{ per meq/L})$$
$$= 3.5 \text{ meq/L}$$

You should practice converting the units of alkalinity between eq/L (or meq/L) and mg/L as $CaCO_3$ until the conversion becomes second nature to you.

13.5 USING THE DEFINITIONS OF ALKALINITY TO SOLVE PROBLEMS

13.5.1 Alkalinity definitions

Key idea: The alkalinity equation is really three equations in one:
$$Alk = (\alpha_1 + 2\alpha_2)C_T + [OH^-] - [H^+]$$
$$C_B - C_A = (\alpha_1 + 2\alpha_2)C_T + [OH^-] - [H^+]$$
$$Alk = C_B - C_A$$

It is very important to see that eq. 13.9 is really three equations in one. First, eq. 13.9 tells you that:

$$Alk = (\alpha_1 + 2\alpha_2)C_T + [OH^-] - [H^+]$$

This is identical to eq. 13.3. Second, eq. 13.9 tells you that:

$$C_B - C_A = (\alpha_1 + 2\alpha_2)C_T + [OH^-] - [H^+]$$

as in eq. 13.8. Finally, eq. 13.9 tells you that:

$$Alk = C_B - C_A$$

Key idea: If you know two of Alk, pH, and C_T, then you can calculate the other parameter easily

This is a new expression and a very powerful way to calculate alkalinity. *The secret to solving alkalinity problems is to select the most appropriate definition from among the three equations in eq. 13.9.* A few examples will illustrate that much effort can be saved by selecting the proper definition.

Example 13.3: **Alkalinity Calculations: Alk = $(\alpha_1 + 2\alpha_2)C_T$ + [OH⁻] – [H⁺]**

Complete the water quality calculations for the following three water samples:

Sample #1: Alk = ?, C_T = 2.1 mM, pH 6.8
Sample #2: Alk = 157 mg/L as $CaCO_3$, C_T = ?, pH 7.2
Sample #3: Alk = 1.5 meq/L, C_T = 20 mg/L as C, pH ?

13.5.2 Alkalinity as $(\alpha_1 + 2\alpha_2)C_T$ + [OH⁻] – [H⁺]

According to this definition, alkalinity is related to two other parameters: pH and C_T. If you know two out of the three parameters (i.e., two of Alk, pH, and C_T), then you can calculate the other one easily. As an example, consider a water with $C_T = 2 \times 10^{-3}$ M and pH 7.5. For $K_{a,1} = 10^{-6.3}$ and $K_{a,2} = 10^{-10.3}$, you can calculate that the alkalinity is 1.9×10^{-3} eq/L or 94 mg/L as $CaCO_3$. What is C_T if the alkalinity of a water is 140 mg/L as $CaCO_3$ (2.8 meq/L) and the pH is 7.8? You can show that:

$$C_T = (Alk - [OH^-] + [H^+])/(\alpha_1 + 2\alpha_2)$$

or $C_T = 2.9 \times 10^{-3}$ M. The more challenging calculation is to find pH by knowing C_T and alkalinity. You can set up a spreadsheet and determine the pH by iteration or use a nonlinear solver. For example, for a water with C_T

Solution:

Sample #1:

At pH 6.8, $\alpha_1 = 0.760$ and $\alpha_2 = 3.02 \times 10^{-4}$, so:

Alk = $(0.760 + 2 \times 3.02 \times 10^{-4})$
$\times (2.1 \times 10^{-3}) + 10^{-7.2} - 10^{-6.8}$
= **1.60×10⁻³ eq/L or 80 mg/L as CaCO₃**

Sample #2:

Alk = (157 mg/L as CaCO₃)/(50 mg as CaCO₃ per meq)
= 3.14×10^{-3} eq/L

Rearranging:

$C_T = ($Alk $- [OH^-] + [H^+])/(\alpha_1 + 2\alpha_2)$
= $(3.14 \times 10^{-3} - 10^{-6.8} + 10^{-7.2})/(0.887 + 2 \times 8.87 \times 10^{-4})$
= **3.53×10⁻³ M**

Sample #3:

Alk = 1.5 meq/L and:
$C_T = (20$ mg/L as C)/(12 mg per mmol)
= 1.67×10^{-3} M

Iterating: **pH 7.24**.

= 1.5×10^{-3} M and Alk = 1.2×10^{-3} eq/L, you should be able to confirm that the pH is about 6.9. Other examples are worked out in Example 13.3.

The expression for Alk in the title of this section emphasizes that Alk is a function of pH and C_T. You could create a three-dimensional plot showing the relationship among Alk, pH, and C_T, but such plots are hard to read. A common approach is to plot lines of equal pH for pairs of Alk and C_T. This was introduced by Kenneth Deffeyes (Deffeyes, 1965), so these plots sometimes are called *Deffeyes diagrams*. An example is shown in Figure 13.9.

Deffeyes diagrams are easy to generate since a plot of Alk versus C_T clearly has a slope of $\alpha_1 + 2\alpha_2$ and an intercept of $[OH^-] - [H^+]$. Note that this intercept is nearly zero, except at extreme pH values (see also Figure 13.9). Another approach is to plot lines of equal Alk for pairs of C_T and pH. This plot is shown in Figure 13.10. You may wish to verify the Alk, pH, and C_T examples in the previous paragraph, using both Figures 13.9 and 13.10.

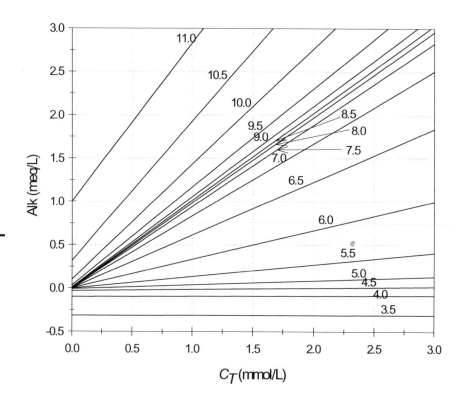

Figure 13.9: Deffeyes Diagram for Alkalinity
(lines are constant pH)

A final lesson from Alk = $(\alpha_1 + 2\alpha_2)C_T + [OH^-] - [H^+]$ concerns the alkalinity near neutral pH. For reasonable values of C_T and pH conditions near neutral pH, $[HCO_3^-]$ is much greater than $[CO_3^{2-}]$, $[OH^-]$, and $[H^+]$.

In other words: $\alpha_1 C_T \gg \alpha_2 C_T$, [OH⁻], and [H⁺]. Thus, for reasonable values of C_T *and* near neutral pH:

$$\text{Alk} \approx [HCO_3^-] = \alpha_1 C_T$$

(for reasonable values of C_T *and* near neutral pH)

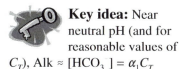

Key idea: Near neutral pH (and for reasonable values of C_T), Alk ≈ [HCO₃⁻] = $\alpha_1 C_T$

This approximation is extremely useful. It allows you to estimate the alkalinity of many natural waters with ease. For example, from Table 13.1, you may estimated that the average alkalinity in the Great Lakes is about [HCO₃⁻] = 1.65 meq/L = 83 mg/L as CaCO₃. The approximation also allows you to estimate C_T. For example, the C_T of a water with pH 7.7 and alkalinity = 125 mg/L as CaCO₃ (= 2.5 meq/L) is about Alk/α_1 = (2.5×10⁻³ eq/L)/(0.959 eq/mol) = 2.6×10⁻³ M. The exact solution for C_T is 2.59×10⁻³ M. C_T values of about 2.6×10⁻³ M also can be estimated from both Figures 13.9 and 13.10.

Example 13.4: **Alkalinity Consumption**

One way to remove ammonium from water is through the process of breakpoint chlorination. The overall reaction is: $2NH_4^+ + 3Cl_2 \rightarrow N_2(g) + 8H^+ + 6Cl^-$. How much alkalinity must be added per mg of NH_4^+-N removed to maintain the pH during breakpoint chlorination?

Solution:
Breakpoint chlorination generates 4 eq of acidity (H⁺) per mole of ammonium consumed or (4 meq Acy/mmol NH_4^+)/(14 mg N per mmol NH_4^+) = 0.286 meq Acy/mg N.

Thus, you need 0.286 meq of Alk per mg of NH_4^+-N removed (about 11 mg Alk as CaCO₃ per mg of NH_4^+-N removed) to maintain the pH.

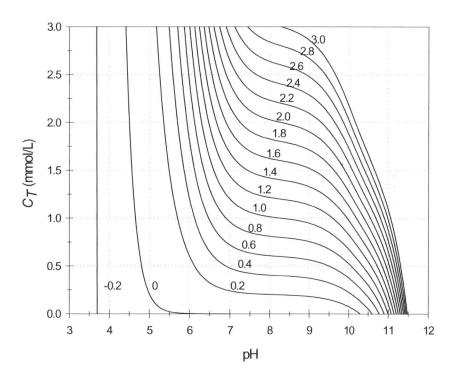

Figure 13.10: Another Approach to Showing the Relationship Among Alkalinity, pH, and C_T
(lines are constant Alk in meq/L)

13.5.3 Alkalinity as $C_B - C_A$
This form of the alkalinity definition is very valuable when individual ion concentrations are known. As an example with a pure solution, what is the

alkalinity of a 1×10^{-3} N NaOH solution? You could calculate the alkalinity the *hard* way:

1. Find the pH
   The charge balance, $[Na^+] + [H^+] = [OH^-]$, and K_W expression allow you to calculate pH 11

2. Find C_T
   $C_T = 0$ from starting materials

3. Use eq. 13.3 to calculate the alkalinity
   Alk $= 1 \times 10^{-3}$ eq/L (from eq. 13.3 or Figures 13.9 or 13.10)

You can calculate the alkalinity much more easily in this case:

$$\text{Alk} = C_B - C_A = [Na^+] = 1 \times 10^{-3} \text{ eq/L}$$

Again, selecting the most appropriate definition of alkalinity saves a lot of computational time and effort. As another example, find the alkalinity of a 1×10^{-3} M $NaHCO_3$ solution. Hard way: find the pH (equilibrium calculation yields pH 8.3), find C_T ($C_T = 1 \times 10^{-3}$ M from starting materials), and use eq. 13.3 (or Figure 13.9 or Figure 13.10) to calculate the alkalinity (Alk $= 1 \times 10^{-3}$ eq/L). Easy way: Alk $= C_B - C_A = [Na^+] = 1 \times 10^{-3}$ eq/L.

The definition of alkalinity as $C_B - C_A$ also is useful with natural waters. From Table 13.1, you could estimate the average alkalinity in the Great Lakes as $C_B - C_A \approx [Na^+] + [K^+] + [Mg^{2+}]^\dagger + [Ca^{2+}] - [SO_4^{2-}]^\dagger - [Cl^-] = 1.64$ meq/L, close to the 1.65 meq/L value estimated in Section 13.4.2.

Another use of the $C_B - C_A$ definition of alkalinity is in calculating the *consumption* of alkalinity from natural and engineered processes. When H^+ is released from a process (i.e., when acidity is released), we say that alkalinity is consumed. Why is alkalinity consumed? The addition of strong acid moves you closer to the H_2CO_3 equivalence point. After the acid is added, the alkalinity is smaller than before the acid is added. In a true sense, alkalinity is consumed.

For example, a commonly used rule of thumb in environmental engineering is that adding alum to raw drinking water for turbidity removal consumes about 0.5 mg of alkalinity as $CaCO_3$ for every mg of alum added.

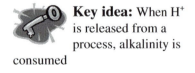 **Key idea:** When H^+ is released from a process, alkalinity is consumed

---

† This equation is written for the data in Table 13.1, where the ion concentrations are in meq/L. If the ion concentrations were in mol/L, the correct equation would be $C_B - C_A \approx [Na^+] + [K^+] + 2[Mg^{2+}] + 2[Ca^{2+}] - 2[SO_4^{2-}] - [Cl^-]$.

Where does this rule of thumb come from? Alum dissolves initially in water to produce Al^{3+} and sulfate: $Al_2(SO_4)_3 \cdot 18H_2O \rightarrow 2Al^{3+} + 3SO_4^{2-} + 18H_2O$. The aquo aluminum ion, Al^{3+}, hydrolyzes to produce aluminum hydroxide solid (see Chapter 19): $Al^{3+} + 3H_2O \rightarrow Al(OH)_3(s) + 3H^+$. Overall, 1 mole of alum (666 g) produces six equivalents of acidity and thus consumes six equivalents of alkalinity, or (6 eq)(50 g Alk as $CaCO_3$ per eq) = 300 g Alk as $CaCO_3$. This means (300 g Alk as $CaCO_3$)/(666 g alum) = 0.45 ≈ 0.5 mg of alkalinity as $CaCO_3$ are consumed per mg alum. Another illustration of alkalinity consumption is given in Example 13.4.

13.5.4 Alkalinity and acidity as conservative properties

As discussed in this chapter, alkalinity can be expressed as $C_B - C_A$. Since strong base cations and strong acid anions are relatively unreactive in water, the alkalinities are additive when two water samples of equal volume are combined. There is an important exception to this rule: we are assuming that no solids precipitate. If solids precipitate, the situation becomes a bit more complex (see Chapter 19).   If two streams with flows Q_1 and Q_2 and alkalinities Alk_1 and Alk_2 combine, the resulting stream has a flow of $Q_1 + Q_2$ (assuming the waters have the same density) and alkalinity = $Alk_1 + Alk_2$ (assuming no solids precipitate).

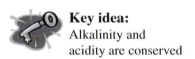

Key idea:
Alkalinity and acidity are conserved

The conservation of alkalinity (and acidity) aids in the solution of water chemistry problems. For example, you can find the alkalinity C_T, and pH of the river formed when two tributaries with the following characteristics combine:

Tributary A: flow = 1500 cfs
Alk = 200 mg/L as $CaCO_3$, and:
pH 8.1

Tributary B: flow = 700 cfs
Alk = 100 mg/L as $CaCO_3$, and:
$C_T = 2.07 \times 10^{-3}$ M

First, you must calculate Alk, C_T, and pH for both tributaries. Using the approach in Section 13.5.2, you can find that C_T for Tributary A is 4.01×10^{-3} M and the pH of Tributary B is 7.7. Alkalinity is conserved, so:

$Q_A Alk_A + Q_B Alk_B = (Q_A + Q_B)Alk$
(if the densities of the waters are similar)

Thus: Alk = 168 mg/L as $CaCO_3$ (3.36 meq/L). *If no inorganic carbon is lost to the atmosphere or precipitates*, then carbon is conserved and:

$Q_A C_{T,A} + Q_B C_{T,B} = (Q_A + Q_B)C_T$ or $C_T = 3.39 \times 10^{-3}$ M

Using the alkalinity equation or Figure 13.9 or Figure 13.10, you can show that the pH of the combined rivers is 8.1. Conservation of alkalinity is shown again in Example 13.5.

13.6 EFFECTS OF OTHER WEAK ACIDS AND BASES ON ALKALINITY

***Example 13.5:* Conservation of Alkalinity**

Acid rain falls on a lake with low Alk. How many inches of rain would have to fall to reduce the pH of the lake to 6.7? The lake characteristics are pH 7.0, Alk = 30 mg/L as $CaCO_3$, mean depth = 10 ft. The acid rain characteristics are pH 4 and $[Cl^-]$ = 1×10^{-4} M.

Solution:
Assume that C_T is unchanged after the acid rain addition, no minerals dissolve, and only rain falling on the lake surface is important (i.e., the alkalinity of the soil neutralizes acid rain falling in the drainage basin). Also, assume no inputs to or outputs from the lake.

The lake alkalinity is (30 mg/L as $CaCO_3$)/(50,000 mg/L as $CaCO_3$ per eq/L) = 6×10^{-4} eq/L. The lake C_T is (Alk − $[OH^-]$ + $[H^+]$)/($\alpha_1 + 2\alpha_2$) = 7.19×10^{-4} M at pH 7.0.

After the acid rain addition, the pH is 6.7 and C_T (assumed constant) = 7.19×10^{-4} M. Thus, the lake Alk after acid rain addition is ($\alpha_1 + 2\alpha_2$)C_T + $[OH^-]$ − $[H^+]$ = 5.14×10^{-4} eq/L.

Alk is a conservative property. Thus:
$$(Alk_{LR})(V_{LR}) = (Alk_L)(V_L) + (Alk_R)(V_R)$$
where V = volume and the subscripts L, R, and LR are lake,

13.6.1 Introduction

For most natural waters, the members of the carbonate family are the dominate weak acids and weak bases. However, on some occasions, other weak acids or bases may be important. Other weak acids or bases can influence the alkalinity by increasing (for weak bases) or decreasing (for weak acids) the acid neutralizing capacity of the water. In such cases, the weak acids and bases must be included in the alkalinity calculation.

How should weak acids and bases be included in the alkalinity equation? The easiest way to include the other species is to return to the derivation of alkalinity as a proton condition. We already have established the endpoint of the alkalinity titration as about pH 4.5. Now, *define the zero level of protons as the dominate species in the acid-base family at pH 4.5* The proton condition becomes:

Σ(species with excess protons over the zero level)(no. of excess protons) =
Σ(species with deficient protons over the zero level)(no. of deficient protons)

Note that this definition works even for carbonate species, OH^-, and H^+. For the carbonate family, H_2CO_3 is the dominate species at pH 4.5 (since $pK_{a,1} = 6.3$) and for the "water family", H_2O is the dominate species at pH 4.5. Other examples are shown in Table 13.2.

Table 13.2: Zero Proton Levels for Common Species at pH 4.5

Family	pK_a Value(s)	Zero Level
$H_3PO_4/H_2PO_4^-/HPO_4^{2-}/PO_4^{3-}$	2.1, 7.2, 12	$H_2PO_4^-$
$H_3BO_3/B(OH)_4^-$	9.3	H_3BO_3
NH_4^+/NH_3	9.3	NH_4^+
Monoprotic acid with $pK_a > 4.5$	>4.5	HA

13.6.2 Example

As an example of the effects of other weak acids and bases on alkalinity, consider the impact of naturally occurring organic acids on the alkalinity of natural waters. These acids are divided into several fractions, primarily fulvic and humic acids. A highly colored water body may have 10 mg/L

rain, and lake after rain addition, respectively. Note that $Alk_R = C_B - C_A = -1 \times 10^{-4}$ eq/L.

Now:

V_L = (mean depth)(surf. area)
V_R = (rainfall rate)(surf. area)

Thus:

(Alk_{LR})(mean depth + rainfall rate) = (Alk_L)(mean depth) + (Alk_R)(rainfall rate)

Or:

rainfall rate
$= (Alk_L - Alk_{LR}) \times$(mean depth)/$(Alk_{LR} - Alk_R)$
$= (6 \times 10^{-4}$ eq/L $-$ 5.14×10^{-4} eq/L)(120 inches)$\times (5.14 \times 10^{-4}$ eq/L $+ 1 \times 10^{-4}$ eq/L)
= 16.7 inches or 42.2 cm of rain

of dissolved organic carbon (DOC). Humic and fulvic acids have many acidic functional groups. At neutral pH, the acidic groups contributing to alkalinity are carboxylic acid groups with concentrations of up to 10 meq per gram of carbon and a typical pK_a between 1.5 and 6. This yields a total acid concentration, A_T, of (10 meq/g C)(10 mg C/L)$(10^{-3}$ g/mg)$= 0.1$ meq/L. In this example, a typical pK_a of 5 will be used. Since the pK_a of the carboxylic acids is greater than the endpoint pH, use HA as the zero proton level for the carboxylate species. Thus, the proton condition is:

$$[H^+] = [HCO_3^-] + [CO_3^{2-}] + [A^-] + [OH^-]$$

The alkalinity equation becomes:

$$Alk = (\alpha_1 + 2\alpha_2)C_T + [OH^-] + [A^-] - [H^+]$$

The concentration of A^- at the endpoint pH is:

$$\alpha_{1,HA}A_T = (10^{-5})(1 \times 10^{-4}\ M)/(10^{-4.5} + 10^{-5}) = 2.4 \times 10^{-5}\ M$$

or 0.024 meq/L of ANC. This is a contribution of only about 1 mg/L of Alk as $CaCO_3$. Since most natural waters have much higher alkalinities from carbonate species, fulvic and humic acids usually do not contribute very much to the alkalinity.

13.7 SUMMARY

In this chapter, water quality parameters were developed to measure the amount of acid or base a water requires to reach specified endpoints. Natural and process waters rarely are pure acid or base solutions. They usually are acid or base solutions that have been partially titrated to some unknown point in their titration curves. For general acids and bases, the acid neutralizing capacity (ANC) is defined to be the amount of strong acid required to reach the zeroth equivalence point. Similarly, base neutralizing capacity (BNC) is the amount strong base required to reach the first equivalence point (for monoprotic acids). If the carbonate system provides the dominate weak acids and bases, then the ANC is called *alkalinity* (titration to $f = 0$ for H_2CO_3) and the BNC is called *acidity* (titration with base to $g = 0$ for Na_2CO_3).

Expressions for alkalinity and acidity can be developed from proton conditions or charge balances. For alkalinity, the resulting expression is called the alkalinity equation: $Alk = C_B - C_A = (\alpha_1 + 2\alpha_2)C_T + [OH^-] - [H^+]$. Alkalinity can be measured by titration to the H_2CO_3 endpoint, about pH 4.5 for most waters. Although alkalinity measures the ANC of a water to this endpoint, other types of alkalinity (e.g., phenolphthalein alkalinity) may be more important for other pH ranges of interest. Common units of alkalinity are eq/L and mg/L as $CaCO_3$: 1 meq/L = 50 mg/L as $CaCO_3$.

In solving problems involving alkalinity, it is important to remember that knowing two of Alk, C_T, or pH allows you to calculate the third parameter. For the problem at hand, carefully decide whether the information you know allows you to more easily calculate Alk from $Alk = C_B - C_A$ or from $Alk = (\alpha_1 + 2\alpha_2)C_T + [OH^-] - [H^+]$. In addition, remember that alkalinity is conserved when waters are mixed (if no solids precipitate).

Weak acids and bases can influence alkalinity and acidity in certain waters. To include the effects of weak acids and bases, define their zero proton level as the species in highest concentration at the endpoint (about pH 4.5 for alkalinity). Then adjust the alkalinity equation to include species with excess or deficient protons compared to the zero proton level.

13.8 PART III CASE STUDY: ACID RAIN

From what you have learned in this chapter, it is apparent that the influence of acid rain on cistern water pH can be determined much more easily by using the concept of alkalinity. Remember that alkalinity is conserved if no solids precipitate.

From Section 12.8, you found that the rainwater has a pH of 4.0, a total carbonate concentration, $C_{T,rain}$, of 8.3×10^{-6} M, and a total strong acid anion concentration, $C_{A,rain}$, of 9.5×10^{-5} eq/L. This means that the alkalinity is $Alk_{rain} = -9.5 \times 10^{-5}$ eq/L. You may wish to confirm that these values are consistent with the alkalinity equation. The river water has a pH of 7.5 and a total carbonate concentration, $C_{T,river}$, of 8.3×10^{-6} M. Using the alkalinity equation, you can verify that the alkalinity = total strong base cation concentration = $Alk_{river} = 1.13 \times 10^{-3}$ eq/L (or about 55 mg/L as $CaCO_3$, fairly low). When a volume v (= 1500 L) of rainwater is mixed with a volume V (= 15,000 L) of water in the cistern, the resulting concentrations are:

$$C_T = (vC_{T,rain} + VC_{T,cistern})/(v + V), \text{ and:}$$
$$Alk = (vAlk_{rain} + VAlk_{cistern})/(v + V)$$

where the subscript "cistern" represents the water in the cistern prior to the rain falling that day. The mass balance on C_T assumes that no inorganic carbon is exchanged with the atmosphere or precipitates.

To calculate the pH, simply apply the alkalinity equation each day. The resulting pH and Alk of the cistern water each day are shown in Figure 13.11 (pH values also were shown in Section 12.8). Note that the alkalinity of the cistern water transitions from the alkalinity of the river water to the alkalinity of the rainwater.

As a result of this analysis, you know why the pH in the cistern water dropped so rapidly: the river water had a very low Alk. The case study shows that the power of alkalinity in solving water quality problems.

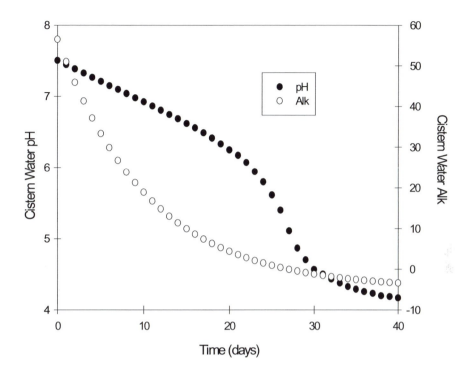

**Figure 13.11: Cistern Water pH and Alkalinity in the
Part III Case Study**

SUMMARY OF KEY IDEAS

- It is possible to probe a system and determine its ability to buffer against changes in pH

- Weak acids (or bases) partially neutralize strong bases (or acids) added to them

- Adding strong acid to a partially titrated weak acid will retrace the titration curve of the weak acid

- ANC and BNC tell you how much added acid or base can be added without significantly changing the pH of the system

- For a monoprotic acid, ANC is the amount of strong acid needed to change $\alpha_1 A_T + [OH^-] - [H^+]$ from its current value to zero, so ANC = $\alpha_1 A_T + [OH^-] - [H^+]$

- For a monoprotic base, BNC is the amount of strong base needed to change $\alpha_0 A_T + [H^+] - [OH^-]$ from its current value to zero, so BNC = $\alpha_0 A_T + [H^+] - [OH^-]$

- The major ions in most fresh waters are Na^+, K^+, Ca^{2+}, Mg^{2+}, SO_4^{2-}, Cl^-, and HCO_3^-

- Alkalinity commonly is used as an indicator of buffering, even though it is not a measure of buffer intensity

- Use caution when accepting alkalinity as a indicator of buffering if the lower pH range of interest is different than 4.3-4.7

- The pH values at the equivalence points for alkalinity and acidity depend on the total carbonate concentration (C_T) of the sample

- Different types of alkalinity are defined for different endpoints

- Alkalinity will be calculated in units of eq/L if C_B and C_A are in units of eq/L

- 1 eq/L Alk is equivalent to 50,000 mg/L Alk as $CaCO_3$

- The alkalinity equation is really three equations in one:
$$Alk = (\alpha_1 + 2\alpha_2)C_T + [OH^-] - [H^+]$$
$$C_B - C_A = (\alpha_1 + 2\alpha_2)C_T + [OH^-] - [H^+]$$
$$Alk = C_B - C_A$$

- If you know two of Alk, pH, and C_T, then you can calculate the other parameter easily

- Near neutral pH (and for reasonable values of C_T), Alk $\approx [HCO_3^-] = \alpha_1 C_T$

- When H^+ is released from a process, alkalinity is consumed

- Alkalinity and acidity are conserved

PROBLEMS

13.1 Using the definitions of alkalinity and acidity, explain why 1 eq of acidity consumes 1 eq of alkalinity.

13.2 An industry wishes to discharge an acidic waste continuously to a river. Describe briefly the information you would require to determine the impact of the waste on the river pH.

13.3 The pH of a lake water is measured in the field to be pH 8.0. A 100 mL sample of the water required 5.0 mL of 0.10 N HCl to titrate it to the H_2CO_3 equivalence point.

A. What is the alkalinity of the water?

B. What are the concentrations of C_T, H_2CO_3, HCO_3^-, and CO_3^{2-} in the lake?

13.4 **Standard Methods** lists the following endpoints for alkalinity titrations, based on the expected alkalinity: pH 4.9, 4.6, and 4.3 for expected alkalinity values of 30, 150, and 500 mg/L as $CaCO_3$, respectively. Verify that these endpoint pH values are reasonable for the anticipated alkalinity values.

13.5 Lake Mendota, located in Madison, Wisconsin, is one of the most thoroughly studied lakes in the world. Average water quality conditions are pH 8.5, Alk = 3.4 meq/L, $[Ca^{2+}]$ = 32 mg/L, $[Mg^{2+}]$ = 32 mg/L, $[SO_4^{2-}]$ = 22 mg/L, $[Cl^-]$ = 27 mg/L, $[Na^+]$ = 1 mg/L, $[K^+]$ = 3.2 mg/L, and total dissolved solids (see Section 2.4.3) = 260 mg/L.

A. Estimate the average C_T in Lake Mendota.

B. Do charges balance? Include all ions you expect in the lake.

C. How does the calculated TDS compare to the measured TDS?

13.6 Lago De Amatitlan (Lake Amatitlan) is a moderate-sized lake in southwestern Guatemala. Typical water quality is as follows (all data are from the International Lake Environment Committee's World Lake Database, at **www.ilec.or.jp/e_index.html**): Ca = 88.3 mg/L, bicarbonate = 410.4 mg/L, K = 7.4 mg/L, and Na = 219.2 mg/L.

A. If the only important unmeasured ion is chloride, estimate the chloride concentration in mg/L.

B. Estimate the alkalinity in mg/L as $CaCO_3$. Is it possible to calculate the pH and C_T of this lake with the information given?

13.7 Using the Deffeyes diagram (Figure 13.9) or Figure 13.10 to estimate the new pH if 0.5 meq/L of alkalinity is added to a water initially at pH 7.2 with an initial alkalinity of 110 mg/L as $CaCO_3$.

13.8 PollutiCon Inc., a large multinational widget manufacturer, wishes to discharge 50,000 gallons of an acidic waste into Lake Fishkill. The waste is 0.1 M sulfuric acid. Lake Fishkill has the following characteristics: alkalinity = 100 mg/L as $CaCO_3$, pH 7.5, mean depth = 7 feet, and surface area = 3 acres. PollutiCon claims that the pH of the lake water after the addition of the acidic waste will not drop below pH 7. You have been hired as a consulting engineer to assess the situation. What will be the pH of the lake water after the addition of the acidic waste? Assume: (1) closed system (no CO_2 lost to the atmosphere), (2) lake volume = (mean depth)(surface area), and (3) the lake has no other inputs or outputs.

13.9 Calculate the alkalinity (in units of both eq/L and mg/L as $CaCO_3$) of solutions containing

A. 2×10^{-3} M $NaHCO_3$

C. 2×10^{-3} M NaCl

B. 2×10^{-3} M Na_2CO_3 and 1×10^{-3} M HCl

D. 2×10^{-3} N H_2SO_4

13.10 If no solids form or precipitate, does the alkalinity of a 1 L sample of a typical natural water increase, decrease, or stay the same if you:

A. Add a small amount of NaCl to it?

B. Add 5 mL of 0.1 M NaOH and 5 mL of 0.1 M H_2SO_4 to it?

C. Titrate it to the H_2CO_3 endpoint?

D. Titrate it to the second equivalence point ($f = 2$)?

13.11 In wastewater treatment plants, ammonium can be oxidized biologically to nitrate. The overall reaction is: $NH_4^+ + 2O_2 \rightarrow NO_3^- + 2H^+ + H_2O$. Based on this reaction, justify the statement that "... for each g of ammonia nitrogen (as N) converted, 7.14 g of alkalinity as $CaCO_3$ will be required" (Metcalf and Eddy, 2003).

13.12 Ammonia also can be removed from wastewater by ammonia stripping. In this process, ammonium is converted to NH_3 by raising the pH to 11 with lime, and ammonia is volatilized into a moving airstream. Create a plot showing the lime dose required (in mg/L) to raise the wastewater pH to 11 as a function of the initial alkalinity of the water (in mg/L as $CaCO_3$) over the normal alkalinity range of wastewaters (50 to 200 mg/L as $CaCO_3$). Lime is $Ca(OH)_2(s)$. Assume all the alkalinity is from carbonate species and the added lime. Also assume that the lime dissolves completely and that the initial pH of the wastewater is 7.5.

13.13 In an anaerobic digester, organics in wastewater sludge are converted into volatile acids (short-chain organic acids), which are subsequently converted into methane and carbon dioxide by a group of organisms called *methogens*. In a typical anaerobic digester(pH 7, Alk = 3000 mg/L as $CaCO_3$, volatile acids concentration = 250 mg/L), how much of the alkalinity is due to the weak volatile acids? Assume that the volatile acids are acetic acid with $pK_a = 4.7$.

13.14 In Figure 13.5, a plot of the pH at the H_2CO_3 equivalence point versus $\log(C_T)$ is a straight line. Find the slope of the line from the alkalinity equation.

13.15 Draw a titration curve like Figures 13.6-13.8 to show the fractions of acidity.

PART IV

Other Equilibria in Homogeneous Aqueous Systems

I shall now attempt to show how, by a
single type of chemical combination,
we may explain the widely varying
phenomena of chemical change.
Gilbert N. Lewis

Here's metal more attractive.
William Shakespeare

CHAPTER 14

Getting Started with Other Equilibria in Homogenous Aqueous Systems

14.1 INTRODUCTION

Acid-base equilibria in Part III introduced you to the concept of a proton donor. Recall that a Brønsted-Lowery acid donates a proton to water. Since acid-base reactions are so important in aquatic chemistry, pH is a logical master variable. However, protons are not the only chemical species that can be moved between species. The other major "currency" in aquatic chemistry is the movement of electrons between species.

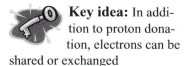 **Key idea:** In addition to proton donation, electrons can be shared or exchanged

As discussed in Chapter 10, electrons can be shared or transferred. The sharing of electrons creates products that are composites of the reactants. This process is shown schematically in Figure 14.1. The formation of composite products from the sharing of electrons is called *complexation*. In some cases, the composite products are not very stable. The free energy of the system may be minimized if the product reverts back to the reactants. For some reactants, free energy will be minimized if electrons are completely transferred. This results in the formation of new species with different oxidation states than the reactants (see Figure 14.1).

The equilibrium tools you have mastered in Part II and honed in Part III will be applied to electron-sharing and electron-transfer equilibria in Part IV of this text. By the conclusion of Part IV, you will be able to describe and calculate species concentrations in realistic models of homogeneous aqueous systems.

14.2 ELECTRON-SHARING REACTIONS

14.2.1 Alfred Werner and molecular compounds

In the early part of the twentieth century, chemists struggled to understand a particular class of compounds. These so-called molecular compounds consisted of a metal and several other groups. One example of interest at the time was $Pt(NH_3)_2Cl_2$. The structure of this compound was not known. Did it form a long chain, such as $Cl-NH_3-Pt-NH_3-Cl$? Were the chlorine atoms associated with ammonia or were they associated with platinum?

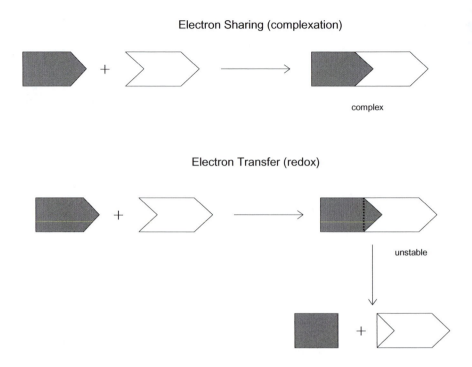

Electron Sharing (complexation)

complex

Electron Transfer (redox)

unstable

Figure 14.1: Electron Sharing and Electron Transfer

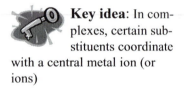 **Key idea**: In complexes, certain substituents coordinate with a central metal ion (or ions)

Alfred Werner (1866-1916) realized that nearly all the molecular compounds could be described by placing the metal ion in a *central location* and having it associate with four or six substituents. The realization that the substituents were associated with the central metal and not with each other was a major breakthrough. It explained many of the observations concerning molecular compounds. His work was so significant that Werner was awarded the Nobel Prize in Chemistry, the first laureate in inorganic chemistry.

Werner used the term "coordination number" for the number of substituents associated with the central metal, a term still used today. We now call molecular compounds *complexes* or *coordination complexes*. The formation of complexes is called *complexation*. Complexation is the subject of Chapter 15.

14.2.2 Sharing of electrons during complexation

Why do certain substituents associate with metals? To answer this question, examine a list of common central metal species (e.g., Cu^{2+}, Hg^{2+}, Zn^{2+}, Fe^{2+}, and Fe^{3+}) and substituents (e.g., Cl^-, OH^-, CN^-, and S^{2-}).

Thoughtful Pause
Based on these list, what do you think is the nature of the interaction
between the metal and substituents?

It appears that the metal ions have a deficiency of electrons (and hence are positively charged) and the substituents have an excess of electrons (and hence are negatively charged). Thus, it makes sense that electrons may be shared during complex formation.

 **Key idea**: Electrons are shared between metals and substituents in complexation

We can now extend Werner's ideas: complexes form when a central metal ion and other species share electrons. This model will go a long way toward explaining why certain complexes form in the aquatic environment. However, the model seems to have a few problems. For example, if electron sharing is so important, why do complexes form with uncharged species such as water, ammonia, and organic acids? The answer is that nitrogen and/or oxygen atoms in these species have *unshared electrons* available to share with the central metal ion. As a result, you may expect that water and OH^- will be important contributors to complex formation in water.

Another potential problem with the electron-sharing idea is a bit more subtle. Remember from freshman chemistry that *all* chemical bonds involve some degree of electron sharing. A *covalent bond* is one in which the electrons are shared fairly equally (that is, the electrons are associated with both atoms in the bond). In an *ionic bond*, the electrons are much more closely associated with one atom (as in Cl in NaCl). If all electrons involve electron sharing, then shouldn't all bond formation be called complexation? More specifically, should the association of H^+ with a conjugate base be called complexation and not acid-base chemistry?

In answer to the first question, the word "complexation" often is limited to describing electron sharing with certain electrons. Thus, the covalent bonds common in organic chemistry (for example, in methane, CH_4) are not referred to as complex formation. However, addressing the second question, we shall find that it is insightful to include H^+ as a metal. In this way, acid-base chemistry is a sort of complexation. This concept is explored further in Chapter 15.

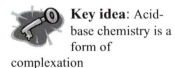 **Key idea**: Acid-base chemistry is a form of complexation

14.2.3 Examples of complexation in the environment

By broadening your view of aquatic chemistry to include electron-sharing reactions, the richness of homogeneous chemistry will become apparent to you in Chapter 15. For example, you will learn about the chemistry of dissolved metals. In natural systems, numerous species react with metals, including the ubiquitous OH^- and naturally occurring acids. Complexation is at the heart of the biochemistry of many vitamins and important enzymes. Complexes also can be formed by reaction between metals and species of anthropogenic origin, such as detergents and cyanide.

Complexation is very useful chemistry. For example, the principles of complexation are used to control metal solubility in industrial processes (e.g., metal plating) and in the laboratory (in metal buffers). In addition, the determination of metal concentrations at very low levels with simple instruments is made possible by highly specific complexation reactions.

14.3 ELECTRON TRANSFER

14.3.1 When electron sharing becomes electron transfer

When the sharing of electrons creates a stable compound, we say that complexation has occurred, and the stable compound is called a *complex*. In many instances, the complex formed from the initial sharing of electrons is not stable. In this case, the electrons may be completely transferred from one species to another.

Consider the reaction of ferric ion with iodide. Initially, a complex is formed:

$$Fe^{3+} + I^- \ = \ FeI^{2+} \hspace{3cm} \text{eq. 14.1}$$

The reaction in eq. 14.1 is complexation equilibrium. It is reasonable to assume that the electrons in iodide are being shared with the ferric ion. The oxidation state of iron remains at +III and the oxidation state of I remains at $-I$ since electrons are merely being shared. The complex reacts with more iodide:

$$FeI^{2+} + I^- \ = \ FeI_2^+$$

Again, this is a simple complexation reaction. However, this intermediate is unstable. The free energy of the system is minimized if the intermediate rearranges:

$$FeI^{2+} \ = \ Fe^{2+} + I_2^- \hspace{3cm} \text{eq. 14.2}$$

Note that the oxidation state of the iron has been reduced from +III to +II, and the average oxidation state of the iodines has been increased from $-I$ to $-\frac{1}{2}$. Thus, in eq. 14.2, electrons are transferred from one chemical species to another. A series of reactions, which started with electrons being shared, ended with electrons being transferred. To complete the reaction scheme, the iodine anion reacts with more ferric ion:

$$I_2^- + Fe^{3+} \ \rightarrow \ I_2 + Fe^{2+}$$

In this final reaction, the oxidation state of the iron has been reduced to +II again, and the average oxidation state of the iodines has been increased further to 0. The overall reaction is:

$$2I^- + 2Fe^{3+} \rightarrow I_2 + 2Fe^{2+} \qquad\qquad \text{eq. 14.3}$$

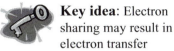

Key idea: Electron sharing may result in electron transfer from one species to another to minimize the free energy of the system

Note that in the overall reaction, no sign of electron sharing is evident. Recall, though, that *electron sharing made electron transfer possible.* Since the FeI_2^+ complex was unstable, free energy was minimized by transferring the electrons from iodine to iron.

Reactions such as eq. 14.3 are called *overall redox reactions.* It is an example of oxidation-reduction chemistry. In eq. 14.3, the ferric ion is said to be *reduced* to ferrous ion, and iodide ion is *oxidized* to iodine. In Chapter 16, the equilibrium aspects of redox reactions will be considered.

14.3.2 Examples of redox reactions in the environment

In many ways, redox chemistry is even more textured than acid-base chemistry. In acid-base chemistry, the Brønsted-Lowery definition allows for only one acceptor of protons, H_2O. In redox chemistry, many different species act as electron acceptors. Thus, many species participate in electron-transfer reactions.

In natural aquatic systems, the cycling of elements between different oxidation states is an important part of chemical ecology. The focus in this text is on chemical processes. However, the importance of redox chemistry in natural systems is best seen by considering the role of electron transfer in biota. Nearly all microorganisms (and all organisms, including humans, that respire) require a *terminal electron acceptor* as the final step in the process of deriving energy from chemical bonds. Organisms undergoing respiration use oxygen as the terminal electron acceptor. Oxygen accepts electrons and protons to form water.

Redox reactions occur in engineered systems as well. Corrosion is the process by which electrons are transferred from a metal in its zero oxidation state to an electron acceptor. Most chemical disinfectants (such as chlorine and ozone) are strong electron acceptors. They disrupt cellular function by accepting electrons from electron-rich sites such as sulfhydryl groups (-SH groups) and double bonds.

14.4 PART IV ROAD MAP

Part IV of this book discusses systems with electron-sharing (complexation) and electron-transfer (redox) equilibria. In Part IV, both equilibrium calculation techniques and qualitative tools will be applied to complexation and redox chemistry. By the conclusion of this part of the text, you should know the effects of electron-sharing and electron-transfer equilibria on homogeneous aquatic systems.

In Chapter 15, the nomenclature of complexation is introduced. The presence of many different elements in complexation equilibrium generally means that multiple mass balances are required. Several different classes of species that react with metals are discussed. By the conclusion of

Chapter 15, you should be able to determine species concentrations in fairly complicated homogeneous aquatic systems.

Chapter 16 provides solution techniques for systems involving electron transfer. The importance of the master variable pe will be emphasized. Graphical solution techniques showing the effects of both pe and pH on speciation will be developed. By the conclusion of Chapter 16, you should be able to answer questions about the dominant chemical species in homogeneous aquatic systems.

14.5 SUMMARY

In this chapter, the fate of electrons in homogeneous aquatic systems was introduced. You found that electrons could be shared or transferred. Electron-sharing reactions with metals results in the formation of complexes. Complexes have one or more central metal ions and one or more species sharing electrons with the metal(s). Many different types of species can form complexes with metals, including common species in water (such as water itself and OH^-).

If electron sharing results in the formation of unstable species, free energy in the system may be minimized by complete transfer of electrons from one chemical species to another. This is called oxidation-reduction (or redox) chemistry. The net result is the change in the oxidation state of elements in the species accepting electrons and in the species donating electrons.

14.6 PART IV CASE STUDY: WHICH FORM OF COPPER PLATING SHOULD YOU USE?

A quick glance around your apartment or office should convince you that hundreds of everyday items are covered with a very thin coat of metal. From doorknobs to razor handles, many metal and plastic objects benefit from metal finishing. In most metal finishing operations, metal in an oxidation state greater than zero is reduced electrochemically to the zero oxidation state on the object to be plated (called the *piece*). This produces a stable metal coating to increase the strength or appearance of the piece.

One of the most common types of metal plating operations is copper plating. Copper can be used to provide a number of finishes, from a useful undercoat for additional plating operations to a bright, final finish. Two types of copper plating processes are common. In copper cyanide plating, the source of copper is $CuCN(s)$. Note that the copper source in copper cyanide plating is $Cu(+I)$. In acid copper plating, the source of copper is $CuSO_4(s)$; copper in the +II oxidation state.

In this case study, the relative merits of copper cyanide and acid copper plating will be examined. The advantages of one plating type over another will be quantified through aqueous equilibrium calculations.

SUMMARY OF KEY IDEAS

- In addition to proton donation, electrons can be shared or exchanged

- In complexes, certain substituents coordinate with a central metal ion (or ions)

- Electrons are shared between metals and substituents in complexation

- Acid-base chemistry is a form of complexation

- Electron sharing may result in electron transfer from one species to another to minimize the free energy of the system

Complexation

15.1 INTRODUCTION

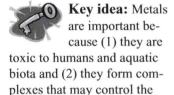

Key idea: Metals are important because (1) they are toxic to humans and aquatic biota and (2) they form complexes that may control the chemistry of the ligands

The world of environmental chemistry is much broader than acids and bases. An important component of the aquatic environment is metals. Metals are worthy of study because of their toxicity to humans and aquatic biota. In addition, metals equilibrate with many common aqueous species (called *ligands*) to form compounds called *complexes*. Complexes may control the chemistry of the ligands.

In this chapter, you will learn about the formation of complexes. In Sections 15.2 and 15.3, complex formation will be discussed on a qualitative basis. You will learn about the roles of both the metal and the ligands. Equilibrium calculation techniques for systems exhibiting complexation will be presented in Section 15.4. A fairly involved system containing acids, bases, metals, and ligands is discussed in Section 15.5. Applications of complexation chemistry are presented in Section 15.6.

15.2 METALS

15.2.1 Introduction

You now have a great deal of experience with acid-base equilibria. So far in this text, acid-base equilibria have been analyzed from the point of view of the acid and conjugate base. For example, the calculation techniques (Chapter 7), alpha values (Chapter 11), and approaches to titration (Chapter 12) have emphasized the role of HA and A^-.

For the moment, focus on H^+ rather than HA or A^-. Usually, we write:

$$HA = A^- + H^+ \qquad\qquad K = K_a$$
$$\text{or: } HA + H_2O = A^- + H_3O^+ \qquad\qquad K = K_a$$

(Recall that the equilibrium constant for the second equilibrium is equal to the equilibrium constant for the first equilibrium because we define K for the equilibrium $H_2O + H^+ = H_3O^+$ to be 1.) To emphasize the fate of H^+, rewrite these equilibria as:

$$H^+ + A^- = HA \qquad\qquad K = 1/K_a$$
$$\text{or: } H_3O^+ + A^- = HA + H_2O \qquad\qquad K = 1/K_a \qquad \text{eq. 15.1}$$

You can rewrite eq. 15.1 slightly as follows:

$$H^+(H_2O) + A^- = HA + H_2O \qquad K = 1/K_a \qquad \text{eq. 15.2}$$

coordination: the association of two chemical species to form a third species

Equation 15.2 shows the fate of H^+ in the $HA/A^-/H_2O$ system: H^+ equilibrates between being associated with water and being associated with A^-. We say that H^+ *coordinates* with H_2O and with A^-.

metal: in aquatic chemistry, the positively charged partner in a coordination reaction

Coordination is the association of two chemical species to form a third species[†]. Coordination involves two partners. The positively charged partner is called a **metal**. The other partner, usually negatively charged or uncharged, is called the **ligand**. Compounds formed by the coordination of a metal and a ligand are called *coordination complexes* or *metal-ligand complexes* or just *complexes*. The formation of coordination complexes is called **complexation**.

ligand: a species that coordinates with a metal

complex: a compound formed by the coordination of a metal and a ligand (also called *coordination complexes* or *metal-ligand complexes*)

Thoughtful Pause
Names the metal(s) and ligand(s) in eq. 15.2.

 **Key idea:** Metals and ligands coordinate to form complexes

In eq. 15.2, the only metal is H^+ and the ligands are H_2O and A^-. Because of its high concentration, water is a very important and common ligand in aquatic systems.

15.2.2 Lewis acids and bases

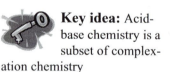 **Key idea:** Acid-base chemistry is a subset of complexation chemistry

If you wished, you could now completely redo Part III of this text and eliminate the words *acid* and *base*. You could say that Part III of this text was devoted to one metal, namely H^+. A variety of ligands was considered, including H_2O, OH^-, Cl^-, SO_4^{2-}, HSO_4^-, and CH_3COO^-. The ligands were Brønsted-Lowry bases, and the metal-ligand complexes (e.g., CH_3COOH) were Brønsted-Lowry acids. It is completely acceptable to think of acid-base chemistry as a subset of metal-ligand chemistry. In fact, in some compilations of equilibrium constants, the K_a values (actually, $1/K_a$ values) are listed for equilibria involving the metal H^+ and the pertinent ligand (conjugate base) as reactants.

[†] The association can be strong or weak. Strong association results in chemical bond formation and is called *inner-sphere coordination*. Weak association may occur if water molecules exist between the coordinating species and is called *ion pair coordination*.

Rather than redefining acids as metal-ligand complexes, you also could redefine metal-ligand complexes as a sort of acid. The great American chemist Gilbert Newton Lewis (1875-1946) formalized this idea through his notation that *electrons are shared between atoms in chemical bonds*. Lewis extended the Brønsted-Lowry definitions of acids and bases to what are now called Lewis acids and bases (Lewis, 1923). A ***Lewis acid*** is a substance that accepts an electron pair. A ***Lewis base*** is a substance that donates an electron pair. Note that Brønsted-Lowry acids and bases also are Lewis acids and bases[†]. For example, the Brønsted-Lowry/Lewis base PO_4^{3-} both accepts a proton and donates (or shares) one or more electron pairs.

It is important to note that the Lewis acid definition is much broader than the Brønsted-Lowry definition. For example, all the transition metals are Lewis acids. The environmentally important transition metals are chromium, manganese, iron, cobalt, nickel, copper, zinc, silver, cadmium, gold, and mercury. Other important Lewis acids in the environment come from magnesium, aluminum, and calcium.

15.2.3 Coordination number
The number of ligands with which a metal coordinates is called the ***coordination number***.

Thoughtful Pause
What is the coordination number for H^+?

H^+ has a coordination number of 1, as it generally coordinates with one ligand [as in $H^+(H_2O) = H_3O^+$]. Most other metals of environmental interest have coordination numbers of 2, 4, or 6. Common metals in the aquatic environment and their coordination numbers are listed in Table 15.1.

The coordination numbers are satisfied at all times in the aquatic environment. This means that uncomplexed metal ions such as Cu^{2+} and Al^{3+} do not exist in the environment. However, *a shorthand notation is used frequently where the ligand water is not written explicitly*. Thus, it is common to write Cu^{2+} as an abbreviation for the species $Cu(H_2O)_4^{2+}$ and Al^{3+} as an abbreviation for the species $Al(H_2O)_6^{3+}$. If the number of ligands is less than the coordination number, you generally can assume that the unwritten ligands are water. Thus, write $ZnOH^+$ as shorthand for $Zn(H_2O)_3OH^+$.

[†] It is easy to see that H^+ is both a Brønsted-Lowry acid and a Lewis acid. It is a bit more difficult to see how other Brønsted-Lowry acids, say HCl, are Lewis acids. In the Lewis acid-base view, HCl is a considered a Lewis acid because it *donates* a Lewis acid, namely, H^+ (March, 1985). Similarly, Cl_2 is a Lewis acid because it donates "Cl^+".

Lewis acid: a substance that accepts an electron pair

Lewis base: a substance that donates an electron pair

Key idea: Brønsted-Lowry acids and bases also are Lewis acids and bases, but the Lewis acid concept is much broader than the Brønsted-Lowry acid concept

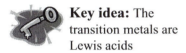

Key idea: The transition metals are Lewis acids

coordination number: the number of ligands with which a metal coordinates

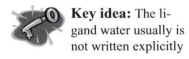

Key idea: The ligand water usually is not written explicitly

**Table 15.1: Coordination Numbers of Common Metals
in the Environment**

Coordination Number	Example Metals[1]
1	H^+, Na^+, K^+
2	Cu^+, Ag^+, Au^+
4^2	Cu^{2+}, Ni^{2+}, Zn^{2+}, Cd^{2+}, Hg^{2+}
6	Al^{3+}, Fe^{2+}, Fe^{3+}

Notes: 1. The metals are written as ions in this table. The ions represent the oxidation states, for example, H(+I), Ni(+II), and Al(+III).
2. The coordination number of four has two geometric configurations: square planar (ligands in the same plane as the metal and at the corners of a square) and tetrahedral (ligands at the corners of a tetradehron with the metal in the center).

polynuclear (multinuclear) complexes: complexes with more than one coordinating metal

Complexes that contain more than one coordinating metal are called *polynuclear* (or *multinuclear*) *complexes*. Complexes with two coordinating metals are called *dimers*, and complexes with three coordinating metals are called *trimers*. One chemical used as a coagulant in drinking water treatment, polyaluminum chloride, is a mixture of polynuclear complexes. One polynuclear complex in polyaluminum chloride is the tridecamer $Al_{13}O_4(OH)_{24}(H_2O)_{12}^{7+}$.

15.3 LIGANDS

Key idea: Ligands donate electron pairs to metals and commonly contain N, O, or S

monodentate ligand: a ligand capable of forming one bond with a metal

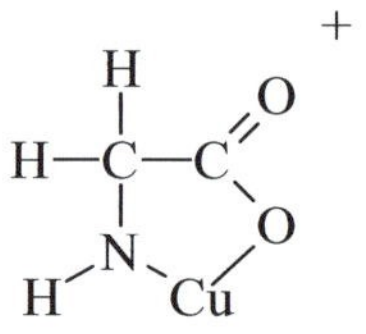

15.3.1 Ligand characteristics and nomenclature

Ligands are Lewis bases and thus donate electron pairs. Thus, ligands must contain atoms with electron pairs to donate. In the aquatic environment, many ligands contain N, O, or S since these atoms have unshared electron pairs in their common oxidation states in the environment. Thus, carboxylic acids (which contain the functional group -COOH or -COO⁻) are common ligands in the aquatic environment. Ammonia can be an important ligand as well.

Recall that the number of ligands with which a metal can form coordination complexes is called the coordination number of the metal. Ligands also are labeled according to how many bonds they can form with metals. A ligand capable of forming one bond (e.g., -COO⁻ and Cl⁻) is called a *monodentate* ligand (from the Latin *dentatus* tooth). Similarly, bidentate ligands can form two bonds. Amino acids contain both -NH₂ and -COOH groups and can be bidentate ligands. The copper(+II)-glycine complex is shown in Figure 15.1. (For clarity, the four water ligands are not shown.)

Figure 15.1: Copper(+II)-Glycine Complex

chelates: complexes formed
with a multidentate ligand

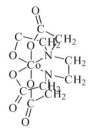

**Figure 15.2: Cobalt-EDTA
Chelate (see also Section
15.3.5)**

Complexes formed with multidentate ligands are called *chelates* or *chelate compounds* (from the Greek *chele* claw). The formation of chelates is called *chelation*. By definition, chelates have a ring structure. For example, the copper(+II)-glycine complex shown in Figure 15.1 is a chelate. Note the presence of a ring (cyclic) structure. Another example of a chelate is shown in Figure 15.2.

15.3.2 Aquo complexes (ligand = H_2O)

As stated in Section 15.2.1, water is an important ligand because water is present at such high concentrations in aqueous systems. The ligand H_2O is given the name *aquo*. If water is the only ligand, then the metal-water complex is called the aquo X ion, where X is the name of the oxidation state of the metal. For example, $Cu(H_2O)_4^{2+}$ is the aquo cupric ion.

The water ligands help explain why some Lewis acids act as Brønsted-Lowry acids - that is, why some Lewis acids donate protons to water. As an example, consider $Al(H_2O)_6^{3+}$. This ion can donate a proton to water:

$$Al(H_2O)_6^{3+} + H_2O = Al(H_2O)_5OH^{2+} + H_3O^+ \qquad \text{eq. 15.3}$$

The complexes in equilibrium expressions such as eq. 15.3 usually are written without the waters. Note that the following equilibria all give the same information and have the same equilibrium constant:

$$Al(H_2O)_6^{3+} + H_2O = Al(H_2O)_5OH^{2+} + H_3O^+$$
$$Al^{3+} + 2H_2O = AlOH^{2+} + H_3O^+$$
$$Al^{3+} + H_2O = AlOH^{2+} + H^+$$

You can "subtract" (i.e., reverse and add) the self-ionization of water from the last version of this equilibrium to obtain:

$$Al^{3+} + OH^- = AlOH^{2+} \qquad \text{eq. 15.4}$$

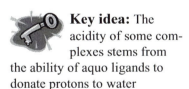

Key idea: The acidity of some complexes stems from the ability of aquo ligands to donate protons to water

Although eq. 15.3 contains the same information as eq. 15.4 (plus the self-ionization of water), the two reactions given a very different feeling for the origin of $AlOH^{2+}$. From eq. 15.4, it appears that $AlOH^{2+}$ comes from the formation of a complex between Al^{3+} and the hydroxide ion. However, from eq. 15.3, it appears that $AlOH^{2+}$ (or $Al(H_2O)_5OH^{2+}$) comes from the fact that the ligand waters are acidic and can donate protons to water.

The acidity of the ligand waters can be estimated. Recall that water itself is a pretty weak acid. Its first pK_a is 14. In contrast, the water molecules coordinated with metal ions are more acidic. For example, the pK_a for $Al(H_2O)_6^{3+}$ (pK for the equilibrium in eq. 15.3) is 5.

Thoughtful Pause

Why are the waters around a metal ion so much more acidic than water itself?

It is likely that the protons in the water ligands are electrostatically repelled by the metal ion, reducing the attraction to the water molecule.

Thoughtful Pause

If the ligand water molecules are acidic, what is the expected pH range where aquo complexes are favored?

In a sense, the aquo complexes are analogous to the most protonated form of a moderately strong acid. In other words, *the aquo complexes are expected to be the dominant form of the metal only at reasonably acidic pH values*.

15.3.3 Hydroxo complexes (ligand = OH⁻)

Hydroxo complexes have OH⁻ as the only ligand(s) other than H_2O. As in the Brønsted-Lowry analogy discussed in Section 15.3.2, hydroxo complexes are analogous to the conjugate bases of moderately strong acids. Recall that the moderately strong acids are the aquo complexes. The water ligands will donate protons to water in a stepwise fashion to form complexes with increasingly larger numbers of OH ligands per mole as the pH increases. As an example, copper(+II) goes from being primarily in the form of $Cu(H_2O)_4^{2+}$ at low pH to the stepwise dominance of $Cu(H_2O)_3OH^+$, $Cu(H_2O)_2(OH)_2^0$, $Cu(H_2O)(OH)_3^-$, and $Cu(OH)_4^{2-}$ as the pH increases. These species are usually written as, respectively, Cu^{2+}, $CuOH^+$, $Cu(OH)_2^0$, $Cu(OH)_3^-$, and $Cu(OH)_4^{2-}$.

A word about the uncharged soluble species such as $Cu(H_2O)_2(OH)_2^0$ $\equiv Cu(OH)_2^0$. The uncharged soluble hydroxo complexes do not form with all metals. When writing these species, it is important to differentiate them from the hydroxide solids. We usually use a superscript zero to indicate that the species are uncharged. You also can label the species as an aqueous (i.e., dissolved) species. Thus, the symbols $Cu(OH)_2^0$, $Cu(OH)_2^0(aq)$, and $Cu(OH)_2(aq)$ all are used to denote the soluble hydroxo-copper(+II) complex with two hydroxy groups.

15.3.4 Other inorganic ligands

In addition to aquo and hydroxo complexes, other inorganic ligands also can be important in natural waters. Several important inorganic ligands are listed in Table 15.2, along with examples of their environmentally

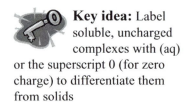 **Key idea:** Label soluble, uncharged complexes with (aq) or the superscript 0 (for zero charge) to differentiate them from solids

important complexes. Oxo complexes are important with several metals. For example, Cr(+IV) complexes with four O^{2-} ligands to form chromate (CrO_4^{2-}) in water. Another common oxo complex is the familiar carbonate: $C(+IV)O(-II)_3^{2-} = CO_3^{2-}$. Bicarbonate is a **mixed ligand complex** - that is, a complex consisting of two or more different ligands. Bicarbonate is an oxohydroxo complex: $C(+IV)O(-II)_2OH(-I)^- = HCO_3^-$.

mixed ligand complex: a complex consisting of two or more different types of ligands

The complexes with bicarbonate and carbonate usually are found in lower concentrations than hydroxo complexes in fresh water. The chloro complexes are very important in seawater. Cyano complexes in the environment typically are found only in waters with high cyanide concentrations, for example, in metal finishing wastewaters.

Table 15.2: Common Inorganic Ligands and Complexes
(Source: Stumm and Morgan, 1996)

Name of Complex	Ligand	Examples
Oxo	O^{2-}	CrO_4^{2-}, HCO_3^-, CO_3^{2-}
Bicarbonato/carbonato	HCO_3^-/CO_3^{2-}	$CoCO_3^0$, $NiCO_3^0$, $CuCO_3^0$, $ZnCO_3^0$, $CdCO_3^0$
Chloro	Cl^-	$NiCl^+$, $ZnCl^+$, $AgCl^0$, $AgCl_2^-$, $CdCl_2^0$, $HgCl_4^{2-}$
Cyano	CN^-	many, e.g., $Ni(CN)_4^{2-}$

15.3.5 Organic ligands

Metal-organic complexes often control the chemistry of many metals in the environment. Although there are many types of organic ligands, three classes of organic ligands play especially important roles in the environment and in methods for the analysis of metals in the environment. The first class is the simple organic acids. Common organic acids in the environment include citric acid (formally, 3-hydroxy-3-carboxypentanedioic acid; see Figure 15.3), the amino acids (see Figure 15.1 for an example), and anionic surfactants such as nitrilotriacetic acid (NTA, formally *N,N-bis*(carboxymethoxy)glycine; see Figure 15.4). Both citric acid and NTA are tridentate ligands.

Another organic acid, EDTA (ethylenediaminetetraacetic acid or ethylenedinitrilotetraacetic acid or, formally, *N,N'*-1,2-ethanediyl*bis*(*N*-(carboxymethyl)glycine), forms very strong complexes with many metals of environmental interest. The structure of EDTA is

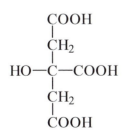

Figure 15.3: Structure of Citric Acid

HOOCCH₂ ⟩ N—CH₂—CH₂—N ⟨ CH₂COOH
HOOCCH₂ ⟩   ⟨ CH₂COOH
. EDTA coordinates with all four acidic oxygen atoms and both nitrogen atoms. Thus, it is a hexadentate

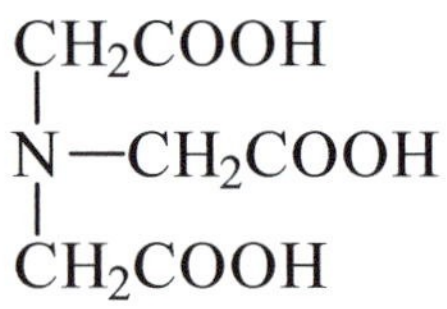

CH$_2$COOH
|
N—CH$_2$COOH
|
CH$_2$COOH

Figure 15.4: Structure of Nitrilotriacetic Acid (NTA)

ligand and can completely coordinate a metal with a coordination number of six. The ligand EDTA forms exceptionally strong complexes. It is used in pollutant analysis methods to bind unwanted metals and is administered in acute metal poisoning.

A second important class of ligands comprises naturally occurring organics in the environment. Litter from plant material is biodegraded by microorganisms into complex, poorly characterized organic matter. In the aquatic environment, the two important organic fractions are called fulvic and humic acids. Fulvic and humic acids are large molecular weight materials with many acidic functional groups. For example, fulvic acid is typically about 40% oxygen and 0.5% nitrogen by weight with a carboxyl acid content of about 6 mmol/g (Thurman and Malcolm, 1983). Thus, fulvic and humic acids are expected to be both Brønsted and Lewis acids.

The third class of organic ligands comprises those important in biochemistry. The classic example is the porphyrin ring, a modification of porphin (see Figure 15.5). One important metal-porphyrin complex is *hemoglobin*, an iron-porphyrin complex. Hemoglobin is responsible for oxygen transport in the blood of vertebrates. (Related compounds include *myoglobin*, which transports oxygen in muscle tissue, and *hemocyanin*, which is responsible for oxygen transport in the blood of some invertebrates.) Recall that iron has a coordination number of six (Table 15.1). Four electron pairs come from the four nitrogen atoms in the porphyrin ring. The iron-porphyrin system lies in a plane. The fifth electron pair comes from the nitrogen in an amino acid (histidine) in hemoglobin that lies outside the iron-porphyrin plane. The sixth electron pair comes from reversible binding to oxygen.

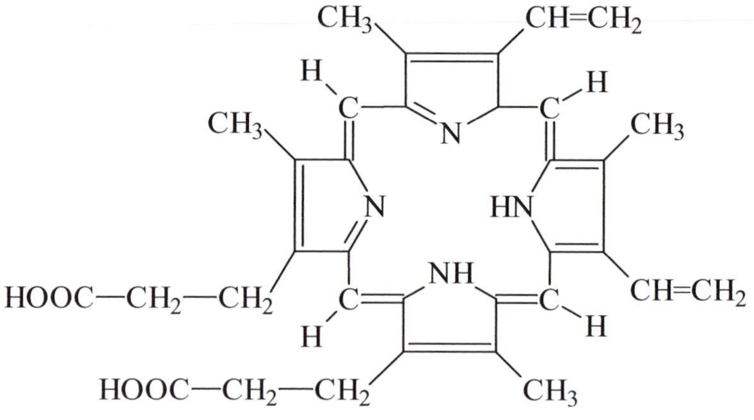

Figure 15.5: Structure of Porphin

Another important metal-porphyrin complex is chlorophyll. Chlorophyll, the principal photoreceptor in green plants, is based on porphyrin complexed with magnesium. Vitamin B_{12} (cobalamin) is based on the complexation of cobalt with a corrin ring (similar to the porphyrin ring). Cobalt is complexed with two other substituents perpendicular to the ring to satisfy its coordination number of six. As a final example, the cytochrome family comprises iron porphyrins used for electron transfer reactions. In the cytochromes, the central iron atom undergoes oxidation-reduction reactions (see Chapter 16).

15.3.6 Effects of pH

Key idea: Metals and H^+ often compete for binding sites with ligands that are Brønsted acids

With substances exhibiting both Brønsted and Lewis acidity, protons (H^+) and other metals (M^{n+}) compete for binding sites. Thus, the ability of fulvic acid humic acids (along with the simple organic acids) to bind metals depends on the pH.

Thoughtful Pause
How does pH affect metal binding?

At low pH, the activity of H^+ is high and H^+ tends to displace metals from the binding sites. For example, acidic rainfall will leach aluminum from soils. Why? The functional groups in soils that act as ligands are also Brønsted acids. At low pH, H^+ replaces Al^{3+} and aluminum is released.

It is important to remember that pH affects the chemistry of almost every metal *and* many ligands. Consider a qualitative description of the complexation of cadmium by acetic acid. At very low pH, acetic acid is protonated and a weaker complexing agent. The $[OH^-]$ is low, so cadmium is mostly in the aquo form. As the pH increases, the cadmium-acetate complex will form. At higher pH values, the ligand OH^- will be in high concentration and the hydroxo complexes will dominate.

15.4 EQUILIBRIUM CALCULATIONS WITH COMPLEXES

15.4.1 Introduction

Key idea: The equilibrium calculations for homogeneous systems involving complexation are set up exactly as with acid-base systems

Equilibrium calculations for homogeneous systems involving complexation are set up exactly as with acid-base systems. To set up the system, write the species list, pertinent equilibria, mass balance equations, and in some cases the electroneutrality equation. After the equilibria are translated into mathematical equations, the n equations in n unknowns (n species concentrations) can be solved for.

Recall from Chapter 6 that mass balances are needed to account for all species except H^+, OH^-, and H_2O. In this vein, systems with complexation usually have one mass balance equation for each metal and one mass balance equation for each ligand. Mass balances are not needed for the

Key idea: Metals that form hydroxo complexes will precipitate under the proper pH and total metal concentrations

ligands OH^- and H_2O. In the typical problem in aquatic chemistry, the total dissolved metal and total dissolved ligand are known.

Before going into the details of equilibrium calculations with complexes, an important caveat: *the techniques presented in this chapter are appropriate for homogeneous systems*. When metals hydrolyze to form aquo and hydroxo complexes, it is very common for solids to form under the proper total metal and pH conditions. For example, Fe(+III) can precipitate as $Fe(OH)_3(s)$ if the pH and $Fe(+III)_T$ are in the proper ranges. Equilibrium calculations in the presence of solids will be covered in Chapter 19. For the moment, *consider homogeneous systems only*. Thus, many of the examples in this chapter are at low total metal concentrations and/or low pH where metal precipitation typically does not occur.

15.4.2 Equilibrium constants for complexation

formation (stability) constant: the equilibrium constant for a complexation equilibrium, written with the complex as a product

The equilibrium constants for complex formation reactions are called *formation constants* or *stability constants*. Remember: a formation constant is just an equilibrium constant. It is an equilibrium constant for a reaction of the form:

$$aM^{n+} + bL^{m-} = M_aL_b^{\,n-m} \qquad K = \text{formation constant}$$

Most formation constants are large. Therefore, it is common to write and compare log K values. Recall that acid dissociation reactions are written in the reverse from of the complexation equilibrium shown above (with $M^{n+} = H^+$ and $K = K_a$). Thus, acid dissociation reactions have small K_a values and it is common to write and compare pK_a values.

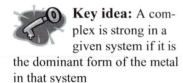

Key idea: A complex is strong in a given system if it is the dominant form of the metal in that system

It is common to say that the complex is *strong* if the formation constant is large. However, you must be careful when deciding whether a formation constant is large or when comparing formation constants. *The magnitude of the formation constant depends on the stoichiometry of the reaction.* For example, try to determine whether the tetrachloro-Hg(+II) complex is strong. The equilibrium constant for $Hg^{2+} + 4Cl^- = HgCl_4^-$ is $K = 10^{+15.2}$. This seems like a very large equilibrium constant, but recall that it has units of M^{-4}. If the Hg^{2+} and Cl^- concentrations could be held magically at 10^{-5} M, then the concentration of the complex would be only $10^{-9.8}$ M[†]. An example of comparing formation constants is given in Example 15.1. Given the potential problems with comparing formation constants, it is more precise to say that *a complex is strong in a given system if it is the dominant form of the metal in that system*.

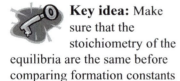

Key idea: Make sure that the stoichiometry of the equilibria are the same before comparing formation constants

---

[†] The phrase "only $10^{-9.8}$ M" is a bit misleading since a concentration of $10^{-9.8}$ M may be significant if the complex participates in important biological reactions.

Example 15.1: Comparing Formation Constants

Which metal forms a stronger complex with cyanide ion: Fe^{3+} or Ni^{2+}?

Solution:
In Table D.1 in Appendix D, the only complex formation equilibria for ferric iron and nickel with CN^- are:

$$Fe^{3+} + 6CN^- = Fe(CN)_6^{3-}$$
$$\log K_1 = 43.6$$

$$Ni^{2+} + 4CN^- = Ni(CN)_4^{2-}$$
$$\log K_2 = 30.2$$

It is not appropriate to simply compare K_1 and K_2, since they have different associated units. To determine which complex is stronger, it is necessary to perform an equilibrium calculation to see whether more $Fe(CN)_6^{3-}$ or more $Ni(CN)_4^{2-}$ is formed (see Problem 15.13).

Recall that acid dissociation constants are given a special symbol, namely, K_a. Similarly, some formation constants also are given special symbols. A common naming convention for formation constants is as follows:

1. Use K_i for stepwise formation equilibria (i.e., when the stoichiometry of the ligand is 1) resulting in the formation of ML_i.

   Example: For $Cu(OH)_2^0 + OH^- = Cu(OH)_3^-$, use the symbol K_3.

2. Use β_i for stepwise formation equilibria (i.e., when the stoichiometry of the ligand is $i \neq 1$).

   Example: For $Cu^{2+} + 3OH^- = Cu(OH)_3^-$, use the symbol β_3.

3. Use β_{nm} for formation of the polynuclear complex M_mL_n.

   Example: For $2Cu^{2+} + 2OH^- = Cu_2OH_2^{2+}$, use the symbol β_{22}.

4. Use $*K_i$, $*\beta_i$, or $*\beta_{nm}$ when the ligand is protonated (including reactions with water, where H_2O is the protonated form of OH^-).

   Example: For $Fe^{3+} + 2H_2O = Fe(OH)_2^+$, use the symbol $*\beta_2$.

For simplicity, only the symbol K_i will be used in this text, with i having no significance and chosen simply to assign a unique symbol to each formation constant. However, you should be aware of the naming convention, as formation constants frequently are compiled by using this system[†]. Remember: K_i, β_i, β_{nm}, $*K_i$, $*\beta_i$, and $*\beta_{nm}$ are just equilibrium constants.

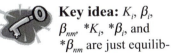 **Key idea:** K_i, β_i, β_{nm}, $*K_i$, $*\beta_i$, and $*\beta_{nm}$ are just equilibrium constants for formation equilibria written in a specific form

15.4.3 Hydroxo complexes
Recall from Section 15.3.2 that the formation of hydroxo complexes stems from the acidity of the aquo ligands. Thus, equilibrium calculations containing only soluble species are similar to acid-base equilibrium calculations. In general, pC-pH diagrams are useful for displaying the results of calculations.

[†] The naming convention has certain advantages. You could write $\beta_2 = 10^{+12.8}$ for Fe^{2+} and NTA and the reader would know you are referring to the formation equilibrium: $Fe^{2+} + 2L^{3-} = FeL_2^{4-}$ (where L^{3-} = unprotonated NTA). Thus, you could tabulate formation constants without writing the equilibria explicitly if you use the naming convention.

As an example, consider the speciation of dissolved aluminum in treated drinking water. Assume that no solid phases exist and that the total soluble aluminum concentration is about 0.5 µg/L as Al. Find the major dissolved aluminum species in tap water (pH 7.5-8) and determine whether the predominant soluble aluminum species changes when consumed tap water hits the stomach (at about pH 2). For the dissolved aluminum system, the pertinent equilibria are:

$$
\begin{aligned}
Al^{3+} + OH^- &= AlOH^{2+} & K_1 &= 10^{+9} \\
AlOH^{2+} + OH^- &= Al(OH)_2^+ & K_2 &= 10^{+8.8} \\
Al(OH)_2^+ + OH^- &= Al(OH)_3^0 & K_3 &= 10^{+7.7} \\
Al(OH)_3^0 + OH^- &= Al(OH)_4^- & K_4 &= 10^{+7.9}
\end{aligned}
$$

Thus:

$$
\begin{aligned}
[AlOH^{2+}] &= K_1[Al^{3+}][OH^-] \\
[Al(OH)_2^+] &= K_1K_2[Al^{3+}][OH^-]^2 \\
[Al(OH)_3^0] &= K_1K_2K_3[Al^{3+}][OH^-]^3 \\
[Al(OH)_4^-] &= K_1K_2K_3K_4[Al^{3+}][OH^-]^4
\end{aligned}
$$

A mass balance on soluble aluminum reveals:

$$
\begin{aligned}
Al(+III)_T &= \text{total soluble aluminum} \\
&= [Al^{3+}] + [AlOH^{2+}] + [Al(OH)_2^+] + [Al(OH)_3^0] \\
&\quad + [Al(OH)_4^-] \\
&= [Al^{3+}](1 + K_1[OH^-] + K_1K_2[OH^-]^2 + K_1K_2K_3[OH^-]^3 \\
&\quad + K_1K_2K_3K_4[OH^-]^4)
\end{aligned}
$$

Or:

$$
\begin{aligned}
Al(+III)_T &= [Al^{3+}](1 + K_1K_W/[H^+] + K_1K_2K_W^2/[H^+]^2 \\
&\quad + K_1K_2K_3K_W^3/[H^+]^3 + K_1K_2K_3K_4K_W^4/[H^+]^4)
\end{aligned}
$$

Here: $Al(+III)_T = (5 \times 10^{-7} \text{ g/L})/(27 \text{ g/mol}) = 1.85 \times 10^{-9}$ M; $pAl(+III)_T = 7.7$.

A pC-pH diagram for the system is shown in Figure 15.6. Clearly, Al^{3+} dominates at low pH and $Al(OH)_4^-$] dominates at high pH. The pC-pH diagram is a little crowded at intermediate pH values. To look for the dominate species, it is convenient to replot the calculated concentrations on a percentage basis, as shown in Figure 15.7. From Figure 15.7, it is easy to see that $Al(OH)_4^-$ will predominate in tap water. As the pH decreases when tap water mixes with gastric fluids, the predominant soluble aluminum species will change to the lower hydroxo complexes (i.e., the complexes with fewer hydroxo ligands). Below about pH 4.5, Al^{3+} is present at the highest concentration.

15.4.4 Other complexes

Complexes with ligands other than hydroxo and oxo ligands require an additional mass balance (or mass balances) to account for the material in the ligand. As an example, consider the speciation of mercury and chloride in a solution where 10^{-3} mole of mercuric chloride, $HgCl_2(s)$, is added to 1 L of water and dissolves completely. Assume the solution has been acidified and the pH is fixed and low enough so that hydroxo complexes are not formed. Because Hg(+II) has a coordination number of 4 (from Table 15.1), you might expect that the mercury-containing species will include Hg^{2+}, $HgCl^+$, $HgCl_2^0$, $HgCl_3^-$, and $HgCl_4^{2-}$.

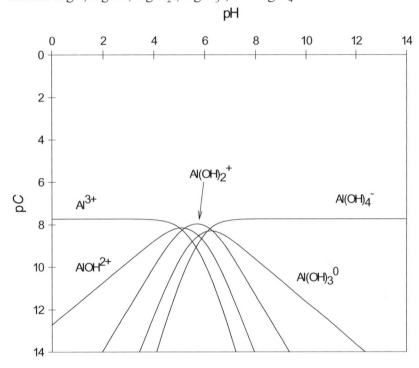

Figure 15.6: pC-pH Diagram for Aluminum
[no solid phase, $Al(+III)_T = 1.85 \times 10^{-9}$ M]

Example 15.2: **Inorganic Complex Formation**

(Note: the conditions and all formation constants in this example were taken from Butler, 2000.)

Solid waste from mining operations (mine tailings) frequently contain various FeS(s) minerals. In the presence of bacteria, water will leach through the tailings to form a very acidic liquid stream containing Fe^{3+} and SO_4^{2-}. The liquid stream is called *acid mine drainage*.

Determine the speciation of ferric iron and sulfur in acid mine drainage if the water is at pH 2 with $Fe(+III)_T = 5 \times 10^{-3}$ M and $S_T = 10^{-2}$ M.

Solution:
The species list includes Fe^{3+}, $Fe(OH)_x^{3-x}$ ($x = 1-4$), $Fe_2(OH)_2^{4+}$, $Fe_3(OH)_4^{5+}$, $FeSO_4^+$, $Fe(SO_4)_2^-$, SO_4^{2-}, and HSO_4^- (in addition to H^+, OH^-, and H_2O, all of which are at fixed activity at pH 2).

The pertinent equilibria are (with log K values):

$Fe^{3+} + OH^- = FeOH^{2+}$ 11.81
$Fe^{3+} + 2OH^- = FeOH_2^+$ 23.3
$Fe^{3+} + 3OH^- = FeOH_3^0$ 28.8
$Fe^{3+} + 4OH^- = FeOH_4^-$ 34.4
$2Fe^{3+} + 2OH^- = Fe_2(OH)_2^{4+}$ 25.1
$3Fe^{3+} + 4OH^- = Fe_3(OH)_4^{5+}$ 49.7
$Fe^{3+} + SO_4^{2-} = FeSO_4^+$ 4.04
$Fe^{3+} + 2SO_4^{2-} = Fe(SO_4)_2^-$ 5.38
$SO_4^{2-} + H^+ = HSO_4^-$ 1.99

The mass balances are:

Thoughtful Pause
What other species should be in the species list?

The species list also should include Cl^-, H^+, OH^-, and H_2O (assuming that the pH is not so low that HCl forms). However, since the pH is assumed to be fixed, $[H^+]$ and $[OH^-]$ are not unknowns. As usual, assume that the activity of water is unity. The species list becomes: Cl^-, Hg^{2+}, $HgCl^+$,

$Fe(+III)_T = [Fe^{3+}] + [FeOH^{2+}]$
$\quad + [FeOH_2^+] + [FeOH_3^0]$
$\quad + [FeOH_4^-]$
$\quad + 2[Fe_2(OH)_2^{4+}]$
$\quad + 3[Fe_3(OH)_4^{5+}] + [FeSO_4^+]$
$\quad + [Fe(SO_4)_2^-]$

$S_T = [SO_4^{2-}] + [HSO_4^-]$
$\quad + [FeSO_4^+] + 2[Fe(SO_4)_2^-]$

Writing each term in the ferric iron mass balance in terms of $[Fe^{3+}]$ and each term in the sulfate mass balance in terms of $[SO_4^{2-}]$:

$Fe(+III)_T = [Fe^{3+}](1 + K_1[OH^-]$
$\quad + K_2[OH^-]^2 + K_3[OH^-]^3$
$\quad + K_4[OH^-]^4 + 2K_5[Fe^{3+}]\times$
$\quad [OH^-]^2 + 3K_6[Fe^{3+}]^2[OH^-]^4$

$\quad + K_7[SO_4^{2-}] + K_8[SO_4^{2-}]^2)$

$S_T = [SO_4^{2-}](1 + K_9[H^+] +$
$\quad K_7[Fe^{3+}] + 2K_8[Fe^{3+}][SO_4^{2-}])$

The system has been reduced to two equations in two unknowns. Solving with $Fe(+III)_T = 5\times10^{-3}$ M, $S_T = 10^{-2}$ M, $[H^+] = 10^{-2}$ M, and $[OH^-] = 10^{-12}$ M (pH 2):

$[Fe^{3+}] = 1.75\times10^{-4}$ M
$[FeOH^{2+}] = 1.13\times10^{-4}$ M
$[FeOH_2^+] = 3.49\times10^{-5}$ M
$[FeOH_3^0] = 1.10\times10^{-11}$ M
$[FeOH_4^-] = 4.39\times10^{-18}$ M
$[Fe_2(OH)_2^{4+}] = 3.85\times10^{-7}$ M
$[Fe_3(OH)_4^{5+}] = 2.68\times10^{-10}$ M
$[FeSO_4^+] = 4.67\times10^{-3}$ M
$[Fe(SO_4)_2^-] = 2.49\times10^{-4}$ M
$[SO_4^{2-}] = 2.44\times10^{-3}$ M
$[HSO_4^-] = 2.38\times10^{-3}$ M

Note that over 98% of the ferric iron is complexed with sulfate, whereas only about 49% of the sulfate is complexed with iron.

$HgCl_2^0$, $HgCl_3^-$, and $HgCl_4^{2-}$.
The pertinent equilibria are:

$$Hg^{2+} + Cl^- = HgCl^+ \qquad K_1 = 10^{+6.74}$$
$$HgCl^+ + Cl^- = HgCl_2^0 \qquad K_2 = 10^{+6.48}$$
$$HgCl_2^0 + Cl^- = HgCl_3^- \qquad K_3 = 10^{+1}$$
$$HgCl_3^- + Cl^- = HgCl_4^{2-} \qquad K_4 = 10^{+1}$$

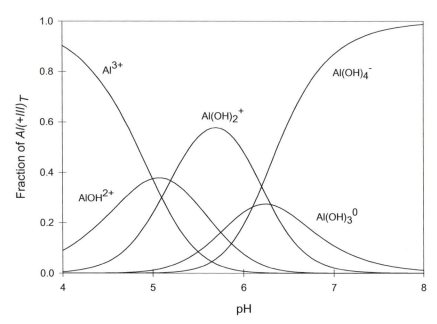

Figure 15.7: Predominance Diagram for Aluminum
[no solid phase, $Al(+III)_T = 1.85\times10^{-9}$ M]

Thus:

$$[HgCl^+] = K_1[Hg^{2+}][Cl^-] \qquad\qquad \text{eq. 15.5}$$
$$[HgCl_2^0] = K_2[HgCl^+][Cl^-] \qquad\qquad \text{eq. 15.6}$$
$$[HgCl_3^-] = K_3[HgCl_2^0][Cl^-] \qquad\qquad \text{eq. 15.7}$$
$$[HgCl_4^{2-}] = K_4[HgCl_3^-][Cl^-] \qquad\qquad \text{eq. 15.8}$$

Mass balances must be written for total dissolved mercury(+II) and total dissolved chloride. The pertinent mass balances are (you may wish to write the mass balances before reading on):

$$Hg(+II)_T = [Hg^{2+}] + [HgCl^+] + [HgCl_2^0] + [HgCl_3^-]$$
$$+ [HgCl_4^{2-}] \qquad\qquad \text{eq. 15.9}$$

$$Cl_T = [Cl^-] + [HgCl^+] + 2[HgCl_2^0] + 3[HgCl_3^-]$$
$$+ 4[HgCl_4^{2-}] \qquad \text{eq. 15.10}$$

Note the stoichiometric coefficients in the chloride mass balance: for example, 1 mole of $HgCl_3^-$ contains 3 moles of chloride. Also, the total chloride concentration is 2×10^{-3} M, and the total mercury(+II) concentration is 1×10^{-3} M if 10^{-3} mole of $HgCl_2(s)$ is added to 1 L of water and dissociates completely.

What about electroneutrality? Through equations 15.5-15.10, you have 6 equations in 6 unknowns. Another equation is not needed. Does this make sense? It is not possible to write the electroneutrality equation, since you do not know how much strong acid was added to reduce the pH.

A useful solution technique with complexation systems is as follows (see also Appendix C, Section C.7).

Key idea: To solve systems involving complexation (1) Solve for each species concentration as a function of the uncomplexed (i.e, aquo complexed) metals and uncomplexed ligands, (2) Substitute these expressions into the mass balances, (3) Rework to form one equation in one unknown and solve for the uncomplexed metal or ligand concentration, and (4) Back-calculate the concentrations of the complexed species

Step 1: Solve for each species concentration as a function of the uncomplexed (i.e, aquo complexed) metal and uncomplexed ligand

In this case, write each species concentration in terms of $[Hg^{2+}]$ and $[Cl^-]$. For the mercuric chloride system:

$$[HgCl^+] = K_1[Hg^{2+}][Cl^-] \qquad \text{eq. 15.11}$$
$$[HgCl_2^0] = K_1K_2[Hg^{2+}][Cl^-]^2 \qquad \text{eq. 15.12}$$
$$[HgCl_3^-] = K_1K_2K_3[Hg^{2+}][Cl^-]^3 \qquad \text{eq. 15.13}$$
$$[HgCl_4^{2-}] = K_1K_2K_3K_4[Hg^{2+}][Cl^-]^4 \qquad \text{eq. 15.14}$$

Step 2: Substitute the expressions from Step 1 into the mass balances

$$Hg(+II)_T = [Hg^{2+}](1 + K_1[Cl^-] + K_1K_2[Cl^-]^2 + K_1K_2K_3[Cl^-]^3 + K_1K_2K_3K_4[Cl^-]^4)$$

$$Cl_T = [Cl^-] + [Hg^{2+}](K_1[Cl^-] + 2K_1K_2[Cl^-]^2 + 3K_1K_2K_3[Cl^-]^3 + 4K_1K_2K_3K_4[Cl^-]^4)$$

Step 3: Rework the mass balances to obtain one equation in one species concentration

Solving the mercury(+II) mass balance for $[Hg^{2+}]$ and substituting into the chloride mass balance:

$$Cl_T = [Cl^-] + Hg(+II)_T(1 + K_1[Cl^-] + 2K_1K_2[Cl^-]^2$$
$$+ 3K_1K_2K_3[Cl^-]^3 + 4K_1K_2K_3K_4[Cl^-]^4)/(K_1[Cl^-]$$
$$+ K_1K_2[Cl^-]^2 + K_1K_2K_3[Cl^-]^3 + K_1K_2K_3K_4[Cl^-]^4)$$

eq. 15.15

Although messy, eq. 15.5 is one equation in one unknown, namely $[Cl^-]$. Solving by iteration, $[Cl^-] = 1.79\times10^{-5}$ M, and from the $Hg(+II)_T$ expression $[Hg^{2+}] = 1.85\times10^{-7}$ M.

Step 4: Back-calculate the concentrations of the complexed species

Thus: $[HgCl^+] = 1.82\times10^{-5}$ M, $[HgCl_2^0] = 9.81\times10^{-4}$ M, $[HgCl_3^-] = 1.76\times10^{-7}$ M, and $[HgCl_4^{2-}] = 3.14\times10^{-11}$ M.

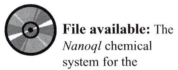

File available: The *Nanoql* chemical system for the Hg(+II)-chloride system may be found on the CD - file location **/Chapter 15/Hg chloride.nqs**

You can verify that the mass balances are satisfied. Note that $HgCl_2^0$ is the dominant species, accounting for 98% of the total dissolved mercury and 98% of the total chloride. The remaining mercury is mostly $HgCl^+$ and the remaining total chloride is mostly Cl^- and $HgCl^+$. The *Nanoql* file for this system is available on the CD at **/Chapter 15/Hg chloride.nqs**.

In this example, the ratio of $Hg(+II)_T:Cl_T$ was fixed because $HgCl_2(s)$ was added to water. If you were to dissolve 10^{-3} M of Hg(+II) into water and add chloride (by, say, adding NaCl(s)), the changes in species concentrations with Cl_T would be as shown in Figure 15.8. (To avoid too many lines in Figure 15.8, the concentrations of Hg^{2+} and Cl^- are shown in the top panel and the other species shown in the bottom panel.)

Thoughtful Pause
Is complexation important at low total chloride concentrations?

As with so many of the *Thoughtful Pauses* in this text, the answer to this question is: It depends. *The importance of complexation depends on whether you are interested in mercury or chloride.* At $Cl_T \approx 1\times10^{-4}$ M, most of the mercury is *uncomplexed* (at low pH) and Hg^{2+} is the dominant mercury species. However, most of the chloride is *complexed* and found in the form of $HgCl^+$. At $Cl_T \approx 1\times10^{-3}$ M, $HgCl^+$ and $HgCl_2^0$ are becoming more important in the mercury mass balance. In addition, they are the dominant chloride-containing species. Thus, most of the mercury is complexed and most of the chloride is complexed. At $Cl_T \approx 1\times10^{-2}$ M, $HgCl_2^0$ is the dominant mercury-containing species and chloride is the chloride-containing species at highest concentration. Here, most of the mercury is complexed, but most of the chloride is not complexed since chloride is present in great excess over mercury. Another worked example of complexation is shown in Example 15.2.

15.5 SYSTEMS WITH SEVERAL METALS AND LIGANDS

15.5.1 Introduction

Natural waters contain many metals and ligands. In principle, equilibrium calculations with very complicated natural or process waters should be conducted with the same steps as outlined in Section 15.4.2: (1) solve for each species concentration as a function of the uncomplexed (i.e, aquo complexed) metals and uncomplexed ligands, (2) substitute these expressions into the mass balances, (3) rework to form one equation in one unknown and solve for the uncomplexed metal or ligand concentration, and (4) back-calculate the concentrations of the complexed species.

15.5.2 Example system

As an example, consider the addition of nitrilotriacetic acid (NTA) to a natural water. NTA (shown in Figure 15.4) is used as a replacement for phosphate in detergents and for treating boiler water (to complex the calcium and magnesium that make up hardness). The World Health Organization has recommended that drinking water contain no more than 200 μg/L of NTA. Our task is to determine the speciation of metals and ligands in a lake at pH 8 if the total soluble calcium concentration (= $Ca(+II)_T$) is 1×10^{-3} M, the total soluble copper concentration (= $Cu(+II)_T$) is 2×10^{-6} M, and the total soluble ferrous iron concentration (= $Fe(+II)_T$) is 2×10^{-6} M[†]. The calculations shall be performed here with two total NTA concentrations (= N_T): a low NTA concentration (10 μg/L ≈ 5×10^{-8} M) and a high NTA concentration (2 mg/L ≈ 1×10^{-5} M).

To simplify the symbols, the symbol L^{3-} will be used to represent the uncomplexed NTA. This system is very complicated.

Thoughtful Pause
What are the uncomplexed Lewis acids and Lewis bases in the system?

The system contains four uncomplexed Lewis acids (H^+, Ca^{2+}, Cu^{2+}, and Fe^{2+}) and three uncomplexed Lewis bases (OH^-, L^{3-}, and H_2O). The pertinent equilibrium are:

[†] To simplify the system a bit, complexation of NTA with Fe(+III) is ignored in this problem.

***Example 15.3:* Complex Formation with More than One Metal**

Return to Example 15.1, where the challenge was to determine whether Fe^{3+} or Ni^{2+} forms the stronger complex with cyanide ion. One way to address this question is to add the same amount of total Fe(+III) and Ni(+II) to a solution containing some total cyanide and calculating the concentrations of the complexes formed. At pH 2, $Fe_T = Ni_T = 1\times10^{-5}$ M, and $CN_T = 1\times10^{-3}$ M, determine which cyanide complex is formed at the highest concentration.

Solution:
The pertinent equilibria are:

$CN^- + H^+ = HCN \quad \log K = 9.2$
$Fe^{3+} + 6CN^- = Fe(CN)_6^{3-}$
$\quad \log K_1 = 43.6$
$Ni^{2+} + 4CN^- = Ni(CN)_4^{2-}$
$\quad \log K_2 = 30.2$

(Note: At pH 2, the hydroxo complexes of iron and nickel can be ignored.)

Writing mass balances:

$Fe_T = [Fe^{3+}] + [Fe(CN)_6^{3-}]$
$Ni_T = [Ni^{2+}] + [Ni(CN)_4^{2-}]$
$CN_T = [CN^-] + [HCN] +$
$\quad 6[Fe(CN)_6^{3-}] +$
$\quad 4[Ni(CN)_4^{2-}]$

Thus:

$$Fe_T = [Fe^{3+}](1 + K_1[CN^-]^6)$$
$$Ni_T = [Ni^{2+}](1 + K_2[CN^-]^4)$$
$$CN_T = [CN^-](1 + K[H^+] +$$
$$K_1[Fe^{3+}][CN^-]^5 +$$
$$K_1[Ni^{2+}][CN^-]^3)$$

Write expressions for $[Fe^{3+}]$ and $[Ni^{2+}]$ from the metal mass balances and substitute them into the cyanide mass balance. The resulting expression is one equation in one unknown (i.e., $[CN^-]$). Solving at pH 2 with the total masses specified:

$$[Fe(CN)_6^{3-}] = 2.5 \times 10^{-23} \text{ M and}$$
$$[Ni(CN)_4^{2-}] = 2.5 \times 10^{-16} \text{ M}$$

Thus, **$Ni(CN)_4^{2-}$ is the stronger complex**.

(Note: Most of the total cyanide is HCN. At pH 2, H^+ outcompetes both Fe^{3+} and Ni^{2+} for cyanide ion. Also note that "strong complex" is used here to refer to complex formation. The bonds in $Fe(CN)_6^{3-}$ are very difficult to break, and thus it is sometimes called a strong cyanide complex.)

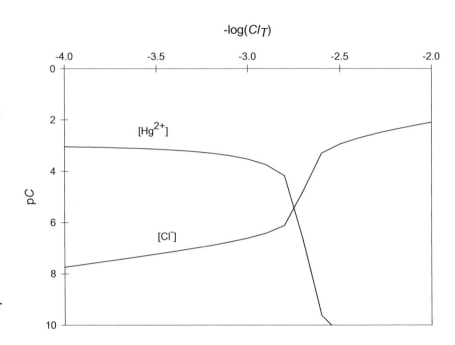

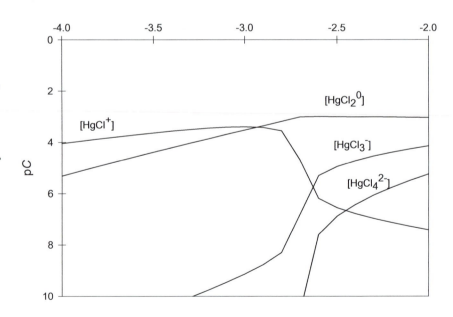

Figure 15.8: pC-pCl$_T$ Diagram for Mercury(+II)
(no solid phases considered)

Complexation with H^+:

$$H^+ + L^{3-} = HL^{2-} \qquad\qquad K_1 = 10^{+10.3}$$
$$H^+ + HL^{2-} = H_2L^- \qquad\qquad K_2 = 10^{+2.9}$$
$$H^+ + H_2L^- = H_3L \qquad\qquad K_3 = 10^{+1.7}$$
$$H^+ + OH^- = H_2O \qquad\qquad K = 1/K_W = 10^{+14}$$

Complexation with Ca^{2+}:

$$Ca^{2+} + L^{3-} = CaL^- \qquad\qquad K_4 = 10^{+7.6}$$
$$Ca^{2+} + 2L^{3-} = CaL_2^{4-} \qquad\qquad K_5 = 10^{+8.9}$$
$$Ca^{2+} + OH^- = CaOH^+ \qquad\qquad K_6 = 10^{+0.64}$$

Complexation with Cu^{2+}:

$$Cu^{2+} + L^{3-} = CuL^- \qquad\qquad K_7 = 10^{+12.9}$$
$$Cu^{2+} + 2L^{3-} = CuL_2^{4-} \qquad\qquad K_8 = 10^{+17.4}$$
$$CuL^- + OH^- = CuOHL^{2-} \qquad\qquad K_9 = 10^{+4.4}$$
$$Cu^{2+} + OH^- = CuOH^+ \qquad\qquad K_{10} = 10^{+6.3}$$
$$Cu^{2+} + 2OH^- = Cu(OH)_2^0 \qquad\qquad K_{11} = 10^{+12.8}$$
$$Cu^{2+} + 3OH^- = Cu(OH)_3^- \qquad\qquad K_{12} = 10^{+14.5}$$
$$Cu^{2+} + 4OH^- = Cu(OH)_4^{2-} \qquad\qquad K_{13} = 10^{+16.4}$$
$$2Cu^{2+} + 2OH^- = Cu_2OH_2^{2+} \qquad\qquad K_{14} = 10^{+17.7}$$

Complexation with Fe^{2+}:

$$Fe^{2+} + L^{3-} = FeL^- \qquad\qquad K_{15} = 10^{+8.3}$$
$$Fe^{2+} + 2L^{3-} = FeL_2^{4-} \qquad\qquad K_{16} = 10^{+12.8}$$
$$H^+ + FeL^- = FeHL^0 \qquad\qquad K_{17} = 10^{+1.9}$$
$$Fe^{2+} + OH^- + L^{3-} = FeOHL^{2-} \qquad\qquad K_{18} = 10^{+11.7}$$
$$Fe^{2+} + OH^- = FeOH^+ \qquad\qquad K_{19} = 10^{+4.5}$$
$$Fe^{2+} + 2OH^- = Fe(OH)_2^0 \qquad\qquad K_{20} = 10^{+7.4}$$
$$Fe^{2+} + 3OH^- = Fe(OH)_3^- \qquad\qquad K_{21} = 10^{+10.0}$$
$$Fe^{2+} + 4OH^- = Fe(OH)_4^{2-} \qquad\qquad K_{22} = 10^{+9.6}$$

The pertinent mass balances are:

$$NTA_T = [L^{3-}] + [HL^{2-}] + [H_2L^-] + [H_3L] + [CaL^-] +$$
$$2[CaL_2^{4-}] + [CuL^-] + 2[CuL_2^{4-}] + [CuOHL^{2-}] +$$
$$[FeL^-] + 2[FeL_2^{4-}] + [FeHL^0] + [FeOHL^{2-}] \quad \text{eq. 15.16}$$

$$Ca(+II)_T = [Ca^{2+}] + [CaL^-] + [CaL_2^{4-}] + [CaOH^+] \qquad\qquad \text{eq. 15.17}$$

$$Cu(+II)_T = [Cu^{2+}] + [CuL^-] + [CuL_2^{4-}] + [CuOHL^{2-}] +$$
$$[CuOH^+] + [Cu(OH)_2^0] + [Cu(OH)_3^-] +$$
$$[Cu(OH)_4^{2-}] + 2[Cu_2(OH)_2^{2+}] \qquad\qquad \text{eq. 15.18}$$

$$Fe(+II)_T = [Fe^{2+}] + [FeL^-] + [FeL_2^{4-}] + [FeHL^0] +$$
$$[FeOHL^{2-}] + [FeOH^+] + [Fe(OH)_2^0] +$$
$$[Fe(OH)_3^-] + [Fe(OH)_4^{2-}] \qquad \text{eq. 15.19}$$

Note that not all the stoichiometric coefficients in the mass balances are equal to one. This system has 29 unknowns (i.e., 29 species concentrations) and 29 equations (23 equilibria, 4 mass balances, and two species with fixed activities: $\{H_2O\} = 1$ and $\{H^+\} \approx [H^+] = 10^{-8}$ M or pH 8.

15.5.3 Manipulating the system

Although this is a big system, approach it just like other complexation equilibria. You must solve for each species concentration as a function of the uncomplexed metals and ligands, then substitute the resulting expressions into the mass balances (eqs. 15.16-15.19). The mass balances then become:

$$NTA_T = [L^{3-}](1 + K_1[H^+] + K_1K_2[H^+]^2 + K_1K_2K_3[H^+]^3 + K_4[Ca^{2+}] +$$
$$2K_5[Ca^{2+}][L^{3-}] + K_7[Cu^{2+}] + 2K_8[Cu^{2+}][L^{3-}] +$$
$$K_7K_9[Cu^{2+}][OH^-] + K_{15}[Fe^{2+}] + 2K_{16}[Fe^{2+}][L^{3-}] +$$
$$K_{15}K_{17}[Fe^{2+}][H^+] + K_{18}[Fe^{2+}][OH^-])$$

$$Ca(+II)_T = [Ca^{2+}](1 + K_4[L^{3-}] + K_5[L^{3-}]^2 + K_6[OH^-])$$

$$Cu(+II)_T = [Cu^{2+}](1 + K_7[L^{3-}] + K_8[L^{3-}]^2 + K_7K_9[L^{3-}][OH^-] +$$
$$K_{10}[OH^-] + K_{11}[OH^-]^2 + K_{12}[OH^-]^3 + K_{13}[OH^-]^4 +$$
$$2K_{14}[Cu^{2+}][OH^-]^2)$$

$$Fe(+II)_T = [Fe^{2+}](1 + K_{15}[L^{3-}] + K_{16}[L^{3-}]^2 + K_{15}K_{17}[H^+][L^{3-}] +$$
$$K_{18}[L^{3-}][OH^-] + K_{19}[OH^-] + K_{20}[OH^-]^2 + K_{21}[OH^-]^3 +$$
$$K_{22}[OH^-]^4)$$

The system has been reduced to four equations in four unknowns: $[L^{3-}]$, $[Ca^{2+}]$, $[Cu^{2+}]$, and $[Fe^{2+}]$. At pH 8 and assuming a dilute solution: $[H^+] = 10^{-8}$ M and $[OH^-] = 10^{-6}$ M. Solving and back-substituting, you can obtain the concentrations shown in Table 15.3. Another example of an equilibrium calculation with more than one metal is shown in Example 15.3.

File available: The spreadsheet for the NTA-Ca(+II)-Cu(+II)-Fe(+II) system may be found on the CD - file location **/Chapter 15/NTA system.xls**

15.5.4 Computer solution

How can you solve such a complicated system? An easy way to calculate equilibrium concentrations is with a spreadsheet. An example spreadsheet is shown in Figure 15.9 and is on the CD at **/Chapter 15/NTA system.xls**. The equilibria and equilibrium constants are listed in columns H, I, and J. For ease of entry, log K values were entered and K values calculated by Excel as $K = 10^{\log K}$. $[H^+]$ and $[OH^-]$ values are listed in cells C10 and C11. The four unknowns are shown in cells B2 through B5. The guesses of their

$-\log$(concentration) and the calculated concentrations ($= 10^{-pC}$) are in columns D and C, respectively. The given total masses are listed in cells C41-C44. The equations for each species are entered in cells C11 through C37. For example, $[CaL_2^{4-}]$ at equilibrium is equal to $K_5[Ca^{2+}][L^{3-}]^2$, so the formula entered in cell C16 is: **=I7*C18*C12^2**. The estimated total masses are calculated by summing the species concentrations with the appropriate stoichiometric coefficients (see eqs. 15.16-15.19).

Table 15.3: Solutions to the NTA Problem
(all concentrations in M)

	Exact Solution		Approximate Solution	
NTA_T:	5×10^{-8} M	1×10^{-5} M	5×10^{-8} M	1×10^{-5} M
$[L^{3-}]$	2.92×10^{-14}	2.00×10^{-10}	2.92×10^{-14}	2.00×10^{-10}
$[HL^{2-}]$	5.83×10^{-12}	3.99×10^{-8}	5.83×10^{-12}	3.98×10^{-8}
$[H_2L^-]$	4.63×10^{-17}	3.17×10^{-13}	4.63×10^{-17}	3.16×10^{-13}
$[H_3L]$	2.32×10^{-23}	1.59×10^{-19}	2.32×10^{-23}	1.58×10^{-19}
$[Ca^{2+}]$	1.00×10^{-3}	9.92×10^{-4}	1.00×10^{-3}	1.00×10^{-3}
$[CaL^-]$	1.16×10^{-9}	7.90×10^{-6}	1.16×10^{-9}	7.94×10^{-6}
$[CaL_2^{4-}]$	6.78×10^{-22}	3.15×10^{-14}	6.79×10^{-22}	3.16×10^{-14}
$[CaOH^+]$	4.37×10^{-9}	4.33×10^{-9}	4.37×10^{-9}	4.37×10^{-9}
$[Cu^{2+}]$	2.05×10^{-7}	1.22×10^{-9}	2.10×10^{-7}	1.25×10^{-9}
$[CuL^-]$	4.76×10^{-8}	1.94×10^{-6}	4.87×10^{-8}	1.99×10^{-6}
$[CuL_2^{4-}]$	4.40×10^{-17}	1.23×10^{-11}	4.50×10^{-17}	1.25×10^{-11}
$[CuOHL^{2-}]$	1.20×10^{-9}	4.87×10^{-8}	1.22×10^{-9}	4.99×10^{-8}
$[CuOH^+]$	4.09×10^{-7}	2.44×10^{-9}	4.18×10^{-7}	2.50×10^{-9}
$[Cu(OH)_2^0]$	1.29×10^{-6}	7.71×10^{-9}	1.32×10^{-6}	7.92×10^{-9}
$[Cu(OH)_3^-]$	6.49×10^{-11}	3.86×10^{-13}	6.63×10^{-11}	3.97×10^{-13}
$[Cu(OH)_4^{2-}]$	5.15×10^{-15}	3.07×10^{-17}	5.27×10^{-15}	3.15×10^{-17}
$[Cu_2(OH)_2^{2+}]$	2.11×10^{-8}	7.48×10^{-13}	2.20×10^{-8}	7.89×10^{-13}
$[Fe^{2+}]$	1.94×10^{-6}	1.87×10^{-6}	1.94×10^{-6}	1.87×10^{-6}
$[FeL^-]$	1.13×10^{-11}	7.45×10^{-8}	1.13×10^{-11}	7.43×10^{-8}
$[FeL_2^{4-}]$	1.04×10^{-20}	4.71×10^{-13}	1.05×10^{-20}	4.69×10^{-13}
$[FeHL^0]$	8.98×10^{-18}	5.91×10^{-14}	8.98×10^{-18}	5.90×10^{-14}
$[FeOHL^{2-}]$	2.84×10^{-14}	1.87×10^{-10}	2.84×10^{-14}	1.87×10^{-10}
$[FeOH^+]$	6.13×10^{-8}	5.90×10^{-8}	6.13×10^{-8}	5.90×10^{-8}
$[Fe(OH)_2^0]$	4.87×10^{-11}	4.69×10^{-11}	4.87×10^{-11}	4.69×10^{-11}
$[Fe(OH)_3^-]$	1.94×10^{-14}	1.87×10^{-14}	1.94×10^{-14}	1.87×10^{-14}
$[Fe(OH)_4^{2-}]$	7.72×10^{-21}	7.43×10^{-21}	7.72×10^{-21}	7.43×10^{-21}

To find the equilibrium concentrations, use a nonlinear tool (such as *Solver*) to adjust the concentrations of $[L^{3-}]$, $[Ca^{2+}]$, $[Cu^{2+}]$, and $[Fe^{2+}]$ until the calculated total masses (cells G12-G15) match the actual total masses

Key idea: Spreadsheet hints: (1) optimize by changing the p(concentration) or log(concentration) values, not the concentration values themselves, and (2) use relative errors in the objective function, not absolute errors

File available: The *Nanoql* chemical system for the NTA-Ca(+II)-Cu(+II)-Fe(+II) system may be found on the CD - file location
/Chapter 15/NTA.nqs

Key idea: To simplify chemical systems, ignore the species in each mass that are expected to be at low concentrations relative to the other species in the mass balance

---

***Example 15.4:* Approximation Method**

Gold can be electroplated onto material in a cyanide-based process. Nickel is added to the plating bath to improve the hardness and brightness of the gold finish. Typical bath conditions for acidic plating are pH 4.2, $Ni(+II)_T = 0.4$ mg/L as Ni, and 8 g/L of potassium gold cyanide, $Au(CN)_2K$.

What fractions of the gold and nickel are complexed with cyanide? At low pH, cyanide forms HCN, which is very toxic. What is the HCN concentration at equilibrium?

(cells F12-F15). To adjust the concentrations, you must devise an objective function. Here, calculate the sum of the squares of the relative errors for each total mass. The square of the relative error is:

$$[(\text{actual total mass} - \text{calculated total mass})/\text{actual total mass}]^2$$

Using *Solver*, minimize (or set equal to zero) the sum of the squares of the relative errors (cell H17) by changing the pC values of $[L^{3-}]$, $[Ca^{2+}]$, $[Cu^{2+}]$, and $[Fe^{2+}]$ (cells F6 through F9). Initial guesses for the pC values of $[L^{3-}]$, $[Ca^{2+}]$, $[Cu^{2+}]$, and $[Fe^{2+}]$ are required, but can be crude.

	A B	C	D	E	F	G	H	I	J	K	L M
1	**Components**	Conc.	pC	Guesses		**Equilibria**		K'	log K'		
2	L^{3-}	2.00E-10	9.70E+00	1.00E+01		$H^+ + L^{3-} = HL^{2-}$		2.00E+10	10.3	K_1	
3	Ca^{2+}	9.92E-04	3.00E+00	4.00E+00		$H^+ + HL^{2-} = H_2L^-$		7.94E+02	2.9	K_2	
4	Cu^{2+}	1.22E-09	8.91E+00	7.00E+00		$H^+ + H_2L^- = H_3L$		5.01E+01	1.7	K_3	
5	Fe^{2+}	1.87E-06	5.73E+00	6.00E+00		$H^+ + OH^- = H_2O$		1.00E+14	14	$1/K_w$	
6						$Ca^{2+} + L^{3-} = CaL^-$		3.98E+07	7.6	K_4	
7						$Ca^{2+} + 2L^{3-} = CaL_2^{4-}$		7.94E+08	8.9	K_5	
8	**Species**	Conc.				$Ca^{2+} + OH^- = CaOH^+$		4.37E+00	0.64	K_6	
9	H_2O	1.00E+00	known			$Cu^{2+} + L^{3-} = CuL^-$		7.94E+12	12.9	K_7	
10	H^+	1.00E-08	known			$Cu^{2+} + 2L^{3-} = CuL_2^{4-}$		2.51E+17	17.4	K_8	
11	OH^-	1.00E-06				$CuL^- + OH^- = CuOHL^{2-}$		2.51E+04	4.4	K_9	
12	L^{3-}	2.00E-10				$Cu^{2+} + OH^- = CuOH^+$		2.00E+06	6.3	K_{10}	
13	HL^{2-}	3.99E-08				$Cu^{2+} + 2OH^- = Cu(OH)_2$		6.31E+12	12.8	K_{11}	
14	H_2L^-	3.17E-13				$Cu^{2+} + 3OH^- = Cu(OH)_3^-$		3.16E+14	14.5	K_{12}	
15	H_3L	1.59E-19				$Cu^{2+} + 4OH^- = Cu(OH)_4^{2-}$		2.51E+16	16.4	K_{13}	
16	Ca^{2+}	9.92E-04				$2Cu^{2+} + 2OH^- = Cu_2OH_2^{2+}$		5.01E+17	17.7	K_{14}	
17	CaL^-	7.90E-06				$Fe^{2+} + L^{3-} = FeL^-$		2.00E+08	8.3	K_{15}	
18	CaL_2^{4-}	3.15E-14				$Fe^{2+} + 2L^{3-} = FeL_2^{4-}$		6.31E+12	12.8	K_{16}	
19	$CaOH^+$	4.33E-09				$H^+ + FeL^- = FeHL^0$		7.94E+01	1.9	K_{17}	
20	Cu^{2+}	1.22E-09				$Fe^{2+} + OH^- + L^{3-} = FeOHL^{2-}$		5.01E+11	11.7	K_{18}	
21	CuL^-	1.94E-06				$Fe^{2+} + OH^- = FeOH^+$		3.16E+04	4.5	K_{19}	
22	CuL_2^{4-}	1.23E-11				$Fe^{2+} + 2OH^- = Fe(OH)_2$		2.51E+07	7.4	K_{20}	
23	$CuOHL^{2-}$	4.87E-08				$Fe^{2+} + 3OH^- = Fe(OH)_3^-$		1.00E+10	10	K_{21}	
24	$CuOH^+$	2.44E-09				$Fe^{2+} + 4OH^- = Fe(OH)_4^{2-}$		3.98E+09	9.6	K_{22}	
25	$Cu(OH)_2$	7.71E-09									
26	$Cu(OH)_3^-$	3.86E-13									
27	$Cu(OH)_4^{2-}$	3.07E-17			**Notes**						
28	$Cu_2(OH)_2^{2+}$	7.48E-13				Problem:		NTA with Ca, Cu, and Fe(II)			
29	Fe^{2+}	1.87E-06				Fitting:		pL, pCa, pCu, pFe			
30	FeL^-	7.45E-08				Obj. function:		Min. sum of squares of rel. error in mass balances			
31	FeL_2^{4-}	4.71E-13				Constraints:		None			
32	$FeHL^0$	5.91E-14									
33	$FeOHL^{2-}$	1.87E-10									
34	$FeOH^+$	5.90E-08									
35	$Fe(OH)_2$	4.69E-11									
36	$Fe(OH)_3^-$	1.87E-14									
37	$Fe(OH)_4^{2-}$	7.43E-21									
38											
39											
40	**Mass Balances**	Added	Calculated	Rel Error^2							
41	N_T	1.00E-05	1.00E-05	1.49E-11							
42	Ca_T	1.00E-03	1.00E-03	3.70E-11							
43	Cu_T	2.00E-06	2.00E-06	4.87E-11							
44	Fe_T	2.00E-06	2.00E-06	1.05E-10							
45			sum	2.06E-10							
46											
47	**Electroneutral**	Calculated									
48	Charge bal.	N/A									
49	Proton cond.										
50											

Figure 15.9: Spreadsheet Solution for NTA Problem

Two useful tricks aid in the spreadsheet solution. First, *optimize by changing the p(concentration) or log(concentration) values, not the concentration values themselves.* This increases the likelihood of converging, since the equilibrium concentration values may vary over many orders of magnitude. Second, use *relative* errors in the objective function, not *absolute* errors. In this problem, the absolute error for the calcium mass balance will overwhelm the other mass balances, since $Ca(+II)_T$ is so much larger than the other total masses. Using relative errors results in more

Solution:
Potassium gold cyanide serves as the source of both gold and cyanide. The total gold(+I) concentration is (8 g/L)/(288.1 g/mol) = 2.8×10^{-2} M. The total cyanide concentration is two times this value or 5.6×10^{-2} M. The total nickel(+II) concentration is $(4 \times 10^{-4} \text{ g/L})/(58.7$ g/mol) = 6.8×10^{-6} M.

Possible gold, nickel, and cyanide species include Au^+, $Au(CN)_2^-$, Ni^{2+}, $NiOH^+$, $Ni(OH)_2^0$, $Ni(OH)_3^-$, $Ni(CN)_4^{2-}$, CN^-, and HCN. The pertinent equilibria are (with log K values):

$Au^+ + 2CN^- = Au(CN)_2^-$	40
$Ni^{2+} + OH^- = NiOH^+$	4.1
$Ni^{2+} + 2OH^- = Ni(OH)_2^0$	9
$Ni^{2+} + 3OH^- = Ni(OH)_3^-$	12
$Ni^{2+} + 4CN^- = Ni(CN)_4^{2-}$	30.2
$CN^- + H^+ = HCN$	9.2

The resulting mass balances are:

$$Au(+I)_T = [Au^+] + [Au(CN)_2^-]$$

$$Ni(+II)_T = [Ni^{2+}] + [NiOH^+] + [Ni(OH)_2^0] + [Ni(OH)_3^-] + [Ni(CN)_4^{2-}]$$

$$CN_T = [CN^-] + [HCN] + 2[Au(CN)_2^-] + 4[Ni(CN)_4^{2-}]$$

An examination of the formation constants for the hydroxo complexes reveals that hydroxo complex formation is expected to be low at pH 4.1. In addition, the concentration of HCN (pK_a 9.2) will be much larger than [CN^-]. Thus, the mass balances become:

equitable weighting of the errors in calculating the total masses.

The system also can be solved by *Nanoql*. The *Nanoql* file is included on the CD at **/Chapter 15/NTA.nqs**.

15.5.5 Approximate solution
The available software allows for the solution of complex systems with relative ease. However, your chemical intuition is sharpened by simplifying the system through approximations. Approximations should be approached in a systematic way. A good approach is to *ignore the species in each mass balance that are expected to be at low concentrations relative to the other species in the mass balance*. For example, in the calcium mass balance, you know that:

$$[CaL^-] = K_4[Ca^{2+}][L^{3-}] = 10^{+7.6}[Ca^{2+}][L^{3-}]$$
$$[CaL_2^{4-}] = K_5[Ca^{2+}][L^{3-}]^2 = 10^{+8.9}[Ca^{2+}][L^{3-}]^2 \text{ and:}$$
$$[CaOH^+] = K_6[Ca^{2+}][OH^-] = 10^{+0.64}[Ca^{2+}][OH^-] = 10^{-5.36}[Ca^{2+}]$$
(at pH 8)

Thoughtful Pause
Which species in the calcium mass balance can be ignored?

Clearly, $[CaOH^+] \ll [Ca^{2+}]$. Also, $[CaL^-]$ and $[CaL_2^{4-}]$ will be much smaller than $[Ca^{2+}]$ as long as $[L^{3-}] \ll 10^{-7.6}$ M. It seems likely that $[L^{3-}]$ will be much smaller than $10^{-7.6}$ M when $L_T = 5 \times 10^{-8}$ M $= 10^{-7.3}$ M. For the moment, assume that $[L^{3-}]$ also will be much smaller than $10^{-7.6}$ M when $L_T = 1 \times 10^{-5}$ M. Under this assumption, the calcium mass balance becomes: $Ca(+II)_T \approx [Ca^{2+}]$.

For the hydroxo-only complexes of copper, $[CuOH^+]$ and $[Cu(OH)_2^0]$ are much larger than $[Cu(OH)_3^-]$ and $[Cu(OH)_4^{2-}]$. The dimer $Cu_2(OH)_2^{2+}$ also can be ignored. Why? Note that $[Cu_2(OH)_2^{2+}] = K_{14}[Cu^{2+}]^2[OH^-]^2 = 10^{+17.7}[Cu^{2+}]^2[OH^-]^2$) is much smaller than $[Cu(OH)_2^0]$ ($= K_{11}[Cu^{2+}][OH^-]^2 = 10^{+12.8}[Cu^{2+}][OH^-]^2$) as long as $[Cu^{2+}] < 10^{-4.9}$ M. Of course, $[Cu^{2+}] < 10^{-4.9}$ M, since $Cu(+II)_T = 10^{-6}$ M. As a general rule, dimers often are present at small concentrations when the total metal concentration is small (see Problem 15.9).

How about the relative concentrations of $[Cu^{2+}]$, $[CuL^-]$ ($= 10^{+12.9}[Cu^{2+}][L^{3-}]$), $[CuL_2^{4-}]$ ($= 10^{+17.4}[Cu^{2+}][L^{3-}]^2$), and $[CuOHL^{2-}]$ ($= 10^{+11.3}[Cu^{2+}][L^{3-}]$ at pH 8)? Of these species, $[CuL_2^{4-}]$ can be ignored relative to $[CuL^-]$. Why? You know $[CuL^-]/[CuL_2^{4-}] = 10^{-4.5}/[L^{3-}]$. This ratio must be greater than one, since $[L^{3-}]$ is less than $10^{-4.5}$ M. Also you know at pH 8 that $[CuL^-]/[CuOHL^{2-}] = 10^{+1.6}$, so $[CuOHL^{2-}]$ can be ignored relative to $[CuL^-]$. The simplified mass balance for cooper

$Au(+I)_T = [Au^+] +$
$\qquad\qquad [Au(CN)_2^-]$
$\qquad = [Au^+](1 +$
$\qquad\qquad K_1[CN^-]^2)$

$Ni(+II)_T \approx [Ni^{2+}] +$
$\qquad\qquad [Ni(CN)_4^{2-}]$
$\qquad = [Ni^{2+}](1 +$
$\qquad\qquad K_5[CN^-]^2)$

$CN_T \approx [HCN] + 2[Au(CN)_2^-]$
$\qquad + 4[Ni(CN)_4^{2-}]$
$\qquad = [CN^-](K_6[H^+] +$
$\qquad\qquad 2K_1[Au^+][CN^-] +$
$\qquad\qquad 4K_5[Ni^{2+}][CN^-]^3)$

Solving for $[Au^+]$ and $[Ni^{2+}]$ and substituting into the cyanide mass balance:

$CN_T \approx [CN^-]\{K_6[H^+] +$
$\qquad 2K_1Au(+I)_T[CN^-]/(1 +$
$\qquad K_1[CN^-]^2) +$
$\qquad 4K_5Ni(+II)_T[CN^-]^3/(1 +$
$\qquad K_5[CN^-]^4)\}$

This is one equation in one unknown. Solving with a spreadsheet: $[CN^-] = 2.5 \times 10^{-16}$ M. Thus:

$[Au^+] = 4.4 \times 10^{-11}$ M
$[Au(CN)_2^-] = 2.8 \times 10^{-2}$ M
$[Ni^{2+}] = 6.8 \times 10^{-6}$ M
$[NiOH^+] = 1.1 \times 10^{-11}$ M
$[Ni(OH)_2^0] = 1.1 \times 10^{-16}$ M
$[Ni(OH)_3^-] = 1.4 \times 10^{-23}$ M
$[Ni(CN)_4^{2-}] = 4.3 \times 10^{-38}$ M
$[HCN] = 3.2 \times 10^{-11}$ M

Checking the assumptions:

1. Ni(+II)-hydroxyo complexes can be ignored in the nickel mass balance.

2. $[CN^-]$ can be ignored relative to [HCN].

The conclusions:

becomes:

$$Cu(+II)_T \approx [Cu^{2+}] + [CuL^-] + [CuOH^+] + [Cu(OH)_2^0]$$

A similar argument for iron (see Problem 15.10) reveals:

$$Fe(+II)_T \approx [Fe^{2+}] + [FeOH^+]$$

What about NTA? The acid-base chemistry is the most straight forward.

Thoughtful Pause

From the acid-base equilibria, which protonated form of NTA is at the highest concentration at pH 8?

NTA is a triprotic acid with pK_a values of 1.7, 2.9, and 10.3. Clearly, the form $[HL^{2-}]$ will be in the highest concentration of the four acid-base forms of NTA. Using arguments similar to those expressed above, it is reasonable that $[CaL^-]$, $[CuL^-]$, and $[FeL^-]$ will be larger than the other NTA-containing complexes with calcium, copper, and iron, respectively. You have no way of comparing $[CaL^-]$, $[CuL^-]$, and $[FeL^-]$ at this point. The simplified mass balance for NTA becomes:

$$NTA_T \approx [HL^{2-}] + [CaL^-] + [CuL^-] + [FeL^-]$$

The approximate mass balances are:

$$NTA_T \approx [HL^{2-}] + [CaL^-] + [CuL^-] + [FeL^-]$$
$$Ca(+II)_T \approx [Ca^{2+}]$$
$$Cu(+II)_T \approx [Cu^{2+}] + [CuL^-] + [CuOH^+] + [Cu(OH)_2^0]$$
$$Fe(+II)_T \approx [Fe^{2+}] + [FeOH^+]$$

Rearranging the latter three expressions to solve for $[Ca^{2+}]$, $[Cu^{2+}]$, and $[Fe^{2+}]$ and substituting into the mass balance on NTA:

$$NTA_T \approx [L^{3-}]\{K_1[H^+] + Ca(+II)_TK_4 + Cu(+II)_TK_7/(1 + K_7[L^{3-}] + K_{10}[OH^-] + K_{11}[OH^-]^2) + Fe(+II)_TK_{15}/(1 + K_{19}[OH^-])\}$$

This is one equation in one unknown and can be solved easily for $[L^{3-}]$. The approximate solution and the exact solution are compared in Table 15.3. The approximate solution is fairly accurate. The assumptions made in this section can be checked in Table 15.3, including the important

Almost all of the gold and almost none of the nickel is complexed with cyanide. Almost all of the cyanide is present as $Au(CN)_2^-$. The HCN concentration is very low.

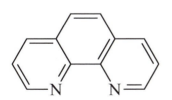

Key idea: The species at the largest concentration in each mass balance usually depends on the total ligand concentration

assumption that $[L^{3-}] \ll 10^{-7.6}$ M. Another example of the method of approximation is given in Example 15.4.

15.5.6 Interpreting the equilibrium calculations

The species at the largest concentration in each mass balance usually depends on the total ligand concentration. Remember that a species may dominate one mass balance, but be a relatively small contributor to another mass balance.

This point is illustrated in the example problem. At the lower NTA concentration, about 95% of the NTA is in the form CuL^-, with about 2% each in CaL^- and $CuOHL^{2-}$. The copper-NTA complex is strong and it dominates the NTA mass balance. About 65% of the dissolved copper is $Cu(OH)_2^0$, about 20% is $CuOH^+$, and about 10% is in the form Cu^{2+}. However, almost all the calcium is in the form Ca^{2+}. The soluble ferrous iron is primarily Fe^{2+} ($\approx 97\%$) and $FeOH^+$ ($\approx 3\%$).

At the higher NTA level, the speciation of the organic ligand changes. Although the copper-NTA complex is strong, the total soluble copper concentration is small. This means that the copper becomes saturated and NTA is available to complex with other metals. As a result, about 80% of the organic ligand is complexed with calcium (as CaL^-) and about 20% with copper (as CuL^-). Since the total NTA concentration is so much smaller than the total calcium concentration, calcium is still found almost completely as Ca^{2+}. Copper, on the other hand, is nearly completely complexed, with 97% of the copper found as CuL^- and 2.5% as $CuOHL_2^-$. Ferrous iron still is mainly found in the aquo form ($\approx 94\%$ as Fe^{2+}), with about 4% as FeL^- and about 3% as $FeOH^+$.

15.6 APPLICATIONS OF COMPLEXATION CHEMISTRY

15.6.1 Complexometric titrations

Titration with a ligand can be used to determine the concentration of certain metals. A good example is the determination of ferrous iron by complexation with 1,10-phenanthroline. The chemical structure of 1,10-phenanthroline is shown in Figure 15.10. This ligand is very specific for Fe^{2+} since the N-N spacing fits Fe^{2+} but not Fe^{3+} or other common metallic ions.

Figure 15.10: Structure of 1,10-Phenanthroline

Thoughtful Pause
What is the stoichiometry of the Fe(+II)-phenanthroline complex?

Each N in 1,10-phenanthroline will share an electron pair with ferrous iron. Since ferrous iron has a coordination number of 6, the expected stoichiometry of the Fe(+II)-phenanthroline complex is 1:3. The aquo and hydroxo complexes of Fe(+II), along with 1,10-phenanthroline, do not absorb light in the visible light wavelength range. However, the Fe(+II)-phenanthroline complex absorbs light strongly at 510 nm. Thus, the concentration of Fe(+II) can be estimated by adding 1,10-phenanthroline to a sample and measuring the absorbance at 510 nm. A quantitative discussion of the 1,10-phenanthroline method is provided in Example 15.5.

Many other metals can be determined by complexometric titration. Examples include the complexation of Cu^{2+} by neocuproine (2,9-dimethyl-1,10-phenanthroline) and the complexation of the hardness cations (Ca^{2+} and Mg^{2+}) with EDTA.

15.6.2 Metal buffers

Synthetic ligands can be used to control the free (i.e., aquo) metal concentrations. For example, EDTA often is found in biological growth media to control the speciation of free metals. Since the free metals usually are more toxic the bacteria than complexed metals, metal toxicity can be controlled by EDTA.

15.7 SUMMARY

Metals are important parts of the aquatic environment because of their toxicity to humans and aquatic biota. Metals form complexes with ligands. Complex formation often controls the chemistry of both the metals and ligands.

In this chapter, you learned that the concept of acids and bases can be extended to include complex formation. Lewis acids and bases are defined to be substances which accept or donate an electron pair. All Brønsted-Lowry acids are Lewis acids. However, the Lewis acid concept is much broader and includes metals as well.

Equilibrium calculations for systems exhibiting complexation are similar to acid-base equilibrium calculations. In general, several mass balances are involved. In most applied problems, the pH is known and thus the electroneutrality equation is not needed. Complexation chemistry can be applied to many areas, including the complexation of metals by naturally-occurring organics, biochemistry, metal plating, analytical chemistry (e.g., complexometric titrations), and metal buffering.

15.8 PART IV CASE STUDY: WHICH FORM OF COPPER PLATING SHOULD YOU USE?

As presented in Section 14.6, the Part IV case study concerns the relative merits of copper cyanide plating and acid copper plating. In this section, the effects of complexation with several metals and ligands on the plating process will be explored.

Metal finishers continually are striving to reduce discharges of potentially harmful chemicals to the environment. As a result, the use of

Example 15.5: Complexometric Titration

Determine the appropriate pH range and appropriate ratio of total 1,10-phenanthroline to total Fe(+II) for the measurement of Fe(+II) by complexation with 1,10-phenathroline.

Solution:

Let P = 1,10-phenanthroline. To measure ferrous iron by complexation, you want the concentration of the complex, $[FeP_3^{2+}]$, to be nearly equal to $Fe(+II)_T$. This means that (1) the predominant complex must be FeP_3^{2+} rather than hydroxo complexes and (2) $[FeP_3^{2+}]$ must be much larger than $[Fe^{2+}]$.

The formation equilibria and constants for the Fe(+II)-hydroxo complexes are listed in Section 15.5.2. With no P:

$$Fe(+II)_T = [Fe^{2+}](1$$
$$+ K_{19}[OH^-]$$
$$+ K_{20}[OH^-]^2$$
$$+ K_{21}[OH^-]^3$$
$$+ K_{22}[OH^-]^4)$$

Solving, $[Fe^{2+}]/Fe(+II)_T$ will be greater than 99.9% if the pH is less than 6.5.

The formation equilibrium for FeP_3^{2+} is:

$$Fe^{2+} + 3P = FeP_3^{2+}; \quad K = 10^{+16}$$

In the presence of P with pH < 6.5:
$$Fe(+II)_T = [Fe^{2+}] + [FeP_3^{2+}]$$
$$= [Fe^{2+}](1 + K[P]^3)$$

$$P_T = [P] + 3[FeP_3^{2+}]$$
$$= [P](1 + 3K[Fe^{2+}][P]^2)$$

cyanide in metal finishing operations is being minimized. In spite of this trend, copper cyanide plating is fairly common. Why? The normal answer to why cyanide is used in any plating process is to increase the solubility of the metal. The chemistry behind this statement will be explored in Chapter 19. However, cyanide is used in copper plating for other reasons as well. Two reasons for the use of cyanide in copper plating will be explored: economics and preservation of the integrity of the piece.

In the discussion of the Part IV case study after Chapter 16, an economic advantage of copper cyanide over acid copper will be quantified. However, the economic advantage disappears if the Cu^+ from copper cyanide is converted into Cu^{2+}. The interconversion of Cu^+ and Cu^{2+} can occur according to the equilibrium:

$$Cu(s) + Cu^{2+} = 2Cu^+ \qquad \log K_1 = -6.1$$

(The equilibrium constant for this equilibrium will be developed in Section 16.10.) Note that since the activity of a pure solid phase such as Cu(s) is 1, then: $[Cu^+]^2/[Cu^{2+}] = 10^{-6.1}$ or $[Cu^{2+}] = 10^{+6.1}[Cu^+]^2$. Since Cu^{2+} and Cu^+ are the only forms of Cu(+II) and Cu(+I), respectively, you can write:

$$Cu(+II)_T = 10^{+6.1}(Cu(+I)_T)^2$$

To prevent this interconversion of copper oxidation states, cyanide is added. How does cyanide help? Cyanide forms complexes with Cu^+ (see equilibria below), but not with Cu^{2+} (see Appendix D).

$$Cu^+ + 2CN^- = Cu(CN)_2^- \quad \log K_2 = 16.3$$
$$Cu^+ + 3CN^- = Cu(CN)_3^{2-} \quad \log K_3 = 21.6$$
$$Cu^+ + 4CN^- = Cu(CN)_4^{3-} \quad \log K_4 = 23.1$$

Performing a mass balance on Cu(+I):

$$Cu(+I)_T = [Cu^+] + [Cu(CN)_2^-] + [Cu(CN)_3^{2-}] + [Cu(CN)_4^{3-}]$$
$$= [Cu^+](1 + K_2[CN^-]^2 + K_3[CN^-]^3 + K_4[CN^-]^4)$$

Thus:

$$[Cu^+] = Cu(+I)_T/(1 + K_2[CN^-]^2 + K_3[CN^-]^3 + K_4[CN^-]^4)$$
$$= Cu(+I)_T/\alpha_{Cu} \quad (\alpha_{Cu} > 1)$$

Recall that Cu^{2+} is still the only form of Cu(+II) in the system. Now the equilibrium expression $[Cu^{2+}] = 10^{+6.1}[Cu^+]^2$ becomes:

$$Cu(+II)_T = 10^{+6.1}[Cu^+]^2 = 10^{+6.1}(Cu(+I)_T)^2/\alpha_{Cu}^2 ; \quad \alpha_{Cu} > 1$$

Or:

$$P_T = [P]\{1 + 3KFe(+II)_T[P]^2/(1 + K[P]^3)\} \qquad (*)$$

From the mass balance on Fe(+II), the complex will be 99.9% of $Fe(+II)_T$ {i.e., $[Fe^{2+}]/Fe(+II)_T = 0.001$} when $1/(1 + K[P]^3) = 0.001$ or $[P] = 4.64 \times 10^{-5}$ M. From (*), when $[P] = 4.64 \times 10^{-5}$ M:

$$P_T = 4.64 \times 10^{-5} + 3.0Fe(+II)_T$$

Or: $P_T/Fe(+II)_T \approx 3.0$.

Thus, adding phenanthroline at a 3:1 molar excess over Fe(+II) at pH less than 6.5 should ensure that the measured response (i.e., $[FeP_3^{2+}]$) is nearly equal to the total Fe(+II).

Thus, by adding cyanide, the concentration of Cu(+II) is reduced as α_{Cu} increases and the interconversion of Cu^+ and Cu^{2+} is minimized.

A second reason for adding cyanide in the copper cyanide process is to avoid the formation of *immersion deposits*. Immersion deposits occur when the oxidized form of the metal to be plated oxidizes the metal in the piece. This process can be described by the following equilibrium:

$$M^{n+} + P(s) = M(s) + P^{n+}$$

where M is the metal to be plated and P is the metal in the piece. Of particular concern with copper is immersion deposits formed during electroplating on zinc. With copper cyanide plating, the pertinent equilibrium is:

$$2Cu^+ + Zn(s) = 2Cu(s) + Zn^{2+}; \quad \log K = 43.4$$

(Again, the equilibrium constant for this equilibrium will be developed in Section 16.10.) At equilibrium in the absence of cyanide: $[Zn^{2+}]/[Cu^+]^2 = 10^{43.4}$ or $[Zn^{2+}] = 10^{43.4}[Cu^+]^2$ or:

$$Zn(+II)_T = 10^{43.4}Cu(+I)_T \qquad \text{eq. 5.20}$$

Clearly, the potential for the oxidation of metallic zinc by Cu^+ is real and an appreciable quantity of Zn^{2+} may be formed at equilibrium.

Again, the picture changes in the presence of cyanide. With cyanide, $[Cu^+] = Cu(+I)_T/\alpha_{Cu}$. Zinc complexes with cyanide as well according to the equilibria:

$$Zn^{2+} + 2CN^- = Zn(CN)_2^0 \quad \log K_5 = 11.1$$
$$Zn^{2+} + 3CN^- = Zn(CN)_3^- \quad \log K_6 = 16.1$$
$$Zn^{2+} + 4CN^- = Zn(CN)_4^{2-} \quad \log K_7 = 19.6$$

From a mass balance on Zn(+II):

$$[Zn^{2+}] = Zn(+II)_T/(1 + K_5[CN^-]^2 + K_6[CN^-]^3 + K_7[CN^-]^4)$$
$$= Zn(+II)_T/\alpha_{Zn}$$

Thus, the equilibrium expression $[Zn^{2+}] = 10^{43.4}[Cu^+]^2$ becomes:

$$Zn(+II)_T = 10^{43.4}(Cu(+I)_T)^2(\alpha_{Zn}/\alpha_{Cu}^2) \qquad \text{eq. 5.21}$$

Note that zinc complexes with cyanide less strongly than Cu(+I). Therefore, $\alpha_{Zn} < \alpha_{Cu}$ and the term $\alpha_{Zn}/\alpha_{Cu}^2$ must be less than one. Thus, comparing eqs. 5.20 and 5.21 with $\alpha_{Zn}/\alpha_{Cu}^2 < 1$, you can conclude that cyanide helps to minimize the formation of immersion deposits.

Further discussion of this case study and the calculation of the pertinent equilibrium constants may be found in Section 16.10. The equilibrium chemistry of copper cyanide baths is discussed in a much more quantitative fashion in Section 23.2, where more realistic plating conditions are explored.

SUMMARY OF KEY IDEAS

- Metals are important because (1) they are toxic to humans and aquatic biota and (2) they form complexes which may control the chemistry of the ligands

- Metals and ligands coordinate to form complexes

- Acid-base chemistry is a subset of complexation chemistry

- Brønsted-Lowry acids and bases also are Lewis acids and bases, but the Lewis acid concept is much broader than the Brønsted-Lowry acid concept

- The transition metals are Lewis acids

- The ligand water usually is not written explicitly

- Ligands donate electron pairs to metals and commonly contain N, O, or S

- The acidity of some complexes stems from the ability of aquo ligands to donate protons to water

- Label soluble, uncharged complexes with (aq) or the superscript 0 (for zero charge) to differentiate them from solids

- Metals and H^+ often compete for binding sites with ligands that are Brønsted acids

- The equilibrium calculations for homogeneous systems involving complexation are set up exactly as with acid-base systems

- Metals which form hydroxo complexes will precipitate under the proper pH and total metal concentrations

- A complex is strong in a given system if it is the dominant form of the metal in that system

- Make sure that the stoichiometry of the equilibria are the same before comparing formation constants K_i, β_i, β_{nm}, $^*K_i$, $^*\beta_i$, and $^*\beta_{nm}$ are just equilibrium constants for formation equilibria written in a specific form

- To solve systems involving complexation (1) Solve for each species concentration as a function of the uncomplexed (i.e, aquo complexed) metals and uncomplexed ligands, (2) Substitute these expressions into the mass balances, (3) Rework to form one equation in one unknown and solve for the uncomplexed metal or ligand concentration, and (4) Back-calculate the concentrations of the complexed species

- Spreadsheet hints: (1) optimize by changing the p(concentration) or log(concentration) values, not the concentration values themselves, and (2) use relative errors in the objective function, not absolute errors

- To simplify chemical systems, ignore the species in each mass that are expected to be at low concentrations relative to the other species in the mass balance

- The species at the largest concentration in each mass balance usually depends on the total ligand concentration

PROBLEMS

15.1 Is CN^- a Brønsted acid? Lewis acid? both? neither? Explain your reasoning.

15.2 Given the formation equilibria below, which metal forms the strongest complex with cyanide? Perform a simple calculation to compare the strengths of the Fe(+III)-CN, Fe(+II)-CN, and Cd(+II)-CN complexes.

$$Fe^{3+} + 6CN^- = Fe(CN)_6^{3-} \quad \log K = 43.6$$
$$Fe^{2+} + 6CN^- = Fe(CN)_6^{4-} \quad \log K = 35.4$$
$$Cd^{2+} + 2CN^- = Cd(CN)_2^0 \quad \log K = 11.1$$

15.3 Using the formation constants in Section 15.5.2, find the concentrations of the hydroxo complexes of copper (including the dimer) in a water at pH 6.0 with total soluble copper equal to 1.0×10^{-6} M.

15.4 Desferrioxamine (structure below) is used to treat iron overload in patients receiving frequent blood transfusions.

A. Looking at the structure of desferrioxamine and with your knowledge of electron sharing, how many moles of desferrioxamine are required to complex one mole of iron?

B. Desferrioxamine binds ferrous iron preferentially over calcium (stability constant $K = 10^{31}$ M for ferrous iron and 10^9 M for calcium). Determine the desferrioxamine concentration in the blood needed to reduce the iron levels to normal values if the total soluble calcium concentration is 1880 mg/L as Ca and the total soluble iron concentration is 3 mg/L as Fe (three times the normal value of 1 mg/L). The pH of blood is 7.3. Use the equilibrium constants from Section 15.5.2 and assume all the iron is ferrous iron. Assume the complexes are 1:1.

15.5 Redo Example 15.2 and calculate the equilibrium concentrations using a spreadsheet, *Nanoql*, and the method of approximation.

15.6 In the acidic gold cyanide plating process described in Example 15.4, 60 g/L of citric acid also is added. Do you expect the citric acid to change the metal and cyanide speciation? Explain qualitatively how the speciation will change.

15.7 A cyanide bath at pH 7 can contain dangerously high levels of HCN. Calculate the change in the HCN concentration in a cyanide bath at pH 7 and $CN_T = 1 \times 10^{-4}$ M if 1×10^{-5} total Ni(+II) is added. Use the equilibrium constants from Example 15.4. To solve this problem, make an assumption about the predominant form of nickel in the system (by looking at the nickel-cyanide stability constant and the total Ni(+II):CN_T ratio). Be sure to check your assumptions.

15.8 Repeat the calculation in Problem 15.7 using no approximations.

15.9 Using copper as an example, explain quantitatively why dimers often are present at small concentrations (when the total metal concentration is small).

15.10 In the NTA complexation example in Section 15.5.5, show that: $Fe(II)_T \approx [Fe^{2+}] + [FeOH^+]$

15.11 Sometimes the addition of chelating agents has unintended consequences. Jensen and Johnson (1990) discuss the reagents added to measure chlorine by the DPD method. Mercuric chloride is added to the pH buffer to complex trace iodide (I⁻), since iodide causes problems in the chlorine analysis. EDTA is added to the pH buffer to complex trace metals which interfere in the chlorine analysis. As a result of EDTA addition, some mercury(+II) is complexed inadvertently and more free iodide is available to cause problems.

A. Given the equilibria in Appendix D and the equilibrium for the formation of HgI^+ below, calculate the equilibrium $[I^-]$ if the test solution is at pH 6.2 and contains 1.1×10^{-8} M total iodide, 1 mg/L total mercury(+II) as $HgCl_2$, and 40 mg/L EDTA as its disodium salt. (This is the solution composition if chlorine is measured by the standard method. Iodide is a contaminant of the phosphate buffer.)

B. What is the equilibrium $[I^-]$ of the test solution in Part A if EDTA is omitted?

$$Hg^{2+} + I^- = HgI^+ \qquad \log K = 12.9$$

15.12 Why is the leaching of trace metals from soils considered one of the impacts of acid rain?

15.13 A community uses a groundwater supply for drinking water. The water is fairly hard, with a hardness of 300 mg/L as $CaCO_3$. To minimize soap scum, the residents wish to add a detergent containing nitrilotriacetate (L^{3-}) to their washing machines. The detergent is 3% L^{3-} by weight. If the average washing machine has a capacity of 40 L, how much detergent should be added to reduce the hardness to 25 mg/L as $CaCO_3$? Assume the hardness is from calcium. The molecular weight of L^{3-} is 188 g/mole.

CHAPTER 16

Oxidation and Reduction

16.1 INTRODUCTION

Acid-base and complexation chemistry describe only some of the interactions that occur between chemical species. In addition to exchanging protons and ligands, aquatic species can exchange electrons. The equilibrium calculation tools developed so far in this text can be applied to electron transfer reactions in a straight forward manner. However, some of the thermodynamic principles and graphical calculation tools presented earlier in the text must be extended to account for the new interactions.

Electron exchange brings with it many new terms. Important concepts in electron exchange chemistry are reviewed in Section 16.2. The techniques you learned to balance chemical reactions will be broadened to electron transfer reactions in Section 16.3. Correctly balanced electron exchange reactions are essential to determining the dose of some chemicals used in treatment processes. In Sections 16.4 and 16.5, a system is developed to quantify the thermodynamic tendency to transfer electrons, just as pK_a and pK_b were developed to quantify the thermodynamic tendency to exchange protons. Equilibrium calculation techniques are presented in Section 16.6. New graphical presentation methods are developed in Section 16.7. In Section 16.8, limitations and strengths of redox equilibrium calculations are discussed.

16.2 A FEW DEFINITIONS

16.2.1 Introduction

Chemical reactions in which electrons are *consumed* are called ***reduction reactions***. In reduction reactions, the free electron is a *reactant* and the oxidation state of at least one reactant is *decreased*. (The procedure for calculating the oxidation states of elements in a chemical species is reviewed in Appendix C, Section C.1.1.) For example, the following reaction is a reduction:

reduction reaction: a reaction in which electrons are consumed (and the oxidation state of one or more reactants is decreased)

$$Cl_2(g) + 2e^- \rightarrow 2Cl^- \qquad \text{eq. 16.1}$$

Note that one of the reactants is the free electron. The overall oxidation state of chlorine gas, $Cl(0)$, is reduced to chloride ion, $Cl(-I)$. Note that both material (i.e., chlorine) and electrons (i.e., charge) are balanced in eq. 16.1. Remember that the oxidation states of reactants are *reduced* in *reduction* reactions.

oxidation reaction: a reaction in which electrons are produced (and the oxidation state of one or more reactants is increased)

Chemical reactions in which electrons are *produced* are called **oxidation reactions**. In oxidation reactions, the free electron is a *product* and the oxidation state of at least one reactant is *increased*. For example, the following reaction is an oxidation:

$$Fe^{2+} \rightarrow Fe^{3+} + e^-$$
eq. 16.2

redox reaction: a reaction in which reduction or oxidation or both take place (i.e., a reaction in which electrons are transferred)

Note that one of the products is the free electron. In addition, the oxidation state of ferrous iron, Fe(+II), is increase to ferric iron, Fe(+III). Any reaction in which reduction or oxidation or both take place is called a **redox reaction** (for reduction-oxidation reaction). Redox reactions are electron transfer reactions.

16.2.2 Half reactions and overall redox reactions

half reaction: a reaction having electrons as either reactants or products

Equations 16.1 and 16.2 have electrons as either reactants or products. Any reaction having electrons as either reactants or products is called a **half reaction**. In aqueous systems under environmental conditions, the free electron does not exist. In other words, all electrons are associated with other chemical species in chemical systems of environmental interest. Thus, redox reactions such as eqs. 16.1 and 16.2 cannot occur in isolation.

overall redox reaction: a reaction in which electrons are transferred but free electrons do not appear as reactants or products

Since half reactions cannot exist in isolation, it is sometimes desirable to write the **overall redox reaction**. In an overall redox reaction, electrons are transferred, but free electrons do not appear as reactants or products. The half reactions are added together in such a way that the electrons cancel out. For example, if your groundwater has a high ferrous iron concentration, the ferrous iron can be oxidized to ferric iron by chlorine and subsequently precipitated. The overall reaction, obtained by adding eqs. 16.1 and 16.2 so that the electrons cancel, is shown in eq. 16.3:

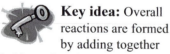 **Key idea:** Overall reactions are formed by adding together half reactions in such a way that the free electrons cancel out

$$2Fe^{2+} + Cl_2(g) \rightarrow 2Fe^{3+} + 2Cl^-$$
eq. 16.3

Note that material and charge both balance in eq. 16.3.

In overall reactions, certain species are given special names. The compound that oxidizes another species is called the **oxidant**. The compound that reduces another species is called the **reductant**. These labels can be confusing at first. In overall reactions, oxidants accept electrons. They oxidize other species and are, in turn, reduced. Reductants donate electrons. They reduce other species and are, in turn, oxidized.

oxidant: a compound that oxidizes another species

reductant: a compound that reduces another species

  Key idea: In an overall reaction, oxidants accept electrons and are reduced, whereas reductants donate electrons and are oxidized

16.2.3 pe revisited

There are many similarities between acid-base chemistry and redox chemistry. One similarity concerns the quantities pe and pH. In Section

pe: a measure of the availability of electrons; pe = -log{e⁻}

1.3, pe was introduced as a master variable in aquatic chemistry. Recall that $pe = -\log\{e^-\}$. As you saw in Section 16.2.1, $\{e^-\}$ at equilibrium is very small in environmental systems. How can pe be useful if e⁻ does not exist?

To see the usefulness of pe, imagine freezing the chemistry of an aquatic system. Although no free electrons would exist, chemical species would "feel" whether electrons were available to be accepted or donated. For example, if electrons are available (i.e., if a significant concentration of strong electron donors, reductants, was present), an oxidant such as dissolved oxygen may be reduced. If electrons are much less available (i.e., if a significant concentration of strong electron acceptors, oxidants, was present), a reductant such as hydrogen sulfide may be oxidized. Thus, pe is a useful concept as a measure of the *availability* of free electrons.

If electrons are readily available, the system behaves as if the activity of the free electron is high; thus, pe is small or even negative. These are called *reducing conditions*. Under reducing conditions, many oxidants will be reduced. In other words, the reduced forms of chemical species will be favored.

If electrons are *not* readily available, the system behaves as if the activity of the free electron is low; thus, pe is large and positive. These are called *oxidizing conditions*. Under oxidizing conditions, many reductants will be oxidized and the oxidized forms of chemical species will be favored. Oxidizing and reducing conditions are summarized in Table 16.1.

16.3 Balancing Redox Reactions

16.3.1 Balancing half reactions

The general procedure for balancing any chemical reaction is given in Appendix C, Section C.1.2. This procedure will be repeated shortly with the example of balancing the reduction of sulfate (SO_4^{2-}) to elemental sulfur S(s) at high pH.

Key idea: To balance a half reaction (1) write the known reactants on the left and known products on the right, (2) adjust stoichiometric coefficients to balance all elements (except H and O), (3) add water (H_2O) to balance the element O, (4) add H⁺ to balance the element H, and (5) add electrons (e⁻) to balance the charge

Table 16.1: Characteristics of Reducing and Oxidizing Conditions

Characteristic	Reducing Conditions	Oxidizing Conditions
pe	low or < 0	high (> 0)
Reactions	oxidants are reduced	reductants are oxidized
Favored species	reduced forms	oxidized forms
Environmental examples	sediments, anaerobic digesters	aerated waters

Example 16.1: **Balancing Half Reactions**

Balance the half reactions for the reduction of hypochlorous acid (HOCl) to chloride and the oxidation of hydrogen sulfide to sulfate at low pH.

Solution:
Following each step in the balancing procedure:

Step 1: Write known participants
$$HOCl \rightarrow Cl^-$$
$$H_2S \rightarrow SO_4^{2-}$$

Step 2: Balance all but H, O
Cl and S balanced in Step 1

Step 3: Add water to balance O
$$HOCl \rightarrow Cl^- + H_2O$$
$$H_2S + 4H_2O \rightarrow SO_4^{2-}$$

Step 4: Add H^+ to balance H
$$HOCl + H^+ \rightarrow Cl^- + H_2O$$
$$H_2S + 4H_2O \rightarrow SO_4^{2-} + 10H^+$$

Step 5: Add e^- to balance charge
$$HOCl + H^+ + 2e^- \rightarrow Cl^- + H_2O$$
$$H_2S + 4H_2O \rightarrow SO_4^{2-} + 10H^+ + 8e^-$$

The balanced half reactions are:

$$\mathbf{HOCl + H^+ + 2e^- \rightarrow Cl^- + H_2O}$$

$$\mathbf{H_2S + 4H_2O \rightarrow SO_4^{2-} + 10H^+ + 8e^-}$$

Thoughtful Pause

How do you know that the conversion of sulfate to elemental sulfur is a reduction?

The oxidation state of S in sulfate is +VI, and the oxidation state of S in elemental is 0. Thus, the oxidation state of sulfur is reduced, and the reaction is a reduction. In the half reaction, electron(s) should appear as reactants. The half reaction balancing procedure is as follows (see also Section 4.7.1 and Appendix C, Section C.1.2):

Step 1: Write the known reactants on the left and known products on the right

In the example: $SO_4^{2-} \rightarrow S(s)$

Step 2: Adjust stoichiometric coefficients to balance all elements (except H and O)

In this case, you need only to balance S, and S is already balanced in the first step.

Step 3: Add water (H_2O) to balance the element O

Thus: $SO_4^{2-} \rightarrow S(s) + 4H_2O$

Step 4: Add H^+ to balance the element H

In the example: $SO_4^{2-} + 8H^+ \rightarrow S(s) + 4H_2O$

Step 5: Add electrons (e^-) to balance the charge

The left side has an overall charge of +6, and the right side has an overall charge of 0. Thus, you must add 6 electrons to the left side to balance the charge: $SO_4^{2-} + 8H^+ + 6e^- \rightarrow S(s) + 4H_2O$

For basic conditions, add the reaction $8H_2O \rightarrow 8H^+ + 8OH^-$ to eliminate H^+. The resulting expression is:

$$SO_4^{2-} + 4H_2O + 6e^- \rightarrow S(s) + 8OH^-$$ eq. 16.4

Note that materials and charge balance and that electrons are reactants, as anticipated. Another illustration of balancing a half reaction is given in Example 16.1.

If materials and charges balance in the final half reaction, then the reaction is balanced. However, a useful check on the charge balance is to make sure that the change in oxidation state of the pertinent species makes sense. In eq. 16.4, H and O are in the +1 and −2 oxidation state always and do not undergo redox chemistry in this reaction. As stated above, the oxidation state of S in sulfate is +VI, and the oxidation state of S in elemental sulfur is 0. Since 1 mole of sulfate is converted to 1 mole of sulfur, you expect eq. 16.4 to require six electrons. *This check on the change in oxidation state should become second nature to you*. It is a useful check to catch errors in half reactions.

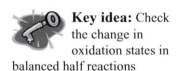

Key idea: Check the change in oxidation states in balanced half reactions

16.3.2 Balancing overall reactions

To balance overall reactions, two steps are required. First, the reduction half reaction and oxidation half reaction should be balanced separately. Second, add the half reactions in such a way that the electrons cancel. As an example, consider the oxidation of hydrogen sulfide to elemental sulfur, S(s), by hypochlorous acid. The hypochlorous acid will be reduced to chloride. Following the protocol for balancing overall reactions (compiled in Appendix C, Section C.1.3):

Key idea: To balance overall reactions: (1) balance the half reactions separately, and (2) add the half reactions to eliminate free electrons

Step 1: Balance the half reactions separately

> The half reactions are balanced in Example 16.1. The resulting half reactions are: $HOCl + H^+ + 2e^- \rightarrow Cl^- + H_2O$ and $H_2S + 4H_2O \rightarrow SO_4^{2-} + 10H^+ + 8e^-$.

Step 2: Add the half reactions to eliminate free electrons

> As written, the reduction half reaction consumes two electrons and the oxidation half reaction produced eight electrons. Thus, to eliminate free electrons, multiply the reduction half reaction by four and add:

$$(HOCl + H^+ + 2e^- \rightarrow Cl^- + H_2O) \times 4$$
$$\underline{+ \ H_2S + 4H_2O \rightarrow SO_4^{2-} + 10H^+ + 8e^-}$$
$$4HOCl + H_2S \rightarrow 4Cl^- + SO_4^{2-} + 6H^+ \qquad \text{eq. 16.5}$$

Example 16.2: **Balancing Overall Reactions**

One measure of the organic content of water is chemical oxygen demand (COD). In the COD test, organic matter is oxidized with dichromate ($Cr_2O_7^{2-}$) at low pH to CO_2, and the dichromate is reduced to Cr^{3+}.

Write the balanced overall reaction for the oxidation of ethanol (CH_3CH_2OH) in the COD test.

Another example of balancing overall reactions is given in Example 16.2.

You may be tempted to shortcut the balancing process and balance the overall reaction in one step. For example, you may say to yourself, Why

Solution:

You must first balance the two half reactions. For chromium species:

Starting point:
$$Cr_2O_7^{2-} \rightarrow Cr^{3+}$$
Balancing Cr:
$$Cr_2O_7^{2-} \rightarrow 2Cr^{3+}$$
Balancing O:
$$Cr_2O_7^{2-} \rightarrow 2Cr^{3+} + 7H_2O$$
Balancing H:
$$Cr_2O_7^{2-} + 14H^+ \rightarrow$$
$$2Cr^{3+} + 7H_2O$$
Balancing charge:
$$Cr_2O_7^{2-} + 14H^+ + 6e^- \rightarrow$$
$$2Cr^{3+} + 7H_2O$$

For ethanol:

Starting point:
$$CH_3CH_2OH \rightarrow CO_2$$
Balancing C:
$$CH_3CH_2OH \rightarrow 2CO_2$$
Balancing O:
$$CH_3CH_2OH + 3H_2O \rightarrow$$
$$2CO_2$$
Balancing H:
$$CH_3CH_2OH + 3H_2O \rightarrow$$
$$2CO_2 + 12H^+$$
Balancing charge:
$$CH_3CH_2OH + 3H_2O \rightarrow$$
$$2CO_2 + 12H^+ + 12e^-$$

Adding:
$$(Cr_2O_7^{2-} + 14H^+ + 6e^- \rightarrow$$
$$2Cr^{3+} + 7H_2O) \times 2$$
$$+ CH_3CH_2OH + 3H_2O \rightarrow$$
$$2CO_2 + 12H^+ + 12e^-$$
_____
$$CH_3CH_2OH + 2Cr_2O_7^{2-} + 16H^+$$
$$\rightarrow 2CO_2 + 4Cr^{3+} + 11H_2O$$

The balanced overall reaction is:
$$CH_3CH_2OH + 2Cr_2O_7^{2-} +$$
$$16H^+ \rightarrow 2CO_2 + 4Cr^{3+} +$$
$$11H_2O$$

not just balance the reaction aHOCl + bH$_2$S + cH$_2$O → dCl$^-$ + eSO$_4^{2-}$ + fH$^+$? There are many ways to try to balance this reaction. Consider the solution: HOCl + H$_2$S + 3H$_2$O → Cl$^-$ + SO$_4^{2-}$ + 9H$^+$. Is this balanced? No, the charges do not balance. How about HOCl + H$_2$S + 3H$_2$O → Cl$^-$ + SO$_4^{2-}$ + 3H$^+$? No, H is not balanced. After playing this game for some time, you will probably realize that it is faster (and less error-prone) to balance the two half reactions separately and then add them to eliminate free electrons.[†]

When two balanced half reactions are added, the resulting overall redox reaction *should* be balanced. It is always wise to check the resulting reaction to make sure: (1) electrons cancel out and do not appear in the final expression, and (2) species are written in terms of their predominant forms under a given set of conditions. For example, eq. 16.5 makes sense below pH 7, where HOCl, H$_2$S, Cl$^-$, and SO$_4^{2-}$ are larger in concentration than their conjugate acid or basic forms (since the pK_a values of HOCl and H$_2$S are 7.5 and 7, respectively). Also, at pH less than 7, it makes sense to write the reaction in terms of H$^+$ rather than OH$^-$.

Thoughtful Pause

What would be the appropriate reactions to express the chemistry in eq. 16.5 at higher pH?

You could find the appropriate expression by starting over (i.e., writing the half reactions at higher pH) or by changing eq. 16.5. Since eq. 16.5 already has been developed, start from it. At higher pH, Cl(+I) is predominantly OCl$^-$, H$_2$S is predominantly HS$^-$ (bisulfide), and it is preferable to write the reaction in terms of OH$^-$ rather than H$^+$. Adding the appropriate reactions:

$$4HOCl + H_2S \rightarrow 4Cl^- + SO_4^{2-} + 6H^+$$
$$(OCl^- + H^+ \rightarrow HOCl) \times 4$$
$$HS^- + H^+ \rightarrow H_2S$$
$$+ H^+ + OH^- \rightarrow H_2O$$
_____
$$4OCl^- + HS^- + OH^- \rightarrow 4Cl^- + SO_4^{2-} + H_2O \qquad \text{eq. 16.6}$$

In this final expression, materials, charges, and oxidation states all balance. All species are the dominant forms at higher pH values.

[†] You could turn this into an algebraic problem. Balances on Cl, S, O, H and charge yield, respectively: $a = d$, $b = e$, $4e = a + c$, $f = a + 2b + 2c$, and $f - d - 2e = 0$. Solving for each coefficient as a function of a: $b = a$, $c = 0$, $d = a$, $e = a/4$, and $f = 3a/2$. Setting $a = 4$, you can obtain $b = 1$, $c = 0$, $d = 4$, $e = 1$, and $f = 6$, as in eq. 16.5.

As in Section 16.3.1, you can double-check the changes in the oxidation state. *The changes in the oxidation states of all elements must balance in an overall reaction.* In eqs. 16.5 and 16.6, the oxidation states of H and O do not change. Chlorine changes from +I in HOCl and OCl⁻ to −I in chloride. This represents an eight-electron change since 4 moles of HOCl (or OCl⁻) are reduced. Sulfur is in the −II oxidation state in H_2S or HS⁻ and is oxidized to the +VI oxidation state in sulfate. Thus, the oxidation of the S in hydrogen sulfide (or bisulfide) also represents an eight-electron change. The changes in the oxidation states balance. Although this process takes many words to describe, you will come to do the oxidation state check in less time in your head than it takes to reread this paragraph.

16.3.3 Disproportionation and comproportionation reactions

There are several important examples in water chemistry in which a single element is present in more than two (usually three) oxidation states in the same overall reaction. In most cases, the element is both oxidized and reduced. If the element is in one oxidation state in the reactants and two or more oxidation states in the products, the reaction is called a **disproportionation reaction**. As an example, consider the dissolution of chlorine gas into water. Molecular chlorine (Cl_2, oxidation state = 0) forms both hypochlorous acid (HOCl, oxidation state = +I) and chloride (Cl⁻, oxidation state = −I) at pH < 7. Using the usual approach for balancing overall reactions, you can show that:

$$Cl_2(g) + H_2O \rightarrow HOCl + Cl^- + H^+ \qquad \text{eq. 16.7}$$

Equation 16.7 is a disproportionation reaction. You may wish to verify that materials, charges, and oxidation states balance in eq. 16.7.

In a **comproportionation reaction**, chemical species containing an element in two or more oxidation states react to form chemical species in one intermediate oxidation state of the element. Equation 16.7, read right to left, is a comproportionation reaction. Another example of a comproportionation reaction is presented in Section 16.3.4.

16.3.4 Practical applications of balancing overall reactions

A balanced overall reaction gives you a great deal of information. In an engineered setting, the stoichiometry of the balanced overall reacion gives chemical doses. For example, in the removal of hydrogen sulfide by chemical oxidation for odor control, a common rule of thumb is that about 9 lb of chlorine (as Cl_2) are required per lb of hydrogen sulfide removed (as S). This value stems directly from the stoichiometry in eqs. 16.5 and 16.7. From eq. 16.7, 1 mole of Cl_2 forms 1 mole of HOCl. From eq. 16.5, 4

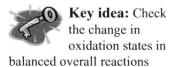

Key idea: Check the change in oxidation states in balanced overall reactions

disproportionation reaction: a reaction in which an element is in one oxidation state in the reactants and two or more oxidation states in the products

comproportionation reaction: a reaction in which an element is in two or more oxidation states in the reactants and one oxidation state in the products

---

Example 16.3: **Determining Dose from Stoichiometry**

Wastewater treatment plants discharging to sensitive watersheds may be required to dechlorinate their effluent. In most cases, a chemical reducing agent is added. Chlorine in most wastewater effluents is in the form of monochloramine, NH_2Cl. Monochloramine is reduced to chloride, and the nitrogen is released as ammonium.

Determine the dose (in lb/d) of sodium bisulfite ($NaHSO_3$) required to dechlorinate 8 mgd of a wastewater containing 2.5 mg/L NH_2Cl as Cl_2 to 0.2 mg/L NH_2Cl as Cl_2. Bisulfite is oxidized to sulfate.

Solution:

To write the overall reaction, you need the two half reactions. The balanced half reactions are:

$$NH_2Cl + 2H^+ + 2e^- \rightarrow NH_4^+ + Cl^-$$
$$HSO_3^- + H_2O \rightarrow SO_4^{2-} + 3H^+ + 2e^-$$

Adding to eliminate free electrons:

$$NH_2Cl + HSO_3^- + H_2O \rightarrow NH_4^+ + Cl^- + SO_4^{2-} + H^+$$

Thus, 1 mole of HSO_3^- is required to reduce 1 mole of NH_2Cl. On a mass basis: [(1 mole HSO_3^-)(104 g/mole $NaHSO_3$)]/[(1 mole NH_2Cl)× (70.9 g/mole Cl_2)] = 1.47 mg $NaHSO_3$ per mg monochloramine as Cl_2. It would require (2.5 – 0.2 mg/L Cl_2)(1.47 mg $NaHSO_3$ per mg monochloramine as Cl_2) = 3.37 mg/L $NaHSO_3$.

For loading calculations:

$$kg/d = Q(m^3/d)C(mg/L), \text{ or:}$$
$$kg/d = 8.34Q(mgd)C(mg/L)$$

where Q = flow, C = concentration, and the density of the fluid is near 1 g/cm³. Thus, the bisulfite requirement is 8.34(8 mgd)(3.37 mg/L) = **225 lb/d NaHSO₃**.

moles of HOCl are required to oxidize 1 mole of H_2S. Thus, the dose requirement is:

$$\begin{aligned} dose &= 4 \text{ moles HOCl/mole } H_2S \\ &= [(4 \text{ moles HOCl})(1 \text{ mole } Cl_2/\text{mole HOCl})(70.9 \text{ g } Cl_2/\text{mole} \\ &\quad Cl_2)]/[(1 \text{ mole } H_2S)(1 \text{ mole S/mole } H_2S)(32 \text{ g S/mole S})] \\ &= 8.9 \text{ g HOCl as } Cl_2 \text{ per g of } H_2S \text{ as S} \\ &= 8.9 \text{ lb HOCl as } Cl_2 \text{ per lb of } H_2S \text{ as S} \end{aligned}$$

Here is a slightly more complicated example. Permanganate (sold in the form of potassium permanganate, $KMnO_4$) is a strong chemical oxidant. It is used to oxidize ferrous iron and reduced manganese, Mn(+II), in groundwater. Iron and manganese can cause problems in treated drinking water. They may be oxidized posttreatment and precipitate as $Fe(OH)_3(s)$ and $MnO_2(s)$, causing red or black spots on clothing in washing machines. Permanganate is used to oxidize Fe^{2+} to Fe^{3+} and Mn^{2+} to $MnO_2(s)$ in the drinking water treatment plant. (The astute reader will note the we add manganese to remove manganese.) Optimization of the permanganate dose is critical because permanganate is relatively expensive and excess permanganate can give the treated water a pinkish hue (a condition called *pink water*, never popular among drinking water consumers). What is the stoichiometric dose of permanganate required to oxidize reduced iron and manganese?

To answer this question, you must write the balanced overall reactions for iron oxidation, manganese oxidation, and manganese reduction. You may wish to do this on your own. First, the balanced half reactions are:

$$Fe^{2+} \rightarrow Fe^{3+} + e^-$$
$$Mn^{2+} + 2H_2O \rightarrow MnO_2(s) + 4H^+ + 2e^-$$
$$MnO_4^- + 4H^+ + 3e^- \rightarrow MnO_2(s) + 2H_2O$$

Adding to eliminate free electrons:

$$\begin{array}{r} (Fe^{2+} \rightarrow Fe^{3+} + e^-)\times3 \\ + \; MnO_4^- + 4H^+ + 3e^- \rightarrow MnO_2(s) + 2H_2O \\ \hline MnO_4^- + 3Fe^{2+} + 4H^+ \rightarrow MnO_2(s) + 3Fe^{3+} + 2H_2O \end{array}$$

And:

$$\begin{array}{r} (Mn^{2+} + 2H_2O \rightarrow MnO_2(s) + 4H^+ + 2e^-)\times3 \\ + \; (MnO_4^- + 4H^+ + 3e^- \rightarrow MnO_2(s) + 2H_2O)\times2 \\ \hline 3Mn^{2+} + 2MnO_4^- + 2H_2O \rightarrow 5MnO_2(s) + 4H^+ \end{array} \qquad \text{eq. 16.8}$$

You may wish to verify the materials, charges, and oxidation states balance in the overall reactions. Note that eq. 16.8 is a comproportionation reaction: $Mn(+II)$ reacts with $Mn(+VII)$ to form the intermediate oxidation state $Mn(+IV)$. The required dose for iron oxidation is:

$$\text{dose} = 1 \text{ mole } MnO_4^-/3 \text{ mole } Fe^{2+}$$
$$= [(1 \text{ mole } MnO_4^-)(158 \text{ g } KMnO_4/\text{mole } KMnO_4)]/$$
$$[(3 \text{ mole } Fe^{2+})(55.8 \text{ g } Fe/\text{mole } Fe)]$$
$$= 0.94 \text{ g } KMnO_4 \text{ per g of Fe or } 0.94 \text{ lb } KMnO_4 \text{ per lb of iron}$$

The required dose for manganese oxidation is:

$$\text{dose} = 2 \text{ mole } MnO_4^-/3 \text{ mole } Mn^{2+}$$
$$= [(2 \text{ mole } MnO_4^-)(158 \text{ g } KMnO_4/\text{mole } KMnO_4)]/$$
$$[(3 \text{ mole } Mn^{2+})(54.9 \text{ g } Mn/\text{mole } Mn)]$$
$$= 1.9 \text{ g } KMnO_4 \text{ per g of Mn or } 1.9 \text{ lb } KMnO_4 \text{ per}$$
$$\text{lb of manganese}$$

Another use of stoichiometry to determine dosing is given in Example 16.3.

16.3.5 How can you tell if electrons are transferred?
In some cases, it is difficult to tell if electrons are transferred in a chemical reaction. In other words, it is sometime difficult to tell if a reaction is a redox reaction or nonredox reaction. Usually, you can identify redox reactions by checking to see if the oxidation states of any element has changed. For example, are electrons transferred in the reaction: $Cl_2(aq) + I^- \rightarrow I_2(aq) + 2Cl^-$? Yes: both chlorine and iodine change oxidation states.

On other occasions, determining whether electron transfer occurs can be more difficult. Consider the chemistry of ozone (O_3). During reactions in water, ozone is converted to oxygen.

Thoughtful Pause

Write the balanced reaction for the conversion of ozone to oxygen

---

Key idea: Exercise care when dealing with oxidation states of O other than $-II$ and of H other than $+I$

If no electrons are transferred, you could write:

$$2O_3 \rightarrow 3O_2 \qquad \text{eq. 16.9}$$

Both species appear to represent oxygen in the zero oxidation state. In fact, ozone is a very strong chemical oxidant and is *reduced* in water to form oxygen (and other species). Knowing that a reduction occurs, you can derive the balanced half reaction (please try this before continuing to read):

$$O_3 + 2H^+ + 2e^- \rightarrow O_2 + H_2O \qquad \text{eq. 16.10}$$

So which reaction is correct: eq. 16.9 or eq. 16.10? In fact, both equations are correct. The half reaction for the reduction of O_3 to O_2 is given by eq. 16.10. If you combine eq. 16.10 with the half reaction for the oxidation of water to O_2 ($H_2O \rightarrow O_2 + 4H^+ + 4e^-$), you obtain eq. 16.9. In other words, eq. 16.9 is an overall reaction. *When dealing with oxidation states of O and H other than $-II$ and $+I$, respectively, exercise great care.*

16.4 WHICH REDOX REACTIONS OCCUR?

16.4.1 Introduction

As the examples presented so far in this chapter show, chemical oxidants have a number of important applications in drinking water and wastewater treatment. In selecting an oxidant to achieve a treatment objective, you want to know if the oxidant can oxidize the pollutant to a given set of oxidation products. In this section, you will develop tools to decide whether one chemical species can oxidize (or reduce) another chemical species to a given oxidation state.

16.4.2 Redox reactions and spontaneity

How can you decide whether one chemical species can oxidize (or reduce) another chemical species? In fact, how do you decide whether any reaction proceeds as written? The answer was discussed in Section 3.5. A reaction proceeds as written (in other words, a reaction is spontaneous) if $\Delta G_{rxn} < 0$. Recall from eq. 3.16 (repeated here for convenience):

$$\Delta G_{rxn} = \Delta G^o_{rxn} + RT \ln \{i\}^{\nu_i} \qquad \text{eq. 16.11}$$

where ν_i are the stoichiometric coefficients (< 0 for reactants and > 0 for products). Recall that ΔG^o_{rxn} is calculated from the partial molar Gibbs free energy of formation values ($\Delta \overline{G}^o_{f,i}$).

An example will clarify the use of eq. 16.11. Suppose you and a fellow student are trying to explain why brominated compounds are found in some treated drinking waters. Engineers usually do not add bromine to water. Perhaps the brominating agent is hypobromous acid (HOBr), potentially formed from the oxidation of bromide (Br^-) by HOCl. For this hypothesis to be plausible, HOCl must be capable of oxidizing bromide to HOBr. You can test this idea by performing the thermodynamic calculation. First, balance the overall reaction. The balanced half reactions are:

$$HOCl + H^+ + 2e^- \rightarrow Cl^- + H_2O$$
$$Br^- + H_2O \rightarrow HOBr + H^+ + 2e^-$$

Example 16.4: Spontaneity

You are working for a start-up environmental technology firm. Your boss is knowledgeable about the use of hydrazine (NH_2NH_2) to reduce trace levels of dissolved oxygen (DO). Hydrazine is oxidized to nitrogen gas, $N_2(g)$, and DO is reduced to water.

Your boss wants you to investigate the possibility of developing a new synthesis method for hydrazine, using the reverse reaction: $N_2(g)$ combining with H_2O to form NH_2NH_2 and $O_2(aq)$.

Is this synthesis reaction possible under atmospheric conditions at room temperature?

Solution:
The half reactions are:
$N_2(g) + 4H^+ + 4e^- \rightarrow NH_2NH_2$
$H_2O \rightarrow O_2(aq) + 4H^+ + 4e^-$

The overall reaction is:
$N_2(g) + H_2O \rightarrow NH_2NH_2 + O_2(aq)$

The pertinent partial molar Gibbs free energy of formation values are +158.64, +16.32, 0, and −237.18 kJ/mol for

NH$_2$NH$_2$, O$_2$(aq), N$_2$(g), and H$_2$O, respectively. Thus, ΔG^o_{rxn} = +412.14 kJ/mol.

At atmospheric conditions, {N$_2$(g)} = 0.79 atm and {H$_2$O} is unity. From the stoichiometry, {NH$_2$NH$_2$} = {O$_2$(aq)} = x. Thus:

$$\Delta G_{rxn} = \Delta G^o_{rxn} + RT[\{NH_2NH_2\} \times \{O_2(aq)\}]/[\{N_2(g)\}\{H_2O\}]$$

Substituting:

$$\Delta G_{rxn} = +412.14 \text{ kJ/mol} + (8.314\times10^{-3} \text{ kJ/mol-}^oK)(298^oK)\ln(x^2/0.79)$$

For ΔG_{rxn} to be negative, x must be less than about 7×10^{-37} M. This is less than one molecule of hydrazine in every 2.5×10^{12} liters of water. (This volume is equivalent to a cube of water about 1.3 km per side.) Tell your boss gently that the synthesis idea is thermodynamically unjustified.

Adding: HOCl + Br$^-$ → HOBr + Cl$^-$. The pertinent partial molar Gibbs free energy of formation values are -79.9, -104.0, -82.2, and -131.3 kJ/mol for HOCl, Br$^-$, HOBr, and Cl$^-$, respectively. Thus:

$$\begin{aligned}\Delta G^o_{rxn} &= (1)(-131.3 \text{ kJ/mol}) + (1)(-82.2 \text{ kJ/mol}) \\ &\quad - (1)(-79.9 \text{ kJ/mol}) - (1)(-104.0 \text{ kJ/mol}) \\ &= -29.6 \text{ kJ/mol}\end{aligned}$$

To calculate the last term in eq. 16.11, you need the concentrations of each species. A common approach in pure chemistry is to assume that each species is present at a concentration of 1 M. This makes the math easier (since the last term in eq. 16.11 becomes zero) but is unrealistic under environmental conditions. If each species was at 1 M, then $\Delta G_{rxn} = -29.6$ kJ/mol < 0 and the reaction is spontaneous as written. Unfortunately, you do not know the concentrations. Reasonable estimates might be:

[HOCl] = 2 mg/L as Cl$_2$ = 2.8×10^{-5} M
[Br$^-$] = 0.1 mg/L = 5.6×10^{-7} M
[HOBr] = 1×10^{-7} M, and:
[Cl$^-$] = 20 mg/L = 5.6×10^{-4} M

With $R = 8.314\times10^{-3}$ kJ/mol-^oK and at 25°C = 298°K, $\Delta G_{rxn} = -26.4$ kJ/mol. Thus, hypochlorous acid is thermodynamically capable of oxidizing bromide to HOBr under the conditions tested. *You do not know if this process is slow or fast, only that the oxidation is possible.* Another determination of whether oxidations or reductions can occur spontaneously is given in Example 16.4.

16.5 REDOX THERMODYNAMICS AND OXIDANT AND REDUCTANT STRENGTH

16.5.1 Introduction
Several oxidants have been referred to in this chapter as strong oxidants. What is meant by a strong oxidant or a strong reductant? Can we develop an approach or scale to rank oxidants (or reductants) by the compounds they can oxidize (or reduce)? Quantitative tools for environmental redox chemistry will be developed in this section to answer such questions.

16.5.2 Electron affinity
How do you judge the strength of an acid? You quantify the strength of an acid by its K_a or pK_a. Recall that the K_a is the equilibrium constant for a specific type of equilibrium. You write:

$$HA + H_2O = A^- + H_3O^+ \quad K = K_a \qquad \text{eq. 16.12}$$

You could also write: $HA = A^- + H^+$; $K = K_a$. Why? You can use the simplified form of the equilibrium because the equilibrium constant of $H^+ + H_2O = H_3O^+$ is *defined* to be 1. In essence, water is selected as a standard proton acceptor. In fact, you can think of eq. 16.12 as a proton transfer equilibria consisting of two proton half reactions: a proton-donating half reaction, $HA = A^- + H^+$, and a proton-accepting half reaction, $H^+ + H_2O = H_3O^+$.

Using this approach with redox equilibria, you could judge oxidant or reductant strength by using a standard form of an equilibrium in which a chemical species acts as an oxidant or reductant. It is common practice to write redox reactions as reductions (i.e., as electron-accepting reactions). Therefore, you need a *standard electron donor*, analogous to the use of water as a standard proton acceptor. The usual standard electron donor is hydrogen gas, which is oxidized to H^+. Thus, the standard form of the equilibrium for determining reductant strength is:

$$X^{m+} + \tfrac{1}{2}H_2(g) = X^{(m-1)+} + H^+ \quad K = K_1$$

This is composed of two half reactions:

$$
\begin{aligned}
X^{m+} + e^- &= X^{(m-1)+} & K &= K_2 \\
\tfrac{1}{2}H_2(g) &= H^+ + e^- & K &= K_3
\end{aligned}
$$

By thermodynamic convention, the K_3 is defined to be 1, just as K for $H^+ + H_2O = H_3O^+$ is defined to be 1. Thus: $K_2 = K_1/K_3 = K_1$.

What is the best measure of reductant strength? You could use the equilibrium constant for equilibrium such as $X^{m+} + e^- = X^{(m-1)+}$ as your measure of reductant strength. However, some reductants accept more than one electron. For example, suppose you write:

$$2X^{m+} + 2e^- = 2X^{(m-1)+} \quad K = K_4$$

Now, $K_4 = K_2^2$ but the oxidant is the same in both equilibria. Therefore, we usually use $(1/n)\log K$ as our measure of reductant strength, where $n =$ number of electrons transferred and $K =$ equilibrium constant for an equilibrium of the form $X^{m+} + ne^- = X^{(m-n)+}$. The expression $(1/n)\log K$ sometimes is called *pe°*.

pe°: a measure of reductant strength and equal to $(1/n)\log K$, where $n =$ number of electrons transferred and $K =$ equilibrium constant for an equilibrium of the form $X^{m+} + ne^- = X^{(m-n)+}$

In many ways, pe° is analogous to pK_a. Remember that pK_a quantifies the thermodynamic tendency of an acid to donate a proton. Thus, pK_a quantifies the thermodynamic tendency of a conjugate base to accept a proton, just as pe° quantifies the thermodynamic tendency of an oxidant to accept an electron. This thermodynamic tendency is called *electron affinity*. Strong oxidants have strong electron affinities and therefore large

K and pe^o values[†]. Similarly, strong reductants have weak electron affinities and therefore small K and small or negative pe^o values.

The symbol pe^o may seem confusing at first because pe^o seems to have little to do with $pe = -\log\{e^-\}$. You can show that for a general redox equilibrium ($X^{m+} + ne^- = X^{(m-n)+}$):

$$K = \log[\{X^{(m-n)+}\}/\{X^{m+}\}] + npe, \text{ or:}$$
$$pe = (1/n)\log K - \log[\{X^{(m-n)+}\}/\{X^{m+}\}], \text{ so:}$$

$$pe = pe^o + \log[\{X^{m+}\}/\{X^{(m-n)+}\}]$$

This leads to another way to look at pe^o: *pe^o is the pe when all species are present at 1 M*. In this way, pe^o is somewhat analogous to ΔG^o_{rxn}, since ΔG^o_{rxn} is the Gibbs free energy of reaction when all species are at 1 M.

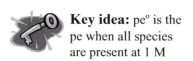

Key idea: pe^o is the pe when all species are present at 1 M

16.5.3 Oxidant strength and cell potentials

There is only one problem with the use of pe^o as a measure of oxidant strength: the K values for redox equilibrium frequently are not known. They are not known, in part, because the oxidation half reaction cannot occur without a corresponding half reaction in which the free electrons are accepted. Not knowing the K values is a big problem. How can you calculate $pe^o = (1/n)\log K$ if you do not know K? One approach is to find ΔG^o_{rxn}. If you know ΔG^o_{rxn}, then (from Section 3.9.2) you can calculate K by:

$$K = \exp(-\Delta G^o_{rxn}/RT)$$

Unfortunately, the ΔG^o_{rxn} also are not known in general. Fortunately, you can measure the energy of the overall reaction by measuring the voltage in an electrochemical cell. How can voltage help? Both voltage and Gibbs free energy are related to the work done by the system. Recall from Section 3.6.5 that electrical work is:

$$\text{Electrical work} = \int(\text{potential difference})dq$$
$$= (\text{potential difference})(\text{total charge transferred})$$
$$= (\text{potential difference})(\text{moles of electrons transferred})(\text{charge per mole of electrons})$$

[†] The relationship between oxidant or reductant strength and K emphasizes that oxidant and reductant are *thermodynamic* concepts. A strong oxidant accepts electrons to a large extent, but not necessarily quickly. The kinetic analogies to oxidant and reductant are *electrophile* and *nucleophile*, respectively. Electrophiles accept electrons rapidly, but not necessarily to a large extent.

faraday (*F*): 1 faraday = 96,845 coulombs per mole of electrons transferred (*F* also has units of joules per mole-volt)

 Key idea: The Gibbs free energy of reaction and the cell potential are related by $\Delta G_{rxn} = -nFE$ (note negative sign)

The potential difference here is the cell potential, *E*. We have been using the symbol *n* to represent the number of moles of electrons transferred. The charge per mole of electrons is a constant: 96,845 coulombs of charge per mole of electrons (96,845 C/mol) = 1 *faraday* = 1*F* (after Michael Faraday, 1791-1867).[†] Based on this analysis, the electric work in an electrochemical cell is related to *nFE*. The cell potential is defined so that the electric work in an electrochemical cell is equal to $-nFE$. Since the chemical work done is ΔG_{rxn}, then:

$$\Delta G_{rxn} = -nFE$$

Note that, by convention, ΔG_{rxn} and *E* are *opposite* in sign. Recall that $\Delta G_{rxn} < 0$ for spontaneous reactions. Thus, $E > 0$ for spontaneous reactions. If all species are at their standard states: $\Delta G_{rxn} = \Delta G^o_{rxn} = -nFE^o$. Here E^o is the cell potential with all species at their standard states. Now you have a way of calculating pe°:

1. Measure the cell potential in an electrochemical cell with all species at their standard states. This is E^o.
2. Calculate $\Delta G^o_{rxn} = -nFE^o$.
3. Calculate $K = \exp(-\Delta G^o_{rxn}/RT)$.
4. Finally, calculate pe° = $(1/n)\log K$.

This procedure relies on measuring the cell potential. To measure the cell potential, an electrochemical cell must be devised in which a standard reductant is oxidized and the oxidant of interest is reduced. Take the example of $Fe^{3+} + e^- \rightarrow Fe^{2+}$. It makes sense to link ferric iron reduction to an oxidation with a known *K* value. We already have defined the equilibrium constant for $\frac{1}{2}H_2(g) = H^+ + e^-$ to be 1. A possible electrochemical cell is shown in Figure 16.1. The hydrogen chamber has $\{H_2(g)\}$ = 1 atm and $\{H^+\}$ = 1 M (i.e., *very* acidic) and is called the *standard* (or *normal*) *hydrogen electrode*. If $\{Fe^{3+}\} = \{Fe^{2+}\}$ = 1 M, then the voltmeter will measure +0.77 V. Thus, E^o = +0.77 V, and:

standard (or *normal*) *hydrogen electrode*: a part of an electrochemical cell where $\{H_2(g)\}$ = 1 atm and $\{H^+\}$ = 1 M

$$\begin{aligned}\Delta G^o_{rxn} &= -nFE^o \\ &= -(1)(96.4\,85\ kJ/mol\text{-}V)(+0.77\ V) \\ &= -74.29\ kJ/mol\end{aligned}$$

So:

$$\begin{aligned}K &= \exp(-\Delta G^o_{rxn}/RT) \\ &= \exp[(+74.29\ kJ/mol)/(8.314\times10^{-3}\ kJ/mol\text{-}°K)(298°K)] \\ &= 1.0\times10^{13}\end{aligned}$$

[†] A volt is defined so that the energy required to move 1 coulomb of charge through 1 volt is 1 joule. Thus, *F* also has units of joules per mol-volt.

Thus: $pe^o = (1/n)\log K = (1/1)\log(1.0 \times 10^{13}) = 13.0$.

Typically, E^o values for redox equilibria (written as reductions) are tabulated rather than pe^o values. Using the equations presented above, it is easy to convert redox thermodynamic data into any form you wish. Interconversion formulas between redox thermodynamic data are listed in Table 16.2. Another example of interconversions between data types is given in Example 16.5.

Example 16.5: **Redox Thermodynamic Data**

Will chlorine gas cause iron to rust spontaneously if all pertinent species are at their standard state?

Solution:
For iron to rust, Fe(s) must be oxidized to Fe^{2+}. The relevant equilibria are:

$Fe^{2+} + 2e^- \rightarrow Fe(s)$
$\qquad E^o = -0.409$ V
$Cl_2(g) + 2e^- \rightarrow 2Cl^-$
$\qquad \log K = 46.0$

Oxygen will rust iron if the overall reaction has $E > 0$ or $E^o > 0$ with all species at their standard states. Thus, we must convert $\log K$ to E^o and then find E^o for the overall reaction. We know:

$E^o = (2.303RT/nF)\log K$
$\quad = (2.303)(8.314$ J/mol-°K$)(298°K)/[(2)(96,485$ V/mol-J$)](46.0)$
$\quad = +1.36$ V

Rearranging the equilibria and adding (E^o in V in parentheses):

$Fe(s) \rightarrow Fe^{2+} + 2e^- (+0.409)$
$+ Cl_2(g) + 2e^- \rightarrow 2Cl^- (+1.36)$
——————————————————
$Cl_2(g) + Fe(s) \rightarrow Fe^{2+} + $
$\qquad 2Cl^-, E^o = +1.77$ V

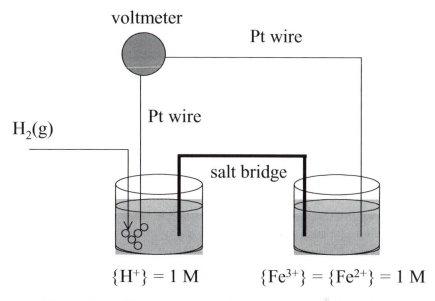

voltmeter

Pt wire

Pt wire

$H_2(g)$

salt bridge

$\{H^+\} = 1$ M $\{Fe^{3+}\} = \{Fe^{2+}\} = 1$ M

Figure 16.1: Electrochemical Cell Showing the Standard Hydrogen Electrode
(the salt bridge completes the circuit and allows electrons to flow)

Table 16.2: Interconversions Between Redox Thermodynamic Data
(tabulated formulas show the row parameter as a function of the column parameter)

	E^o	ΔG^o_{rxn}	$\log K$	pe^o
$E^o =$	–	$-\Delta G^o_{rxn}/nF$	$(2.303RT/nF)\log K$	$(2.303RT/F)pe^o$
$\Delta G^o_{rxn} =$	$-nFE^o$	–	$-2.303RT\log K$	$-2.303nRTpe^o$
$\log K =$	$nFE^o/2.303RT$	$-\Delta G^o_{rxn}/2.303RT$	–	npe^o
$pe^o =$	$FE^o/2.303RT$	$-\Delta G^o_{rxn}/2.303nRT$	$(1/n)\log K$	–

Note: For the reaction: ox + ne^- = red
Also note that $\Delta G_{rxn} = -nFE$
$2.303RT/nF \approx 0.059/n$ volts $\approx 59/n$ mV at 25°C

$E^o > 0$, **so chlorine gas will cause iron to rust spontaneously if $Cl_2(g)$, $Fe(s)$, H^+, Fe^{2+}, and Cl^- are at their standard state**

Key idea: With reactions listed as reductions, strong oxidants appear as reactants in reactions with large positive pe^o values and strong reductants appear as products in reactions with large negative pe^o values

A few environmentally important redox reactions, listed as usual as reductions, are given in Table 16.3 along with pertinent thermodynamic data. You may wish to verify a few of the calculations. Table 16.3 is ordered by pe^o values. Strong oxidants are the reactants in reactions near the bottom of the table (e.g., oxygen and ozone). Strong reductants, for example $Zn(s)$, are products in reactions at the top of the table.

Table 16.3: Common Environmental Redox Reactions[1]

Reaction	E^o (V)	log$K$	pe^o
$Zn^{2+} + 2e^- = Zn(s)$	-0.763	-25.8	-12.9
$Fe^{2+} + 2e^- = Fe(s)$	-0.409	-13.8	-6.92
$2H^+ + 2e^- = H_2(g)$	$\equiv 0$	0.0	0.0
$S(s) + 2H^+ + 2e^- = H_2S$	$+0.14^2$	4.8	2.4
$Cu^{2+} + e^- = Cu^+$	$+0.158$	2.7	2.7
$Cu^{2+} + 2e^- = Cu(s)$	$+0.340$	11.5	5.7
$SO_4^{2-} + 8H^+ + 6e^- = S(s) + 4H_2O$	$+0.36^2$	36.2	6.03
$Fe^{3+} + e^- = Fe^{2+}$	$+0.770$	13.0	13.0
$NO_3^- + 2H^+ + 2e^- = NO_2^- + H_2O$	$+0.84^2$	28.3	14.15
$O_2(g) + 4H^+ + 4e^- = 2H_2O$	$+1.229$	83.1	20.78
$MnO_2(s) + 4H^+ + 2e^- = Mn^{2+} + 2H_2O$	$+1.21$	40.9	20.5
$Cl_2(g) + 2e^- = 2Cl^-$	$+1.36$	46.0	23.0
$O_3(g) + 2H^+ + 2e^- = O_2(g) + H_2O$	$+2.09$	70.7	35.3

Note: 1. All at 25°C; E^o data are from Bard and Faulkner (1980), with log$K$ and pe^o values calculated.

2. The pe^o value is from Stumm and Morgan (1996), with E^o and log$K$ calculated.

16.5.4 How is the cell potential related to species activities?

From Section 16.5.3, you know that ΔG_{rxn} is related to the cell potential, E, by: $\Delta G_{rxn} = -nFE$ or $E = -\Delta G_{rxn}/nF$. Since ΔG_{rxn} is a function of species activities, it makes sense that the E also should depend on the species concentrations. As an example, consider a general reduction in which an oxidized species accepts n electrons to be reduced to a reduced form: red + ne^- = ox. Using hydrogen gas as the electron donor: red + $(n/2)H_2(g)$ = ox + nH^+. The Gibbs free energy of reaction for this equilibrium is (see Section 3.8.3):

$$\Delta G_{rxn} = \Delta G_{rxn}^o + RT \ln \frac{\{ox\}\{H^+\}}{\{red\}\{H_2(g)\}^{n/2}}$$

eq. 16.13

For the standard hydrogen electrode: $\{H^+\} = 1$ M and $\{H_2(g)\} = 1$ atm. Substituting these values and dividing both sides of eq. 16.13 by $-nF$:

$$E = E^o + \frac{RT}{nF} \ln \frac{\{ox\}}{\{red\}}$$ eq. 16.14

Or:

$$E = E^o + \frac{2.303RT}{nF} \log \frac{\{ox\}}{\{red\}}$$

where $2.303RT/nF = 0.059/n$ V $= 59/n$ mV at 25°C.

Nernst equation: an equation relating the cell potential to the activities of the electron-transferring species

Equation 16.14 is called the ***Nernst equation*** (after Walther Hermann Nernst, 1864-1941; see the *Historical Note* at the end of the chapter). The Nernst equation shows the dependence of the cell potential on the activities of the species undergoing redox reactions. For example, the Nernst equation for the reduction of ferric iron to ferrous iron ($Fe^{3+} + e^- = Fe^{2+}$) is:

$$E = E^o + (RT/F)\ln[\{Fe^{2+}\}/\{Fe^{3+}\}]$$

Although the Nernst equation usually is written as shown in eq. 16.14, do not forget that it comes from the expression for the Gibbs free energy of reaction. Thus, the numerator in the logarithm argument is really the products of the activities of the products, each raised to their stoichiometric coefficients and the denominator is the products of the activities of the reactants (excluding e^-), each raised to their stoichiometric coefficients. By convention, the reaction must be written as a reduction. For example, the Nernst equation for the reduction of dissolved oxygen to water ($O_2(aq) + 4H^+ + 4e^- = 2H_2O$) is:

$$E = E^o + (RT/F)\ln[\{H_2O\}^2/\{O_2(aq)\}\{H^+\}^4]$$

16.6 ALGEBRAIC EQUILIBRIUM CALCULATIONS IN SYSTEMS UNDERGOING ELECTRON TRANSFER

16.6.1 Manipulating half reactions

Your experiences so far in aquatic chemistry have convinced you of the need to add, reverse, and multiply equilibria by constants to get them in the form you need. Tracking the thermodynamic parameters as you manipulate redox equilibria can be challenging. The most common form of thermodynamic data listed for redox reactions is E^o. Thus, it is important to practice manipulating E^o and E values as you manipulate equilibria. Two skills are particularly important: tracking E as you multiply redox equilibria by a positive constant and tracking E as you add redox equilibria.

As an example of tracking E as you multiply redox equilibria by a positive constant, compare the following two half reactions (written as equilibria):

$$\text{Equilibrium 1:} \quad Fe^{3+} + e^- = Fe^{2+} \qquad E = E_1$$
$$\text{Equilibrium 2:} \quad 2Fe^{3+} + 2e^- = 2Fe^{2+} \qquad E = E_2$$

Thoughtful Pause
How does E_2 compare with E_1?

From the Nernst equation:

$$E_1 = E_1^o + (RT/F)\ln[\{Fe^{2+}\}/\{Fe^{3+}\}] \text{ and:}$$
$$E_2 = E_2^o + (RT/2F)\ln[\{Fe^{2+}\}^2/\{Fe^{3+}\}^2]$$

We know $E_1^o = E_2^o$, since the E^o values are the cell potentials when all species are at their standard states (1 M for dissolved species). Note also that:

$$\begin{aligned}
E_2 &= E_2^o + (RT/2F)\ln[\{Fe^{2+}\}^2/\{Fe^{3+}\}^2] \\
&= E_1^o + (RT/2F)\ln[\{Fe^{2+}\}/\{Fe^{3+}\}]^2 \\
&= E_1^o + (RT/F)\ln[\{Fe^{2+}\}/\{Fe^{3+}\}] \\
&= E_1
\end{aligned}$$

 Key idea: The cell potential does not change if you multiply the half reaction by a positive constant

Thus, *the cell potential does not change if you multiply the half reaction by a positive constant.* You also can reach this conclusion by converting the cell potentials to Gibbs free energy values (see Problem 16.7).

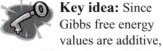 **Key idea:** Since Gibbs free energy values are additive, when adding redox equilibria convert E^o values to ΔG_{rxn}^o, add the ΔG_{rxn}^o values, and then reconvert to E^o

When adding redox equilibria, always come back to the truism that *Gibbs free energy values are additive.* Thus, the most error-free way to manipulate E^o when adding redox equilibria is to *convert E^o values to ΔG_{rxn}^o, add the ΔG_{rxn}^o values, and then reconvert to E^o.* An example will make this process clearer. Using the data given below, decide whether molecular chlorine is capable of oxidizing ferrous iron to ferric iron (if all species are at 1 M) and find ΔG_{rxn}^o, E^o, and K for the overall reaction. The thermodynamic data are:

$$Fe^{3+} + e^- = Fe^{2+} \qquad E_1^o = +0.77 \text{ V}$$
$$Cl_2 + 2e^- = 2Cl^- \qquad E_2^o = +1.39 \text{ V}$$

Calculating the standard Gibbs free energies of reaction: $\Delta G^o_{rxn,1} = -FE_1^o$ and $\Delta G^o_{rxn,2} = -2FE_2^o$. We know the iron reaction should be reversed (written as an oxidation) and multiplied by 2. Reversing the reaction will multiply E_1^o by -1. Multiplying the stoichiometric coefficients by 2 will have no effect on E^o, but will double ΔG^o_{rxn} (since $\Delta G^o_{rxn} = -nFE^o$). Thus:

$$2Fe^{2+} = 2Fe^{3+} + 2e^- \quad E_3^o = -0.77 \text{ V}, \Delta G^o_{rxn,3} = -2FE_3^o$$
$$Cl_2 + 2e^- = 2Cl^- \quad E_2^o = +1.39 \text{ V}, \Delta G^o_{rxn,2} = -2FE_2^o$$

Adding the half reactions: $2Fe^{2+} + Cl_2 = 2Fe^{3+} + 2Cl^-$. Adding the standard Gibbs free energies of reaction:

$$\Delta G^o_{rxn} = \Delta G^o_{rxn,3} + \Delta G^o_{rxn,2}$$
$$= -2F(E_3^o + E_2^o) = -2(96.845 \text{ kJ/mol-V})(-0.77 \text{ V} + 1.39 \text{ V})$$
$$= -120.1 \text{ kJ/mol}$$

Thus: $E^o = -\Delta G^o_{rxn}/2F = +0.62$ V and $K = \exp(-\Delta G^o_{rxn}/RT) = 1.1 \times 10^{21}$.

Thoughtful Pause
Is the reaction as written favorable if all species are at 1 M?

Since ΔG^o_{rxn} is less than zero (or, equivalently, $E^o > 0$), the oxidation of ferrous iron by molecular chlorine is thermodynamically favorable if all species are present at 1 M.

16.6.2 Calculating activities as a function of pe and pH
The master variables in most aquatic systems are pH and pe: pH controls the acid-base chemistry and pe controls the redox chemistry. It is useful to manipulate equilibria to show the chemical speciation at any pe and pH value.

As an example, consider the equilibrium representing the oxidation of ammonium to nitrate. This is an important environmental process called *nitrification*. Writing the equilibrium as usual as a reduction:

$$NO_3^- + 10H^+ + 8e^- = NH_4^+ + 3H_2O, E^o = +0.88$$

Key idea: To calculate activities as functions of pe and pH, rearrange the mathematical form of the equilibrium to obtain an equation of the form pe = a + bpH +$c$log(ratio of species activities)

For this equilibrium: $\log K = (nFE^o/2.303RT)E^o = 119.2$ and $K = 10^{+119.2}$. Therefore:

$$K = \{H_2O\}^3\{NH_4^+\}/(\{NO_3^-\}\{H^+\}^{10}\{e^-\}^8) \text{ or:}$$

$$\log K = 3\log\{H_2O\} + \log\{NH_4^+\} - \log\{NO_3^-\}$$
$$- 10\log\{H^+\} - 8\log\{e^-\}$$

***Example 16.6:* pe, pH, and Species Activities**

What form of sulfur dominates in a bog water at pH 4 and pe = -3?

Solution:
As a start, compare the lowest two oxidation states of sulfur: S(-II) and S(0). The redox equilibrium is:

$$S(s) + 2H^+ + 2e^- \rightarrow H_2S$$
$$\log K = 4.8$$

Thus:
$$2pe = 4.8 - 2pH + \log[\{S(s)\}/\{H_2S\}]$$

At pH 4 and pe = -3:
$$\log[\{S(s)\}/\{H_2S\}] = -2.8$$

Thus, H_2S predominates.

If H_2S predominates over $S(s)$, then higher oxidation states such as sulfate are not expected to be important. You can use the equilibria in Table 16.3 to confirm this point. Adding the S equilibria in Table 16.3:

$$SO_4^{2-} + 12H^+ + 8e^- \rightarrow$$
$$H_2S + 4H_2O, \log K = 41$$

Thus:

$$8pe = 41 - 12pH + \log[\{SO_4^{2-}\}/\{H_2S\}]$$

At pH 4 and pe = -3:

$$\log[\{SO_4^{2-}\}/\{H_2S\}] = -17$$

Thus, H_2S predominates.

In a bog water at pH 4 and pe = -3, H_2S predominates.

Using the definitions of pH and pe and assuming that the activity of water is unity:

$$\log K = 119.2 = \log(\{NH_4^+\}/\{NO_3^-\}) + 10pH + 8pe, \text{ or:}$$

$$pe = 14.9 - 5/4pH - 1/8\log(\{NH_4^+\}/\{NO_3^-\}) \qquad \text{eq. 16.15}$$

Equation 16.15 can be used to determine how the speciation of nitrogen varies with pH and pe at equilibrium. For example, consider water in an oxidizing environment (pe = +10) at neutral pH (pH 7). Under these conditions:

$$\log(\{NH_4^+\}/\{NO_3^-\}) = -8[pe - 14.9 + (5/4)pH]$$
$$= -8[10 - 14.9 + (5/4)7]$$
$$= -30.8$$

or $\{NH_4^+\}/\{NO_3^-\} = 10^{-30.8}$. In an oxidizing environment at neutral pH, N(+V) is greatly favored over N(-III). This makes sense: the more oxidized form is favored in an oxidizing environment.

Thoughtful Pause

What is the ratio of species activities at pe = -5 and pH 7?

At pe = -5 and pH 7, $\{NH_4^+\}/\{NO_3^-\} = 10^{+89.2}$. In other words, in a reducing environment at neutral pH, the reduced form, N(-III) is greatly favored over the oxidized form, N(+V).

Equation 16.15 also can be used to show how pe and pH are related at a given ratio of the reduced and oxidized forms of chemical species. We often examine the relationship between pe and pH when the activities of the reduced and oxidized forms are equal. For the nitrification example, pe = 14.9 - 5/4pH when $\{NH_4^+\} = \{NO_3^-\}$. This approach allows you to calculate the pe. For a water at pH 6.5 where nitrogen chemistry controls the pe and $\{NH_4^+\} = \{NO_3^-\}$, pe = 14.9 - (5/4)(6.5) = +6.78. Another illustration of quantifying the relationship among pe, pH, and species activities is given in Example 6.6.

16.6.3 Application: Potentiometric electrodes and the measurement of pH
The Nernst equation shows that the cell potential is related to species activities. Cell potentials caused by species activities are called *Nernst potentials*. One example of a Nernst potential is the cell potential

generated with two solutions of differing compositions that are separated by a thin membrane. If one solution is of known composition, the activity of the species of interest can be determined in the other solution through measurement of the Nernst potential. In general:

$$E = E^o + (RT/nF)\ln[\{X\}_{side\ 1}/\{X\}_{side\ 2}]$$

A common example of the use of Nernst potentials in chemical analysis is the pH probe. The working (i.e., $\{H^+\}$ sensing) electrode is a thin glass membrane shaped like a bulb and developed by Arnold O. Beckman in the 1920s. The inside of the bulb is filled with a strong acid of known $\{H^+\}$. The sample is the other (or outside) solution. A salt bridge (see Figure 16.1) and reference electrode complete the circuit. Thus:

$$\begin{aligned}
E &= E^o + (RT/nF)\ln[\{H^+\}_{outside}/\{H^+\}_{inside}] \\
&= E^o - (RT/nF)\ln\{H^+\}_{inside} + (RT/nF)\ln\{H^+\}_{outside}, \text{ or, since } n=1: \\
E &= E^{o\prime} - 2.303(RT/F)\text{pH}_{sample} \qquad\qquad \text{eq. 16.16}
\end{aligned}$$

where $E^{o\prime} = E^o - (RT/F)\ln\{H^+\}_{inside}$.

The pH probe must be calibrated. *Calibration* is the process of determining the values of the adjustable variables so that the response (E here) can be related to the quantity of interest (pH_{sample} here). To calibrate the probe, the pH of a solution of known $\{H^+\}$, called a *standard*, is measured.

Thoughtful Pause
How many standards must be measured?

As eq. 16.16 states, the cell potential is linearly related with the pH of the standard or sample. Thus, you must measure the pH of at least two standards to determine the slope ($= -2.303RT/F$) and intercept ($= E^{o\prime} = E^o - (RT/F)\ln\{H^+\}_{inside}$).[†] The Nernst equation gives you a way to measure pH and tells you the minimum number of standards that must be measured to calibrate a pH probe.

[†] You might wonder why standards are even necessary since it appears that the slope and intercept both can be calculated from temperature, constants, and E^o. In fact, the intercept also comprises other terms (including the liquid-junction potential, a potential difference caused by the different rates of ion migration across an interface). In practice, pH is operationally defined from pH measurements by the equation: $\text{pH(sample)} = \text{pH(standard)} + (F/2.303RT)(E_{sample} - E_{standard})$.

16.7 GRAPHICAL REPRESENTATIONS OF SYSTEMS UNDERGOING ELECTRON TRANSFER

16.7.1 Introduction

Since pe and pH are master variables, it is desirable to be able to show how speciation changes with both pe and pH. To show species concentrations as a function of both pe and pH would require a three-dimensional plot, not impossible, of course, but a bit confusing.

To simplify the plots, two approaches have been taken to show the effects of redox conditions on speciation. First, the pH is fixed and pe is used as the master variable. The resulting plots are called ***pC-pe diagrams***, with −log(concentration) on the *y*-axis and pe on the *x*-axis. The p*C*-pe diagrams are exactly analogous to the p*C*-pH diagrams of Chapter 8. The second approach is called a ***pe-pH diagram***. In a pe-pH diagram, pe is plotted on the *y*-axis and pH on the *x*-axis. Lines are drawn on the plot to indicate conditions where two species have the same activity. Before discussing each diagram in more detail, the redox limits of water must be established.

pC-pe diagram: a graphical representation of redox equilibria with −log(concentration) on the *y*-axis and pe on the *x*-axis

pe-pH diagram: a graphical representation of redox equilibria with pe on the *y*-axis and pH on the *x*-axis and lines drawn to indicate conditions where two species have the same activity

16.7.2 The redox limits of water

What happens to water when the oxidizing or reducing conditions are too extreme? Under highly oxidizing conditions, water is oxidized to oxygen gas. Under highly reducing conditions, water is reduced to hydrogen gas. How high or low does the pe have to be before water is oxidized or reduced? Not surprisingly, the answer depends on the pH. For the oxidation of water:

$$O_2(g) + 4H^+ + 4e^- = 2H_2O \qquad \log K = 83.1$$

With the activity of water at unity and the activity of oxygen gas expressed by its partial pressure, you can show that $pe = 20.78 - pH + 1/4\log P_{O_2}$.

If the total pressure of the system is 1 atm, then the largest possible value of P_{O_2} is 1 atm. Thus:

$$pe = 20.78 - pH \qquad \text{(for a pure } O_2 \text{ atmosphere)}$$

A similar equation can be developed for the reduction of water to hydrogen gas. The redox equilibrium is:

$$2H_2O + 2e^- = H_2(g) + 2OH^- \qquad \log K = -28^\dagger$$

[†] The redox equilibrium for the reduction of water can be obtained by adding the following two reactions: $2H^+ + 2e^- = H_2(g)$ ($K \equiv 1$ or $\log K = 0$) and $2H_2O = 2H^+ + 2OH^-$ ($\log K = 2\log K_W = -28$).

Again, for $\{H_2O\} = 1$ M, $\{H_2(g)\} = P_{H_2} = 1$ atm, and as $\log\{OH^-\} = \log K_W + pH = -14 + pH$ at $25°C$:

$$pe = -pH \qquad \text{(for a pure } H_2 \text{ atmosphere)}$$

We now have two equations that define the redox stability of water:

$$pe = 20.78 - pH \text{ and:}$$
$$pe = -pH$$

Key idea: The redox and pH ranges of water are intertwined and given by the equations: pe = 20.78 – pH and pe = –pH

We are only interested in the region between these two extremes (at 1 atm and $25°C$). Note that *the redox stability of water depends on the pH.* What does this mean? First, it means that we must revisit our pC-pH diagrams. The pC-pH diagrams are valid only for a specific pe (or range of pe values). In a perfect world, pC-pH diagrams should be labeled with the pe range for which they are valid. In addition, the pH range in a pC-pH diagram depends on pe. Previously, we have considered the pH range of water to be 0 to 14. Now you know that the pH range of water depends on the pe. For pH = 0 to 14, the pe must be in the range of 0 to +6.78 to satisfy the redox stability of water. Thus, all our previous pC-pH diagrams with x-axes extending from 0 to 14 should be amended to say that they are valid only if pe = 0 to +6.78. Similar statements hold true for other pH ranges: for example, pH 4 to 10 corresponds to pe = –4 to +10.78. (In general, pH = min to max corresponds to pe = –min to 20.78 – max.)

Second, the dependence of the redox stability range of water on pH means that pC-pe and pe-pH diagrams are valid only for specific ranges of pe and pH. For pC-pe diagrams (which are drawn for a specific pH), this means that you must choose a pe range (x-axis range) corresponding to the stability of water at that pH value. For pe-pH diagrams, the standard approach is to draw the lines for the stability of water on the diagram.

Both pC-pe and pe-pH diagrams will be discussed in more detail below. For each type of diagram, three systems will be presented: (1) $H_2(g)/O_2(g)/H_2O$, (2) Fe^{3+}/Fe^{2+}, and (3) NO_3^-/NH_4^+ system from Section 16.6.2.

16.7.3 pC-pe diagrams

To draw a pC-pe diagram, the pH is fixed and pC for each species is plotted as a function of pe. For the $H_2(g)/O_2(g)/H_2O$ system, the pertinent equilibria are:

$$O_2(g) + 4H^+ + 4e^- = 2H_2O \qquad \log K = 83.1$$
$$2H_2O + 2e^- = H_2(g) + 2OH^- \qquad \log K = -28$$

Expressing each species concentration in terms of pe:

$$pP_{O_2} = 83.12 - 4pH - 4pe, \text{ and:}$$

$$pP_{H_2} = 2pH + 2pe$$

The pC-pe diagram for the $H_2(g)/O_2(g)/H_2O$ system is shown at pH 7 in Figure 16.2. At lower pH, the lines for $H_2(g)$ and $O_2(g)$ are both displaced to the right.

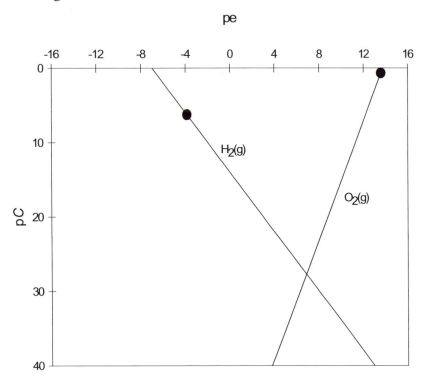

Figure 16.2: pC-pe Diagram for $H_2(g)/O_2(g)/H_2O$ at pH 7

The pC-pe diagram for the $H_2(g)/O_2(g)/H_2O$ system is not very useful, but it does point out an important fact about some redox systems. How would you determine the pe of water at pH 7 in equilibrium with the atmosphere? In the atmosphere, the partial pressures of $H_2(g)$ and $O_2(g)$ are 5×10^{-7} and 0.209 atm, respectively. Applying these data to Figure 16.2, you can obtain *two* equilibrium pe values (indicated by the solid circles in Figure 16.2).[†] How can this be? Clearly, the $H_2(g)/O_2(g)/H_2O$ system at pH 7 is **not** in equilibrium with the atmosphere. In fact, many redox systems are not in equilibrium. This idea will be explored further in Section 16.8.1.

[†] The equilibrium pe value of 13.6, calculated from the partial pressure of oxygen in the atmosphere, corresponds to a hydrogen partial pressure of about 10^{-41} atm. This is about 1000 molecules of hydrogen gas in the *entire* atmosphere - not a very reasonable result.

A more meaningful pC-pe diagram can be obtained with the Fe^{3+}/Fe^{2+} system. The pertinent equations for constructing the pC-pe diagram are:

Equilibrium: $Fe^{3+} + e^- = Fe^{2+}$, $\log K = 13.0$ or $K = \{Fe^{2+}\}/\{Fe^{3+}\}\{e^-\}$
Mass balance: $Fe_T = [Fe^{2+}] + [Fe^{3+}]$

Solving each species activity in terms of $\{e^-\}$ (assuming activities and concentrations are equal):

$$\{Fe^{2+}\} = \{e^-\}Fe_T/(1/K + \{e^-\}) \text{ and } \{Fe^{3+}\} = (Fe_T/K)/(1/K + \{e^-\})$$

These equations are analogous to alpha values with acid-base chemistry. The pC-pe diagram at $Fe_T = 1 \times 10^{-4}$ M is plotted in Figure 16.3 (assuming that the pH is low enough that hydroxo complexes can be ignored). Note that pe = $\log K = pe^o = 13$ at $\{Fe^{2+}\} = \{Fe^{3+}\}$ (dotted line in Figure 16.3), showing the analogy between pe^o and pK_a, discussed in Section 16.5.2.

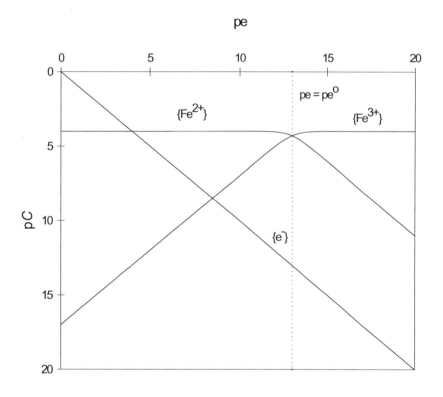

Figure 16.3: pC-pe Diagram for the Fe^{3+}/Fe^{2+} System at Low pH

As a final example of pC-pe diagrams, consider the NO_3^-/NH_4^+ system discussed in Section 16.6.2. At pH 7, the predominate acid-base forms of $N(+V)$ and $N(-III)$ are NO_3^- and NH_4^+, respectively. The pertinent equations for constructing the pC-pe diagram are:

Equilibrium: $NO_3^- + 10H^+ + 8e^- = NH_4^+ + 3H_2O$, $\log K = 119.2$
Mass balance: $N_T = [NO_3^-] + [NH_4^+]$

Solving each species activity in terms of $\{e^-\}$ (assuming activities and concentrations are equal):

$\{NH_4^+\} = \{e^-\}^8 Fe_T/(1/K' + \{e^-\}^8)$ and:
$\{NO_3^-\} = (Fe_T/K')/(1/K' + \{e^-\}^8)$

where $K' = K\{H^+\}^{10} = 10^{49.2}$ at pH 7. The pC-pe diagram at $N_T = 1 \times 10^{-3}$ M is plotted in Figure 16.4. Note that at $\{NO_3^-\} = \{NH_4^+\}$, pe = (1/8)$\log K'$ = pe$^\circ$ at pH 7 = 6.15. The conditions calculated in Section 16.6.2 at pH 7 can be evaluated using Figure 16.4. It is clear from the pC-pe diagram that nitrate dominates at pe = +10 and pH 7 and that ammonium dominates at pe = −5 and pH 7.

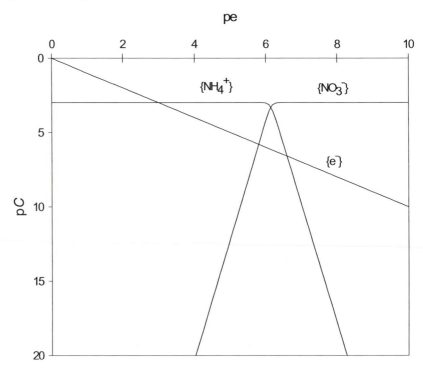

Figure 16.4: pC-pe Diagram for the NO_3^-/NH_4^+ System at pH 7

16.7.4 pe-pH diagrams

Recall that in pe-pH diagrams, lines are drawn to indicate conditions where *two species have the same activity*. You can plot the water stability conditions as lines on a pe-pH diagram to indicate the possible oxidation states of H and O under environmental conditions, as in Figure 16.5. The top line in Figure 16.5 indicates where $\{O_2(g)\} = \{H_2O\}$, and the bottom line indicates where $\{H_2(g)\} = \{H_2O\}$. It is customary to label each line to show the pe region where a given species predominates. Thus, for example, we put the label for $O_2(g)$ at higher pe values than the line indicating where $\{O_2(g)\} = \{H_2O\}$. Since a fluid with pH < 0 or pH > 14 is not called water, water is defined thermodynamically (at 1 atm and 25°C) by the parallelogram in Figure 16.5.

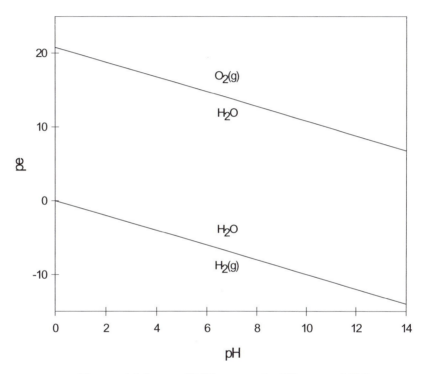

Figure 16.5: pe-pH Diagram for Water at 25°C
(top line: partial pressure of O_2 = 1 atm
bottom line: partial pressure of H_2 = 1 atm)

Now you can simply plot other pe-pH relationships on top of the pe-pH diagram for water (just as we plotted other pC-pH relationships on top of the pC-pH diagram for water in Chapter 8). The general approach to constructing simple pe-pH diagrams is to write equations of the form pe = apH + b for every equilibrium involving the species of interest. For electron transfer reactions with no proton transfer (i.e., no acid-base chemistry), $a = 0$. Thus, the line of equal activities on a pe-pH diagram is a line of slope = 0 at pe = pe° = $(1/n)\log K$.

As an example, consider the reduction of ferric iron: $Fe^{3+} + e^- = Fe^{2+}$, $\log K = 13.0$. Writing the expression for K in terms of species activities and taking logs:

$$pe = 13.0 - \log(\{Fe^{2+}\}/\{Fe^{3+}\})$$

For $\{Fe^{2+}\} = \{Fe^{3+}\}$, $pe = 13.0$. This line is plotted in Figure 16.6. (For systems containing species other than just water, it is common practice to write the water stability lines as dashed lines). Clearly, this pe–pH diagram is valid at low pH only, since the hydroxo complexes must be included at higher pH values.

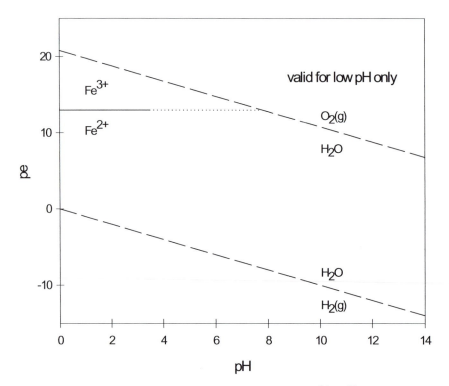

Figure 16.6: pe-pH Diagram for Fe^{3+}/Fe^{2+}

For NH_4^+/NO_3^- chemistry, we know: $pe = 14.9 - 5/4\,pH$ when the ratio $\{NH_4^+\}/\{NO_3^-\}$ is equal to 1. The resulting pe-pH diagram is shown in Figure 16.7. The conditions calculated in Section 16.6.2 also are shown in Figure 16.7. Point A is at $pe = +10$ and pH 7, where nitrate dominates. Point B is at $pe = -5$ and pH 7, where ammonium dominates. Point C is at pe +6.78 and pH 6.5, where the nitrate and ammonium activity are equal. Note that point C lies on the line of equal activity.

The pe-pH diagram gives a *qualitative* sense of which species dominates. For example, as pe increases, the oxidized forms become more dominant. However, be careful not to overinterpret pe-pH diagrams. It is not possible by inspection to see that the ammonium activity is over 89 orders of magnitude larger than the nitrate activity at point B (in Figure 16.7) under equilibrium conditions, as calculated in Section 16.6.2. The magnitude of the ratio of activities depends on the number of electrons transferred. You can show that the log of the ratio of activities is equal to the number of electrons transferred multiplied by the vertical distance between the condition of interest and the line of equal activity (see Problem 16.8).

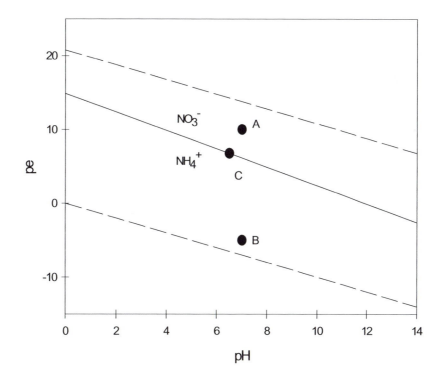

Figure 16.7: pe-pH Diagram for Nitrate/Ammonium

16.7.5 More complex pe-pH diagrams: the example of carbon chemistry

When several species are involved, pe-pH diagrams can become fairly complex. As an example, consider the chemistry of carbon. For this example, consider the $+IV$ and $-IV$ oxidation states of carbon. (In doing so, we neglect the rich redox chemistry of organic carbon.) The steps in constructing a more complex pe-pH diagram will be presented here, using dissolved inorganic carbon as an example (see also Appendix C, Section C.4.4).

Step 1: Make a species list, including species from all reasonable oxidation states

For this system, consider only carbon in the highest oxidation state, C(+IV), and carbon in the lowest oxidation state, C(−IV). The species list in a closed system would include H_2CO_3, HCO_3^-, CO_3^{2-}, and aqueous CH_4 (methane).

Step 2: Carefully consider which chemical processes should be included

 Key idea: To create a pe-pH diagram (1) Make a species list, including species from all reasonable oxidation states; (2) Carefully consider which chemical processes should be included; (3) List the equilibria describing the chemical processes listed in Step 2; (4) Write the expressions for the K values in terms of species activities, then set the activity of water to unity and set the ratio of activities of redox pairs equal to one; (5) Plot the lines generated in Step 4 on a pe-pH diagram; and (6) Erase line segments in the pe-pH diagram that are thermodynamically impossible

It is useful to make a list of how each species in the species list could be related to every other species. You know that $H_2CO_3^*$, HCO_3^-, and CO_3^{2-} are related by proton transfer (acid-base chemistry). Each of these species also can be reduced to methane. A reasonable list of chemical processes is:

1. $H_2CO_3^*$ deprotonation to HCO_3^-
2. HCO_3^- deprotonation CO_3^{2-}
3. $H_2CO_3^*$ reduction to CH_4
4. HCO_3^- reduction to CH_4
5. CO_3^{2-} reduction to CH_4

Step 3: List the equilibria describing the chemical processes listed in Step 2

Tables of redox and acid-base equilibria (see Table 16.3 and Appendix D) provide the following possible equilibria for inclusion in the chemical model:

$$H_2CO_3^* = HCO_3^- + H^+ \qquad \log K_1 = -6.3$$
$$HCO_3^- = CO_3^{2-} + H^+ \qquad \log K_2 = -10.3$$
$$1/8\,H_2CO_3^* + H^+ + e^- = 1/8\,CH_4(aq) + 3/8\,H_2O$$
$$\log K_3 = +2.71$$

In this list, K_i corresponds to process i. By combining the K_1 and K_3 equilibria, we can describe process 4. Process 5 can be described by combining the K_1, K_2, and K_3 equilibria:

$$1/8\,HCO_3^- + 9/8\,H^+ + e^- = 1/8\,CH_4(aq) + 3/8\,H_2O$$
$$\log K_4 = +3.50$$

$$1/8CO_3^{2-} + 5/4H^+ + e^- = 1/8CH_4(aq) + 3/8H_2O$$
$$\log K_5 = +4.79$$

Step 4: Write the expressions for the K values in terms of species activities. Set the activity of water to unity and set the ratio of activities of redox pairs equal to 1.

For the equilibria above:

$$pH = 6.3 - \log(\{H_2CO_3^*\}/\{HCO_3^-\}) = 6.3 \quad \text{eq. 16.17}$$
$$pH = 10.3 - \log(\{HCO_3^-\}/\{CO_3^{2-}\}) = 10.3 \quad \text{eq. 16.18}$$
$$pe = 2.71 - pH + 1/8\log(\{H_2CO_3^*\}/\{CH_4(aq)\}$$
$$= 2.71 - pH \quad \text{eq. 16.19}$$
$$pe = 3.50 - 9/8pH + 1/8\log(\{HCO_3^-\}/\{CH_4(aq)\})$$
$$= 3.50 - 9/8pH \quad \text{eq. 16.20}$$
$$pe = 4.79 - 5/4pH + 1/8\log(\{CO_3^{2-}\}/\{CH_4(aq)\})$$
$$= 4.79 - 5/4pH \quad \text{eq. 16.21}$$

Step 5: Plot the lines generated in Step 4 on a pe-pH diagram

The equations are plotted in Figure 16.8. Blindly plotting the equations did not result in a clear delineation of the pe-pH regions where each species predominates. We have more work to do.

Step 6: Erase line segments in the pe-pH diagram that are thermodynamically impossible

This step requires you to examine each line to determine over what pe and pH range (if any) it makes sense. The dotted lines in Figure 16.9 represent species interconversions that do not make thermodynamic sense. To determine if a line (or portion of a line) should be erased, it is important to determine the pe and pH values where redox or acid-base transitions occur. For example, we know that $\{H_2CO_3^*\} = \{HCO_3^-\}$ at 6.3. From eqs. 16.19 and 16.20 at pH 6.3, the pe must be -3.59 if $\{H_2CO_3^*\} = \{HCO_3^-\} = \{CH_4\}$. Thus, pe $= -3.59$ and pH 6.3 denote the "corner" of the $H_2CO_3^*/HCO_3^-/CH_4$ transition, where all three species activities are equal. At pH < 6.3 and pe > 2.71 − pH, $H_2CO_3^*$ dominates. To prove this point, simply pick a few pe and pH values and substitute them into the eqs. 16.17-16.21. As a result, erase the lines for processes 4 and 5 at pH < 6.3 and pe > -3.59. Similarly, $\{HCO_3^-\} = \{CO_3^{2-}\} = \{CH_4\}$ at pH 10.3 and pe $= -8.09$ from eqs. 16.18, 16.20, and 16.21. Thus, erase the lines for processes 3 and 4 at pH > 10.3 and at pe > -8.09. By similar reasoning, only process 4 is appropriate between pH 6.3 and 10.3. Finally, the lines delineating the C(+IV) species (processes 1 and 2) are not appropriate in the pe region where methane dominates. The final pe-pH diagram is shown in Figure 16.10.

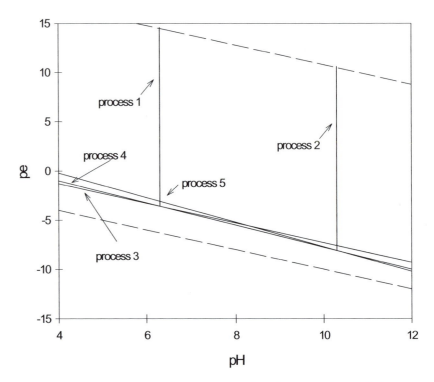

Figure 16.8: pe-pH Diagram for C(+IV)-C(−IV): All Lines

16.7.6 More complex pe-pH diagrams: the example of nitrogen chemistry

As another example of a complex pe-pH diagram, reconsider the nitrate/ammonium pe-pH diagram in Figure 16.7. This pe-pH diagram is incomplete since several acid-base and redox forms of nitrogen are missing. You can construct a more realistic pe-pH diagram for the nitrogen system.

Species: The common oxidation states of inorganic nitrogen-containing species in natural waters include N(+VII), N(+V), N(0), and N(−III). The species list would include NO_3^-, NO_2^-, NH_4^+, and NH_3. We shall ignore dissolved nitrogen gas because it is inert. Also, we can ignore HNO_3 because its pK_a is about −1. To simplify the system a bit, we have ignored HNO_2 ($pK_a \approx 3.3$). Nitrous acid decomposes in water at low pH.

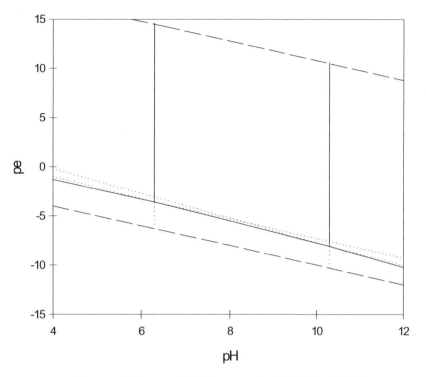

Figure 16.9: pe-pH Diagram for C(+IV)-C(-IV)
(dotted lines are discontinued)

Processes: In the nitrogen system, NO_3^- clearly could be reduced to NO_2^-. At this point, we do not know if NO_3^- could be reduced directly to NH_4^+ or NH_3, so we better not exclude these transformations. Nitrite might to reduced to ammonium at low pH (where ammonium may be expected to predominate) or reduced ammonia at high pH (where ammonia may be expected to predominate). Ammonium and ammonia are interconverted by proton transfer. Thus, a reasonable list of chemical processes is:

1. NO_3^- reduction to NO_2^-
2. NO_3^- reduction directly to NH_4^+ at lower pH
3. NO_3^- reduction directly to NH_3 at higher pH
4. NO_2^- reduction to NH_4^+ at lower pH
5. NO_2^- reduction to NH_3 at higher pH
6. NH_4^+ deprotonation to NH_3

The next step is the generation of a list of equilibria:

$$\tfrac{1}{2}NO_3^- + H^+ + e^- = \tfrac{1}{2}NO_2^- + \tfrac{1}{2}H_2O \qquad \log K_1 = +14.15$$
$$\tfrac{1}{8}NO_3^- + \tfrac{5}{4}H^+ + e^- = \tfrac{1}{8}NH_4^+ + \tfrac{3}{8}H_2O \quad \log K_2 = +14.90$$

$$1/6NO_2^- + 4/3H^+ + e^- = 1/6NH_4^+ + 1/3H_2O \ \log K_4 = +15.14$$
$$NH_4^+ = NH_3 + H^+ \qquad\qquad\qquad \log K_6 = -9.3$$

The equilibria were numbered so that the K_i equilibrium corresponds to process i. We can generate an equilibrium for process 3 by adding the K_1 and 1/8 times the K_6 equilibria, and we can generate an equilibrium for process 5 by adding the K_4 and 1/6 times the K_6 equilibria. The final list of equilibria is:

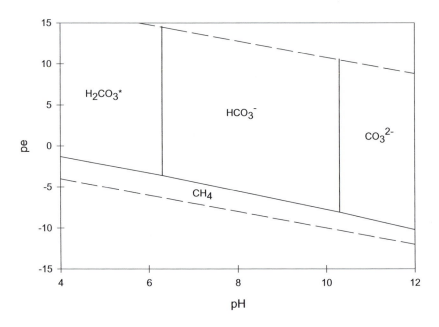

Figure 16.10: pe-pH Diagram for C(+IV)-C(–IV): Final Version

$$½NO_3^- + H^+ + e^- = ½NO_2^- + ½H_2O \qquad \log K_1 = +14.15$$
$$1/8NO_3^- + 5/4H^+ + e^- = 1/8NH_4^+ + 3/8H_2O \ \log K_2 = +14.90$$
$$1/8NO_3^- + 9/8H^+ + e^- = 1/8NH_3 + 3/8H_2O \ \log K_3 = +13.74$$
$$1/6NO_2^- + 4/3H^+ + e^- = 1/6NH_4^+ + 1/3H_2O \ \log K_4 = +15.14$$
$$1/6NO_2^- + 7/6H^+ + e^- = 1/6NH_3 + 1/3H_2O \ \log K_5 = +13.59$$
$$NH_4^+ = NH_3 + H^+ \qquad\qquad\qquad \log K_6 = -9.3$$

The equilibria must be translated into pe-pH equations:

$$pe = 14.15 - pH \qquad (\text{from } NO_3^-/NO_2^-)$$
$$pe = 14.90 - 5/4pH \quad (\text{from } NO_3^-/NH_4^+)$$
$$pe = 13.74 - 9/8pH \quad (\text{from } NO_3^-/NH_3)$$
$$pe = 15.14 - 4/3pH \quad (\text{from } NO_2^-/NH_4^+)$$

$pe = 13.59 - 7/6pH \quad (\text{from } NO_2^-/NH_3)$

$pH = 9.3 \qquad\qquad (\text{from } NH_4^+/NH_3)$

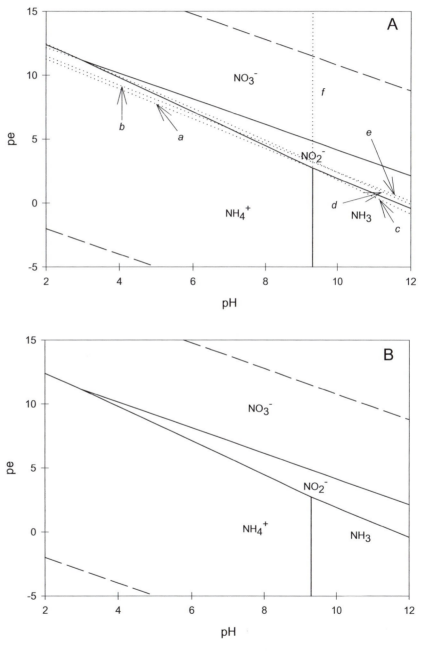

Figure 16.11: pe-pH Diagram for Ammonium/Nitrate
(A: Lines revised, B: Final version)

Now a draft pe-pH diagram can be made (Figure 16.11A). Several line segments do not make sense. For example, process 3 (NO_3^-/NH_3) and process 5 (NO_2^-/NH_3) make no sense at pH < 9.3. Therefore, the portion of the lines below pH 9.3 is erased (lines a and b in Figure 16.11A). Similarly, process 2 (NO_3^-/NH_4^+) and process 4 (NO_2^-/NH_4^+) make no sense at pH > 9.3 and their lines (lines c and d in Figure 16.11A) are erased above pH 9.3. Also, leaving some processes (i.e., lines) in the pe-pH diagram leads to a nonsensical ordering of species from high pe to low pe. For example, consider the reduction of nitrate to ammonia (process 3) above pH 9.3. If this process (line e) were in place, ammonia would be dominant at a pe higher than nitrite. This does not make sense, and so the remainder of line e is erased. By this reasoning, process 6 (NH_4^+/NH_3) does not make sense at pe values greater than 2.74, where nitrite equilibrates with N(–III). Thus, line f is erased at higher pe values. Process 2 (NO_3^-/NH_4^+) is a little more complicated. It can be ignored at pH > 2.88 since nitrite equilibrates with ammonium in that pH range. The final pe-pH diagram is shown in Figure 16.11B.

16.8 APPLYING REDOX EQUILIBRIUM CALCULATIONS TO THE REAL WORLD

16.8.1 Are redox systems really at equilibrium?

Systems in which electrons are transferred are not as easy to deal with as systems in which protons are transferred. One difficulty in applying redox equilibrium calculations to the real world is that redox reactions can be slow. As a result, redox systems may not be in equilibrium.

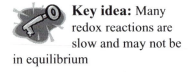

Key idea: Many redox reactions are slow and may not be in equilibrium

You saw an example of the nonequilibrium nature of some redox reactions in Section 16.7.3. In that section, it was shown that water cannot be in redox equilibrium with both the oxygen gas and hydrogen gas in the atmosphere at pH 7. Another classic example of nonequilibrium conditions is chlorine chemistry. The pe-pH diagram for chlorine at a total chlorine concentration of 1×10^{-3} M is shown in Figure 16.12 (see Problem 16.9).

Thoughtful Pause
From Figure 16.12, what form of chlorine is thermodynamically stable?

From Figure 16.12, you can see that the only form of chlorine that is thermodynamically stable is chloride. Is this true? Yes: at equilibrium, chloride predominates. However, the reduction reactions of various oxidation states of chlorine are fairly slow. Aqueous Cl(+I) solutions are fairly stable in the dark. (Otherwise, chlorine bleach, concentrated OCl⁻ solutions, would be unavailable in the stores.)

16.8.2 Measuring pe

Another challenge to applying redox equilibrium calculations to the real world is that it is very difficult to measure pe. In proton transfer (acid-base) chemistry, the master variable, pH, is easy to measure. In electron transfer (redox) chemistry, the master variable, pe, is not easy to measure. A full discussion of the reasons behind the difficulties in measuring pe are beyond the scope of this text. However, problems with impurities and nonequilibrium conditions greatly limit our ability to measure a thermodynamically reliable pe value in natural waters.

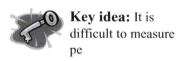

Key idea: It is difficult to measure pe

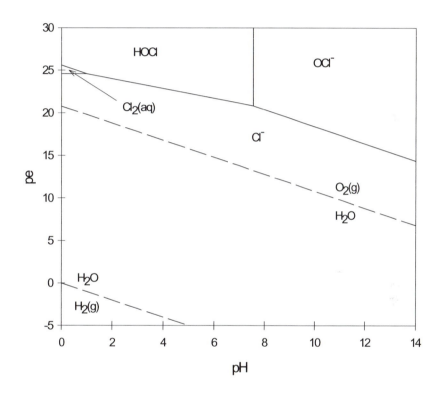

Figure 16.12: pe-pH diagram for Chlorine
(total chlorine concentration = $Cl_T = 1 \times 10^{-3}$ M)

16.8.3 The value of redox equilibrium calculations

Although pe cannot be measured accurately and easily, redox equilibrium calculations still play an important role in the modeling of aquatic chemistry. Why? In acid-base systems, acid/conjugate base pairs are usually related by the exchange of only one or two protons. Thus, the slopes of the lines in pC-pH diagrams are small and the speciation can change over a wide range of pH. In redox systems, redox pairs often are related by the exchange of many electrons. Thus, the slopes of the lines in pC-pe diagrams can be quite large (see Figure 16.4 for an example). As a

Key idea: Redox equilibrium calculations play an important role in chemical modeling, especially in determining the predominant species in a system

result, speciation often changes only over a small pe range. In other words, once you are away from pe°, certain species can be ignored with little error. Thus, redox equilibrium calculations play an important role in chemical modeling, especially in determining the *predominant* species in a system.

16.9 SUMMARY

With the introduction of electron transfer in this chapter, you have now been introduced to the major types of equilibria in homogeneous systems. In this chapter, you learned about oxidants (species that accept electrons to become reduced) and reductants (species that donate electrons to become oxidized). Reactions in which electrons are transferred are called redox reactions and may be half reactions (if electrons appear in the reaction) or overall redox reactions (if one or more oxidation states change but electrons do not appear as reactants or products). Redox (electron transfer) chemistry has many analogies with acid-base (proton transfer) chemistry.

Chemical reactions and chemical equilibria must be balanced to be useful. In Section 16.3, you honed your skills on balancing half reactions and overall reactions. To balance an overall reaction, add the half reactions in such a way that electrons cancel. Correctly balanced electron exchange reactions allow for the determination of oxidant and reductant doses in treatment processes.

In Sections 16.4 and 16.5, a system was developed to quantify the thermodynamic tendency to transfer electrons. You learned about several thermodynamic parameters, including the cell potential (linked to species activities through the Nernst equation) and pe° (equal to the equilibrium constant for a reduction reaction divided by the number of electrons transferred). In fact, pe° is analogous with pK_a. The thermodynamic parameters lead to the calculation of species activities at equilibrium through combinations of redox equilibria and mass balances.

New graphical presentation methods were developed to display the results of redox equilibrium calculations. A key to both types of plots is that the redox stability of water (with respect to oxygen and hydrogen gas) depends on the pH. Two important plot types are (1) pC-pe diagrams, with $-\log$(concentration) on the y-axis and pe on the x-axis, and (2) pe-pH diagrams, with pe plotted on the y-axis and pH on the x-axis. In pe-pH diagrams, lines are drawn to indicate conditions where two species have the same activity.

Finally, the limitations and value of redox equilibrium calculations were reviewed. Limitations include the nonequilibrium state of some redox systems [e.g., $H_2(g)$ in the atmosphere in non-equilibrium with water and the aqueous chlorine system] and the inability to measure pe reliably and easily. In the end, though, redox equilibrium calculations are valuable because often only one oxidation state predominates under a given set of conditions.

16.10 PART IV CASE STUDY: WHICH FORM OF COPPER PLATING SHOULD YOU USE?

Recall that the Part IV case study concerns the relative merits of copper cyanide plating, where CuCN(s) is the copper source, and acid copper plating, where $CuSO_4$(s) is the copper source. After completing Chapter 16, you should be thinking about how the redox chemistry of copper may influence the choice of plating process.

It is of interest to ask whether copper cyanide offers a savings in electrical usage over acid copper. Electrical usage stems directly from redox stoichiometry. For example, how much electricity does it take to reduce Cu(+I) to Cu(0)? The electric requirement is quantified in electroplating by the *electrochemical equivalent*, the number of grams of a metal reduced per amp-hour of reduction. How do you calculate the electrochemical equivalent? Recall that Faraday's constant, F, is equal to 96,485 coulombs of charge per mole. One coulomb is 1 amp-second. Thus, F = 96,485 A-s/mol. A one-electron process would require $1F$ of charge, whereas a two-electron process requires $2F$ of charge. Thus, the electrochemical equivalent is given by:

$$\text{electrochemical equivalent (g/A-hr)} = (MW)(3600 \text{ s/hr})/(nF)$$

where MW = molecular weight (in g/mol) = 63.55 g/mol for copper, n = number of electrons transferred, and F = 96,485 A-s/mole. For copper cyanide plating [$Cu^+ + e^- \rightarrow Cu$(s)], n = 1, while n = 2 for acid copper plating [$Cu^{2+} + 2e^- \rightarrow Cu$(s)]. Thus, the electrochemical equivalent is 2.4 g/A-hr for copper cyanide plating and 1.2 g/A-hr for acid copper plating. Therefore, one of the advantages of copper cyanide plating is that it uses less electrical power than acid copper plating.

Another use of redox chemistry in analyzing copper plating is in determining the pertinent equilibrium constants. For examining the "loss" of Cu(+I) to Cu(+II) (see Section 15.8), the appropriate equilibrium is:

$$Cu(s) + Cu^{2+} = 2Cu^+ \qquad\qquad \text{eq. 16.21}$$

(Note that the forward reaction is a comproportionation reaction and the reverse reaction is a disproportionation reaction.) For the avoidance of immersion deposits with copper cyanide plating (see Section 15.8), the pertinent equilibrium is:

$$2Cu^+ + Zn(s) = 2Cu(s) + Zn^{2+} \qquad\qquad \text{eq. 16.22}$$

To determine the equilibrium constants, it is necessary to start with the reactions for which we have thermodynamic data. For many redox reactions, this means starting with the half reactions. The important half reactions are listed below:

$$Cu^+ + e^- = Cu(s) \quad E_1^o = +0.52 \text{ V}$$
$$Cu^{2+} + e^- = Cu^+ \quad E_2^o = +0.16 \text{ V}$$
$$Zn^{2+} + 2e^- = Zn(s) \quad E_3^o = -0.76 \text{ V}$$

To find the equilibrium constant for the equilibrium in eq. 16.21, it is necessary to add the following two half reactions:

$$Cu(s) = Cu^+ + e^- \quad E_4^o = ?$$
$$Cu^{2+} + e^- = Cu^+ \quad E_2^o = +0.16 \text{ V}$$

Clearly, the E_4^o equilibrium is the reverse of the E_1^o equilibrium. Remember that reversing reactions changes the sign of E^o. Therefore, $E_4^o = -0.52$ V. Remember that the safest way to manipulate redox equilibria is to convert the E^o values of the original equilibria to ΔG_{rxn}^o, add the ΔG_{rxn}^o values, and reconvert to E^o. Recall that:

$$\Delta G_{rxn}^o = -FE^o$$

Thus: $\Delta G_{rxn,4}^o = -FE_4^o$ and $\Delta G_{rxn,2}^o = -FE_2^o$, so:

$$\Delta G_{rxn}^o = -FE_4^o - FE_2^o = -F(E_4^o + E_2^o)$$

Now recall that:

$$\log K = -\Delta G_{rxn}^o/(2.303RT)$$
$$= +(F/2.303RT)(E_4^o + E_2^o)$$
$$= (-0.52 \text{ V} + 0.16 \text{ V})/(0.059 \text{ V})$$
$$= -6.1$$

Thus: K for the equilibrium in eq. 16.21 is equal to $\{Cu^+\}^2/\{Cu^{2+}\} = 10^{-6.1}$.

To find the equilibrium constant for the equilibrium in eq. 16.22, add the following two half reactions:

$$2Cu^+ + 2e^- = 2Cu(s) \quad E_5^o = ?$$
$$Zn(s) = Zn^{2+} + 2e^- \quad E_6^o = ?$$

Recall that multiplying a reaction by a positive constant does not affect E^o, so $E_5^o = E_1^o = +0.52$ V. The E_6^o equilibrium is the reverse of the E_3^o equilibrium, so $E_6^o = -E_3^o = +0.76$ V. Converting to ΔG_{rxn}^o and adding, ΔG_{rxn}^o for $2Cu^+ + Zn(s) = 2Cu(s) + Zn^{2+}$ is equal to $-2F(E_5^o + E_6^o)$. Thus:

$$\log K = -\Delta G_{rxn,}^o/(2.303RT)$$
$$= +(2)(F/2.303RT)(E_5^o + E_6^o)$$
$$= 2(+0.52 \text{ V} + 0.76 \text{ V})/(0.059 \text{ V})$$
$$= 43.4$$

The equilibrium expression for the equilibrium in eq. 16.22 is $\{Zn^{2+}\}/\{Cu^{+}\}^{2} = 10^{43.4}$, since the activities of the pure solids are unity.

SUMMARY OF KEY IDEAS

- Overall reactions are formed by adding together half reactions in such a way that the free electrons cancel out

- In an overall reaction, oxidants accept electrons and are reduced, whereas reductants donate electrons and are oxidized

- To balance a half reaction (1) write the known reactants on the left and known products on the right, (2) adjust stoichiometric coefficients to balance all elements (except H and O), (3) add water (H_2O) to balance the element O, (4) add H^+ to balance the element H, and (5) add electrons (e^-) to balance the charge

- Check the change in oxidation states in balanced half reactions

- To balance overall reactions (1) balance the half reactions separately, and (2) add the half reactions to eliminate free electrons

- Check the change in oxidation states in balanced overall reactions

- Exercise care when dealing with oxidation states of O other than $-II$ and of H other than $+I$

- pe^o is the pe when all species are present at 1 M

- The Gibbs free energy of reaction and the cell potential are related by $\Delta G_{rxn} = -nFE$ (note negative sign)

- With reactions listed as reductions, strong oxidants appear as reactants in reactions with large positive pe^o values and strong reductants appear as products in reactions with large negative pe^o values

- The cell potential does not change if you multiply the half reaction by a positive constant

- Since Gibbs free energy values are additive, when adding redox equilibria convert E^o values to ΔG^o_{rxn}, add the ΔG^o_{rxn} values, and then reconvert to E^o

- To calculate activities as a function of pe and pH, rearrange the mathematical form of the equilibrium to obtain an equation of the form pe $= a + b$pH $+c$log(ratios of species activities)

- Measured cell potentials and the Nernst equation can be used to develop methods to measure species activities

- The redox and pH ranges of water are intertwined and given by the equations $pe = 20.78 - pH$ and $pe = -pH$

- To create a pe-pH diagram (1) Make a species list, including species from all reasonable oxidation states; (2) Carefully consider which chemical processes should be included; (3) List the equilibria describing the chemical processes listed in Step 2; (4) Write the expressions for the K values in terms of species activities, then set the activity of water to unity and set the ratio of activities of redox pairs equal to one; (5) Plot the lines generated in Step 4 on a pe-pH diagram; and (6) Erase line segments in the pe-pH diagram that are thermodynamically impossible

- Many redox reactions are slow and may not be in equilibrium

- It is difficult to measure pe

- Redox equilibrium calculations play an important role in chemical modeling, especially in determining the predominant species in a system

HISTORICAL NOTE: WALTHER HERMANN NERNST

Walther Nernst

Many students of chemistry learn the name Nernst upon their introduction to electrochemistry. Yet few appreciate Nernst's great contributions to physical chemistry, thermodynamics, photochemistry, and industrial chemistry. Born in West Prussia in 1864, Nernst studied in several of the great universities of central Europe, including the Universities of Zurich, Berlin, Graz, and Würzburg. His first academic appointment, at Leipzig University, created a sort of late nineteenth-century Dream Team of physical chemists: Nernst, Ostwald, van't Hoff (see Chapter 22), and Arrhenius (see Chapter 22).

Although his early work was in electrochemistry, Nernst's contributions span physical chemistry and beyond. In 1906, he developed the Third Law of Thermodynamics (see Chapter 3), opening up a new degree of precision in thermodynamic calculations. He collaborated with Otto Haber on the important industrial syntheses of the day. Outside of chemistry, Nernst contributed practical designs of many items, including a replacement for the carbon filament in the light bulb, the use of nitrous oxide to enhance the performance of the internal combustion engine, and an electric piano. A prolific writer of textbooks, he was awarded the Nobel Prize in Chemistry in 1920.

PROBLEMS

16.1 Chlorine exists in seven oxidation states. Examples include (in alphabetical order):

chlorate ClO_3^-
chloride Cl^-
chlorine Cl_2

chlorine dioxide ClO_2
chlorite ClO_2^-
hypochlorous acid $HOCl$

perchlorate ClO_4^-

A. Determine the oxidation state of each compound and order them from the most negative oxidation state to the most positive oxidation state.

B. Write a balanced half reaction for the conversion of each oxidation state to the next highest oxidation state at low pH.

16.2. Balance the following overall reactions:

A. $CH_3OH + O_2 \rightarrow CO_2 + H_2O$ (at high pH)
(oxidation of methanol with oxygen)

B. $HS^- + MnO_4^- \rightarrow S(s) + MnO_2(s)$ (at high pH)
(oxidation of bisulfide with permanganate)

C. IO_3^- (iodate) $+ I^-$ (iodide) $\rightarrow I_2$ (iodine) (at low pH)
(chemistry used to standardize iodate solutions)

16.3 Dissolved oxygen, $O_2(aq)$ or DO, can be measured by a wet-chemical technique called the *Winkler method*. The Winkler method consists of three steps outlined below. For each step, write the balanced overall reaction.

A. In the first step, NaOH and a great deal of Mn^{2+} are added to the sample. The Mn^{2+} is oxidized to $MnO(OH)_2(s)$ by $O_2(aq)$ at high pH, and $O_2(aq)$ is reduced to H_2O.

B. In the second step, acid and a great deal of I^- are added. The $MnO(OH)_2(s)$ oxidizes I^- to I_2 and is reduced to Mn^{2+} at low pH.

C. In the third step, I_2 is titrated with thiosulfate ($S_2O_3^{2-}$). During the titration, I_2 is reduced to I^- and $S_2O_3^{2-}$ is oxidized to $S_4O_6^{2-}$ (tetrathionate).

D. Add together the overall reactions in parts A through C to develop the total overall reaction for the Winkler method. In the standard method, it is stated that 1 mL of 0.025 M $Na_2S_2O_3$ titrant corresponds to 1 mg/L DO in a 200 mL sample. Is this information consistent with your total overall reaction?

16.4 Hydrogen peroxide (H_2O_2) is a good chemical oxidant. You may have used hydrogen peroxide as a topical disinfectant and noticed the evolution of oxygen bubbles.

    A. If the oxidation state of H in hydrogen peroxide is +I, what is the oxidation state of O?

    B. Write the half reactions and overall reaction for the conversion of hydroxide peroxide to molecular oxygen and water.

    C. Is the overall reaction a disproportionation reaction, a comproportionation reaction, or neither a disproportionation nor a comproportionation reaction?

16.5 A drinking water treatment plant is using ozone for disinfection. Plant personnel want to use ozone also to oxidize reduced manganese, Mn^{2+}, to $MnO_2(s)$.

    A. Find the required ozone dose (in g O_3 per g Mn^{2+} oxidized) if ozone is reduced to oxygen.

    B. Plant personnel added ozone to their water and noticed a pink water condition. What was the minimum ratio of O_3 to Mn^{2+} that they were using?

16.6 The following questions concern the COD test, introduced in Example 16.2.

    A. In the COD test, excess dichromate is added. The dichromate remaining after oxidizing the organics is titrated with a ferrous iron salt. Write the balanced overall reaction for the reduction of dichromate to Cr^{3+} (while Fe^{2+} is oxidized to Fe^{3+}).

    B. Comparing the redox chemistry of dichromate and oxygen, verify the statement that " each milliliter of a 0.25 N solution of dichromate is equivalent to 2 mg of oxygen" (Sawyer and McCarty, 1978). Note: a 1 N solution here means one mole/L of electrons available.

16.7 Following the example in Section 16.6.1, convert cell potentials to free energies to show that the cell potential does not change if you multiply a half reaction by a positive constant.

16.8 Show that at any point on a pe-pH diagram, the log of the ratio of activities of the oxidized and reduced forms is equal to the number of electrons transferred multiplied by the signed vertical distance to the line of equal activity.

16.9 Reproduce the pe-pH diagram for the aqueous chlorine system shown in Figure 16.12 (for $Cl_T = 5 \times 10^{-3}$ M). The pertinent equilibria are:

$$HOCl + H^+ + e^- \; = \; \tfrac{1}{2}Cl_2(aq) + H_2O \quad \log K = 26.6$$

$$\tfrac{1}{2}Cl_2(aq) + e^- = Cl^- \qquad\qquad \log K = 23.6$$
$$HOCl = OCl^- + H^+ \qquad\qquad \log K = -7.54$$

Note: because of mass balance considerations, $[HOCl] = \tfrac{1}{2}Cl_T$ and $[Cl_2(aq)] = \tfrac{1}{4}Cl_T$ at the $Cl_2(aq)$-HOCl boundary and $[Cl^-] = \tfrac{1}{2}Cl_T$ and $[Cl_2(aq)] = \tfrac{1}{4}Cl_T$ at the $Cl_2(aq)$-Cl^- boundary.

16.10 In Section 16.7.3, it was suggested that alpha values for redox reactions could be developed by analogy with the acid-base alpha values. Develop redox alpha values for the following system:

$$ox + ne^- + mH^+ = red$$
$$C = \text{total concentration} = [ox] + [red]$$

The alpha values should satisfy the equations: $\{ox\} = \alpha_0' C$ and $\{red\} = \alpha_1' C$. Check your answer with the ammonium/nitrate example in Section 16.6.2. In that example, $\alpha_0' = \alpha_1'$ at pH 6.5 when pe $= +6.78$.

Heterogeneous Systems

Johnny, finding life a bore,
Drank some H_2SO_4.
Johnny's father, an M.D.,
Gave him $CaCO_3$.
Now he's neutralized, it's true,
But he's full of CO_2.
Philip Howard

A field of water betrays the spirit that is in the air. It is
continually receiving new life and motion from above. It is
intermediate in its nature between land and sky.
Henry David Thoreau

CHAPTER 17

Getting Started with Heterogeneous Systems

17.1 INTRODUCTION

In the first four parts of this text, you have been introduced to the major types of chemical equilibria in homogeneous aquatic systems. Homogeneous systems have only one phase. Thus, exchange of mass with gas and/or solid phases was not allowed. The restriction to one phases limits the applicability of the tools you have learned.

A quick glance at any surface water will tell you that exchange with the atmosphere may be possible. Many small ponds and shallow rivers exhibit reasonably large surface-area-to-volume ratios, thus aiding mass transfer. The equilibration of chemical species between the gas and liquid phases, called *gas-liquid equilibrium*, often sets limits on aqueous phase concentrations. Gas-liquid equilibria are the focus of Chapter 18.

Chemical species also can move between solid and aqueous phases. Recall from Section 12.1 that the composition of the early oceans was determined by the leaching of minerals into acidic runoff. This is an example of *dissolution*. The chemical process of forming solid phases from dissolved species is called *precipitation*. At equilibrium, the rates of dissolution and precipitation are equal. The equilibrium state is called *solid-liquid equilibrium* and is the topic of Chapter 19.

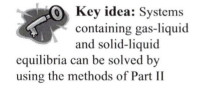 **Key idea:** Systems containing gas-liquid and solid-liquid equilibria can be solved by using the methods of Part II

Although you will be learning about new types of equilibria in this part of the text, it is important to recognize that no new solution techniques are needed. *Systems containing gas-liquid and solid-liquid equilibria can be solved by the same approaches that you already have mastered.* You still will be developing species lists, writing equilibrium expressions, and conjuring up mass balances. In Part V, you simply will be adding new chemical species and new equilibria to the mix. In fact, in some cases, equilibrium calculations with systems containing gas-liquid and solid-liquid equilibria are easier than with homogeneous systems. Why? As you will see, many heterogeneous systems have species with fixed concentrations.

17.2 EQUILIBRIUM EXCHANGE BETWEEN GAS AND AQUEOUS PHASES

17.2.1 Introduction

Some dissolved species are *volatile* (from the Latin *volare* to fly). Volatile compounds may partition into a gas phase in contact with the water. The gas phase may be the atmosphere or gas bubbles. You will discover in Chapter 18 that at equilibrium, the gas phase activity and aqueous phase activity of the equilibrating species are related through an equilibrium constant. This is analogous to the way that the activities of an acid and its conjugate base are related through an equilibrium constant.

Dissolved species that partition into a contacting gas phase are called *dissolved gases*. Dissolved gases play a critical role in the function of ecosystems. One of the most important dissolved gases in natural waters is *dissolved oxygen*. Dissolved oxygen is used as a terminal electron acceptor by a number of aquatic organisms, including fish and aerobic bacteria. Dissolved carbon dioxide, as you will see, helps to buffer pH. Oxygen and carbon dioxide dissolved in the blood play similar roles in mammals and other terrestrial organisms.

17.2.2 Open systems

Some environmental systems have gas phases that serve as extremely large reservoirs of chemical species. The classic example is equilibration with the atmosphere. However, even gas phases with smaller volumes than the atmosphere can contain large amounts of gaseous species. For example, air contains about 20.9% oxygen by volume, or about $(0.209 \text{ L } O_2/\text{L air})(1.331 \text{ g } O_2/\text{L } O_2) = 0.278 \text{ g } O_2/\text{L air}$ at 20°C. Water in equilibrium with the atmosphere has a dissolved oxygen concentration at 20°C of about 9 mg/L $= 0.009 \text{ g } O_2/\text{L water}$. Thus, as long as the ratio of the water volume to air volume in the system is less than about 0.1:1, the mass of oxygen in the water at equilibrium will represent a small (<0.3%) fraction of the mass of oxygen in the air.[†] As a result, gas phase concentrations sometimes can be considered constant even though partitioning into water occurs.

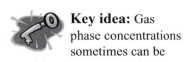 **Key idea:** Gas phase concentrations sometimes can be considered constant, as long as the ratio of the water volume to air volume in the system is small enough

---

Thoughtful Pause

What are the implications of constant gas phase concentrations for equilibrium calculations?

Recall that gas phase and liquid phase concentrations are related through an equilibrium constant. Thus, constant gas phase concentrations should result in constant dissolved gas concentrations. Having species with fixed concentrations generally makes equilibrium calculations easier.

[†] For carbon dioxide, a similar calculation reveals that the mass in the water at equilibrium is <0.2% of the mass in the air, as long as the ratio of the water volume to air volume in the system is less than about 0.1:1.

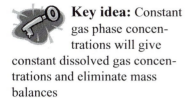

Key idea: Constant gas phase concentrations will give constant dissolved gas concentrations and eliminate mass balances

What about mass balances? Of course, the total mass of equilibrating species must remain constant *for the system as a whole*. Now, however, the control volume includes both gas and liquid phases. Since the gas phase sometimes (but not always) contains a large amount of the equilibrating species, the mass balance is not very helpful in determining the concentrations of dissolved gases. Thus, mass balances usually are not used for equilibrating species.

17.3 EQUILIBRIUM EXCHANGE BETWEEN SOLID AND AQUEOUS PHASES

17.3.1 Introduction

Some dissolved species react to form insoluble complexes. In this way, solid phases are generated and chemical species partition between solid and liquid phases. The formation of solid phases is called *precipitation*. At equilibrium, the rate of precipitation is balanced by the rate of *dissolution*. Dissolution is the partitioning of components in solid phases to the aqueous phase. You have seen an example of dissolution in Section 12.1, where the solubilization of minerals by acidic rainfall to create the early oceans was discussed. The equilibrium partitioning of chemical species between solid and liquid phases is called *solid-liquid equilibrium*.

Solid-liquid equilibria are important in aquatic systems because the presence of solid phases greatly influences the concentrations of metals and ligands in equilibrium with the solids. The discussion of complexation in Chapter 15 was of necessity limited to low pH and low total metal concentrations. Why? At higher pH and higher total metal concentrations, metals may precipitate. Once the solid phase is generated, solid-liquid equilibria come into play and both metal and ligand concentrations are affected. The total *dissolved* concentrations of many metals are small because solid phases containing the metal are formed. Environmental engineers and scientists frequently take advantage of solid-liquid equilibrium chemistry in engineered systems to control the concentrations of metals.

Precipitation also affects ligand concentrations. Thus, solid-liquid equilibrium chemistry can be used in engineered systems to control the concentrations of problematic ligands such as phosphate.

17.3.2 Implications for equilibrium calculations

Similar to gas-liquid exchange (Section 17.2.1), the solid phase activity and aqueous phase activity of the equilibrating species are related at equilibrium through an equilibrium constant. The usual choice of standard states is that the activity of a pure solid is equal to 1 (see Section 4.3). Thus, the solid phase activity is fixed. As discussed in Section 17.2.2, a fixed activity of a nonaqueous species will fix the concentrations of equilibrating species and affect the choice of mass balances. In general, equilibrium calculations in the presence of solid phases are simplified.

17.3.3 Surfaces

Precipitation creates an interface between the solid and aqueous phases. Chemical species bound to the solid side of the interface can interact with dissolved species. Some of these interactions are modeled as complex-formation reactions (called *surface complexation*). You will discover in Chapter 19 that your knowledge of complexation, combined with electrostatic corrections for interactions between charged dissolved species and the charged surface, will allow you to determine quantitatively how surfaces affect the concentrations of dissolved species.

17.4 PART V ROAD MAP

Part V of this book discusses gas-liquid and solid-liquid equilibria. These two types of equilibria are the final major equilibrium classes in aquatic systems. By the conclusion of this part of the text, you will be able to develop fairly realistic chemical and mathematical models for aquatic systems.

In Chapter 18, the equilibrium partitioning of species between gas and liquid phases is explored. Consideration will be given to equilibrium partitioning with both water and pure solvents as liquid phases. New equilibrium constants will be introduced to allow for the calculation of dissolved species concentrations. Special attention will be given to dissolved carbon dioxide. By the conclusion of Chapter 18, you should be able to determine species concentrations at equilibrium in systems having both gas and liquid phases.

In Chapter 19, the equilibrium partitioning of species between solid and liquid phases is explored. Again, a new type of equilibrium constant will be introduced to allow for the calculation of dissolved species concentrations. Special attention will be given to solid-liquid equilibria involving calcium carbonate. By the conclusion of Chapter 19, you should be able to determine species concentrations at equilibrium in systems having gas, aqueous, and solid phases.

17.5 SUMMARY

In this chapter, the implications of including gas and/or solid phases in equilibrium systems were introduced. You learned that systems containing gas-liquid and solid-liquid equilibria can be solved by the same approaches that you already have mastered.

The concept of the dissolved gas was introduced in this chapter. You learned that some species equilibrate between gas and liquid phases. In some cases, the gas phase contains a large amount of the equilibrating species and serves as an infinite reservoir of material. In such cases, the gas phase concentration is fixed. This, in turn, will set the concentration of the dissolved gas and eliminate the corresponding mass balance.

Some dissolved species form insoluble complexes and therefore equilibrate between solid and liquid phases. The usual choice of standard state fixes the activity of the solid phase. This will influence the concentrations of the metals and ligands in equilibrium with the solid phase. In addition, precipitation creates an interface. This introduces new species to the system (surface-bound species), which can form complexes with dissolved species. This is another way that precipitation affects speciation.

17.6 PART V CASE STUDY: THE KILLER LAKES

On the late Thursday evening of August 21, 1986, the villagers living near Lake Nyos, a crater lake in northwestern Cameroon, heard a strange series of rumbling noises. Some villagers reported later that they felt a warm sensation and lost consciousness. The lucky ones awoke 6 to 36 hours later to a scene of unimaginable horror. Surveying the scene, they found 1,700 townspeople and over 3,000 cattle dead. No insects or birds were seen. Strangely, plant life was unaffected but the oil lamps had all gone out.

What caused this terrible and bizarre loss of life? Investigations after the tragedy revealed that the area around the lake had been flooded with a deadly cloud of carbon dioxide. Where did the carbon dioxide come from?

Lake Nyos receives huge inputs of carbon dioxide gas from neighboring volcanos. When the dissolved gas pressure exceeds the weight of the overlying water, the lake can become unstable. In the 1986 incident, the supersaturated bottom waters rose and degassed catastrophically when they reached the surface. Essentially, the supersaturated gases equilibrated with the atmosphere very quickly. Carbon dioxide gas is heavier than water and the evolved gases moved downhill rapidly. A 50-meter-thick gas cloud spread for 25 km at a speed of up to 45 mph (72 kph).

Lake Nyos was not the first Cameroonian lake to degas with tragic consequences. On August 16, 1984, 37 people died when Lake Monoun (another deep crater lake in Cameroon) degassed. In this case study, you will investigate the gas-liquid and solid-liquid water chemistry of Lakes Monoun and Nyos. In addition, you will learn about efforts to ensure that the lakes do not kill again.

SUMMARY OF KEY IDEAS

- Systems containing gas-liquid and solid-liquid equilibria can be solved by using the methods of Part II

- Gas phase concentrations sometimes can be considered constant, as long as the ratio of the water volume to air volume in the system is small enough

- Constant gas phase concentrations will give constant dissolved gas concentrations and eliminate mass balances

CHAPTER 18

Gas-Liquid Equilibria

18.1 INTRODUCTION

In this chapter, the equilibration of chemical species between gas and liquid phases will be explored. In doing so, the applicability of your aquatic chemistry knowledge will be expanded greatly. By the conclusion of this chapter, you will be able to calculate the concentrations of pollutants in both gas and liquid phases after equilibrium partitioning. Of particular importance, you will be able to determine the effects of equilibration with the atmosphere on important chemical species such as dissolved oxygen and dissolved carbon dioxide.

In Section 18.2, you will learn about the equilibrium laws governing gas-liquid partitioning - Raoult's Law and Henry's Law. These principles bring with them two new kinds of equilibrium constants: vapor pressure and Henry's Law constants.

Equilibrium calculation techniques with gas-liquid equilibria will be developed in Section 18.3. The calculation method depends on whether the gas phase has a small volume or whether the gas phase is large and can be considered an infinite reservoir of material. Calculations involving equilibrium with the atmosphere often are simpler than homogeneous calculations since the species in the gas phase fixes the concentration of the aqueous phase species with which it is in equilibrium. Some dissolved species hydrolyze. In such cases, you will learn how the gas solubility depends on pH.

Section 18.4 is devoted to dissolved carbon dioxide. The solubility of dissolved carbon dioxide as a function of pH will be explored. In addition, the effects of equilibration with the atmosphere on total dissolved carbonate (C_T), pH, and alkalinity will be investigated.

18.2 RAOULT'S LAW AND HENRY'S LAW

18.2.1 Ideal substances

Chemical species can equilibrate between gas and liquid phases. If you set out a dish of a pure organic solvent, say acetone or methyl ethyl ketone, you will soon smell the chemical in the gas phase. At equilibrium, the gas phase and pure liquid phase concentrations of the organic species will not change over time.

Now consider for a moment a mixture of two substances, A and B. If you change the composition of the mixture (say by increasing the amount

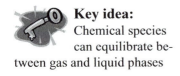

Key idea:
Chemical species can equilibrate between gas and liquid phases

vapor pressure of substance i
(P_i^o): the partial pressure of i
over a pool of pure liquid i

of B relative to A), the concentrations of A and B in the gas phase above the liquid mixture will change at equilibrium. If A and B behave in an *ideal* fashion (i.e., if neither A nor B affects the partitioning behavior of the other species), then doubling the concentration of B in the liquid phase will double the concentration of B in the gas phase.

This statement can be shown graphically. For convenience, use *partial pressures* as gas phase concentrations (P_i = partial pressure of substance i) and *mole fractions* for liquid phase concentrations (x_i = mole fraction of substance i). (Concentration units were reviewed in Chapter 2.) If A and B are ideal and do not interact, then the partial pressures will change as a function of mole fraction as indicated in Figure 18.1. The partial pressures of pure A (left side of Figure 18.1) and pure B (right side of Figure 18.1) are given special names. They are called the **vapor pressure** of A and B. The vapor pressure of substance i, denoted P_i^o, is the partial pressure of i over a pool of pure liquid i. Note also that P_A^o and P_B^o are the slopes of the lines in Figure 18.1 since the x-axis runs from 0 to 1.

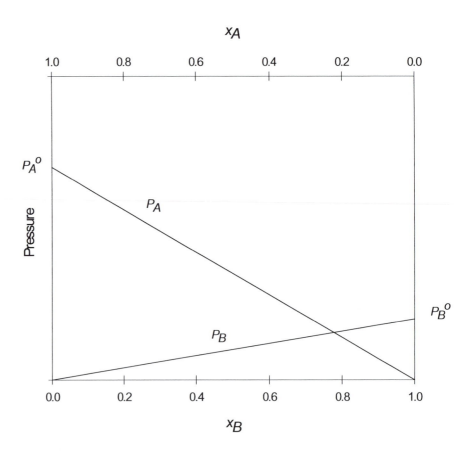

**Figure 18.1: Partial Pressures as a Function of Mole Fractions
in an Ideal Mixture**

The straight lines (i.e., lines of constant slope) in Figure 18.1 tell you that the ratio of the concentration in the gas phase to the concentration in the liquid phase is constant regardless of the composition. In other words: $[A(g)]/[A(l)] = slope = constant = P_A^o$, and $[B(g)]/[B(l)] = slope = constant = P_B^o$. [Note the use of the label (l) to denote a pure liquid phase.] These mathematical equations look suspiciously like equilibrium expressions:

$$A(l) = A(g) \quad K = slope = P_A^o = [A(g)]/[A(l)]$$
$$B(l) = B(g) \quad K = slope = P_B^o = [B(g)]/[B(l)]$$

In other words, straight lines on a plot of partial pressure versus aqueous concentration indicates a constant equilibrium relationship between concentrations in the gas and liquid phases.

18.2.2 Nonideal substances

Of course, many substances of environmental interest *do* interact and are *nonideal*. Thus, a plot of the partial pressures against mole fractions does not yield straight lines. The deviations away from ideality can yield larger or smaller partial pressures than the ideal case. A typical example is shown in Figure 18.2. Note that the partial pressures of A and B do not change linearly with composition.

The nice equilibrium analogy appears to fall apart: the ratio of the gas phase concentration to the liquid phase concentration (i.e., the slope) is no longer constant. In other words, the ratio of the phase concentrations (and thus the equilibrium constant) depends on the composition.

Can you develop equilibrium relationships from Figure 18.2? Yes, if you confine yourself to very small mole fractions (near 0.0) and very large mole fractions (near 1.0). Notice that both A and B exhibit fairly constant slopes at x_A (or x_B) near zero and at x_A (or x_B) near 1. This is emphasized in Figure 18.3, where the limiting slopes are indicated by dashed lines.

18.2.3 Equilibrium relationships at limiting concentrations

Finally, pretend that substance A is a pollutant and substance B is water. The left side of Figure 18.3 now represents nearly pure A. and the right side of Figure 18.3 represents nearly infinite dilution. These are the two extremes of interest to environmental engineers and scientists. We want to know how pollutants partition into the gas phase under each extreme case. For example, if you are determining workers' exposure to a solvent through

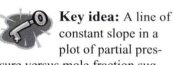

Key idea: A line of constant slope in a plot of partial pressure versus mole fraction suggests a constant equilibrium relationship between concentrations in the gas and liquid phases

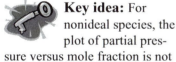

Key idea: For nonideal species, the plot of partial pressure versus mole fraction is not linear, suggesting that the equilibrium relationship between concentrations in the gas and liquid phases varies with composition

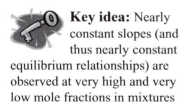

Key idea: Nearly constant slopes (and thus nearly constant equilibrium relationships) are observed at very high and very low mole fractions in mixtures

the air, you need to quantify the gas phase concentration in equilibrium with pure solvent ($x_A = 1$, $x_B = 0$). Or perhaps you are concerned about the partitioning of a solvent into air in the unsaturated zone of soil. If the solvent is less dense than water (a so-called *LNAPL* - light nonaqueous phase liquid), it will float on top of the saturated zone and partition into air found in the interstices of soil particles. These cases are shown by the extreme left side of Figure 18.3. The appropriate equilibrium relationship is: $[A(g)]/[A(l)] = P_A^o$ at $x_A \approx 1$.

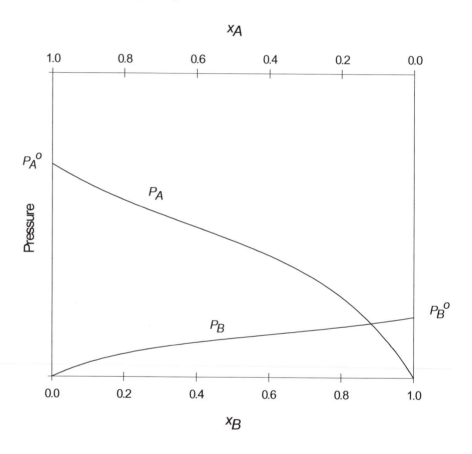

Figure 18.2: Partial Pressures as a Function of Mole Fractions in a NonIdeal Mixture

On the other hand, you may be interested in very small pollutant concentrations as well. Many volatile substances are only marginally soluble in water. In other words, their aqueous concentrations are very small. Since the concentration of water in water is very high [= (1000 g/L)/(18 g/mol) = 55.56 M], the mole fraction of many pollutants is very small, as shown in the extreme right side of Figure 18.3. The appropriate equilibrium relationship is: $[A(g)]/[A(l)]$ = right-hand slope at $x_A \approx 0$.

The two equilibrium relationships have names. For pure or nearly pure substances, the gas phase concentration is proportional to its mole fraction. The proportionality constant is the vapor pressure. This is called ***Raoult's Law***, developed in 1884 by François-Marie Raoult (1830-1901). Formally:

Raoult's Law: for pure or nearly pure substances, the gas phase concentration is proportional to its mole fraction with the proportionality constant equal to the vapor pressure

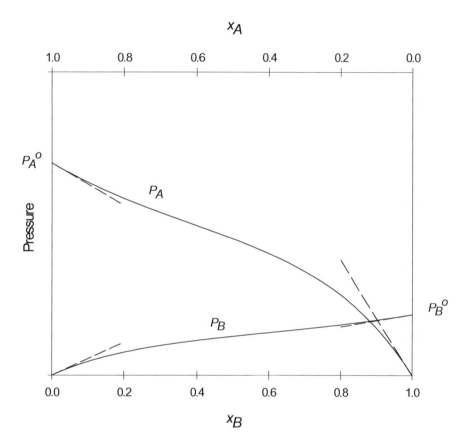

Figure 18.3: Partial Pressures as a Function of Mole Fractions in a Non-Ideal Mixture
(linearity at the extreme mole fractions emphasized)

$$\lim_{x_i \to 1} \frac{\{i(g)\}}{\{i(l)\}} = P_i^{\,o} \qquad\qquad \text{eq. 18.1}$$

Henry's Law: for nearly infinitely dilute substances, the gas phase concentration is proportional to its mole fraction

For almost infinitely dilution solutions, the gas phase concentration also is proportional to its mole fraction, but with a different proportionality constant. This is called ***Henry's Law***, developed in 1803 by the British chemist and physician William Henry (1774-1836). The proportionality

Henry's Law constant: the proportionality constant between gas phase and liquid phase concentrations in dilute solutions obeying Henry's Law

constant is the ***Henry's Law constant*** (or Henry's constant), denoted K_H or H. For dilute aqueous solutions, one way to write Henry's Law is:

$$\lim_{x_i \to 0} \frac{\{i\,(g)\}}{\{i\,(aq)\}} = K_{H,i} \qquad \text{eq. 18.2}$$

Although activities are used in eqs. 18.1 and 18.2, concentrations can be substituted if the gas phase and aqueous phase are very dilute[†]. Tables of vapor pressures and Henry's Law constants may be found in Appendix D.

18.2.4 Units of the Henry's Law constant

Key idea: Henry's Law constants are just equilibrium constants and thus thermodynamic functions

It is important to understand that the Henry's Law constant is *just an equilibrium constant*. As such, it is a thermodynamic property and depends on temperature, pressure, and the concentration of nonparticipating species. In this text, the symbol K_H will be used to emphasize that the Henry's Law constant is just an equilibrium constant.

There are two aspects of Henry's Law constants that you must understand before using the constant in an equilibrium calculation: the form of the equilibrium and the units of the Henry's Law constant. Remember that acid dissociation and complex formation constants are listed for equilibria written in a specific form. However, there is no standardized form for gas-liquid equilibria. Sometimes the equilibria are written with aqueous species as reactants: A(aq) = A(g). On other occasions, the equilibria are written with aqueous species as products: A(g) = A(aq). Of course, the units for the equilibrium constants for these two equilibria are different. *Always make sure you know the equilibrium corresponding to a given Henry's Law constant.*

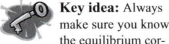

Key idea: Always make sure you know the equilibrium corresponding to a given Henry's Law constant and the units of the Henry's Law constant

Even if you know the equilibrium corresponding to a Henry's Law constant, it is critical that you know the *units* of the Henry's Law constant. For the equilibrium written as A(g) = A(aq), there are two common sets of units for K_H: mol/L-atm and dimensionless. Clearly, the units of mol/L-atm correspond to aqueous concentrations in mol/L and gas phase concentrations in atm. An equilibrium constant with units of mol/L-atm can be made dimensionless by multiplying by RT. In other words:

Key idea: Take care with dimensionless Henry's Law constants

$$K_H(\text{mol/L-atm})RT = K_H(\text{dimensionless})$$

[†] Henry believed that the constant proportionality of the gas phase concentration and mole fraction existed over a wide range of mole fractions. When chemist and burner-developer Robert Wilhelm Bunsen (1811-1899) showed that Henry's Law was only true at very small mole fractions and pressures, Henry apparently was so distraught that he took his own life.

Example 18.1: **Converting Units of Henry's Law Constants**

To test a new device for removing ammonia from water, you need to know about the equilibrium partitioning of ammonia between gas and water. In a book, you see two values of the Henry's Law constants for ammonia: 0.0175 atm-L/mol and 1400 (no units given). You know both values are for 25°C. What equilibria do the two Henry's Law constants correspond to?

Solution:
You must start with the constant with units since the dimensionless constant gives no clues as to the corresponding equilibrium. With units of atm-L/mol, the gas phase species must be a product and the aqueous phase species must be a reactant. Thus:

$$NH_3(aq) = NH_3(g)$$
$$K_H = 0.0175 \text{ atm-L/mol}$$

The dimensionless value for the equilibrium written in this way is, from Table 18.2: K_H/RT, or: $(0.0175 \text{ atm-L/mol})/[(0.082057 \text{ L-atm/mol-°K})(298°K)] = 7.16 \times 10^{-4}$.

This value is not near 1400. Thus, the dimensionless Henry's Law constant for the *reverse* equilibrium, $NH_3(aq) = NH_3(g)$, should be evaluated. The dimensionless Henry's Law constant for the reverse equilibrium is $1/7.16 \times 10^{-4} = 1400$. Therefore, the dimensionless Henry's Law constant 1400 is for the equilibrium written as:

The most appropriate units for R here are 0.082057 L-atm/mol-°K, with T in °K. The Henry's Law constant for the equilibrium A(aq) = A(g) sometimes is expressed in atmospheres. Interconversions between common units for Henry's Law constants are listed in Tables 18.1 and 18.2.

Table 18.1: Interconversions Between Units for Henry's Law Constants for the Equilibrium A(g) = A(aq)
(tabulated formulas show the row parameter as a function of the column parameter)

	K_H (mol/L-atm)	K_H ()	K_H (1/atm)
K_H (mol/L-atm) =	–	K_H ()/RT	55.56K_H (1/atm)
K_H () =	RTK_H (mol/L-atm)	–	55.56RTK_H (1/atm)
K_H (1/atm) =	K_H (mol/L-atm)/55.56	K_H ()/(55.56RT)	–

Note: $R = 0.082057$ L-atm/mol-°K, T in °K. K_H () means dimensionless K_H. The number 55.56 has units of mol/L and is the molar concentration of water (see Section 18.2.3).

Dimensionless Henry's Law constants are especially tricky. If you see a Henry's Law constant for oxygen of 1.26×10^{-3} mol/L-atm, you know that the corresponding equilibrium must be $O_2(g) = O_2(aq)$. However, if you see a Henry's Law constant for oxygen of 3.08×10^{-2} (dimensionless), *you do not know the form of the corresponding equilibrium*. You have no units to help you. [You can verify that $0.0308 = (0.00126 \text{ mol/L-atm})RT$ at 25°C, and thus the Henry's Law constant of 3.08×10^{-2} also corresponds to the equilibrium $O_2(g) = O_2(aq)$.] Another illustration of manipulating the units of Henry's Law constants is given in Example 18.1.

Table 18.2: Interconversions Between Units for Henry's Law Constants for the Equilibrium A(aq) = A(g)
(tabulated formulas show the row parameter as a function of the column parameter)

	K_H (atm-L/mol)	K_H ()	K_H (atm)
K_H (atm-L/mol) =	–	RTK_H ()	K_H (atm)/55.56
K_H () =	K_H (atm-L/mol)/RT	–	K_H (atm)/(55.56RT)
K_H (atm) =	55.56K_H (atm-L/mol)	55.56RTK_H ()	–

Notes: $R = 0.082057$ L-atm/mol-°K, T in °K. K_H () means dimensionless K_H. The number 55.56 has units of mol/L and is the the molar concentration of water (see Section 18.2.3).

$$NH_3(aq) = NH_3(g)$$
$$K_H = 1400 \text{ (di-mensionless)}$$

Since this text is focused on dilute aqueous systems, Raoult's Law will not be used very often. You should note from eq. 18.1 that the vapor pressure will have units of atm if the gas phase activity is in atm and the liquid phase activity is in mole fraction (dimensionless). From eq. 18.1, you can see that vapor pressure can be thought of as just an equilibrium constant. As a result, it is a thermodynamic function.

As an example of the use of vapor pressures, note that the vapor pressure of toluene is 0.0289 atm at 20°C, and thus the partial pressure of toluene in equilibrium with pure liquid toluene is (0.0289 atm)(1.0 mole fraction) = 0.0289 atm at 20°C. Another instance of using vapor pressures is shown in Example 18.2.

18.3 EQUILIBRIUM CALCULATIONS INVOLVING GAS-LIQUID EQUILIBRIA

18.3.1 Introduction

Open systems contain one more species (the gas phase species) and one more equation (from the mathematical form of the Henry's Law equilibrium) than the corresponding closed system. Thus, if the closed system consists of n equations in n unknowns, the open system with a non-hydrolyzing gas should consist of $n+1$ equations in $n+1$ unknowns and should be solvable.

The first step in performing equilibrium calculations with gas-liquid partitioning is to identify the species that partitions between the gas and liquid phases. *Under most environmental conditions, the partitioning species is uncharged.* Why? Recall that water forms aquo complexes with most ions. As a result, charged species have a high affinity for water and very low affinity for the gas phase. Thus, you should expect NH_3 to volatilize but not NH_4^+. You also might expect H_2S to partition into the gas phase but not HS^- or S^{2-}.

The second step in performing equilibrium calculations is to identify the type of open system. Open systems consist of two types. The most common type of open system is equilibrium with a very large volume of gas. A common example is equilibration with the atmosphere. In this case, partitioning of species to and from the gas phase does not change the gas phase concentration (i.e., the species partial pressure) appreciably. Equilibrium calculations with open systems containing a large gas volume are very simple because the partial pressure of the partitioning species is constant. From Henry's Law, *the concentration of the aqueous species in equilibrium with the gas phase (i.e., the dissolved gas) also must be fixed.* In other words, the gas phase is considered to be an infinite reservoir of the species of interest, and *a mass balance on the species of interest is not required.* This is analogous to acid-base chemistry, where water is considered to be an infinite reservoir of H^+ and OH^-.

A second type of open system contains a very small gas volume. For example, in a closed tank with head space, a volatile pollutant in water may

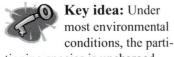

Key idea: Under most environmental conditions, the partitioning species is uncharged

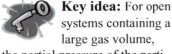

Key idea: For open systems containing a large gas volume, the partial pressure of the partitioning species is constant so the dissolved gas concentration must be fixed and a mass balance on the species of interest is not required

equilibrate with a relatively small gas volume. In this case, the partitioning process may very well change the gas phase concentration.[†] To account for this change, you must perform a mass (or mole) balance on the partitioning species. Examples of the mass balance will be given in Sections 18.3.2 and 18.3.3.

18.3.2 Nonhydrolyzing dissolved gases

Nonhydrolyzing gases do not react with water when dissolved in water. The equations in the chemical system depend on whether the gas phase is large enough so that the partial pressure is fixed. For equilibration with the atmosphere (or other large gas phase), no mass balance on the partitioning species is required. As an example of the equilibrium calculation in this case, determine the concentration of the major species in water at 25°C and 1 atm total pressure in equilibrium with the atmosphere. Assume that the major gases, $N_2(g)$ and $O_2(g)$, are non-hydrolyzing. Proceed as usual: make a species list and then list the equilibria, mass balances, charge balance, and species of fixed activities. (You may wish to try this on your own before reading on.)

Species list: H_2O, H^+, OH^-, $N_2(g)$, $O_2(g)$, $N_2(aq)$, and $O_2(aq)$

Equilibria:

$$H_2O = H^+ + OH^- \qquad K = K_W = 10^{-14}$$
$$N_2(g) = N_2(aq) \qquad K = K_{H,N_2} = 6.61 \times 10^{-4} \text{ mol/L-atm}$$
$$O_2(g) = O_2(aq) \qquad K = K_{H,O_2} = 1.26 \times 10^{-3} \text{ mol/L-atm}$$

Or:

$$K_W = [H^+][OH^-]/[H_2O]$$
$$K_{H,N_2} = [N_2(aq)]/P_{N_2}$$
$$K_{H,O_2} = [O_2(aq)]/P_{O_2}$$

Mass balances:

None - H_2O is an infinite reservoir of H^+ and OH^-, and the atmosphere is an infinite reservoir of $N_2(g)$ and $O_2(g)$

Charge balance:

$$[H^+] = [OH^-]$$

Other:

---

nonhydrolyzing gas: a gas that does not react with water after dissolution

---

Example 18.2: **Vapor Pressure Calculations**

What is the mass concentration of water vapor in a room at 25°C, where the air is in equilibrium with a bowl of liquid water?

Solution:
The vapor pressure of water is 23.8 mm Hg at 25°C or (23.8 mm Hg)/(760 mm Hg/atm) = 3.13×10^{-2} atm. From the ideal gas law: $PV = nRT$, so:

$$
\begin{aligned}
n/V &= P/RT \\
&= (3.13 \times 10^{-2} \text{ atm})/ \\
&\quad [(0.082057 \text{ L-atm/mol-}°K)(298°K)] \\
&= 1.28 \times 10^{-3} \text{ mol/L}
\end{aligned}
$$

or: $(1.28 \times 10^{-3} \text{ mol/L})(18 \text{ g/mol})$ = 23 mg/L.

Thus, air in equilibrium with liquid water at 25°C contains about 23 mg/L of water.

[†] According to some definitions, a water-gas system with a very small gas volume is not considered to be open. The term *open system* is used here to indicate that species can partition into a gas phase.

Dilute solution so that Henry's Law holds and concentrations and activities are interchangeable
Activity of water is unity, $P_{N_2} = 0.79$ atm, and $P_{O_2} = 0.209$ atm

From the K_W expression and the charge balance, you know that $[H^+] = [OH^-] = 10^{-7}$ M (pH 7). From the K_{H,N_2} expression and the fixed partial pressure of $N_2(g)$:

$$[N_2(aq)] = K_{H,N_2}P_{N_2}$$
$$= (6.61 \times 10^{-4} \text{ mol/L-atm})(0.79 \text{ atm}) = 5.22 \times 10^{-4} \text{ M}$$

From the K_{H,O_2} expression and the fixed partial pressure of $O_2(g)$:

$$[O_2(aq)] = K_{H,O_2}P_{O_2}$$
$$= (1.26 \times 10^{-3} \text{ mol/L-atm})(0.209 \text{ atm}) = 2.63 \times 10^{-4} \text{ M}$$

The concentrations of the major species in the system are: $[H^+] = [OH^-] = 10^{-7}$ M, $P_{N_2} = 0.79$ atm, $P_{O_2} = 0.209$ atm, $[N_2(aq)] = 2.63 \times 10^{-4}$ M, and $[O_2(aq)] = 2.63 \times 10^{-4}$ M. You can verify that this dissolved oxygen concentration is equal to 8.4 mg/L, oxygen saturation at 25°C.

What if the volume of gas is small? You can solve the system again for a small gas volume, say a 2 L bottle containing 1.5 L of water and 0.5 L of air. The chemical system is the same, except that *you must now include mass balances on nitrogen and oxygen*. Initially, the water contains no dissolved nitrogen or oxygen. The air contains $[N_2(g)]V_{air}$ moles of nitrogen and $[O_2(g)]V_{air}$ moles of oxygen, where V_{air} = volume of air. To obtain units of moles of gas, you must convert partial pressure to mol/L of gas. From the ideal gas law: $P/RT = n/V$. Thus, divide the partial pressure by RT to get units of mol/L of gas and add the following two mass balances to the chemical system:

$$\begin{aligned} N_T &= N_2 \text{ initially in the gas phase} + N_2 \text{ initially in} \\ &\quad \text{the aqueous phase} \\ &= P_{N_2,i}V_{air}/RT + [N_2(aq)]_i V_{water} \\ &= (0.79 \text{ atm})(0.5 \text{ L})/[(0.082057 \text{ L-atm/mol-°K})(298°K)] + 0 \\ &= 0.0162 \text{ mol} \end{aligned}$$

where $P_{N_2,i}$ = initial partial pressure of nitrogen gas, $[N_2(aq)]_i$ = initial dissolved nitrogen gas concentration = 0, and V_{water} = volume of water. Also:

$$\begin{aligned} O_T &= O_2 \text{ initially in the gas phase} + O_2 \text{ initially in} \\ &\quad \text{the aqueous phase} \\ &= P_{O_2,i}V_{air}/RT + [O_2(aq)]_i V_{water} \\ &= (0.209 \text{ atm})(0.5 \text{ L})/[(0.082057 \text{ L-atm/mol-°K})(298°K)] + 0 \end{aligned}$$

Example 18.3: **Nonhydrolyzing Dissolved Gases**

Professor Whoops makes up a standard solution containing 10 mg/L of tetrachloromethane (carbon tetrachloride, CCl_4). The good professor carefully pours the solution into two 40 mL vials. One vial is filled completely and sealed. The second vial accidently includes a 1 mL gas bubble before it is sealed. What are the tetrachloromethane concentrations in the aqueous phases of the two vials at equilibrium?

Solution:
The first vial was sealed and therefore should contain 10 mg/L of CCl_4 at equilibrium since no gas phases exist. For the other vial, you must combine a mass balance with the Henry's Law expression. The Henry's Law equilibrium is:

$$CCl_4(aq) = CCl_4(g)$$
$$K_H = 24.0 \text{ atm-L/mol}$$
$$\text{at 25°C}$$

For the small gas volume in the second vial, do a mass balance on tetrachloromethane. The total moles of CCl_4, C_T, is $(1 \times 10^{-2} \text{ g/L})(0.039 \text{ L})/(153.8 \text{ g/mol}) = 2.54 \times 10^{-6}$ mol. The mass balance is:

$$C_T = P_C V_{air}/RT + P_C V_{water}/K_H$$

(Note that this equation differs from that in the text because the gas phase species is a product in the equilibrium here.)

Solving with $V_{air} = 0.001$ L and $V_{water} = 0.039$ mL: $P_C = 1.52 \times 10^{-3}$ atm and $[CCl_4(aq)] = P_C/K_H = 6.34 \times 10^{-5}$ M or 9.75 mg/L.

Thus, the presence of a 1 mL air bubble will reduce the tetrachloromethane concentration in the standard by about **2.5%, from 10 mg/L to 9.75 mg/L.**

$$= 0.00427 \text{ mol}$$

where $P_{O_2,i}$ = initial partial pressure of oxygen gas and $[O_2(aq)]_i$ = initial dissolved oxygen gas concentration = 0. The acid-base chemistry is the same as in the large gas phase case, so: $[H^+] = [OH^-] = 10^{-7}$ M. Substituting the Henry's Law expressions into the mass balances:

$$N_T = P_{N_2} V_{air}/RT + K_{H,N_2} P_{N_2} V_{water} = 0.0162 \text{ mol}$$
$$O_T = P_{O_2} V_{air}/RT + K_{H,O_2} P_{O_2} V_{water} = 0.00427 \text{ mol}$$

Solving:

$$P_{N_2} = N_T/(V_{air}/RT + K_{H,N_2} V_{water}) = 0.756 \text{ atm}$$
$$[N_2(aq)] = K_{H,N_2} P_{N_2} = 4.99 \times 10^{-4} \text{ M}$$
$$P_{O_2} = O_T/(V_{air}/RT + K_{H,O_2} V_{water}) = 0.191 \text{ atm, and:}$$
$$[O_2(aq)] = K_{H,O_2} P_{O_2} = 2.41 \times 10^{-3} \text{ M}$$

In the system with a small gas volume, the partial pressures of the gases are smaller and the dissolved concentrations are larger than in the case of equilibrium with the atmosphere. Another equilibrium calculation with nonhydrolyzing dissolved gases is given in Example 18.3.

18.3.3 Hydrolyzing dissolved gases

hydrolyzing dissolved gas: a gas that reacts with water after dissolution

Key idea: For hydrolyzing dissolved gases, the dissolved gas is usually the uncharged member of an acid-base family

Key idea: For hydrolyzing dissolved gases in equilibrium with the atmosphere, the partial pressure of the gas phase species is constant and so the concentration of the dissolved gas species is fixed and independent of pH

Hydrolyzing dissolved gases react with water after dissolution. In the most common cases, *the dissolved gas is the uncharged member of an acid-base family*. Examples include $H_2S(aq)$ [in equilibrium with $H_2S(g)$ and also HS^- and S^{2-}], $H_2CO_3^*$ [in equilibrium with $CO_2(g)$ and also HCO_3^- and CO_3^{2-}], and $NH_3(aq)$ [in equilibrium with $NH_3(g)$ and also NH_4^+]. As stated in Section 18.3.1, chemical systems with dissolved gases have one more species and one more equation than the corresponding closed chemical system. Equilibrium calculations with hydrolyzing dissolved gases follow the usual steps: make a species list, list the equilibria, list the mass balances, form the charge balance, and list species of fixed activities.

For equilibrium with the atmosphere or other large gas volume, the partial pressure of the gas phase species is assumed to be constant. From Henry's Law, *this fixes the concentration of the dissolved gas species*. This simple statement has important consequences. The concentration of the dissolved gas species is fixed and thus *its concentration is independent of pH*. For example, the $H_2S(aq)$ concentration becomes independent of pH if water is in equilibrium with gas containing a fixed partial pressure of $H_2S(g)$. As the pH increases, the $H_2S(aq)$ concentration remains constant, but the concentrations of HS^- and S^{2-} increase. Thus, the total dissolved sulfide concentration, $[H_2S(aq)] + [HS^-] + [S^{2-}]$, increases with pH. As with nonhydrolyzing dissolved gases, a mass balance on the partitioning species is not needed because the atmosphere is an infinite reservoir of material.

This behavior can be demonstrated with a simple example. Find the equilibrium pH and draw a pC-pH diagram for water in equilibrium with air containing $P_{H_2S} = 0.001$ atm. Following the usual procedure:

Species list: H_2O, H^+, OH^-, $H_2S(aq)$, HS^-, S^{2-}, and $H_2S(g)$

Equilibria:

$$H_2O = H^+ + OH^- \qquad K = K_W = 10^{-14}$$
$$H_2S(aq) = H^+ + HS^- \qquad K = K_{a,1} = 10^{-7}$$
$$HS^- = H^+ + S^{2-} \qquad K = K_{a,2} = 10^{-14}$$
$$H_2S(g) = H_2S(aq) \qquad K = K_H = 10^{-0.98} \text{ mol/L-atm}$$

Or:

$$K_W = [H^+][OH^-]/[H_2O]$$
$$K_{a,1} = [H^+][HS^-]/[H_2S(aq)]$$
$$K_{a,2} = [H^+][S^{2-}]/[HS^-]$$
$$K_H = [H_2S(aq)]/P_{H_2S}$$

Mass balances:

   None; H_2O is an infinite reservoir of H^+ and OH^- and the atmosphere is an infinite reservoir of $H_2S(g)$

Charge balance:

   $$[H^+] = [HS^-] + 2[S^{2-}] + [OH^-]$$

Other:

   Dilute solution so that Henry's Law holds and concentrations and activities are interchangeable
   Activity of water is unity
   $P_{H_2S} = 0.001$ atm

From the Henry's Law expression, $[H_2S(aq)]$ is fixed and equal to $K_H P_{H_2S}$. Since $[H_2S(aq)]$ is fixed, it makes sense to express the other dissolved sulfide species in terms of $[H_2S(aq)]$. From the $K_{a,1}$ and $K_{a,2}$ expressions:

$$[HS^-] = K_{a,1}[H_2S(aq)]/[H^+] = K_{a,1}K_H P_{H_2S}/[H^+] \text{ and:}$$
$$[S^{2-}] = K_{a,2}[HS^-]/[H^+] = K_{a,1}K_{a,2}K_H P_{H_2S}/[H^+]^2$$

Thus, each species can be expressed as a function of $[H^+]$ and constants (including P_{H_2S}):

$$[OH^-] = K_W/[H^+]$$
$$[H_2S(aq)] = K_H P_{H_2S} = 10^{-3.98} \text{ M}$$
$$[HS^-] = K_{a,1}K_H P_{H_2S}/[H^+]$$

$$[S^{2-}] = K_{a,1}K_{a,2}K_H P_{H_2S}/[H^+]^2$$

Thoughtful Pause
How can you find the equilibrium pH?

The pC-pH for the H_2S system in equilibrium with $P_{H_2S} = 0.001$ atm is shown in Figure 18.4. To find the equilibrium, use the charge balance:

$$[H^+] = [HS^-] + 2[S^{2-}] + [OH^-]$$

Key idea: Dissolved gases that represent the most protonated form of the acid-base family are more soluble at higher pH, and dissolved gases that represent the least protonated form of the acid-base family are more soluble at lower pH

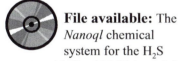

File available: The *Nanoql* chemical system for the H_2S system with fixed $H_2S(g)$ partial pressure may be found on the CD, file location **/Chapter 18/H2S fixed.nqs**

Example 18.4: **Hydrolyzing Dissolved Gases**

You are in charge of production at a bleach manufacturing plant. You want to produce bleach at the highest concentration possible to minimize shipping costs. In your production line, you make bleach by equilibrating chlorine gas ($P_{Cl_2} = 1 \times 10^{-3}$ atm) with water. What should be the pH of the water to maximize the concentration of total dissolved chlorine? The total dissolved chlorine (in the +1 oxidation

Either graphically or algebraically, you can find that the equilibrium pH is 5.55 (be sure to check any assumptions you made in using the charge balance).[†]

The pC-pH diagram in Figure 18.4 is very different from the acid-base pC-pH diagrams you have seen in closed systems. First, the species lines in Figure 18.4 have constant slope. Their slopes reflect the difference in the number of protons between the species of interest and the dissolved gas species. The lines do not curve at the pK_a values, as with closed acid-base systems. Second, the total dissolved sulfide concentration increases with pH. (The total dissolved sulfide is plotted as a thick line in Figure 18.4.) Since $[H_2S(aq)]$ is constant and independent of pH, more hydrogen sulfide dissolves to maintain the equilibrium with the conjugate bases at higher pH. In general, dissolved gases that represent the most protonated form of the acid-base family are more soluble at higher pH. Another example of equilibrium calculations with water in equilibrium with the atmosphere is given in Example 18.4.

Systems with large gas volumes are solved easily by equilibrium calculation software. For example, to solve the open H_2S system with *Nanoql*, simply make the species $H_2S(g)$ a fixed species with a fixed concentration (partial pressure) of 0.001 atm (see the accompanying CD at **/Chapter 18/H2S fixed.nqs**).

If the gas volume is small, then you must add a mass balance for the partitioning species. For example, suppose you have developed a reactor to remove ammonia from water. The 10 L reactor is charged with 0.1 L of water to be treated, then sealed, and mixed vigorously to equilibrate the ammonia with the trapped air. What is the removal efficiency of the reactor at pH 11 if the initial total N concentration is 20 mg/L as N in the water and the air initially has no ammonia? How does the removal efficiency vary with pH if the pH is fixed by a buffer? Setting up the system:

[†] In fact, you can estimate the equilibrium pH in your head. From the pC-pH diagram, the equilibrium pH is about where $[H^+] = [HS^-]$ (here, $[S^{2-}]$ and $[OH^-]$ are negligible). Thus: $[H^+] \approx K_{a,1}K_H P_{H_2S}/[H^+]$ or $[H^+]^2 \approx K_{a,1}K_H P_{H_2S} = 10^{-10.98}$ M, so the equilibrium pH is about $10.98/2 = 5.5$.

state) is [HOCl] + [OCl⁻].
(Note: A qualitative version of
this question was asked in Example 4.2.)

Solution:

$Cl_2(g)$ dissolves to form $Cl_2(aq)$,
which disproportionates to
HOCl and chloride (Cl⁻). HOCl
is in equilibrium with OCl⁻.
The pertinent equilibria are:

$$H_2O = H^+ + OH^-$$
$$K_W = 10^{-14}$$
$$Cl_2(aq) = HOCl + Cl^-$$
$$+ H^+$$
$$K = 10^{-3.3}$$
$$HOCl = H^+ + OCl^-$$
$$K_a = 10^{-7.5}$$
$$Cl_2(aq) = Cl_2(g)$$
$$K_H' = 0.0853 \text{ atm-}$$
$$\text{m}^3/\text{mol}$$

Rewriting the Henry's Law
equilibrium in a more convenient form:

$$Cl_2(g) = Cl_2(aq)$$
$$K_H = 0.0117 \text{ mol/L-atm}$$

Thus:

$$Cl_2(aq) = K_H P_{Cl_2} \text{ and:}$$
$$[HOCl] =$$
$$K[Cl_2(aq)]/([Cl^-][H^+])$$

Now, both chloride and Cl(+I)
are formed from the disproportionation of dissolved chlorine gas, so:

$$[Cl^-] = [HOCl] + [OCl^-]$$

Also:

$$[OCl^-] = [HOCl]K_a/[H^+]$$

Substituting and solving for
[HOCl]:

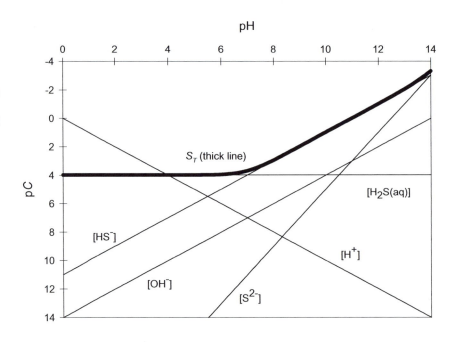

Figure 18.4: pC-pH for the H₂S System in Equilibrium with $P_{H_2S} = 0.001$ atm

Species list: H_2O, H^+, OH^-, $NH_3(aq)$, NH_4^+, and $NH_3(g)$

Equilibria:

$H_2O = H^+ + OH^-$	$K = K_W = 10^{-14}$
$NH_4^+ = H^+ + NH_3(aq)$	$K = K_a = 10^{-9.3}$
$NH_3(g) = NH_3(aq)$	$K = K_H = 57 \text{ mol/L-atm}$

Or:

$$K_W = [H^+][OH^-]/[H_2O]$$
$$K_a = [H^+][NH_3]/[NH_4^+]$$
$$K_H = [NH_3(aq)]/P_{NH_3}$$

Mass balances:

$$N_T = \text{NH}_3 \text{ initially in the gas phase} + \text{N}(-\text{III}) \text{ initially}$$
$$\text{in the aqueous phase}$$
$$= P_{NH_3,i}V_{air}/RT + \text{N}(-\text{III})_i V_{water}$$
$$= 0 + (0.02 \text{ g/L})(8 \text{ L})/(14 \text{ g/mol}) = 1.43 \times 10^{-4} \text{ mol}$$

$$[HOCl] = \sqrt{\frac{KK_H P_{Cl_2}}{[H^+] + K_a}}$$

With these expressions for [HOCl] and [OCl⁻], you can show that above pH about 9 (where pH > pK_a and [H⁺] << K_a):

$$Cl(+I)_T \approx (KK_a K_H P_{Cl_2})^{1/2}/[H^+]$$

Thus, the dissolved Cl(+I) concentration increases by about tenfold with each unit increase in pH above about pH 9. Thus, the **chlorine solutions will be more concentrated at higher pH**.

Note: Household bleach is sold at about 5% chlorine and pH about 12.

where $P_{NH_3,i}$ = initial partial pressure of ammonia = 0 and N(−III)$_i$ = initial total dissolved ammonia.

> Charge balance:
>   Not needed, since [H⁺] is fixed
>
> Other:
>   Dilute solution so that Henry's Law holds and concentrations
>   and activities are interchangeable
>   Activity of water is unity

At equilibrium: [NH₃(aq)] = $K_H P_{NH_3}$. Inserting this equation into the mass balance and noting that [NH₄⁺] = [H⁺][NH₃]/K_a:

$$N_T = P_{NH_3} V_{air}/RT + K_H P_{NH_3}(1 + [H^+]/K_a)V_{water}$$

At pH 11, P_{NH_3} = 2.30×10⁻⁵ atm, so [NH₃(aq)] = $K_H P_{NH_3}$ = 1.31×10⁻³ M, [NH₄⁺] = [H⁺][NH₃]/K_a = 2.61×10⁻⁵ M, and the total dissolved ammonia concentration is 1.3×10⁻³ M or 18.7 mg/L as N. This is only a 6.5% removal, not very good.

Thoughtful Pause
Does the removal of ammonia increase at higher or lower pH values?

The removal should increase at *higher* pH values since more of the total dissolved ammonia is in the form of NH₃(aq) and available for partitioning into the gas phase. At pH 7, only 0.04% of the total dissolved ammonia is removed, whereas at pH 13, 6.6% is removed. Note that in the case of ammonia, the dissolved gas is the least protonated form of the acid-base family and is more soluble at lower pH.

18.4 DISSOLVED CARBON DIOXIDE

18.4.1 How should you write dissolved carbon dioxide equilibria?
The partitioning of carbon dioxide between aqueous and gas phases is of particular interest since the carbonic acid system is the predominant weak acid system in natural waters. It is a simple matter to write the Henry's Law expression for carbon dioxide:

$$CO_2(g) = CO_2(aq) \qquad\qquad K_H'$$

The species $CO_2(aq)$ reacts with water to form H_2CO_3:

$$CO_2(aq) + H_2O = H_2CO_3 \qquad K \approx 1.5 \times 10^{-3}$$

***Example 18.5:* Dissolved Carbon Dioxide**

Cola drinks are strongly buffered by phosphoric acid at about pH 2. An unopened cola drink container has a partial pressure of carbon dioxide equal to about 5 atm. What is the total dissolved carbonate in an unopened cola drink and in a drink that has been equilibrated with the atmosphere?

Solution:
For the unopened drink: $C_T = K_H P_{CO_2}/\alpha_0$, where $K_H = 10^{-1.5}$ mol/L-atm, $P_{CO_2} = 5$ atm, and $\alpha_0 \approx 1$ at pH 2. Thus, $C_T = 0.158$ **M in the unopened bottle**.

For the opened and equilibrated drink: $C_T = K_H P_{CO_2}/\alpha_0$, where: $K_H = 10^{-1.5}$ mol/L-atm, $P_{CO_2} = 10^{-3.5}$ atm, and $\alpha_0 \approx 1$ at pH 2. Thus, $C_T = 1\times10^{-5}$ **M in the opened bottle**. The C_T value in the unopened bottle is almost 16,000 times greater than the C_T value in the opened and equilibrated bottle.

Recall from Section 11.4.2 that the species CO_2(aq) and H_2CO_3 are nearly indistinguishable. Remember that the symbol $[H_2CO_3{}^*]$ is used to represent the sum of the concentrations of CO_2(aq) and H_2CO_3. Thus, the Henry's Law expression for carbon dioxide usually is written:

$$CO_2(g) + H_2O = H_2CO_3{}^* \qquad K_H = 10^{-1.5} \text{ mol/L-atm}$$

18.4.2 Equilibrium calculations with dissolved carbon dioxide

You can draw the pC-pH diagram for water in equilibrium with carbon dioxide in the atmosphere by using exactly the same approach as with hydrogen sulfide in Section 18.3.3. The partial pressure of carbon dioxide in the atmosphere is about $10^{-3.5}$ atm. The pertinent equilibrium constants are: $pK_{a,1} = 6.3$, $pK_{a,2} = 10.3$, and $K_H = 10^{-1.5}$ mol/L-atm. For $P_{CO_2} = 10^{-3.5}$ atm, the pC-pH diagram is given in Figure 18.5.

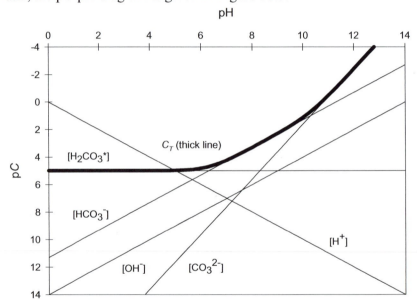

Figure 18.5: pC-pH for the $H_2CO_3{}^*$ System in Equilibrium with $P_{CO_2} = 10^{-3.5}$ atm

There are several important implications of the pC-pH diagram in Figure 18.5. First, carbon dioxide is very soluble at high pH. This can be a problem in some applications. For example, in metal finishing, some plating baths are maintained at high pH to minimize the volatilization of HCN. Carbon dioxide dissolves to a large extent in these baths, resulting in the precipitation of calcium carbonate in the baths. The high solubility of carbon dioxide at high pH can be used to your advantage as well. Environmental microbiologists sometimes use pollutants with radioactively labeled carbon atoms to determine the fate of the pollutant during biodegradation. If the pollutant is mineralized to carbon dioxide, then the released $CO_2(g)$ can be trapped in an alkaline solution since carbon dioxide is so soluble at high pH.

You can use the equilibria to write a very simple expression for the total dissolved carbonate in water in equilibrium with the atmosphere. From Henry's Law: $[H_2CO_3^*] = K_H P_{CO_2}$. From the acid-base equilibria: $[H_2CO_3^*] = \alpha_0 C_T$. Thus:

$$C_T = K_H P_{CO_2} / \alpha_0 \qquad \text{eq. 18.3}$$

You can plot eq. 18.3 to obtain the dotted line in Figure 18.5. Note that the C_T line looks like an inverted α_0 line - it has a slope of zero at pH $< pK_{a,1} = 6.3$, *increases* with a slope of one at pH values between $pK_{a,1}$ and $pK_{a,2}$, and *increases* with a slope of 2 at pH $> pK_{a,2} = 10.3$.

Second, the $CO_2(g)$-H_2O system is a good model for clean water. You know that the pH of water in a closed system is 7.0 at 25°C. What is the pH of otherwise clean water in equilibrium with the atmosphere? To determine the equilibrium pH, apply the charge balance to Figure 18.5. The charge balance is: $[H^+] = [HCO_3^-] + 2[CO_3^{2-}] + [OH^-]$. From the figure, this occurs at about $[H^+] \approx [HCO_3^-]$. From the footnote in Section 18.3.3: $[H^+]^2 \approx K_1 K_H P_{CO_2} = 10^{-11.3}$ M, so the equilibrium pH is about 11.3/2 = 5.6-5.7. This is a pretty good model for clean rain since rain has a significant time to equilibrate with the atmosphere. Because equilibration with the atmosphere results in slightly acidic rain, the term *acid rain* is used to refer to rain with pH less than about 5.6. Another equilibrium calculation with dissolved carbon dioxide may be found in Example 18.5.

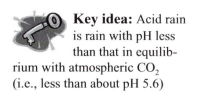 **Key idea:** Acid rain is rain with pH less than that in equilibrium with atmospheric CO_2 (i.e., less than about pH 5.6)

18.4.3 Dissolved carbon dioxide and alkalinity

In the previous section, the pH of clean water in equilibrium with the atmosphere was calculated to be about 5.6.

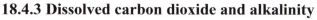

Thoughtful Pause

What is the alkalinity of clean water in equilibrium with the atmosphere?

You can determine the alkalinity in two ways: by logic or by calculation. Logically, you know that H_2CO_3 is at the zero level of alkalinity. Why? First, Alk = $C_B - C_A$ and both C_B and C_A are zero. Second, recall from Section 13.4.2 that alkalinity is the amount of strong acid required to make a water equivalent to an H_2CO_3 solution. Thus, adding or subtracting H_2CO_3 or $CO_2(aq)$ or $H_2CO_3^*$ does not affect the alkalinity (but may affect both pH and C_T). Clean water not in equilibrium with the atmosphere clearly has $Alk = 0$. Therefore, if only $CO_2(g)$ is exchanged, the Alk should remain zero: starting with zero alkalinity and adding or subtracting zero alkalinity should leave you with zero alkalinity.

You also can calculate the alkalinity. From the charge balance (namely, $[H^+] = [HCO_3^-] + 2[CO_3^{2-}] + [OH^-]$), you know:

$$[HCO_3^-] + 2[CO_3^{2-}] + [OH^-] - [H^+] = 0$$

Rewriting with alpha values: $(\alpha_1 + 2\alpha_2)C_T + [OH^-] - [H^+] = 0$. However, $(\alpha_1 + 2\alpha_2)C_T + [OH^-] - [H^+] = Alk$. Thus: $Alk = 0$.

How does equilibration with the atmosphere affect other systems? To investigate this question, consider two solutions in closed systems (i.e., *not* in equilibrium with the atmosphere): 2×10^{-3} M $NaHCO_3$ and 1×10^{-3} M Na_2CO_3.

Thoughtful Pause

What are the values of C_T, Alk, and pH for closed solutions of 2×10^{-3} M $NaHCO_3$ and 1×10^{-3} M Na_2CO_3?

In your head (from the composition and $Alk = C_B - C_A$), you should be able to show that the C_T values are 2×10^{-3} M and 1×10^{-3} M for the bicarbonate and carbonate solutions, respectively, and that both solutions have alkalinity values equal to 2×10^{-3} eq/L. From sketches of pC-pH diagrams, you can find equilibrium pH values of 8.3 for 2×10^{-3} M $NaHCO_3$ and 10.7 for 1×10^{-3} M Na_2CO_3, respectively.

Now, how do C_T, Alk, and pH change if the solutions are allowed to equilibrate with the atmosphere? The alkalinity calculation is straightforward: *if only $CO_2(g)$ is exchanged, then the alkalinity remains constant.*[†] Thus, upon equilibration with the atmosphere, the alkalinity for both solutions is 2×10^{-3} eq/L. How about pH and C_T? As you have done many times, the equilibrium pH is determined where the charge balance is satisfied, that is, where:

$$[H^+] + [Na^+] = [HCO_3^-] + 2[CO_3^{2-}] + [OH^-], \text{ or:}$$
$$[Na^+] = (\alpha_1 + 2\alpha_2)C_T + [OH^-] - [H^+]$$

From eq. 18.3, you know that $C_T = K_H P_{CO_2}/\alpha_0$. Thus:

$$[Na^+] = (\alpha_1 + 2\alpha_2)(K_H P_{CO_2}/\alpha_0) + [OH^-] - [H^+]$$

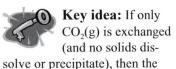 **Key idea:** If only $CO_2(g)$ is exchanged (and no solids dissolve or precipitate), then the alkalinity remains constant

---

[†] Strictly speaking, the alkalinity remains constant when only $CO_2(g)$ is exchanged only if no solids dissolve or precipitate. Precipitating solids could remove C_B or C_A and thus affect the alkalinity.

For both solutions, $[Na^+] = 2 \times 10^{-3}$ M. If you assume that you are near neutral pH (where $[OH^-]$, $[H^+]$, and α_2 are negligible), then:

$$[Na^+] \approx \alpha_1 K_H P_{CO_2}/\alpha_0$$

For any diprotic acid: $\alpha_1/\alpha_0 = K_{a,1}/[H^+]$ exactly (see Section 11.5.2). Thus:

$$[Na^+] \approx K_{a,1}K_H P_{CO_2}/[H^+] \text{ or } [H^+] = K_{a,1}K_H P_{CO_2}/[Na^+]$$

For $K_{a,1} = 10^{-6.3}$, $K_H = 10^{-1.5}$, $P_{CO_2} = 10^{-3.5}$ atm, and $[Na^+] = 2 \times 10^{-3}$ M, you can show: $[H^+] = 10^{-11.3}/10^{-2.7} = 10^{-8.6}$ or about pH 8.6. You can verify that the assumptions are valid at pH 8.6. (Without assumptions, the calculated pH is 8.58.) At this pH, $\alpha_0 = 5.12 \times 10^{-3}$ and $C_T = K_H P_{CO_2}/\alpha_0 = 1.95 \times 10^{-3}$ M. Thus, for *both* solutions upon equilibration with the atmosphere: $C_T = 1.95 \times 10^{-3}$ M, Alk $= 2 \times 10^{-3}$ eq/L, and pH ≈ 8.6.

This is a remarkable result. The two solutions had the same alkalinity and different pH and C_T values as closed systems. After they were equilibrated with the atmosphere, they became identical. The bicarbonate solution lost a little C_T to the atmosphere, and consequently its pH went up slightly. The carbonate solution gained a significant amount of C_T from the atmosphere, and consequently its pH went down significantly.

You can generalize these results by noting that the general charge balance is: Alk $= (\alpha_1 + 2\alpha_2)C_T + [OH^-] - [H^+]$. For waters without solids in equilibrium with the atmosphere:

$$\text{Alk} = (\alpha_1 + 2\alpha_2)(K_H P_{CO_2}/\alpha_0) + [OH^-] - [H^+] \qquad \text{eq. 18.4}$$

If the pH is near neutral and the alkalinity is not zero:

$$[H^+] \approx K_{a,1}K_H P_{CO_2}/\text{Alk} = 10^{-11.3}/\text{Alk} \qquad \text{eq. 18.5}$$

Or:

$$\text{pH} \approx 11.3 + \log(\text{Alk in eq/L}) \qquad \text{eq. 18.6}$$

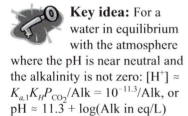

Key idea: For a water in equilibrium with the atmosphere where the pH is near neutral and the alkalinity is not zero: $[H^+] \approx K_{a,1}K_H P_{CO_2}/\text{Alk} = 10^{-11.3}/\text{Alk}$, or pH $\approx 11.3 + \log(\text{Alk in eq/L})$

oversaturation: the concentration of the dissolved gas is greater than that in equilibrium with the gas phase

These equations aid in the calculation of pH in natural waters. Suppose you are evaluating a groundwater for use as a drinking water source. The groundwater has a pH of 5.3 and Alk $= 50$ mg/L as $CaCO_3$. You suspect that the relatively low pH is due to **oversaturation** with carbon dioxide. This means that the $H_2CO_3^*$ concentration in the groundwater is greater than that in equilibrium with the atmosphere (where the CO_2 partial pressure is $10^{-3.5}$ atm). You believe that the groundwater can be aerated (i.e., equilibrated with the atmosphere) to bring the pH up to a reasonable value without chemical addition. Can you test these ideas quantitatively? For the groundwater:

$$\text{Alk} = (\alpha_1 + 2\alpha_2)(K_H P_{CO_2}'/\alpha_0) + [\text{OH}^-] - [\text{H}^+]$$

where P_{CO_2}' is the partial pressure of $CO_2(g)$ with which the groundwater is in equilibrium. Solving for P_{CO_2}' at pH 5.3 and Alk = 50 mg/L as $CaCO_3$ = 1×10^{-3} eq/L: $P_{CO_2}' = 10^{-0.50}$ atm. Thus, the groundwater is indeed oversaturated with carbon dioxide. Upon equilibration with the atmosphere, the alkalinity will remain constant (if no solids form or dissolve) and P_{CO_2} will decrease to $10^{-3.5}$ atm. From eq. 18.6, you can estimate the pH rapidly:

$$\text{pH} \approx 11.3 + \log(\text{Alk in eq/L}) = 11.3 - 3 = 8.3$$

A more careful calculation reveals a pH of 8.29. Thus, equilibration with the atmosphere would raise the pH up to about 8.3. Another illustration of the relationships among C_T, Alk, and pH may be found in Example 18.6.

18.5 SUMMARY

In this chapter, the equilibration of chemical species between the gas and liquid phases was explored. The aqueous species in equilibrium with the gaseous species is called the dissolved gas. The equilibrium laws governing gas-liquid partitioning (Raoult's Law and Henry's Law) were presented. The resulting constants (vapor pressure and Henry's Law constants) *are just equilibrium constants*. Remember that Henry's Law constants can be expressed with gases as either reactants or products and can be written with several sets of units (including no units).

Equilibrium calculation techniques with gas-liquid equilibria were developed. Two cases were considered: small gas volumes (in which case, mass balances are needed) and very large gas volumes (such as the atmosphere, where the gas phase is considered to be an infinite reservoir of material and some mass balances are not needed). In cases involving equilibrium with the atmosphere, the species in the gas phase fixes the concentration of the dissolved gas species. The dissolved gas species usually is uncharged (e.g., NH_3, not NH_4^+). Another way of categorizing gas-liquid equilibria is whether or not the dissolved species reacts with water. If the dissolved species hydrolyzes, then the gas solubility will depend on pH.

One of the most important hydrolyzing dissolved gases in natural systems is dissolved carbon dioxide. The solubility of dissolved carbon dioxide increases with increasing pH. Thus, C_T is not constant but depends on the equilibrium pH.

18.6 PART IV CASE STUDY: THE KILLER LAKES

The natural tragedies at Lakes Monoun and Nyos were described in Section 17.6. Unfortunately, future catastrophic degassing events could occur since

Example 18.6: Dissolved Carbon Dioxide and C_T, pH, and Alk

A lake is at pH 7.7 and in equilibrium with the atmosphere. What are the C_T, alkalinity, and pH after 5×10^{-4} M of dissolved carbon dioxide are consumed through algal photosynthesis? Repeat the calculation where nitrate also is consumed as the nitrogen source as the algae growth.

Solution:

If the lake water is in equilibrium with the atmosphere, then $C_T = K_H P_{CO_2}/\alpha_0$. At pH 7.7, $\alpha_0 = 0.0382$ so: $C_T = 2.62\times10^{-4}$ M. From eq. 18.4, Alk = 2.53×10^{-4} eq/L (or 13 mg/L as $CaCO_3$).

If only carbon dioxide is consumed during photosynthesis and no solids precipitate, then the Alk remains constant. Of course, C_T will decrease and pH will increase until the lake reequilibrates with the atmosphere. Upon reequilibration, the initial conditions are reestablished:

pH 7.7, $C_T = 2.62\times10^{-4}$ M, and Alk = 2.53×10^{-4} eq/L

If nitrate is consumed, the balanced equation is: $106CO_2 + 16NO_3^- + HPO_4^{2-} + 122H_2O + 18H^+ \rightarrow C_{106}H_{263}O_{110}N_{16}P + 138O_2$, where the products are a model of algae and oxygen. Thus, 18/106 mole of H^+ are consumed per mole of CO_2 consumed. Photosynthesis will consume $(18/106)(5\times10^{-4}$ M$) = 8.49\times10^{-5}$ eq/L of Alk, leaving 2.53×10^{-4} eq/L $- 8.49\times10^{-5}$ eq/L $= 1.68\times10^{-4}$ eq/L of Alk.

carbon dioxide continues to accumulate in the lower waters of the two lakes. International efforts to prevent future disasters have focused on controlled degassing. With your knowledge of gas-liquid equilibria, you should be able to estimate when the lakes will again reach the point of instability.

It is believed that the lakes were completely degassed in the violent events of the 1980s. Thus, carbon dioxide accumulation essentially restarted from the last catastrophic degassing events. The pertinent data for the lakes are summarized in Table 18.3.

Table 18.3: Characteristics of Lakes Nyos and Monoun

Parameter	Lake Nyos	Lake Monoun
Last degassing	1986	1984
Surface area (km²)	1.4	0.8
Depth (m)	208	96
CO_2 accumulation rate (mol/yr)[1]	2x20⁸	1.7×10^7
pH at bottom[2]	4.81	5.63

Notes: 1. From Kling et al., 1994
2. From Kling et al., 1999 - collected Nov., 1999

The CO_2 accumulation rates in Table 18.3 are assumed to be valid for the bottom 10 m of the lake. Thus, the volume of interest is the volume in the bottom 10 m of each lake. This is a volume of $(1.4\ km^2)(10^6\ m^2/km^2)(10\ m)$, or $1.4\times10^7\ m^3$ for Lake Nyos and $(0.8\ km^2)(10^6\ m^2/km^2)(10\ m)$ or $8\times10^6\ m^3$ for Lake Monoun.

Violent turnover is expected when the CO_2 pressure equals the hydrostatic pressure. The hydrostatic pressure is given by $\rho g z$, where ρ = water density (1 kg/m³), g = gravitational acceleration (= 9.8 m/s²), and z = depth at the bottom of the lake (in m). With the units indicated, the hydrostatic pressure will be calculated in Pascals (Pa). Recall that 1 atm is 101,325 Pa. The hydrostatic pressures at the bottoms of Lakes Nyos and Monoun can be calculated to be 20.1 and 9.3 atm, respectively.

From eq. 18.3: $C_T = K_H P_{CO_2}/\alpha_0$. If the pH values listed in Table 18.3 measured at the lake bottom do not change much over time (a crude assumption at best) and the equilibrium constants are not significantly affected by pressure, then the C_T values required to obtain P_{CO_2} = bottom hydrostatic pressure can be calculated for each lake. The values are reported in Table 18.4. The time to required to reach the critical C_T values is obtained by:

time to reach critical C_T (yr) =
(critical C_T in mol/L)(volume in bottom 10 m in L)/(CO_2 accumulation rate in mol/yr)

The new pH can be determined from eq. 18.5:

$$[H^+] \approx K_{a,1}K_H P_{CO_2}/\text{Alk}$$
$$= 10^{-11.3}/2.36\times10^{-4}$$

or pH 7.52. The new C_T is 1.77×10^{-4} M from $C_T = K_H P_{CO_2}/\alpha_0$.

In summary, if nitrate is the nitrogen source, then the new conditions will be:

pH 7.52, $C_T = 1.77\times10^{-4}$ M, and Alk = 1.68×10^{-4} eq/L

The pH, C_T, and Alk decrease because of the H^+ production from nitrogen assimilation.

Based on this very limited analysis, the estimated times to reach the critical C_T values are summarized in Table 18.4. Employing a much more sophisticated analysis, Kling and coworkers (1994) estimated that critical conditions would be reached in Lake Nyos in 30 ± 8 years from 1994 and in 7 ± 2 years from 1994 in Lake Monoun. It is hoped that controlled degassing of the lakes will minimize future threats.

Table 18.4: Calculated Values for Lakes Nyos and Monoun

Parameter	Lake Nyos	Lake Monoun
Volume in bottom 10 m (m³)	1.4×10^7	8×10^6
Volume in bottom 10 m (L)	1.4×10^{10}	8×10^9
Hydrostatic pressure at bottom (= P_z, atm)	20.1	9.3
C_T required for $P_{CO_2} = P_z$ (M)	0.66	0.36
Time required for $P_{CO_2} = P_z$ (yr)	≈170	≈46
Time required after 1994	≈34	≈160

SUMMARY OF KEY IDEAS

- Chemical species can equilibrate between gas and liquid phases

- A line of constant slope in a plot of partial pressure versus mole fraction suggests a constant equilibrium relationship between concentrations in the gas and liquid phases

- For nonideal species, the plot of partial pressure versus mole fraction is not linear, suggesting that the equilibrium relationship between concentrations in the gas and liquid phases varies with composition

- Nearly constant slopes (and thus nearly constant equilibrium relationships) are observed at very high and very low mole fractions in mixtures

- Henry's Law constants are just equilibrium constants and thus thermodynamic functions

- Always make sure you know the equilibrium corresponding to a given Henry's Law constant and the units of the Henry's Law constant

- Take care with dimensionless Henry's Law constants

- Under most environmental conditions, the partitioning species is uncharged

- For open systems containing a large gas volume, the partial pressure of the partitioning species is constant so the dissolved gas concentration must be fixed and a mass balance on the species of interest is not required

- For hydrolyzing dissolved gases, the dissolved gas is usually the uncharged member of an acid-base family

- For hydrolyzing dissolved gases in equilibrium with the atmosphere, the partial pressure of the gas phase species is constant and so the concentration of the dissolved gas species is fixed and independent of pH

- Dissolved gases that represent the most protonated form of the acid-base family are more soluble at higher pH, and dissolved gases that represent the least protonated form of the acid-base family are more soluble at lower pH

- Acid rain is rain with pH less than that in equilibrium with atmospheric CO_2 (i.e., less than about pH 5.6)

- If only $CO_2(g)$ is exchanged (and no solids precipitate), then the alkalinity remains constant

- For a water in equilibrium with the atmosphere where the pH is near neutral and the alkalinity is not zero:
  $[H^+] \approx K_{a,1} K_H P_{CO_2}/\text{Alk} = 10^{-11.3}/\text{Alk}$ or pH $\approx 11.3 + \log(\text{Alk in eq/L})$

PROBLEMS

18.1 Which is the most appropriate equilibrium model for the following situations: Raoult's Law, Henry's Law, or neither Raoult's nor Henry's Law?

A. You are concerned about the air quality after a barrel of nitrobenzene was spilled on a factory floor.

B. You want to model the gas phase methanol concentrations above a vat containing a 50:50 methanol:water mixture.

C. You are asked to assess workers' exposure to airborne chlorophenol in a manhole, where chlorophenol is present at trace levels in the wastewater.

18.2 Ozone is generated from a oxygen gas to produce a gas stream with a 2% ozone concentration, by volume. The ozone/oxygen gas stream is at 1 atm total pressure. It is bubbled through a solution at 25°C until ozone equilibrates between the gas and liquid streams. What will be the equilibrium concentration of ozone in solution, in mg/L? Assume the pH is low enough that ozone does not react appreciably with water.

18.3 The concentration of dissolved oxygen in equilibrium with the atmosphere is 11.28 mg/L at 10°C. Calculate the Henry's Law constant at 10°C, using all three sets of units presented in Table 18.2.

18.4 Another measure of gas solubility at equilibrium is the *Bunsen absorption coefficient*, defined to be the volume of a gas that is soluble in a volume of fluid (at 0°C and 1 atm pressure of the gas). Devise the conversion factor for converting Henry's Law constant in atm-L/mol (at 0°C and 1 atm pressure) to the Bunsen absorption coefficient.

18.5 The maximum contaminant level (MCL) for chloroform ($CHCl_3$) in drinking water is 100 µg/L. The occupational exposure level (OEL) for chloroform in air is 10 ppm by weight ($= 2.54 \times 10^{-6}$ atm chloroform). If treated water in an enclosed drinking water reservoir is at the MCL for chloroform, is the OEL for chloroform in air exceeded? Assume that the air volume of the enclosed reservoir is very large.

18.6 As discussed in Section 18.4.2, the pH of otherwise clean water in equilibrium with the atmosphere is about 5.6. This calculation assumes an atmospheric carbon dioxide concentration of 300 ppm $= 3 \times 10^{-4}$ atm $= 10^{-3.5}$ atm. As a result of the combustion of fossil fuels and other factors, the atmospheric CO_2 concentration is increasing. What is the pH of clean water in equilibrium with the atmosphere if the atmospheric CO_2 concentration increases to 350 ppm? to 600 ppm?

18.7 A groundwater containing 0.5 mg/L as S of total dissolved sulfide ($= [H_2S(aq)] + [HS^-] + [S^{-2}]$) at pH 7.5 is aerated to volatilize H_2S. If the H_2S concentration in the atmosphere in contact with the aerated grroundwater is 0.75 ppm by volume and the total system pressure is 1 atm, what will be the total sulfide concentration in the water after equilibrium has been reached? Assume the pH of the water remains constant. Is this a reasonable assumption?

18.8 Ammonia stripping refers to the volatilization of ammonia from water.

    A. Ammonia stripping is found to be most efficient at pH greater than 11. Why is ammonia stripping most efficient at high pH?

    B. If the effluent standard is 1.5 mg/L as N for the total ammonia concentration in solution (i.e., $[NH_3]$ + $[NH_4^+]$), what is the partial pressure of ammonia in the stripping gas in equilibrium with the waste-water?

18.9 Sketch a pC-pH diagram for the concentration of $NH_3(aq)$ in equilibrium with an atmosphere containing $10^{-2.5}$ atm of ammonia.

18.10 The following questions concern the relationships between Alk, C_T, P_{CO_2}, and pH:

A. A groundwater with an alkalinity of 200 mg/L as $CaCO_3$ is in equilibrium with 0.02 atm of CO_2. Find the pH and C_T of the water.

B. If the groundwater in Part A is contaminated by 1×10^{-3} eq/L of mineral acidity (i.e., C_A), what is the new pH of the groundwater? Does C_T increase, decrease, or stay the same?

C. If the groundwater in Part A is aerated by equilibration with the atmosphere, what is the resulting pH and C_T?

18.11 Professor Whoops carefully collected a water sample from the field in a closed sample container. The sample was brought back to the lab and its pH measured at 7.0. Professor Whoops then stirred 100 mL of the sample rapidly and began to measure its alkalinity by titrating with 0.1 N H_2SO_4. After adding one drop (0.05 mL) of the acid, to the good professor's astonishment, the pH *increased*. Later, Professor Whoops determined the alkalinity of the sample to be 10 mg/L as $CaCO_3$. (Assume no solids were present in the sample.)

A. What was the pH of the sample pH after stirring and adding 0.05 mL of the acid?

B. Why did the pH of the sample increase after Professor Whoops added the acid? (Hint: calculate $C_T = [H_2CO_3] + [HCO_3^-] + [CO_3^{2-}]$ before and after acid addition and stirring.)

C. Professor Whoops took another portion of the original sample, stirred it rapidly, and measured its pH *without adding any acid*. What was the final pH?

18.12 From Example 18.6, you learned that photosynthesis may change the C_T, pH, and Alk if the Alk is altered by production or consumption of H^+.

A. Calculate the new C_T, pH, and Alk for the lake water in Example 18.6 if ammonium (rather than nitrate) is the nitrogen source during photosynthesis. The pertinent reaction is: $106CO_2 + 16NH_4^+ + HPO_4^{2-} + 108H_2O \rightarrow C_{106}H_{263}O_{110}N_{16}P + 107O_2 + 14H^+$. Assume that 5×10^{-4} M of dissolved carbon dioxide is consumed.

B. Calculate the new C_T, pH, and Alk for the lake water in Example 18.6 if ammonium is *produced* and 5×10^{-4} M of dissolved carbon dioxide are *produced* during respiration. The pertinent reaction for respiration with ammonium production is the reverse of the reaction in Part A.

18.13 Human blood typically has a pH of 7.4, total dissolved carbonate (C_T) $= 2.5\times10^{-2}$ M, and is at 37°C. With what partial pressure of carbon dioxide is blood in equilibrium? (At 37°C: $pK_1 = 6.2$, $pK_2 = 10.2$, and $K_H = 10^{-1.6}$ mol/L-atm).

18.14 Equation 18.6 appears to be a very general result for natural waters. Discuss the conditions under which eq. 18.6 is valid.

CHAPTER 19

Solid-Liquid Equilibria

19.1 INTRODUCTION

In this chapter, the equilibration of chemical species between solid and liquid phases will be explored. Solid-liquid equilibria, also called *precipitation-dissolution equilibria*, represent the final major type of chemical equilibrium to be discussed in this text. By the conclusion of this chapter, you will be able to calculate the concentrations of aqueous species in equilibrium with pure solid phases and solid surfaces. Of particular importance is the solid calcium carbonate.

In Section 19.2, you will learn about the thermodynamics of solid-liquid equilibria. In particular, you will see that a common thermodynamic convention greatly simplifies equilibrium calculations with systems containing a pure solid phase. Equilibrium calculation techniques with solid-liquid equilibria will be developed in Section 19.3.

In Section 19.4, two factors affecting the solubility of metals will be explored. First, equilibrium partitioning in the presence of more than one solid phase will be discussed. Second, the effects of additional ligands on metal solubility will be explored. Section 19.5 discusses the solubility of calcium carbonate. Equilibration with calcium carbonate influences the dissolved total carbonate and calcium concentrations in water. Thus, inclusion of equilibrium with calcium carbonate will allow you to develop and test more realistic models for natural waters and process waters in engineered systems. Finally, Section 19.6 discusses the partitioning of dissolved species between water and surfaces.

19.2 SATURATION AND THE ACTIVITY OF PURE SOLIDS

19.2.1 Introduction

precipitation: the formation of insoluble complexes from dissolved species

dissolution: the formation of dissolved species from insoluble complexes

Certain metals and ligands can form complexes that are insoluble in water under a given set of temperature and pressure conditions. Insoluble complexes become solid phases, and *precipitation*[†] occurs. As an example, sodium ions and chloride ions can form the solid sodium chloride: $Na^+(aq) + Cl^-(aq) \rightarrow NaCl(s)$. The reverse reaction (i.e., the formation of soluble metal and ligand species from solids) is called *dissolution* (from the Latin *dis* + *solvere* to loosen or break up): $NaCl(s) \rightarrow Na^+(aq) + Cl^-(aq)$.

[†] The word "precipitation" comes from the Latin *pre* before + *caput* head; literally, "before the head" or the act of throwing down headlong. By analogy, a chemical is deposited downward during precipitation.

Can you consider precipitation and dissolution reactions to be in equilibrium? In other words, is it acceptable to write the reaction NaCl(s) = Na⁺ + Cl⁻ with an equilibrium constant K? [Here, as usual, species are assumed to be dissolved in water until otherwise indicated and the (aq) label is omitted.] You can write the opposing reactions as an equilibrium if they occur fast enough. For example, you will learn in Section 22.2 that the proton transfer reactions occurring in acid-base chemistry are fast. Thus, equilibrium models are appropriate for describing acid-base chemistry under environmentally important conditions in most cases. *Precipitation reactions may be fast or slow and cannot always be assumed to be at equilibrium.* For the moment, an equilibrium model will be used to describe precipitation and dissolution reactions. Nonequilibrium reactions in general will be examined in Chapter 22.

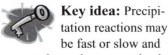

Key idea: Precipitation reactions may be fast or slow and cannot always be assumed to be at equilibrium

solubility product (symbol: K_{s0}): equilibrium constant for an equilibrium written with the solid phase as a reactant and the metals and ligands as products in their uncomplexed forms

It is common to write the precipitation/dissolution equilibria with the solid phase as a reactant and the metals and ligands as products in their *uncomplexed* forms. For equilibria of this form, the equilibrium constant is given a special name. It is called the *solubility product* and given the symbol K_{s0} (pronounced "kay ess zero"). The "s" in K_{s0} indicates that a solid phase is involved and the zero (**not** the letter O) indicates that the metal and ligand are uncomplexed. Remember that K_{s0} *is just another equilibrium constant.* Thus, you can write, for example:

$$NaCl(s) = Na^+ + Cl^- \qquad K = K_{s0}, \text{ or:}$$
$$CaCO_3(s) = Ca^{2+} + CO_3^{2-} \qquad K = K_{s0}$$

Note that the metal and ligand must be uncomplexed to use the K_{s0} label. The equilibrium constant for the equilibrium $CaCO_3(s) + H^+ = Ca^{2+} + HCO_3^-$ should not be called K_{s0}. Other K_{s0} values are listed in Appendix D. In general:

$$M_xL_y(s) = xM^{y+} + yL^{x-} \qquad K = K_{s0} \qquad \text{eq. 19.1}$$

In the last example:

$$K_{s0} = \{M^{y+}\}^x\{L^{x-}\}^y/\{M_xL_y(s)\} \qquad \text{eq. 19.2}$$

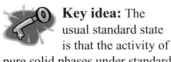

Key idea: The usual standard state is that the activity of pure solid phases under standard thermodynamic conditions is unity

19.2.2 Activity of pure solid phases

Recall from Section 3.6.3 that we usually pick standard states so that *the activity of pure solid phases under standard thermodynamic conditions is unity.* Thus, for the equilibrium in eq. 19.1, the expression for K_{s0} from eq. 19.2 becomes:

$$K_{s0} = \{M^{y+}\}^x\{L^{x-}\}^y \qquad \text{eq. 19.3}$$

The choice of setting the activity of *pure solids* equal to unity for pure solids has profound implications on equilibrium calculations. In fact, as you will see shortly, the choice makes equilibrium calculations much easier.

Thoughtful Pause

Why does the choice of setting the activity of pure solids equal to unity for pure solids make equilibrium calculations easier?

Key idea: The activity of pure solids is usually either zero (if it does not exist) or one (if it does exist)

Under standard thermodynamic conditions, *pure* solids can take only two values of activity: zero or unity. If the pure solid does not exist, then its activity is zero and its precipitation equilibrium (eq. 19.1) is meaningless. If the pure solid *does* exist, then its activity is unity and its solubility product takes the form indicated in eq. 19.3. In a sense, the activity of pure solids is binary: it takes on values of zero or one.

What if the solid is not chemically pure? In many cases in the environment, the solid phase produced may not be of uniform chemical composition. For example, in the precipitation of aluminum as aluminum hydroxide during coagulation in drinking water treatment, the solid phase may not be the pure solid gibbsite[†], α-Al(OH)$_3$(s). In this case, activity of the solid phase may not be exactly unity. In general, solubility constants are defined operationally for different solids. In other words, sometimes the activity of the solid is assumed to be unity in equilibrium calculations, and the value of the equilibrium constant reflects the thermodynamic activity of the solid.

19.2.3 Oversaturation and undersaturation

saturation: the condition in which the product of the free metal and ligand activities are *equal* to the solubility product (K_{s0} value)

How can you tell if the solid phase exists at equilibrium? If the solid phase exists, then the product of the free metal and ligand activities should equal the solubility product. This condition is called *saturation*. If the product of the free metal and ligand activities is less than the solubility product, then the solid phase should not form at equilibrium and the solution is said to be *undersaturated*. The solid phase will not exist at equilibrium in undersaturated solutions.

undersaturation: the condition in which the product of the free metal and ligand activities are *less* than the solubility product (K_{s0} value)

What about the case where the product of the free metal and ligand activities is greater than the solubility product? This type of solution is said to be *oversaturated*.

oversaturation: the condition in which the product of the free metal and ligand activities are *greater* than the solubility product (K_{s0} value)

[†] Lest the reader think that *all* important parameters in aquatic chemistry are named after J. Willard Gibbs, rest assured that gibbsite was named after the American mineralogist and collector Col. George Gibbs (1776-1833) in 1822 (17 years before the thermodynamicist Gibbs was born).

Thoughtful Pause
Are oversaturated solutions at equilibrium?

Oversaturated solutions are not at equilibrium since the product of the free metal and ligand activities should equal the solubility product if the solid phase exists at equilibrium. However, because the formation of some solid phases can be slow, oversaturated solutions can exist as transient systems approaching equilibrium.

19.3 EQUILIBRIUM CALCULATIONS INVOLVING SOLID-LIQUID EQUILIBRIA

19.3.1 Introduction

Example 19.1: **Solubility Calculations with Nonhydrolyzing Metals and Ligands**

A metal finishing process uses a silver plating bath. Because of the value of silver, it is desirable to precipitate silver out of the plant wastewater for metals recovery. How many grams of NaCl must be added per liter of wastewater to ensure that the soluble silver concentration is less than 1 µg/L at low pH?

Solution:
At low pH, you can assume that the only important soluble silver species is Ag^+. You want the total soluble silver concentration (= $[Ag^+]$) to be less than 1 µg/L = $(1 \times 10^{-6}$ g/L)/(107.9 g/mol) = 9.3×10^{-9} M. From the solubility product: $[Ag^+][Cl^-] = K_{s0} = 1.8 \times 10^{-10}$ M^2, so:

$[Cl^-] = 1.8 \times 10^{-10}$ $M^2/9.3 \times 10^{-9}$
M = 1.9×10^{-2} M

Therefore, 1.9×10^{-2} M = $(1.9 \times 10^{-2}$ M)(58.5 g/mol) = **1.1 g/L of NaCl(s) is required.**

You probably have some experience in performing equilibrium calculations with solids. In freshman chemistry courses, it is quite common to see examples of equilibrium calculations with solids that dissolve to form metals and ligands that themselves do not react appreciably with water. A favorite example is silver chloride, AgCl(s). The solubility product of silver chloride is about 1.8×10^{-10}. In other words:

$$AgCl(s) = Ag^+ + Cl^- \qquad K_{s0} = 1.8 \times 10^{-10}$$

A typical problem is as follows: find the concentrations of Ag^+ and Cl^- in a solution containing a lump of AgCl(s). *If* the solid phase exists, then $[Ag^+][Cl^-] = 1.8 \times 10^{-10}$ (if concentrations and activities are interchangeable). The charge balance is $[H^+] + [Ag^+] = [OH^-] + [Cl^-]$. *If* that $[H^+]$ and $[OH^-]$ can be ignored, then the charge balance is $[Ag^+] = [Cl^-]$. Thus: $[Ag^+]^2 = 1.8 \times 10^{-10}$ M^2 and $[Ag^+] = [Cl^-] = 1.3 \times 10^{-5}$ M. Another illustration is shown in Example 19.1.

Such simple examples do little to aid in your understanding of solid-liquid equilibria. For more complicated systems, you already know the drill in setting up chemical equilibria: make a species list and then list the equilibria, mass balances, charge balance, and species of fixed activities. Consider two chemical systems containing zinc that are identical, except that one system contains a solid phase and the other system does not. In the first hypothetical system, all the zinc is soluble. For example, consider a solution where 10^{-6} mole of zinc are added to 1 L of water. In the other system, many moles of zinc are added to 1 L of water and $Zn(OH)_2$(s) exists. How will the usual set of equations for the two systems differ?

The chemical systems will be written in their full glory in Section 19.3.2. For now, just consider the numbers of equations and unknowns. Assume for the moment that the system without $Zn(OH)_2$(s) contains n species. You should be confident that you can find n equations to solve the system. The system with $Zn(OH)_2$(s) will contain $n+1$ species: all the

Note: AgCl(s) is a very insoluble solid. Also, at this high NaCl dose, activities and concentrations are not interchangeable and the calculation should be refined to take into account the effects of other ions (see Section 21.2).

soluble species, plus $Zn(OH)_2(s)$. Thus, the system with $Zn(OH)_2(s)$ must have one more equation. This creates a problem: the system with $Zn(OH)_2(s)$ actually has *two* more equations.

Thoughtful Pause
What are the two additional equations in the $Zn(OH)_2(s)$ system?

The system with the solid phase will have the solubility product expression (in this case: $\{Zn^{2+}\}\{OH^-\}^2/\{Zn(OH)_2(s)\} = K_{s0}$) and it will have the expression $\{Zn(OH)_2(s)\} = 1$. Comparing the systems, you can see that the system with the solid has $n+1$ unknowns but $n+2$ equations. To look at it another way, you have introduced a new species (i.e., the solid phase) and immediately set its activity to unity (as we did earlier with water), but you also added a new constraint: the solubility constraint (here, that $\{Zn^{2+}\}\{OH^-\}^2 = K_{s0}$). How can this situation be resolved?

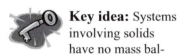 **Key idea:** Systems involving solids have no mass balances on the dissolved metals or ligands that precipitate in the solid phase(s)

The answer is that an equation must be eliminated. What is new about the system containing the solid phase? If the solid phase forms, then some of the metal (here, some of the zinc) and some of the ligand is removed from solution. Thus, the mass balance on *soluble* species is no longer valid. For the soluble-species-only system, you can certainly write a mass balance on soluble zinc species:

$$Zn(+II)_T = 1 \times 10^{-6} \text{ M} = [Zn^{2+}] + [ZnOH^+] + ...$$

In the system containing $Zn(OH)_2(s)$, this statement is no longer true: $[Zn^{2+}] + [ZnOH^+] + ...$ must be *less* than the added $Zn(+II)_T$, since some zinc precipitates. Thus, you lose the mass balance on soluble zinc and the system with the solid becomes at least possible to solve, with $n+1$ unknowns and $n+1$ equations.

19.3.2 Example: Soluble zinc without $Zn(OH)_2(s)$
Having whet your appetite in the previous section, we will now solve the two zinc systems (with and without the solid zinc hydroxide). Consider first the system without solid-phase zinc (you may wish to try this on your own before reading further):

Species list:
 H_2O, H^+, OH^-, Zn^{2+}, $ZnOH^+$, $Zn(OH)_2^0$, $Zn(OH)_3^-$, and $Zn(OH)_4^{2-}$
 (For simplicity, ignore here the dimers Zn_2OH^{3+} and $Zn_2(OH)_6^{2-}$ since they are found in low concentrations relative to the other species under the conditions of this problem.)

Equilibria:

$$H_2O = H^+ + OH^- \qquad K_W = 10^{-14}$$
$$Zn^{2+} + OH^- = ZnOH^+ \qquad K_1 = 10^{+5.0}$$
$$Zn^{2+} + 2OH^- = Zn(OH)_2^0 \quad K_2 = 10^{+11.1}$$
$$Zn^{2+} + 3OH^- = Zn(OH)_3^- \quad K_3 = 10^{+13.6}$$
$$Zn^{2+} + 4OH^- = Zn(OH)_4^{2-} \quad K_4 = 10^{+14.8}$$

Or:

$$K_W = [H^+][OH^-]/[H_2O]$$
$$K_1 = [ZnOH^+]/([Zn^{2+}][OH^-])$$
$$K_2 = [Zn(OH)_2^0]/([Zn^{2+}][OH^-]^2)$$
$$K_3 = [Zn(OH)_3^-]/([Zn^{2+}][OH^-]^3)$$
$$K_4 = [Zn(OH)_4^{2-}]/([Zn^{2+}][OH^-]^4)$$

Mass balance:

$$Zn(+II)_T = [Zn^{2+}] + [ZnOH^+] + [Zn(OH)_2^0] + [Zn(OH)_3^-] + [Zn(OH)_4^{2-}]$$
$$= 1 \times 10^{-6} \text{ M}$$

Charge balance:

If you knew the source of zinc [say, $ZnCl_2(s)$] which you could assume dissolved completely, then you could write a charge balance: $[H^+] + 2[Zn^{2+}] + [ZnOH^+] = [OH^-] + [Zn(OH)_3^-] + 2[Zn(OH)_4^{2-}] + [Cl^-]$. From a mass balance on chloride: $[Cl^-] = 2Zn(+II)_T = 2 \times 10^{-6}$ M. In this problem, we will only construct a pC-pH diagram and the charge balance is not necessary.

Other:

Dilute solution so that concentrations and activities are interchangeable

Activity of water is unity

Assume no solids are present

This type of system (complexation with no solids) is very familiar to you from Chapter 15.4. Remember from Section 15.4.3 that it is easier to solve the system if you express the concentration of each species in the mass balance in terms of the concentration of *one* species (for example, Zn^{2+}). Thus:

$$Zn(+II)_T = [Zn^{2+}](1 + K_1[OH^-] + K_2[OH^-]^2 + K_3[OH^-]^3 + K_4[OH^-]^4)$$
$$= 1 \times 10^{-6} \text{ M}$$

Therefore, you can easily solve for $[Zn^{2+}]$ and then all the other zinc-containing species. The resulting pC-pH diagram is shown in Figure 19.1.

The pC-pH diagram offers no surprises. Each species plots as a curved line on the pC-pH diagram. You can see the transition from Zn^{2+} to $ZnOH^+$ to $Zn(OH)_2^0$ to $Zn(OH)_3^-$ to $Zn(OH)_4^{2-}$ as the pH increases. The total soluble zinc is 10^{-6} M, of course, since the mass balance on total soluble zinc was used in the equilibrium calculations.

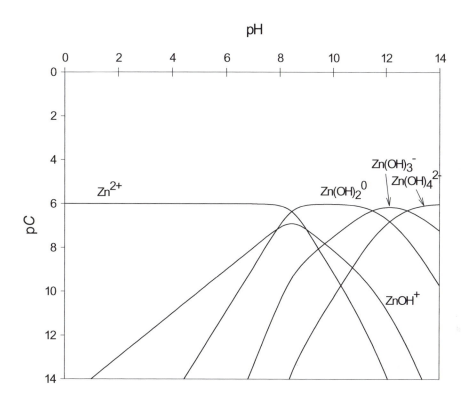

Figure 19.1: Zinc Hydrolysis in the Absence of $Zn(OH)_2(s)$
$[Zn(+II)_T = 10^{-6}$ M; H^+ and OH^- omitted for clarity)]

In solving the system, you have *assumed* that $Zn(OH)_2(s)$ does not precipitate.

Thoughtful Pause

How can you *prove* that $Zn(OH)_2(s)$ does not precipitate at $Zn(+II)_T = 1 \times 10^{-6}$ M?

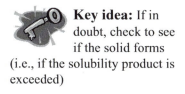

Key idea: If in doubt, check to see if the solid forms (i.e., if the solubility product is exceeded)

The solid phase will not precipitate at equilibrium if the solution remains undersaturated, that is, if $[Zn^{2+}][OH^-]^2 < K_{s0}$. To test whether the solubility product is exceeded, the value of $[Zn^{2+}][OH^-]^2$ is plotted as a function of pH in Figure 19.2. Note that $[Zn^{2+}][OH^-]^2$ is less than $K_{s0} = 10^{-15.55}$ at all pH values. Thus, the assumption is fine: $Zn(OH)_2(s)$ does not precipitate at $Zn(+II)_T = 1 \times 10^{-6}$ M.

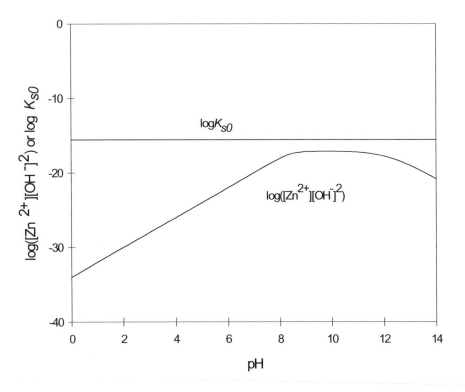

Figure 19.2: Undersaturation of the Zn System at All pH Values
$[Zn(+II)_T = 1 \times 10^{-6}$ M]

19.3.3 Example: Precipitation of zinc as $Zn(OH)_2(s)$

How does the system change if the total added zinc is increased significantly and solid zinc hydroxide forms? The new system, *assuming the solid exists*, is as follows:

Species list:
As before [H_2O, H^+, OH^-, Zn^{2+}, $ZnOH^+$, $Zn(OH)_2^0$, $Zn(OH)_3^-$, and $Zn(OH)_4^{2-}$] but now including $Zn(OH)_2(s)$.

Equilibria:
As before, but now including:

$$Zn(OH)_2(s) \; = \; Zn^{2+} + 2OH^- \qquad\qquad K_{s0} = 10^{-15.55}$$

Or:

The equations given earlier plus:
$$K_{s0} = \{Zn^{2+}\}\{OH^-\}^2/\{Zn(OH)_2(s)\}$$
With the other information listed below, this becomes:
$$K_{s0} = [Zn^{2+}][OH^-]^2$$

Mass balance:
> Now, there is no mass balance on zinc since some zinc precipitates.

Charge balance:
> Again, you could write a charge balance if you knew the source of zinc. No charge balance is needed for a pC-pH diagram.

Other:
> Dilute solution so that concentrations and activities are interchangeable
> Activities of water and $Zn(OH)_2(s)$ are unity
> $Zn(OH)_2(s)$ exists

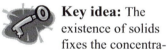

solubility constraint: the mathematical form of the solubility product, where the concentration of the metal ion is expressed in terms of K_{s0} and the concentration of ligand species

How can you solve this system? The mathematical version of the solubility product (namely, $K_{s0} = [Zn^{2+}][OH^-]^2$) puts a constraint on the possible values of $[Zn^{2+}]$. You know $[Zn^{2+}] = K_{s0}/[OH^-]^2 = K_{s0}[H^+]^2/K_W^2$. Equations of this form sometimes are called *solubility constraints*. Just as Henry's Law constrains (or fixes) the concentration of the aqueous species in equilibrium with the gas phase, so also *solids constrain the concentration of the metal in equilibrium with the solid phase* (and sometimes the ligand as well). How can you calculate the concentrations of the other zinc-containing species? The relationships between $[Zn^{2+}]$ and the other zinc-containing species are the same as in the soluble case. For example, the equilibrium $Zn^{2+} + OH^- = ZnOH^+$ still holds, so the following is still true: $[ZnOH^+] = K_1[Zn^{2+}][OH^-]$. Thus, once $[Zn^{2+}]$ is calculated at a given pH by the solubility constraint, the concentrations of all the other zinc-containing species can be determined easily. You may wish to try this on your own for the $Zn(OH)_2(s)$ system. Equilibrium calculations of this type lend themselves nicely to solution by use of a spreadsheet.

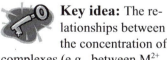

Key idea: The existence of solids fixes the concentration of certain metal species through the solubility constraint (at fixed ligand concentration)

Key idea: The relationships between the concentration of complexes (e.g., between M^{2+} and MOH^+) is the same whether or not solids exist

The pC-pH diagram for the $Zn(OH)_2(s)$ system is given in Figure 19.3. The pC-pH diagram for the $Zn(OH)_2(s)$ system looks very different from the pC-pH diagram for soluble zinc (Figure 19.1). The species concentration lines plot at straight lines. Why? Recall that:

$$[Zn^{2+}] = K_{s0}/[OH^-]^2 = K_{s0}[H^+]^2/K_W^2$$
$$\text{or: } pZn^{2+} = p(K_{s0}/K_W^2) + 2pH$$

Thus, the $[Zn^{2+}]$ line on the pC-pH diagram should have a slope of $+2$ and a y-intercept of $p(K_{s0}/K_W^2) = 15.55 - 2(14) = -12.45$. The other zinc hydroxo complexes also will plot as straight lines. For example, recall that:

$$[ZnOH^+] = K_1[Zn^{2+}][OH^-] = K_1K_W[Zn^{2+}]/[H^+] = K_1K_{s0}[H^+]/K_W$$

Thus: $pZnOH^+ = p(K_1K_{s0}/K_W) + pH$, and the $[ZnOH^+]$ line on the pC-pH diagram should have a slope of $+1$ and a y-intercept of $p(K_1K_{s0}/K_W) = -5 + 15.55 - 14 = -3.45$. Similarly, you may wish to confirm that the $Zn(OH)_2^0$ line has a slope of 0 and an intercept of $p(K_2K_{s0}) = 4.45$, the $Zn(OH)_3^-$ line has a slope of -1 and an intercept of $p(K_3K_{s0}K_W) = 15.95$, and the $Zn(OH)_4^{2-}$ line has a slope of -2 and an intercept of $p(K_4K_{s0}K_W^2) = 28.75$.

 Key idea: Each point on the total soluble metal line represents the total soluble metal concentration in equilibrium with the solid at a given pH (and other total ligand concentrations)

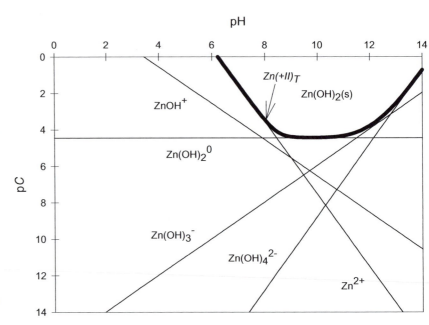

Figure 19.3: Zinc Speciation in Equilibrium with Zn(OH)$_2$(s)

Example 19.2: pH of Minimum Metal Solubility

Personnel at a hazardous waste treatment, storage, and disposal facility (TSDF) wish to remove cobalt from a liquid hazardous waste by precipitation as $Co(OH)_2(s)$. In what pH range should the precipitation process operate to produce a supernatant with a total soluble cobalt concentration less than 0.01 mg/L?

Solution:
The pertinent equilibria are:

$H_2O = H^+ + OH^-$ $K_W = 10^{-14}$
$Co^{2+} + OH^- = CoOH^+$
    $K_1 = 10^{+4.3}$
$Co^{2+} + 2OH^- = Co(OH)_2^0$
    $K_2 = 10^{+9.2}$
$Co^{2+} + 3OH^- = Co(OH)_3^-$
    $K_3 = 10^{+10.5}$
$Co(OH)_2(s) = Co^{2+} + 2OH^-$
    $K_{s0} = 10^{-15.7}$

Another interesting observation is that *the total soluble zinc concentration changes with pH*. The total soluble zinc concentration can be calculated by adding up the concentrations of the individual zinc-containing species (again, a spreadsheet is useful). The line for $pZn(+II)_T$ is shown as the thick black line in Figure 19.3. Make sure you realize the importance of the total soluble zinc line. *Each point on the line represents the total soluble zinc concentration in equilibrium with Zn(OH)$_2$(s) at a given pH.* For example, at pH 7, if the solid exists then the total soluble zinc concentration will be about 10^{-2} M. Zinc hydroxide is very soluble at pH

The total soluble cobalt concentration, $Co(+II)_T$, is given by:

$$Co(+II)_T = [Co^{2+}] + [CoOH^+]$$
$$+ [Co(OH)_2^0] +$$
$$[Co(OH)_3^-]$$

$$= [Co^{2+}](1 + K_1[OH^-]$$
$$+ K_2[OH^-]^2 +$$
$$K_3[OH^-]^3)$$

If cobalt precipitates, then the $Co(OH)_2$(s) solid phase exists and $[Co^{2+}] = K_{s0}/[OH^-]^2$. Thus:

$$Co(+II)_T = (K_{s0}/[OH^-]^2)(1 +$$
$$K_1[OH^-] + K_2[OH^-]^2$$
$$+ K_3[OH^-]^3)$$

Using a spreadsheet (or setting $dCo(+II)_T/d[OH^-] = 0$), you can find that the minimum total soluble cobalt concentration is about 3.3×10^{-7} M (0.019 mg/L) at about pH 10.9. **Thus, precipitation of cobalt as Co(OH)$_2$(s) cannot be used to reduce the residual soluble cobalt concentration to less than 0.01 mg/L.**

 Key idea: Many metals are soluble at both low and high pH, with minimum solubility at an intermediate pH value

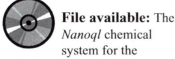 **File available:** The *Nanoql* chemical system for the Zn(+II) system in the presence of Zn(OH)$_2$(s) may be found on the CD, file location **/Chapter 19/zinc.nqs**

7, dissolving to form a lot of soluble zinc (mostly in the form of Zn^{2+}, from Figure 19.3). Similarly, the solid is pretty soluble at pH 13, dissolving to form again about 10^{-2} M of soluble zinc [mostly in the form of $Zn(OH)_4^{2-}$ at pH 13].

Thoughtful Pause

At what pH is the solubility of Zn(OH)$_2$(s) minimized?

The solubility of Zn(OH)$_2$(s) is minimized at about pH 10 (more accurately, pH 9.76). The minimum total soluble zinc is about 3.7×10^{-5} M. If you wanted to remove zinc from solution by precipitation at Zn(OH)$_2$(s), you could raise the pH to 9.76. The remaining (or residual) total soluble zinc would be about 3.7×10^{-5} M (or about 2.4 mg/L as Zn).

Note also in Figure 19.3 that the $Zn(+II)_T$ line divides the pC-pH diagram into two regions. At $pZn(+II)_T$ and pH conditions on or inside the line (at the Zn(OH)2(s) label), the solid will form. At $pZn(+II)_T$ and pH conditions outside the line, the solid will not form and zinc will remain soluble.

Many other metals follow the pattern of zinc. They are soluble at low pH (where species with *fewer* OH ligands than in the solid dominate) and soluble at high pH (where species with *more* OH ligands than in the solid dominate). Their minimum solubility lies at an intermediate pH value. The example of cobalt solubility is given in Example 19.2.

Systems where the solid phase is known to exist are solved easily by equilibrium calculation software. For example, with *Nanoql*, the solid phase activity is fixed at 1 in the calculation. The *Nanoql* file for zinc in the presence of Zn(OH)$_2$(s) may be found at **/Chapter 19/zinc.nqs**.

19.3.4 Example: Speciation of zinc where the solid phase may or may not exist

In the example in Section 19.3.2, you determined the zinc speciation at very low total zinc in the absence of a solid phase. In the example in Section 19.3.3, you found the zinc speciation and total soluble zinc concentration at very high total zinc where the presence of Zn(OH)$_2$(s) was assumed. What if you are interested in an intermediate zinc concentration in which you do not know whether or not the solid phase exists?

As an example, explore the speciation of zinc when 10^{-3} mole of zinc is added to 1 liter of water. An easy approach to solving for the zinc speciation is to start with Figure 9.3. Note that the total soluble zinc concentration exceeds 10^{-3} M at pH values less than about 7.7 and greater than about 12.7. Since it is impossible for the total soluble zinc concentration to exceed the added zinc concentration, the solid phase must not exist at pH values less than about 7.7 and greater than about 12.7. In those pH

Key idea: If the solid phase is assumed to exist under certain conditions, but the calculated total soluble metal concentration exceeds the available metal concentration, then the solid phase does not exist under those conditions

regions, you must rework the problem as a homogeneous system with no solid phase. The resulting pC-pH diagram is shown in Figure 9.4.

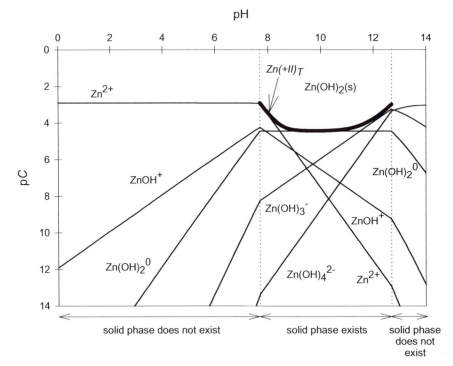

Figure 9.4: Zinc Speciation with 10^{-3} M Total Zinc
[Zn(OH)$_2$(s) exists at 7.74 < pH < 12.69]

19.4 FACTORS AFFECTING METAL SOLUBILITY

19.4.1 Introduction

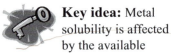

Key idea: Metal solubility is affected by the available metal concentration, pH, types of solid phases potentially formed, and ligand concentrations

The only factors affecting metal solubility discussed so far in this chapter are the total available metal concentration and pH. You know pH plays an important role with hydroxide solids such as Zn(OH)$_2$(s). Other kinds of solids can be important in natural and engineered systems, including solids formed with sulfide, phosphate, and carbonate. Thus, it is possible that more than one solid phase can form. In Section 19.4.2, the effects of multiple solid phases on metal solubility will be explored.

In Section 19.3.3, it was pointed out that some hydroxide solids dissolve at high pH because of the large [OH$^-$]. This concept can be expanded. In fact, the presence of any additional ligand may affect metal solubility. In Section 19.4.3, the effects of multiple ligands on metal solubility will be explored.

19.4.2 Multiple solid phases

Is it possible for two or more solid phases to exist in equilibrium? How do you decide which solid phases form and which solid phase controls the solubility of a metal? To answer these questions, consider the example of ferrous iron and its two common solid forms, $Fe(OH)_2(s)$ and $FeCO_3(s)$. Suppose that, under a given set of conditions (i.e., pH and total carbonate concentration), $Fe(OH)_2(s)$ alone produces a total Fe(+II) concentration, $Fe(+II)_T$, of 4.8×10^{-5} M, and $FeCO_3(s)$ alone produces a $Fe(+II)_T$ of 8.0×10^{-2} M. Under these conditions, $Fe(OH)_2(s)$ is *much less* soluble than $FeCO_3(s)$. Why? $Fe(OH)_2(s)$ produces much less $Fe(+II)_T$. Thus, $Fe(OH)_2(s)$ will form and $FeCO_3(s)$ will not form under these conditions. In general, *the controlling solid phase is the solid phase producing the smallest equilibrium total metal concentration.* In other words, the least soluble solid phase controls solubility.

Now we can decide whether it is possible for two or more solid phases to exist in equilibrium. *Two or more solid phases exist in equilibrium under the same set of conditions if they produce the same equilibrium total metal concentration.* Why? If one solid produces a smaller equilibrium total metal concentration than the other solid, the first solid will exist and the second solid will not. Only if they produce the same equilibrium total metal concentration will more than one solid phase exist under one set of conditions.

To determine which solid phase controls the solubility of a metal, construct lines of total soluble metal, assuming that only one solid phase exists at a time. For ferrous iron, the pertinent equilibria are:

Key idea: The controlling solid phase is the solid phase producing the *smallest* equilibrium total metal concentration: the *least* soluble solid phase controls solubility

Key idea: Two or more solid phases exist in equilibrium under the same set of conditions if they produce the *same* equilibrium total metal concentration

***Example 19.3:* Multiple Solid Phases**

For the Fe(+II) example in the text, find the pH where the transition between the controlling solids occurs.

Solution:
A simple way to find the transition point is to construct an equilibrium between the two solid phases. Reversing eq. 19.5 and adding it to eq. 19.4:

$Fe(OH)_2(s) + CO_3^{2-} = FeCO_3(s) + 2OH^-$
$K = K_{s0,1}/K_{s0,2} = 10^{-4.4}$

Since the activities of the pure solid phases are 1:

$K = [OH^-]^2/[CO_3^{2-}]$
$= [OH^-]^2/(\alpha_2 C_T)$

$$\begin{array}{ll}
H_2O \ = \ H^+ + OH^- & K_W = 10^{-14} \\
Fe^{2+} + OH^- \ = \ FeOH^+ & K_1 = 10^{+4.5} \\
Fe^{2+} + 2OH^- \ = \ Fe(OH)_2^{\,0} & K_2 = 10^{+7.4} \\
Fe^{2+} + 3OH^- \ = \ Fe(OH)_3^{\,-} & K_3 = 10^{+11}
\end{array}$$

In this case:

$$Fe(+II)_T = [Fe^{2+}] + [FeOH^+] + [Fe(OH)_2^{\,0}] + [Fe(OH)_3^{\,-}]$$
$$= [Fe^{2+}](1 + K_1[OH^-] + K_2[OH^-]^2 + K_3[OH^-]^3)$$

If you pretend for the moment that only $Fe(OH)_2(s)$ exists, then include the equilibrium:

$$Fe(OH)_2(s) \ = \ Fe^{2+} + 2OH^- \qquad K_{s0,1} = 10^{-15.1} \qquad \text{eq. 19.4}$$

Thus: $[Fe^{2+}] = K_{s0,1}/[OH^-]^2$, and:

$$Fe(+II)_T = (K_{s0,1}/[OH^-]^2)(1 + K_1[OH^-] + K_2[OH^-]^2 + K_3[OH^-]^3)$$

The resulting $Fe(+II)_T$ is shown in Figure 19.5A. If you pretend for the moment that only $FeCO_3(s)$ exists, then use the K_W, K_1, K_2, and K_3 equilibria plus the equilibrium:

Thus, the transition pH is where:

$K = K_{s0,1}/K_{s0,2}$
$= 10^{-4.4}$
$= [OH^-]^2/\alpha_2 C_T$
$= [H^+]^2/\alpha_2 C_T K_w^2$

At high pH, $\alpha_2 \approx K_2'/([H^+] + K_2')$, where: $K_2' = 10^{-10.3}$. Thus:

$K = K_w^2([H^+] + K_2')/([H^+]^2 K_2' C_T)$

or: $[H^+]^2 K_2' C_T K - K_w^2 [H^+] - K_w^2 K_2' = 0$

Solving with the quadratic equation at $C_T = 10^{-4}$ M: $[H^+] = 5.47 \times 10^{-10}$ M or **pH 9.26**.

$$FeCO_3(s) = Fe^{2+} + CO_3^{2-} \qquad K_{s0,2} = 10^{-10.7} \qquad \text{eq. 19.5}$$

Thus: $[Fe^{2+}] = K_{s0,2}/[CO_3^{2-}] = K_{s0,2}/(\alpha_2 C_T)$, and

$$Fe(+II)_T = [K_{s0,2}/(\alpha_2 C_T)](1 + K_1[OH^-] + K_2[OH^-]^2 + K_3[OH^-]^3)$$

Here, C_T is the total carbonate concentration. The resulting $Fe(+II)_T$ for $C_T = 10^{-4}$ M is shown in Figure 19.5B.

At what pH are both solids present? The key pH is the pH where both solids give the same $Fe(+II)_T$. This occurs where $K_{s0,1}/[OH^-]^2 = K_{s0,2}/[CO_3^{2-}]$, or pH 9.26 at $C_T = 10^{-4}$ M. Another way to determine the pH where both solids exist is explored in Example 19.3. The overall $Fe(+II)_T$ is shown in Figure 19.5C. Note that $FeCO_3(s)$ controls at pH < 9.26 since it yields the smallest $Fe(+II)_T$.

19.4.3 Effects of other ligands on metal solubility

Suppose that the concentration of a free metal ion is controlled by one solid. Does the soluble metal concentration increase or decrease if a different ligand is added to the system? As an example, suppose that the solubility of Cu(+II) is controlled by $Cu(OH)_2(s)$. The question before you is whether or not the total soluble Cu(+II) concentration will increase or decrease if, say, chloride is added to the system.

In the most general case, you must solve the new equilibrium system to assess whether $Cu(+II)_T$ increases [as $Cu(OH)_2(s)$ dissolves] or decreases [as Cu(+II) precipitates] upon addition of chloride. If we narrow the problem a bit, then you can answer the question at hand without doing any calculations. Suppose that only a small amount of chloride is added so that the pH and total ion concentration do not change appreciably. Also, assume that so little chloride is added that no new copper-containing *solid* species form. In the absence of chloride, the total soluble copper is likely made up of four species:

$$Cu(+II)_T = [Cu^{2+}] + [CuOH^+] + [Cu(OH)_2^0] + [Cu(OH)_4^{2-}]$$

[For simplicity, the dimer $Cu_2(OH)_4^0$ and bicarbonate/carbonate complexes are ignored here.] You can rewrite this sum by expressing each species concentration in terms of $[Cu^{2+}]$ and remembering that $[Cu^{2+}] = K_{s0}/[OH^-]^2$ if $Cu(OH)_2(s)$ controls the Cu(II) solubility:

$$Cu(+II)_T = (K_{s0}/[OH^-]^2)(1 + K_1[OH^-] + ...)$$

If chloride is added, then a new soluble Cu(+II) species is introduced, namely, $CuCl^-$. This species equilibrates with Cu^{2+} as follows:

$$Cu^{2+} + Cl^- = CuCl^+ \qquad K_{Cl} = 10^{+0.5}$$

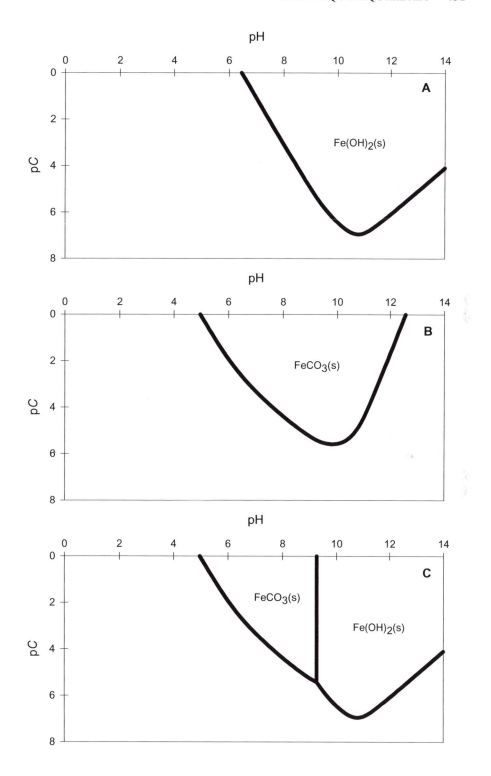

Figure 19.5: Total Soluble Fe(II) in Equilibrium with Fe(OH)$_2$(s) (A), FeCO$_3$(s) (B), and Both Solids (C)

The expression for total soluble copper becomes:

$$Cu(+II)_T = [Cu^{2+}] + [CuOH^+] + [Cu(OH)_2^0] + [Cu(OH)_4^{2-}]$$
$$+ [CuCl^+]$$
$$= (K_{s0}/[OH^-]^2)(1 + K_1[OH^-] + ... + K_{Cl}[Cl^-])$$

Thus, if $Cu(OH)_2(s)$ still controls the copper solubility and the pH does not change, the total soluble copper concentration in the presence of chloride must be *larger* than the total soluble copper concentration in the absence of chloride because a new species has been added. In general, *adding a new ligand increases the solubility of metals* (assuming that the same solid phase controls before and after ligand addition and the pH and other water quality characteristics do not change appreciably).

 Key idea: Adding a new ligand usually increases the solubility of metals, assuming that the same solid phase controls before and after ligand addition and the pH and other water quality characteristics do not change appreciably

Of course, you can show quantitatively that ligands increase metal solubility. As an example, consider the use of cyanide to increase the solubility of metals in metal plating operations. What is the effect on copper solubility at pH 8 if 1×10^{-4} M of total cyanide is added? In the absence of cyanide, the pertinent equilibria are:

$$
\begin{array}{ll}
H_2O = H^+ + OH^- & K_W = 10^{-14} \\
Cu^{2+} + OH^- = CuOH^+ & K_1 = 10^{+6.3} \\
Cu^{2+} + 2OH^- = Cu(OH)_2^0 & K_2 = 10^{+11.8} \\
Cu^{2+} + 4OH^- = Cu(OH)_4^{2-} & K_4 = 10^{+16.4} \\
Cu(OH)_2(s) = Cu^{2+} + 2OH^- & K_{s0} = 10^{-19.3}
\end{array}
$$

In metal plating, it is desirable to maintain the highest possible dissolved metal concentration. This means operating at the solubility limit. The total soluble copper concentration is:

$$Cu(+II)_T = [Cu^{2+}] + [CuOH^+] + [Cu(OH)_2^0] + [Cu(OH)_4^{2-}]$$

If $Cu(OH)_2(s)$ controls the copper solubility, then:

$$Cu(+II)_T = [Cu^{2+}](1 + K_1[OH^-] + K_2[OH^-]^2 + K_4[OH^-]^4)$$
$$= (K_{s0}/[OH^-]^2)(1 + K_1[OH^-] + K_2[OH^-]^2 + K_4[OH^-]^4)$$
$$= 1.8 \times 10^{-7} \text{ M at pH 8}$$

Thus, only about 1.8×10^{-7} M $= 12$ µg/L of total copper can be maintained at pH 8. If cyanide is added, the following equilibria also must be considered:

$$
\begin{array}{ll}
Cu^{2+} + 2CN^- = Cu(CN)_2^0 & K_{2,CN} = 10^{+16} \\
Cu^{2+} + 3CN^- = Cu(CN)_3^- & K_{3,CN} = 10^{+21.6} \\
Cu^{2+} + 4CN^- = Cu(CN)_4^{2-} & K_{4,CN} = 10^{+23.1}
\end{array}
$$

$$HCN \ = \ CN^- + H^+ \qquad\qquad K_a = 10^{-9.2}$$

In the presence of cyanide, the total soluble copper concentration is:

$$Cu(+II)_T \ = [Cu^{2+}] + [CuOH^+] + [Cu(OH)_2^0] + [Cu(OH)_4^{2-}]$$
$$+ [Cu(CN)_2^0] + [Cu(CN)_3^-] + [Cu(CN)_4^{2-}]$$

Again, if $Cu(OH)_2(s)$ controls the copper solubility, then:

$$Cu(+II)_T \ = [Cu^{2+}](1 + K_1[OH^-] + K_2[OH^-]^2 + K_4[OH^-]^4 +$$
$$K_{2,CN}[CN^-]^2 + K_{3,CN}[CN^-]^3 + K_{4,CN}[CN^-]^4)$$

$$= (K_{s0}/[OH^-]^2)(1 + K_1[OH^-] + K_2[OH^-]^2 + K_4[OH^-]^4 +$$
$$K_{2,CN}[CN^-]^2 + K_{3,CN}[CN^-]^3 + K_{4,CN}[CN^-]^4)$$

Writing a mass balance for cyanide:

$$CN_T \ = [CN^-] + [HCN] + 2[Cu(CN)_2^0] + 3[Cu(CN)_3^-]$$
$$+ 4[Cu(CN)_4^{2-}]$$
$$= [CN^-](1 + [H^+]/K_a + 2K_{2,CN}[Cu^{2+}][CN^-] +$$
$$3K_{3,CN}[Cu^{2+}][CN^-]^2 + 4K_{4,CN}[Cu^{2+}][CN^-]^3)$$

If no cyanide-containing species precipitate, then $CN_T = 1 \times 10^{-4}$ M. At pH 8, $[OH^-] = 10^{-6}$ M and $[Cu^{2+}] = K_{s0}/[OH^-]^2 = 10^{-7.3}$ M. Solving the cyanide mass balance for $[CN^-]$ (by using a spreadsheet and guesses or *Solver*): $[CN^-] = 2.84 \times 10^{-7}$ M. Thus, $Cu(+II)_T = 4.53 \times 10^{-5}$ M or 2.9 mg/L of dissolved copper. Note that copper is about 250 times more soluble in the presence of 1×10^{-4} M of total cyanide at pH 8. In the presence of 1×10^{-3} M of total cyanide at pH 8, copper is about 2400 times more soluble than in the absence of cyanide.

The addition of cyanide also has changed the speciation of dissolved copper. In the absence of cyanide, the dissolved copper is primarily $CuOH^+$, Cu^{2+}, and $Cu(OH)_2^0$. In the presence of cyanide, the dissolved copper is primarily $Cu(CN)_3^-$ and $Cu(CN)_2^0$.

19.5 SOLUBILITY OF CALCIUM CARBONATE

19.5.1 Introduction

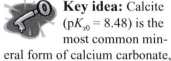

Key idea: Calcite ($pK_{s0} = 8.48$) is the most common mineral form of calcium carbonate, $CaCO_3(s)$

Calcium carbonate, $CaCO_3$, is an important solid phase in natural waters. Calcium carbonate has two common mineral forms, calcite (from the Latin *calx* lime) and aragonite (named after the Spanish province, Aragón, where it was first found). Calcium carbonate is the main constituent of several rocks, including limestone (at least 50% $CaCO_3$), marble (recrystallized calcium carbonate), and chalk. The pK_{s0} values for the mineral forms are

8.48 for calcite and 8.34 for aragonite. Calcite is the most common mineral form.

Calcium carbonate is important in aquatic chemistry because it influences several common water quality parameters.

Thoughtful Pause

What common water quality parameters are influenced by Ca^{2+} and CO_3^{2-}?

Calcium carbonate dissolves to form calcium and carbonate (eq. 19.6):

$$CaCO_3(s) = Ca^{2+} + CO_3^{2-} \; ; \; K_{s0} = 10^{-8.48} \text{ (calcite)} \qquad \text{eq. 19.6}$$

From eq. 19.6, you can see that calcium carbonate serves as a source of both alkalinity (from the cation Ca^{2+}) and C_T. How does Ca^{2+} affect alkalinity? Recall from Section 13.3 that alkalinity comes from a charge balance. Thus, any cations or anions (that do not cancel each other out) will influence alkalinity. Remember also that pH, alkalinity, and C_T are

hardness (or *total hardness*): the sum of the concentrations of the divalent cations in water (usually about equal to the sum of the calcium and magnesium ions)

interrelated (see Section 13.5). Thus, you might expect that the dissolution or precipitation of calcium carbonate also may affect the master variable pH.

The calcium ion also affects hardness. **Hardness** (also called **total hardness**) is the sum of the concentrations of the divalent cations in water, principally calcium ion and magnesium ion. Thus, in most waters: hardness $\approx [Ca^{2+}] + [Mg^{2+}]$. Hardness typically is expressed in units of mg/L as $CaCO_3$, with 1 mol of Ca^{2+} and 1 mol of Mg^{2+}, each equivalent to 1 mol of $CaCO_3$. The hardness from calcium is called *calcium hardness*, and the hardness from magnesium is called *magnesium hardness*.

carbonate (or temporary) hardness: hardness from calcium carbonate (called *temporary* because it can be removed by merely rasing the pH)

Hardness from calcium carbonate is called **carbonate** or **temporary hardness**. It is *temporary* because it can be removed by merely raising the pH.

Thoughtful Pause

Why can temporary hardness be removed by raising the pH?

Key idea: The carbonate hardness is (1) equal to the Alk if Alk < total hardness and (2) equal to the total hardness otherwise

If Ca^{2+} comes from calcium carbonate, then the C_T must be at least as large as the calcium ion concentration. Thus, raising the pH will convert the C_T to carbonate and allow for precipitation of most of the calcium as calcium carbonate. For waters near neutral pH, Alk $\approx [HCO_3^-] \approx C_T$ (see Section 13.4). Thus, the carbonate hardness is equal to the alkalinity, if Alk < total hardness and if all parameters are expressed in the same units (usually mg/L as $CaCO_3$). If the alkalinity is greater than the total hardness, then the carbonate hardness is equal to the total hardness.

Hardness also can come from calcium sulfate, $CaSO_4(s)$, often found in the form of the mineral gypsum. Hardness from calcium sulfate is called **noncarbonate** or **permanent hardness**. It is called *permanent* because it cannot be removed by merely raising the pH. If Ca^{2+} comes from calcium sulfate, then the C_T may not be at least as large as the calcium ion concentration. Thus, raising the pH, although it will convert the C_T to carbonate, may not provide enough carbonate to precipitate most of the calcium as calcium carbonate.

noncarbonate (or permanent) hardness: hardness from calcium sulfate (called *permanent* because it cannot be removed by merely raising the pH)

19.5.2 Equilibrium calculations with calcium carbonate

In many natural systems, calcium carbonate solid exists. If calcium carbonate exists, then its solubility and the resulting dissolved calcium concentration are pH dependent. You can show the pH dependence easily. From eq. 19.6: $[Ca^{2+}][CO_3^{2-}] = K_{s0} = 10^{-8.48}$. Thus:

$$[Ca^{2+}] = K_{s0}/[CO_3^{2-}] = K_{s0}/(\alpha_2 C_T)$$

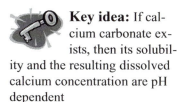

Key idea: If calcium carbonate exists, then its solubility and the resulting dissolved calcium concentration are pH dependent

If C_T is fixed, you can calculate $[Ca^{2+}]$ as a function of pH (remember that for the carbonate system, $pK_{a,1} = 6.3$ and $pK_{a,2} = 10.3$). The calcium solubility is shown in Figure 19.6, where the solid line is the total soluble calcium concentration.[†]

Note the changes in slope in the total soluble calcium line at pH 6.3 and 10.3 (caused by the changes in the speciation of dissolved carbonate). You can interpret Figure 19.6 as you have done previously. "Inside" the line (i.e., at larger pH and smaller pC values than points on the line), calcium carbonate precipitates. For example, you can see that calcium carbonate would precipitate at pH 8 and $[Ca^{2+}] = 10^{-2}$ M (point A, Figure 19.6), but it would not precipitate at pH 7 and $[Ca^{2+}] = 10^{-4}$ M (point B, Figure 19.6).

Figure 19.6 tells you that calcium carbonate is very soluble at low pH. Why? At low pH, the carbonate ion concentration is small. Therefore, $CaCO_3(s)$ dissolves to increase the concentration of Ca^{2+} and maintain the product $[Ca^{2+}][CO_3^{2-}]$ at $K_{s0} = 10^{-8.48}$. This phenomenon explains a number of observations in the environment. For example, acid rain is known to dissolve limestone monuments in a sort of titration (see Section 12.1 for examples). The low pH of the acid rain solubilizes $CaCO_3(s)$. On the other hand, $CaCO_3(s)$ precipitates at higher pH values. An example can be found in groundwater remediation. Environmental engineers sometimes use reactive barriers to treat groundwater passively. In some reactive barriers, zero-valent iron, $Fe(0)$, is used as an electron source (i.e., a reducing agent) to reduce chlorinated organics. As a result of the redox chemistry, the pH increases. Since some groundwaters can have appreciable calcium concentrations from equilibrium with $CaCO_3(s)$, an increase in the pH results in calcium precipitation and may clog the reactive barrier.

Key idea: Calcium carbonate is very soluble at low pH

[†] In this calculation, the species $CaOH^+$ has been ignored. $CaOH^+$ is present in significant concentrations relative to $[Ca^{2+}]$ only above about pH 12.

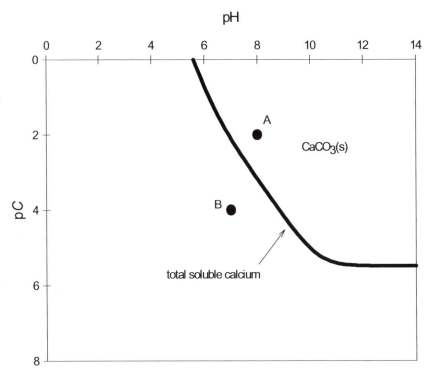

Figure 19.6: Total Dissolved Calcium in the Presence of CaCO₃(s)

***Example 19.4:* Langlier Index**

One way to determine whether a water is undersaturated, saturated, or supersaturated with respect to calcium carbonate is to calculate the *Langlier index* (also called the LI, saturation index, or SI). The Langlier index is the difference between the observed pH and the pH at saturation for a closed system at the observed alkalinity and calcium ion concentration ($LI = pH_{obs} - pH_{sat}$).

Calculate the Langlier index for a water with Alk = 250 mg/L as $CaCO_3$, $[Ca^{2+}]$ = 50 mg/L as Ca, and pH 7.15. Will $CaCO_3(s)$ tend to dissolve or precipitate in this water?

Solution:

For this water, Alk = 5 meq/L, and $[Ca^{2+}]$ = 1.25 mM. To find pH_{sat}, note that $[Ca^{2+}] = K_{s0}/[CO_3^{2-}] = K_{s0}/(\alpha_2 C_T)$, so $\alpha_2 = K_{s0}/([Ca^{2+}]C_T)$. From the alkalinity expression:

$$Alk = (\alpha_1 + 2\alpha_2)C_T + [OH^-] - [H^+]$$

Thus:

$$C_T = (Alk - [OH^-] + [H^+])/(\alpha_1 + 2\alpha_2)$$

Combining:

$$\alpha_2 = K_{s0}(\alpha_1 + 2\alpha_2)/\{[Ca^{2+}](Alk - [OH^-] + [H^+])\} \quad (*)$$

19.5.3 Water in equilibrium with calcium carbonate and the atmosphere

What if the C_T of the water is not fixed at some arbitrary value, but rather the water is in equilibrium with both calcium carbonate and the atmosphere? This is an excellent model for large, pristine lakes. Remarkably, you can find the equilibrium pH under these conditions with pen and paper (without even using a calculator).

For a large, pristine lake, assume that no ions are present other than those resulting from the dissolution of calcium carbonate and carbon dioxide gas. The charge balance is:

$$[H^+] + 2[Ca^{2+}] = [HCO_3^-] + 2[CO_3^{2-}] + [OH^-]$$

If the pH is nearly neutral (an assumption we must check later), then $[H^+]$, $[CO_3^{2-}]$, and $[OH^-]$ can be ignored. Thus: $2[Ca^{2+}] \approx [HCO_3^-]$. Your job is to find expressions for $[Ca^{2+}]$ and $[HCO_3^-]$.

If calcium carbonate exists, then: $[Ca^{2+}] = K_{s0}/[CO_3^{2-}] = K_{s0}/(\alpha_2 C_T)$, as before. Now, however, C_T is fixed by the atmosphere. From Chapter 18: $[H_2CO_3^*] = \alpha_0 C_T = K_H P_{CO_2}$. Thus:

At pH values above 6.3, $\alpha_2 \approx K_2/(K_2 + [H^+])$ and $\alpha_1 + \alpha_2 \approx 1$. Thus, (*) simplifies to:

$$\frac{K_2}{K_2+[H^+]} = \frac{K_{s0}\left(1+\dfrac{K_2}{K_2+[H^+]}\right)}{[Ca^{2+}]\left(Alk - \dfrac{K_W}{[H^+]}+[H^+]\right)}$$

This equation can be solved iteratively for $[H^+]$ and thus pH_{sat}. For the water in question, pH_{sat} is calculated to be 7.02 (pH > 6.3, so the assumption holds). Thus: LI = 7.15 – 7.02 = +0.13. The water is supersaturated with calcium carbonate (which raises the pH above pH_{sat}), and thus calcium carbonate will precipitate.

In general, it is desirable to have a slightly positive LI in a drinking water distribution system so that a protective layer of calcium carbonate will form on the pipes and corrosion will be inhibited.

To summarize: **LI = +0.13 and calcium carbonate will precipitate**.

$$C_T = K_H P_{CO_2}/\alpha_0 \text{ and } [Ca^{2+}] = K_{s0}\alpha_0/(\alpha_2 K_H P_{CO_2})$$

From Section 11.5, recall that for any diprotic acid: $\alpha_0/\alpha_2 = [H^+]^2/K_{a,1}K_{a,2}$ Thus:

$$[Ca^{2+}] = K_{s0}[H^+]^2/(K_1 K_2 K_H P_{CO_2})$$

What about $[HCO_3^-]$? You know $[HCO_3^-] = \alpha_1 C_T$ and $C_T = K_H P_{CO_2}/\alpha_0$. Therefore: $[HCO_3^-] = \alpha_1 K_H P_{CO_2}/\alpha_0$. Since $\alpha_1/\alpha_0 = K_{a,1}/[H^+]$, then:

$$[HCO_3^-] = K_H P_{CO_2} K_1/[H^+]$$

Where does all of this leave us? The charge balance is approximated by $2[Ca^{2+}] \approx [HCO_3^-]$ (if near neutral pH). Substituting in the expressions for $[Ca^{2+}]$ and $[HCO_3^-]$:

$$2K_{s0}[H^+]^2/(K_1 K_2 K_H P_{CO_2}) \approx K_H P_{CO_2} K_1/[H^+]$$

Simplifying:

$$[H^+]^3 \approx K_1^2 K_H^2 P_{CO_2}^2 K_2/(2K_{s0})$$

At $P_{CO_2} = 10^{-3.5}$ atm:

$$\begin{aligned}[H^+]^3 &\approx [(10^{-6.3}\text{ M})(10^{-1.5}\text{ M/atm})(10^{-3.5}\text{ atm})]^2 \times \\ &\quad (10^{-10.3}\text{ M})/(2\times10^{-8.48}\text{ M}^2) \\ &= 10^{-24.42}\text{ M}^3/2 \\ &= 10^{-24.72}\text{ M}^3\end{aligned}$$

So: $[H^+] = 10^{-24.72/3}$ M $= 10^{-8.24}$ M or about pH 8.24. (The solution without approximations also rounds to 8.24 because $[H^+]$, $[CO_3^{2-}]$, and $[OH^-]$ are indeed small compared to $[Ca^{2+}]$ and $[HCO_3^-]$.) For this model, you can show that: $C_T = 8.84\times10^{-4}$ M and $[Ca^{2+}] = 4.41\times10^{-4}$ M. The alkalinity (from Alk $= 2[Ca^{2+}]$) is equal to 8.83×10^{-4} eq/L or 44 mg/L as $CaCO_3$. Another equilibrium calculation with calcium carbonate is given in Example 19.4.

19.6 SUMMARY OF THE ACID-BASE CHEMISTRY OF NATURAL WATERS

19.6.1 Introduction

With the addition of calcium carbonate chemistry, you now can create more realistic models of the aquatic environment. In this section, all the possible chemistries of aquatic systems will be reviewed, including open and closed systems, systems with and without calcium carbonate, and systems with and without strong acid anions and strong base cations other than the calcium ion.

This review will take you from the simplest system (closed to the atmosphere, no solids, and no ions other than H^+ and OH^-, i.e., pure water) to a general model for surface waters.

19.6.2 Closed systems with no solids

For this type of system, the charge balance can be rearranged to: $C_B - C_A = (\alpha_1 + 2\alpha_2)C_T + [OH^-] - [H^+]$, where Alk $= C_B - C_A$. Numerical answers can be obtained if the alkalinity is known. For example, if Alk $= C_B - C_A = 0$, then the pH is a function of C_T and can be obtained with the charge balance and a pC-pH diagram. For the special case where Alk $= 0$ and $C_T = 0$, then the charge balance becomes: $[H^+] = [OH^-]$ or pH 7 at 25°C. For the general case, see Figures 13.8 and 13.9 in Chapter 13.

19.6.3 Closed systems with CaCO₃(s) but no other solids

There are two cases in this type of system. In the simplest case, no C_B or C_A exists other than Ca^{2+}. Actually, we need not be so restrictive. All that needs to be assumed in this case is that C_A is equal to the sum of all the strong base cations except the calcium ion. It is useful to define $C_B' =$ the sum in eq/L of all strong base cations except calcium ion $= C_B - 2[Ca^{2+}]$. Thus, for this case: $C_B' - C_A = 0$.

In this case, the charge balance becomes:

$$2[Ca^{2+}] = (\alpha_1 + 2\alpha_2)C_T + [OH^-] - [H^+]$$

From the solubility constraint: $[Ca^{2+}] = K_{s0}[CO_3^{2-}] = K_{s0}/(\alpha_2 C_T)$. In this case, $[Ca^{2+}] = C_T$ since all calcium and all carbonate comes from CaCO₃(s). Why? The system is closed to the atmosphere and contains no solids other than CaCO₃(s). Thus: $[Ca^{2+}] = C_T = K_{s0}/(\alpha_2 C_T)$. Solving for C_T: $C_T = [Ca^{2+}] = (K_{s0}/\alpha_2)^{1/2}$. Thus, the charge balance becomes:

$$2(K_{s0}/\alpha_2)^{1/2} = (\alpha_1 + 2\alpha_2)(K_{s0}/\alpha_2)^{1/2} + [OH^-] - [H^+]$$

Solving: pH 9.9, $C_T = [Ca^{2+}] = 1.1 \times 10^{-4}$ M, and Alk $= 2[Ca^{2+}] = 2.2 \times 10^{-4}$ eq/L. This is a model for pristine groundwaters.

As you know, most groundwaters do not have pH values near 9.9. This means that most groundwaters *do* have some additional acid inputs and thus some extra C_A. Therefore, we must consider a second case, namely, a closed system with CaCO₃(s) (but no other solids) that *does* have other C_B or C_A. In other words, $C_B' - C_A \neq 0$. In this case, the charge balance becomes:

$$C_B' - C_A + 2[Ca^{2+}] = (\alpha_1 + 2\alpha_2)C_T + [OH^-] - [H^+]$$

Now, in this case, $[Ca^{2+}]$ may or may not be equal to C_T, since there may be sources of dissolved carbonate and/or calcium other than CaCO₃(s). In any

$C_B' =$ the sum in eq/L of all strong base cations except calcium ion $= C_B - 2[Ca^{2+}]$

event, the solubility constraint still holds if $CaCO_3(s)$ controls the solubility of calcium: $[Ca^{2+}] = K_{s0}/(\alpha_2 C_T)$. Thus:

$$C_B' - C_A + 2K_{s0}/(\alpha_2 C_T) = (\alpha_1 + 2\alpha_2)C_T + [OH^-] - [H^+]$$

This is as far as you can get without knowing $C_B' - C_A$ and C_T.

In general, pH, C_T, $[Ca^{2+}]$, and Alk are functions of $C_B' - C_A$. The relationships among pH, C_T, and Alk in the absence of $CaCO_3(s)$ were given in Chapter 13. Plots of $[Ca^{2+}]$ and $C_B' - C_A$ as functions of pH and C_T in the presence of $CaCO_3(s)$ are shown in Figure 19.7.

Figure 19.7 appears a bit confusing at first but can be useful. In the absence of $CaCO_3(s)$, you know from Chapter 13 that specifying two of pH, C_T, and Alk will allow you to calculate the third parameter. In the presence of $CaCO_3(s)$, specifying two of pH, C_T, Alk, and $[Ca^{2+}]$ will allow you to calculate the other two parameters. In addition, you have a choice of information about alkalinity - you can specify Alk and $[Ca^{2+}]$, Alk and $C_B' - C_A$, or $[Ca^{2+}]$ and $C_B' - C_A$.

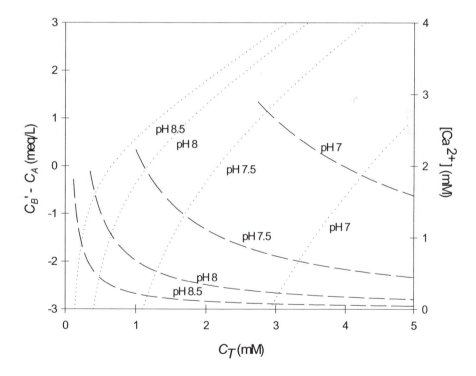

Figure 19.7: Calcium Ion Concentration (dashed lines) and $C_B' - C_A$ (dotted lines) as Functions of pH and C_T for Closed Systems with $CaCO_3(s)$ but No Other Solids
(analogous to the Deffeyes diagram in Figure 13.8)

An example will help clarify the use of Figure 19.7. Suppose you analyze a groundwater sample and find that it has pH 7 and $[Ca^{2+}] = 80$ mg/L as Ca = 2 mM. You can use Figure 19.7 to help characterize the water (assuming the water is in equilibrium with calcium carbonate). Extend a line parallel to the x-axis from $[Ca^{2+}] = 2$ mM to the dashed line at pH 7 (see Figure 19.8). The corresponding C_T is about 4 mM. The value of $C_B' - C_A$ can be found by extending a line parallel to the x-axis from $C_T = 4$ mM and the dotted pH 7 line to the left y-axis. The corresponding value of $C_B' - C_A$ is about -0.8 meq/L.

Thoughtful Pause
What does the negative value of $C_B' - C_A$ mean?

The negative value of $C_B' - C_A$ means that the sum of the strong acid anions, C_A, exceeds the sum of the strong base cations other than the calcium ion, C_B'. The Alk is $C_B' - C_A + 2[Ca^{2+}] \approx -0.8$ meq/L + (2 meq/mmol)(2 mM) = 3.2 meq/L or 160 mg/L as $CaCO_3$. Thus, the groundwater (if in equilibrium with $CaCO_3(s)$) has the following water quality characteristics: pH 7, $[Ca^{2+}] = 2$ mM (or 80 mg/L as Ca or 200 mg/L as $CaCO_3$), $C_T = 4$ mM, Alk = 3.2 meq/L (160 mg/L as $CaCO_3$), and $C_B' - C_A = -0.8$ meq/L (-40 mg/L as $CaCO_3$). The groundwater has an excess of strong acid anions over strong base cations other than the calcium ion.

19.6.4 Open systems with no solids
For this type of system, the general charge balance equation is $C_B - C_A = (\alpha_1 + 2\alpha_2)C_T + [OH^-] - [H^+]$. In this case, the total dissolved carbonate is fixed by the atmosphere and $C_T = K_H P_{CO_2}/\alpha_0$ (see Section 18.4). Thus: $C_B - C_A = (\alpha_1 + 2\alpha_2)(K_H P_{CO_2}/\alpha_0) + [OH^-] - [H^+]$. Consider the special case where $C_B = C_A = 0$. This is a model for pristine rainwater (see Section 18.4.2). In this case: $0 = (\alpha_1 + 2\alpha_2)(K_H P_{CO_2}/\alpha_0) + [OH^-] - [H^+]$. For $P_{CO_2} = 10^{-3.5}$ atm, you can calculate: pH 5.65, $C_T = 1.22 \times 10^{-5}$ M, and Alk = 0 eq/L.

19.6.5 Open systems where CaCO₃(s) exists
This is the most general system. If there are no C_B or C_A except Ca^{2+} from $CaCO_3(s)$ (or, more generally, if $C_B' - C_A = 0$), then the charge balance becomes: $2[Ca^{2+}] = (\alpha_1 + 2\alpha_2)C_T + [OH^-] - [H^+]$. The atmospheric carbon dioxide fixes the concentration of $H_2CO_3^*$ and $C_T = K_H P_{CO_2}/\alpha_0$. From the solubility constraint: $[Ca^{2+}] = K_{s0}/(\alpha_2 C_T)$. Combining these two expressions: $[Ca^{2+}] = K_{s0}\alpha_0/(\alpha_2 K_H P_{CO_2})$. The charge balance becomes:

$$2K_{s0}\alpha_0/(\alpha_2 K_H P_{CO_2}) = (\alpha_1 + 2\alpha_2)(K_H P_{CO_2}/\alpha_0) + [OH^-] - [H^+]$$

If $P_{CO_2} = 10^{-3.5}$ atm, then you can show (Section 19.5.3): pH 8.24, $C_T = 8.84 \times 10^{-4}$ M, $[Ca^{2+}] = 4.41 \times 10^{-4}$ M, and Alk = 8.83×10^{-4} eq/L.

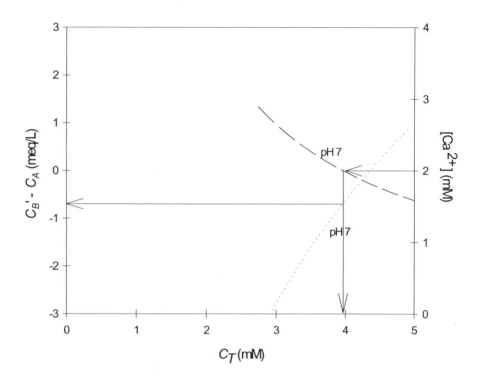

Figure 19.8: Detail of Figure 19.7 for the Example in the Text

In the most general case, $C_B' - C_A \neq 0$. Thus:

$$C_B' - C_A + 2[Ca^{2+}] = (\alpha_1 + 2\alpha_2)C_T + [OH^-] - [H^+]$$

In this case, pH, C_T, $[Ca^{2+}]$, and Alk are functions of $C_B' - C_A$.

The results for several of the cases discussed above are summarized in Table 19.1. The calculated values in Table 19.1 are useful in making assessments of water quality. Consider the groundwater discussed in Section 19.6.3. Its water quality was: pH 7, $[Ca^{2+}] = 2$ mM, $C_T = 4$ mM, and Alk = 3.2 meq/L. Of the choices in Table 19.1, this groundwater is modeled most closely by system 2. If you compare with system 2 in Table 19.1 [closed, $CaCO_3(s)$ exists, and no strong base cations or strong base ions other than calcium ion], it is clear that the groundwater in question has a lower pH, larger calcium ion concentration, larger alkalinity, and larger C_T than system 2.

Thoughtful Pause
Can you explain qualitatively why the groundwater has a lower pH and larger [Ca^{2+}], alkalinity, and C_T than system 2?

The groundwater *must* have an excess of strong acid anions over strong base cations other than the calcium ion (in other words, $C_B' - C_A < 0$). The presence of strong acid anions would reduce the pH. This in turn would dissolve more calcium carbonate and increase the alkalinity, calcium ion concentration, and total dissolved carbonate concentration.

As another example, consider Lake Ontario. Its typical pH, [Ca^{2+}], and Alk values are 8.0-8.3, 40 mg/L, and 95 mg/L as CaCO$_3$, respectively. A good model for Lake Ontario is an open system with CaCO$_3$(s) or system 4. The system 4 pH, [Ca^{2+}], and Alk values are (from Table 19.1) 8.2, 18 mg/L, and 44 mg/L as CaCO$_3$, respectively. Recall that system 4 assumed that $C_B' - C_A = 0$. Clearly for Lake Ontario, $C_B' - C_A < 0$. This would lower the pH and dissolve additional calcium, thereby increasing the alkalinity. You can show from Figure 19.7 that $C_B' - C_A \approx -1.5$ meq/L for Lake Ontario. These examples show that comparisons of real waters with the idealized cases in Table 19.1 can reveal qualitative information about water quality.

Table 19.1: Summary of Calculated Values ($P_{CO_2} = 10^{-3.5}$ atm)

System	pH	C_T (M)	Alk (eq/L)	[Ca^{+2}] (M)
1. Closed, no solids, $C_B - C_A = 0$ (Section 19.6.2)	7.0	0	0	0 if $C_B = 0$
2. Closed, CaCO$_3$(s) exists, $C_B' - C_A = 0$ (Section 19.6.3)	9.9	1.1×10^{-4}	2.2×10^{-4}	1.1×10^{-4}
3. Open, no solids, $C_B - C_A = 0$ (Section 19.6.4)	5.7	1.2×10^{-3}	0	0 if $C_B = 0$
4. Open, CaCO$_3$(s) exists, $C_B' - C_A = 0$ (Section 19.6.5)	8.2	8.8×10^{-4}	8.8×10^{-4}	4.4×10^{-4}

19.7 AQUATIC CHEMISTRY OF SURFACES

19.7.1 Introduction

surface: an interface between a solid phase and a fluid (such as water or air)

The precipitation of solids creates particles, and with particles come surfaces. A *surface* is the interface between a solid phase and a fluid such as water or air. Surfaces provide sites for chemical interactions with dissolved species.

Many chemical processes other than precipitation generate particles and surfaces. The most important process is erosion. Erosion transmits enormous quantities of minerals to the oceans through streams. The total suspended sediment flux to the oceans is about 1.4×10^{16} g/year (Fyfe, 1987), a mass of material each year equivalent to a cube of soil about 19 kilometers on a side. Other sources of surfaces besides erosion include algae, bacteria, and colloidal material such as aquatic humic and fulvic acids. Colloids are particles with characteristic lengths less than about 1 μm.

Solids with very small diameters can exhibit large surface areas. For example, 1 gram of nonporous colloidal particles with a diameter of 1 μm and a density of 2.5 g/cm^3 would contain about 2.4 m^2 of surface area. Porous materials, on the other hand, can have incredibly large surface areas. Activated carbon can exhibit greater than 1,000 m^2 per gram of material: about 1½ football fields of surface area in a mass equivalent to a U.S. nickel.

Key idea: Surfaces provide sites for chemical interaction with dissolved species

Dissolved chemical species can interact in a number of ways. For example, pollutants may bind irreversibly to surfaces or decompose at high-energy surface sites. The focus in this section will be on *reversible interactions* between dissolved species and surfaces. Such interactions can be analyzed by using equilibrium tools.

adsorption: the accumulation of mass on a surface

19.7.2 A simple approach to surface complexation

adsorbate: the substance being adsorbed

adsorbent: the solid phase onto which the adsorbate is being adsorbed

There are several types of reversible interactions between dissolved species and surfaces in water. In general, the accumulation of mass on a surface is called *adsorption*. The substance being adsorbed is called the *adsorbate*. The solid phase onto which the adsorbate is adsorbed is called the *adsorbent*.

surface complexation: interactions between species in solution and surfaces to form surface-bound complexes (analogous to homogeneous complexation)

When species in solution interact with surfaces to form surface-bound complexes, the form of interaction is called *surface complexation*. Surface complexation is analogous to homogeneous complexation discussed in Chapter 15. For example, you know that copper complexes with the hydroxide ion in part as follows: $Cu^{2+} + OH^- = CuOH^+$. By analogy, aquo cupric ion can complex with a surface site, as shown in eq. 19.7:

$$Cu^{2+} + \equiv SOH = \equiv SOCu^+ + H^+ \qquad K \qquad \qquad \text{eq. 19.7}$$

In eq. 19.7, the symbol "$\equiv$S" is used to indicate a surface, and "$\equiv$SOH" represents a surface oxide site. You can express the concentrations of species containing surfaces in units of moles of surface species per liter of water. Thus, the statement that "$[\equiv SOCu^+] = 10^{-4}$ M" means that 10^{-4} mole of the surface species $\equiv SOCu^+$ are found per liter of water. These concentration units also are analogous to the homogeneous case.

On the simplest level, you might apply the normal procedure for equilibrium calculations to systems containing complexing surfaces. For example, suppose that you expose a surface, $\equiv SOH$, to water containing copper. To make this example more straightforward, assume that copper does not precipitate as $Cu(OH)_2(s)$ and that the pH is low enough so that aqueous phase copper-hydroxy complexes can be ignored. The chemistry of the site may be more complicated than shown in eq. 19.7. Suppose the site also donates a proton to water: $\equiv SOH = \equiv SO^- + H^+$. In other words, the surface possesses acid-base chemistry. In general, many surfaces in the environment exhibit acid-base chemistry. The equilibrium system becomes (as usual, you may wish to try this on your own before reading further):

Species list:
   H_2O, H^+, OH^-, Cu^{2+}, $\equiv SOH$, $\equiv SOCu^+$, and $\equiv SO^-$

Equilibria:
$$H_2O = H^+ + OH^- \qquad K_W = 10^{-14}$$
$$Cu^{2+} + \equiv SOH = \equiv SOCu^+ + H^+ \qquad K$$
$$\equiv SOH = \equiv SO^- + H^+ \qquad K_a$$

Or:

$$K_W = [H^+][OH^-]/[H_2O]$$
$$K = [\equiv SOCu^+][H^+]/([Cu^{2+}][\equiv SOH])$$
$$K_a = [\equiv SO^-][H^+]/[\equiv SOH]$$

Mass balances:
   Mass balances are required on copper and the sites.
   For copper: $Cu(+II)_T = [Cu^{2+}] + [\equiv SOCu^+]$
   For sites: $S_T = [\equiv SOH] + [\equiv SOCu^+] + [\equiv SO^-]$

Charge balance:
   $$2[Cu^{2+}] + [\equiv SOCu^+] + [H^+] = [\equiv SO^-] + [OH^-]$$

Other:
   Dilute solution so that concentrations and activities are interchangeable
   Activity of water is unity

Thoughtful Pause
Is this system solvable?

The system has six unknowns ($[H^+]$, $[OH^-]$, $[\equiv SOCu^+]$, $[Cu^{2+}]$, $[\equiv SOH]$, and $[\equiv SO^-]$) and six linearly independent equations. Thus, a solution should be possible. In fact, the solution method is similar to that employed in homogeneous complexation (see Chapter 15): express both mass balances is terms of the concentration of two species. In the example above, you can show that:

$$Cu(+II)_T = [Cu^{2+}] + [\equiv SOCu^+]$$
$$= [Cu^{2+}](1 + K[\equiv SOH]/[H^+]) \qquad \text{eq. 19.8}$$
$$S_T = [\equiv SOH] + [\equiv SOCu^+] + [\equiv SO^-]$$
$$= [\equiv SOH](1 + K[Cu^{2+}]/[H^+] + K_a/[H^+]) \qquad \text{eq. 19.9}$$

To construct a pC-pH diagram, $[Cu^{2+}]$ and $[\equiv SOH]$ can be calculated at any pH from eqs. 19.8 and 19.9[†]. After a little algebra, you can combine eqs. 19.8 and 19.9 to write one equation in either $[Cu^{2+}]$ or $[\equiv SOH]$. For example, you can combine eqs. 19.8 and 19.9 to show:

$$K[Cu^{2+}]^2 + \{[H^+] + K_a + KS_T - KCu(+II)_T\}[Cu^{2+}]$$
$$- ([H^+] + K_a)Cu(+II)_T = 0$$

If $K = 10^{-1.5}$, $K_a = 10^{-8}$, $Cu(+II)_T = 1 \times 10^{-6}$ M, and $S_T = 1 \times 10^{-4}$ M, then at pH 6: $[Cu^{2+}] = 2.4 \times 10^{-7}$ M, $[\equiv SOCu^+] = 7.6 \times 10^{-7}$ M, $[\equiv SOH] = 9.8 \times 10^{-5}$ M, and $[\equiv SO^-] = 9.8 \times 10^{-7}$ M. Note that the surface greatly influences the fate of the copper. About 76% of the copper will be bound to the surface in this example. However, because the available copper concentration is so low, the copper does not exert very much influence on the surface. About 98% of the surface is in the form $\equiv SOH$ and thus complexed with H^+, not Cu^{2+}.

If pH is varied, then the resulting pC-pH diagram is shown in Figure 19.9. Does the pC-pH diagram make *qualitative* sense? As the pH increases from low pH conditions, the surface becomes deprotonated. Aquo cupric ion (Cu^{2+}) begins to compete with H^+ for surface sites. Since the total available copper is only 10^{-6} M, the copper-surface complex levels out at 10^{-6} M.

[†] If you wish to determine the equilibrium pH, simply follow the usual procedure and find the point on the pC-pH diagram where the charge balance is satisfied. To solve the system algebraically, use eqs. 19.8 and 19.9 along with the equilibria to express each species concentration in terms of $[H^+]$ (as was done for $[Cu^{2+}]$ in the text). Insert the expressions for each species concentration as a function of $[H^+]$ into the charge balance, and then solve for $[H^+]$.

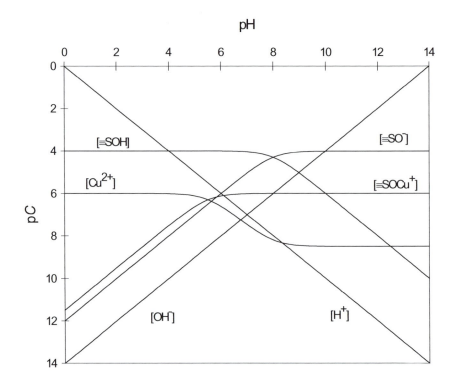

Figure 19.9: p*C*-pH Diagram for Copper in the Presence of a Complexing Surface
(aqueous phase hydroxo-copper complexes and copper precipitates ignored)

adsorption edge plot: a plot of the surface-bound metal concentration (as a percentage of the total metal) as a function of pH

It is common to plot only the surface-bound metal concentration as a function of pH. Usually, the surface-bound metal concentration is expressed as a percentage of the total metal. Such a plot, called an ***adsorption edge plot***, is shown in Figure 19.10. Note that the bound copper concentration is very sensitive to pH over a narrow pH range. The bound copper concentration increases from 10% to 90% over two pH units (about pH 5 to 7). This is a relatively sharp adsorption edge.

19.7.3 Langmuir isotherm
The simple equilibrium approach to surface complexation can be applied to the adsorption of chemical species onto solid surfaces. Consider the adsorption of an uncharged pollutant P onto a surface $\equiv$S. An example might be adsorption of chloroform onto activated carbon. For simplicity, assume that the pH is fixed so that the acid-base chemistry of the surface does not change. Applying the general equilibrium calculation approach:

Species list:
    $\equiv$S, P, and $\equiv$SP

Equilibria:

$$P + \equiv S \;=\; \equiv SP \quad K$$

Or:

$$K = [\equiv SP]/[P][\equiv S]$$

***Example 19.5:* Langmuir Isotherm**

For the adsorption of phenol onto a particular type of activated carbon, the following Langmuir isotherm constants were found: $b = 26.2$ L/mg and $q_{max} = 8.1$ mg/g. Find the aqueous equilibrium concentration of phenol if 100 mg of the activated carbon and 20 mg of phenol are combined in 1 L of water.

Solution:
From a mass balance on phenol:
(q_e)(mass adsorbent) + (C_e)(vol. of water) = mass added. Thus:

$$(0.1 \text{ g})q_e + (1 \text{ L})C_e = 20 \qquad (*)$$

Here, q_e is in mg/g and C_e is in mg/L. Also:
$q_e = q_{max}bC_e/(1 + bC_e)$

From (*): $q_e = 200 - 10C_e$. Substituting into the Langmuir isotherm:

$$200 - 10C_e = q_{max}bC_e/(1 + bC_e)$$

Solving for C_e: **$C_e = 19.2$ mg/L and $q_e = 8.1$ mg of phenol per g of activated carbon**. Clearly, this carbon does not adsorb phenol very well.

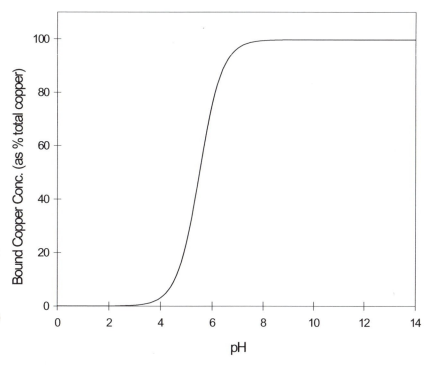

Figure 19.10: Adsorption Edge for Copper (details in the text)

Mass balances:
Mass balances are required on P and the sites.
For P: $P_T = [P] + [\equiv SP]$
For sites: $S_T = [\equiv S] + [\equiv SP]$

In adsorption, we are interested in the relationship between the adsorbed species concentration and the dissolved species concentration. Thus, we seek a relationship between $[\equiv SP]$ and $[P]$. From the equilibrium expression: $[\equiv SP] = K[P][\equiv S]$. From the mass balance on sites: $[\equiv S] = S_T - [\equiv SP]$. Combining: $[\equiv SP] = K[P](S_T - [\equiv SP])$ or:

$$[\equiv SP] = K[P]S_T/(1 + K[P]) \qquad \text{eq. 19.10}$$

isotherm: a quantitative relationship between the mass adsorbed per mass of adsorbate and the dissolved species concentration at constant temperature

Adsorption equilibrium frequently are expressed as the relationship between the mass adsorbed per mass of adsorbate and the dissolved species mass concentration. A relationship between the mass adsorbed per mass of adsorbate and the dissolved species concentration is called an **isotherm** (so named because the relationship is defined for a constant temperature). If the molecular weight of P is M_P and the mass concentration of the adsorbate is M, then the mass adsorbed per mass of adsorbate is $[\equiv SP]M_P/M$. Multiplying both sides of eq. 19.10 by M_P/M, the isotherm becomes:

$$[\equiv SP]M_P/M = K'[P]_m S_T'/(1 + K'[P]_m) \qquad \text{eq. 19.11}$$

where $K' = K/M_P$, $S_T' = S_T M_P/M =$ mass of sites as P per mass of adsorbate, and $[P]_m =$ mass concentration of P $= [P]M_P$. Equations 19.10 and 19.11 are called the **Langmuir isotherm** (after Irving Langmuir, 1881-1957).

Many forms of the Langmuir isotherm are in use. It is common to introduce the following symbols:

Langmuir isotherm: an isotherm given by:
$q_e = q_{max}bC_e/(1 + bC_e)$
(terms defined in the text)

$q_e =$ mass adsorbed per mass of adsorbate $= [\equiv SP]M_P/M$
$C_e =$ pollutant concentration (usually in mass units) $= [P]_m$
$q_{max} =$ mass of sites (as P) per mass of adsorbate $= S_T'$, and:
$b =$ a measure of the strength of adsorption $= K' = K/M_P$

Thus, the Langmuir isotherm becomes:

$$q_e = \frac{q_{max}bC_e}{1 + bC_e}$$

The use of the Langmuir isotherm is illustrated in Example 19.5.

19.7.4 Correction for surface charge: an advanced topic[†]

The simple equilibrium approach discussed in Sections 19.7.2 and 19.7.3 has a serious flaw. The approach fails to take into account the effect of the surface charge on complexation. Many surfaces in natural waters are negatively charged. The charge stems from ionized weak acids at the surface or, in the case of clays, for example, replacement of one metal (e.g., Si^{4+}) with a metal containing a lower positive charge (e.g., Al^{3+}).

[†] The correction of surface equilibria for surface charge is an advanced topic and is covered here only in minimum detail. For more information, see, for example, Stumm and Morgan (1996).

You might expect that the affinity of ions in solution for a charged surface will be influenced by the surface charge. For example, a metal cation such as Cu^{2+} may have an attractive electrostatic interaction with a surface exhibiting a high negative charge. The complexation of H^+ or other metals onto the surface will affect the surface charge and thus the affinity of ions for the surface. In other words, *the surface complexation equilibrium constants are not constant but depend on the state of the surface*. The surface complexation equilibrium constants in Sections 19.7.2 and 19.7.3 ignored the effects of surface charge. These are sometimes called ***intrinsic equilibrium constants*** (or K_{int}) because they represent the intrinsic chemical interaction that is independent of electrostatics. In other words, K_{int} can be calculated strictly from the intrinsic Gibbs free energy. With a charged surface, the Gibbs free energy is the sum of two terms: the intrinsic Gibbs free energy and the free energy associated with electrostatics:

intrinsic equilibrium constants: equilibrium constants uncorrected for electrostatic effects (symbol: K_{int})

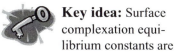 **Key idea:** Surface complexation equilibrium constants are not constant but depend on the surface charge (which, in turn, depends on the degree of surface complexation)

$$\Delta G^o = \Delta G^o_{int} + \Delta G^o_{electro} \qquad \text{eq. 19.12}$$

The charged surface generates an electric potential difference at the surface (i.e., a voltage difference), ψ_0. The electrostatic portion of the Gibbs free energy is the work required to move the ions across the potential difference ψ_0 and can be calculated by:

$$\Delta G^o_{electro} = \Delta Z F \psi_0 \qquad \text{eq. 19.13}$$

where ΔZ = change in the surface charge for a given equilibrium and F = Faraday's constant = 96,485 C/mol. Combining eqs. 19.12 and 19.13:

apparent equilibrium constants: equilibrium constants corrected for electrostatic effects (symbol: K_{app})

$$K_{app} = \exp(-\Delta G^o/RT) = K_{int}\exp(-\Delta Z F \psi_0/RT)$$

where K_{app} is the ***apparent equilibrium constant***.

If you want to correct for surface charge, you need to use K_{app} values. This means you need a way to calculate the surface potential ψ_0. Unfortunately, we cannot measure ψ_0. You can, however, calculate ψ_0 from the surface charge. Several models of surface charge exist. A common model is the ***Guoy-Chapman diffuse layer model***.[†] In this model, the charge distribution near a surface in treated as two layers called the ***electric double layer*** (EDL). The EDL is envisioned as a fixed charge on the surface with a diffuse charge penetrating into the water surrounding the surface.

Guoy-Chapman diffuse layer model: a model for the charge distribution at and near a surface, with a fixed charge on the surface and a diffuse charge penetrating into the water surrounding the surface

electric double layer (EDL): a conceptualization of the charge distribution around a charged surface as consisting of two charged layers

According to the Guoy-Chapman diffuse layer model, the surface charge at 25°C in the presence of c mol/L of a 1:1 electrolyte such as NaCl can be calculated by:

[†] The diffuse layer model was developed independently by Georges Gouy (1854-1926) in 1910 and David Leonard Chapman (1869-1958) in 1913.

$$\text{surface charge (in C/m}^2) = 0.1174c^{\frac{1}{2}}\sinh(19.46\psi_0) \qquad \text{eq. 19.14}$$
$$(\psi_0 \text{ in V})$$

Recall that $\sinh(x) =$ hyperbolic sine of $x = \frac{1}{2}(e^x - e^{-x})$.

How does this equation help? The surface charge can be calculated by the excess of positive charges over negative charges at the surface. For the example in Section 19.7.2:

$$\text{surface charge} = (F/s)([\equiv SOCu^+] - [\equiv SO^-]) \qquad \text{eq. 19.15}$$

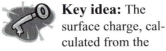 **Key idea:** The surface charge, calculated from the excess of charges at the surface, can be used to calculate the surface potential

where $s =$ surface concentration in m^2/L. Thus, ψ_0 can be determined by eq. 19.14 after calculating the surface charge from eq. 19.15.

To illustrate the effects of surface charge, the problem in Section 19.7.2 will be solved again. Everything stays the same as in Section 19.7.2, except for the equilibria involving surfaces. For these equilibria:

$$Cu^{2+} + \equiv SOH \; = \; \equiv SOCu^+ + H^+ \qquad K_{app} = K_{int}\exp(+F\psi_0/RT)$$
$$\equiv SOH \; = \; \equiv SO^- + H^+ \qquad K_{a,app} = K_{a,int}\exp(-F\psi_0/RT)$$

The signs in the exponents stem from the fact that the surface gains one positive charge per mole in the first reaction and loses one charge per mole in the second reaction. You can calculate ψ_0 from eqs. 19.14 and 19.15. Assuming $s = 10$ m^2/L, a medium of 0.1 M NaCl, and using the same conditions as in Section 19.7.2 ($K_{app} = K = 10^{-1.5}$ and $K_{a,app} = K_a = 10^{-8}$), you can obtain at pH 6: surface charge $= -0.93$ C/m^2, $\psi_0 = -0.20$ V, $[Cu^{2+}] = 1.0\times10^{-6}$ M, $[\equiv SOCu^+] = 4.9\times10^{-11}$ M, $[\equiv SOH] = 3.8\times10^{-6}$ M, and $[\equiv SO^-] = 9.6\times10^{-5}$ M.[†] (To accomplish this calculation, you must iterate since the surface charge is a function of surface species concentrations.)

19.8 SUMMARY

In this chapter, you learned about the final major type of reactions in aquatic systems, the equilibration of chemical species between solid and liquid phases. The symbol K_{s0} is used to represent the equilibrium constant for precipitation-dissolution equilibria written in a certain form (with metals and ligands as uncomplexed products). Equilibrium calculations are simplified by the choice of unit activity for pure solids. Similar to gas-liquid equilibria, the presence of a solid phase fixes the concentration of one or more dissolved species.

[†] The inclusion of surface charge effects had a dramatic impact on the equilibrium concentrations in this example. Part of the reason for the large effects is the relatively high surface charge. The large negative surface charge came about here in part because the only positively charged surface species considered was $\equiv SOCu^+$. More realistic acid-base surface chemistry can be included (see Problem 19.15).

In addition to the total metal concentration and pH, two factors were found to affect the solubility of metals. First, equilibrium partitioning may occur in the presence of more than one solid phase. In this case, the solid giving the lowest total dissolved concentration will control the solubility. Second, additional ligands generally increase the solubility of metals.

Special emphasis was placed in this chapter on the solubility of calcium carbonate. Calcium carbonate affects alkalinity, hardness, pH, and C_T. Inclusion of equilibrium with calcium carbonate allowed for the development of realistic models of the chemistry of natural waters and process waters in engineered systems.

The partitioning of dissolved species between water and surfaces also was discussed in this chapter. Two equilibrium models (surface complexation and Langmuir isotherm) were presented. The inclusion of electrostatic effects allows for the development of more realistic models of surface-dissolved species interactions.

19.9 PART IV CASE STUDY: THE KILLER LAKES

In Section 17.6, you were introduced to the lethal release of carbon dioxide from two Cameroonian lakes, Lake Nyos and Lake Monoun. With your new knowledge of water chemistry, take a moment to contemplate whether the measured water quality of the lakes makes sense.

Water quality data are summarized in Table 19.2. To analyze the system, return to the system discussed in Section 19.6.5: an open system in the presence of $CaCO_3(s)$. Although the bottom waters are not open to the atmosphere, they may likely be in equilibrium with bubbles of carbon dioxide. The charge balance for this system is:

$$\text{Alk} = C_B{}' - C_A + 2K_{s0}\alpha_0/(\alpha_2 K_H P_{CO_2})$$
$$= (\alpha_1 + 2\alpha_2)(K_H P_{CO_2}/\alpha_0) + [OH^-] - [H^+]$$

Table 19.2: Water Quality at the Bottom of Lakes Nyos and Monoun

Parameter	Data Source	Lake Nyos	Lake Monoun
Measured pH	Kling et al., 1999	4.81	5.63
Measured Alk (mg/L as $CaCO_3$)	Kling et al., 1999	1713	4016
Estimated P_{CO_2} (atm)	Section 18.6	20.1	9.3
Calc. pH	calculated	5.0	5.7
Calc. $[Ca^{2+}]$ (mg/L as $CaCO_3$)	calculated	1792	151
Calc. $C_B{}' - C_A$ (mg/L as $CaCO_3$)	calculated	– 79	3865

Using the critical P_{CO_2} values estimated in Section 18.6 and the measured total alkalinity values, you can calculate the pH at the bottom of each lake. The calculated values are listed in Table 19.2.

The pH can be approximated fairly reasonably by considering the water in equilibrium with carbon dioxide at the high pressures expected in the lake and also in equilibrium with calcium carbonate. Lake Monoun water requires a large concentration of excess cations to be fitted by this model.

SUMMARY OF KEY IDEAS

- Precipitation reactions may be fast or slow and cannot always be assumed to be at equilibrium

- The usual standard state is that the activity of pure solid phases under standard thermodynamic conditions is unity

- The activity of pure solids is usually either zero (if it does not exist) or one (if it does exist)

- Systems involving solids have no mass balances on the metals or ligands that precipitate in the solid phase(s)

- If in doubt, check to see if the solid forms (i.e., if the solubility product is exceeded)

- The existence of solids fixes the concentration of certain metal species through the solubility constraint (at fixed ligand concentration)

- The relationships between the concentration of complexes (e.g., between M^{2+} and MOH^+) is the same whether or not solids exist

- Each point on the total soluble metal line represents the total soluble metal concentration in equilibrium with the solid at a given pH (and other total ligand concentrations)

- Many metals are soluble at both low and high pH, with minimum solubility at an intermediate pH value

- If the solid phase is assumed to exist under certain conditions but the calculated total soluble metal concentration exceeds the available metal concentration, then the solid phase does not exist under those conditions

- Metal solubility is affected by the available metal concentration, pH, types of solid phases potentially formed, and ligand concentrations

- The controlling solid phase is the solid phase producing the *smallest* equilibrium total metal concentration: the *least* soluble solid phase controls solubility

- Two or more solid phases exist in equilibrium under the same set of conditions if they produce the *same* equilibrium total metal concentration

- Adding a new ligand usually increases the solubility of metals, assuming that the same solid phase controls before and after ligand addition and the pH and other water quality characteristics do not change appreciably

- Calcite (pK_{s0} = 8.48) is the most common mineral form of calcium carbonate, $CaCO_3(s)$

- The carbonate hardness is (1) equal to the Alk if Alk < total hardness and (2) equal to the total hardness otherwise

- If calcium carbonate exists, then its solubility and the resulting dissolved calcium concentration are pH dependent

- Calcium carbonate is very soluble at low pH

- Surfaces provide sites for chemical interaction with dissolved species

- Surface complexation equilibrium constants are not constant but depend on the surface charge (which, in turn, depends on the degree of surface complexation)

- The surface charge, calculated from the excess of charges at the surface, can be used to calculate the surface potential

PROBLEMS

19.1 The hypolimnetic (bottom) waters of a eutrophic lake contain 0.4 mg/L as S of total dissolved sulfide. The water is in equilibrium with $ZnS(s,\alpha)$.

A. Calculate the pK_{s0} value for $ZnS(s,\alpha)$.

B. What is the concentration of Zn^{2+} at pH 7.2?

C. What is the concentration of total soluble zinc at pH 7.2?

19.2 A municipal wastewater at pH 7.5 contains 6 mg/L as P of total orthophosphate (i.e., $P(+V)$). The engineering staff wants to precipitate the phosphate to achieve a residual orthophosphate concentration of 0.5 mg/L as P using ferrous iron. (For Parts A through C, ignore complexation/precipitation reactions between ferrous iron and ligands other than PO_4^{3-}. Assume that the Fe^{2+}-PO_4 solid phase has the same composition as vivianite, but is 10^7 times more soluble than vivianite.)

A. How much ferrous iron must be added to satisfy the stoichiometry of the precipitation reaction? (See Appendix D for the appropriate equilibria.)

B. How much ferrous iron must be added to maintain equilibrium between the residual phosphate and the solid phase?

C. The required dose is the sum of the ferrous iron doses calculated in Parts A and B. What is the required dose? How much total orthophosphate would have been removed if only the dose based on stoichiometry (i.e., your answer to Part A) had been added?

D. Under the water quality conditions stated in the problem, what solid controls the solubility: $Fe(OH)_2(s)$ or $Fe_3(PO_4)_2(H_2O)_8$?

19.3 A student is studying a hazardous waste stream containing cadmium, and she needs to know the solubility product for $CdCO_3(s)$. She adds some $CdCO_3(s)$ to water and measures the pH as pH 8.1. The solid does not dissolve completely. If the suspension is in equilibrium with the atmosphere ($P_{CO_2} = 10^{-3.5}$ atm), what is the K_{s0} for $CdCO_3(s)$? Compare your answer with the K_{s0} for $CdCO_3(s,Otavite)$ ($\log K_{s0} = -12.1$ from Appendix D). Hint: perform a charge balance and think about the alkalinity equation in an open system.

19.4 For certain divalent metals (M^{2+}), the major hydroxo complexes are MOH^+, $M(OH)_2^0$, and $M(OH)_3^-$. For these metals, you can write the following equilibria:

$$
\begin{aligned}
M^{2+} + OH^- &= MOH^+ & K_1 \\
M^{2+} + 2OH^- &= M(OH)_2^0 & K_2 \\
M^{2+} + 3OH^- &= M(OH)_3^- & K_3 \\
M(OH)_2(s) &= M^{2+} + 2OH^- & K_{s0}
\end{aligned}
$$

A. Show that the pH of minimum solubility for these metals is approximated by $14 + (pK_3 - pK_1)/2$. Hint: write an expression for the total soluble metal as a function of $[OH^-]$. Then differentiate it with respect to $[OH^-]$ and set the resulting expression equal to zero.

B. Compare the exact pH of minimum solubility with the approximation for the following metals (numbers following are $\log K_1$, $\log K_2$, $\log K_3$, and $\log K_{s0}$): Fe^{2+} (4.5, 7.4, 11, and -15.1), Co^{2+} (4.3, 9.2, 10.5, and -15.7), Ni^{2+} (4.1, 9, 12, and -17.2), Pb^{2+} (6.3, 10.9, 13.9, and -15.3), and Hg^{2+} (10.6, 21.8, 20.9, and -25.4).

19.5 Why do ligands generally *increase* the solubility of metals that are in equilibrium with solids? Why do metals typically show high aqueous solubility at low pH *and* high pH?

19.6 The solubility of many metals is increased in the presence of chloride. Plot the total soluble mercury concentration as a function of total chloride concentration at pH 7.7 if the solubility of mercury is controlled by mercuric oxide. Do you expect much increase in mercury solubility at chloride concentrations typical of fresh waters (say, 20 mg/L)? (Note: at relatively low chloride concentrations and pH 7.7, only the first chloro and hydroxo complexes of mercury need be considered. See Appendix D for equilibrium constants.)

19.7 Cyanide sometimes is added to increase the solubility of metals in metal plating operations. A metal plating facility faces the following two discharge regulations: total dissolved nickel must be less than or equal to 0.5 mg/L as Ni ($= 1.0 \times 10^{-5}$ M), and total dissolved cyanide must be less than or equal to 0.5 mg/L as CN. The plant personnel wish to add as much cyanide as possible to increase plating efficiency. They decide to remove nickel in the waste stream by precipitation. To what pH must the waste stream be adjusted to precipitate nickel and meet the effluent limits on both nickel and cyanide? (Ignore the formation of $NiCO_3(s)$ and $Ni_4OH_4^{2-}$.)

19.8 Describe in words how you determine which of two solids controls the solubility of a metal at equilibrium.

19.9 The following questions concern the solubility of manganese. Ignore Mn dimers. Assume the water contains 2×10^{-3} M of total dissolved carbonate (C_T).

A. Draw the solubility diagram for Mn(+II) as a function of pH if $Mn(OH)_2(s)$ controls the Mn solubility and $MnCO_3(s)$ does not exist. Be sure to include the Mn-CO_3 and Mn-HCO_3 complexes.

B. Draw the solubility diagram for Mn(+II) as a function of pH if $Mn(OH)_2(s)$ does not exist and $MnCO_3(s)$ controls the solubility. Use $K_{s0,CO_3} = -10.5$. Be sure to include the hydroxo manganese complexes.

C. Draw the solubility diagram for Mn(+II) as a function of pH, assuming both solids exist (and $C_T = 2 \times 10^{-3}$ M). Show the regions where $MnCO_3(s)$ controls the Mn(+II) solubility and where $Mn(OH)_2(s)$ controls the Mn(+II) solubility.

19.10 The pH of a lake is 6.0, and the total dissolved carbonate is 1×10^{-3} M. The total soluble copper is 1.0×10^{-6} M. Assume that C_T is constant and the total soluble copper consists only of Cu^{2+} and $CuOH^+$. Consider only the solids $Cu(OH)_2(s)$ and $Cu_2(OH)_2(CO_3)(s,$ Malachite$)$.

A. Which solid, if any, controls the solubility of copper in the lake?

B. A lake liming project in an adjacent watershed raises the pH of the lake to 7.0. Which solid, if any, now controls the solubility of copper in the lake? What is the total soluble copper now?

C. If the pH of the lake is raised to 8.0, which solid (if any) controls the solubility of copper in the lake and what is the total soluble copper?

19.11 Manganese can form two common solids in natural waters, $MnO_2(s)$ and $Mn(OH)_2(s)$. For equilibrium constants, see Appendix D and Table 16.3.

A. Which of these solids controls the manganese solubility at pH 8 and pe = 7?

B. Sketch a pe-pH diagram for the equilibrium between $MnO_2(s)$ and $Mn(OH)_2(s)$.

19.12 A groundwater (closed system) in contact with $CaCO_3(s)$ has an alkalinity of 100 mg/L as $CaCO_3$ and a pH of 6.8.

A. *Without doing any calculations* (using only the text), how do you know that this water contains alkalinity (in addition to Ca^{2+})? Calculate the value of the alkalinity other than Ca^{2+}.

B. What is the equilibrium concentration of Ca^{2+} in the water?

C. Calculate the value of the alkalinity other than Ca^{2+}.

D. The water is exposed to the atmosphere ($P_{CO_2} = 10^{-3.5}$ atm). What are the new values of the Ca^{2+} concentration, pH, concentration of total carbonate species (C_T), and alkalinity?   Explain qualitatively why each of the four parameters increased or decreased when the system was equilibrated with the atmosphere.

E. The treatment plant operators are concerned about the low alkalinity of the water after it has been brought to the surface. They add enough NaOH to the water to increase the alkalinity back to 100 mg/L as $CaCO_3$. How much NaOH did they add (in mg/L)? What are the new values of pH, C_T, and $[Ca^{2+}]$? (Note: Assume the water is still in equilibrium with the atmosphere.)

19.13 A water is found to have the following characteristics: pH 7.8, $[Ca^{2+}] = 8\times10^{-4}$ M, $[Mg^{2+}] = 2.5\times10^{-4}$ M, and Alk = 75 mg/L as $CaCO_3$.

A. Calculate the calcium hardness, magnesium hardness, total hardness, carbonate hardness, and non-carbonate hardness (all in mg/L as $CaCO_3$).

B. What is the C_T of the water?

C. To what pH does the water have to be adjusted to reduce the total hardness to 50 mg/L as $CaCO_3$? Assume that the pH adjustment does not affect C_T. Also assume that magnesium does not precipitate as $Mg(OH)_2(s)$ (then check this assumption).

19.14 A worker was instructed to dose 3 kg of $CaCl_2$ into a pond during the winter for ice control. Because of a tragic mislabeling of the bags, the worker accidentally added 3 kg of $CdCl_2$ into the 80,000 L pond. The water in the pond had a pH of 6.9, and a total carbonate concentration (C_T) of 2×10^{-3} M. Does any precipitate form? If so, which one? How much was formed? What is the remaining total soluble cadmium concentration in the pond? Assume no interactions with the atmosphere.

19.15 The point of zero charge (pH_{zpc}) is defined to be the pH where the charge on the surface is zero. (The pH_{zpc} corresponds to a proton condition of zero at the surface, if the zero level of protons is $\equiv$SOH.) Estimate the pH_{zpc} of hematite (α-Fe_2O_3, abbreviated $\equiv$FeOH). In your estimate, do not correct for electrostatics. Is the charge on hematite positive or negative at pH 7.5? At 9.0? For hematite:

$$\equiv FeOH_2^+ \; = \; \equiv FeOH + H^+ \qquad K_1 = 10^{-7.25}$$
$$\equiv FeOH \; = \; \equiv FeO^- + H^+ \qquad K_2 = 10^{-9.25}$$

19.16 For the adsorption of DDT onto activated carbon, the following Langmuir isotherm constants were found: $b = 303$ L/mg and $q_{max} = 41.3$ mg/g. Find the aqueous equilibrium concentration of DDT if 100 mg of activated carbon and 1 μg of DDT are combined in 1 L of water.

Beyond Dilute Solutions at Equilibrium

Now this is not the end. It is not even the beginning of the end. But it is, perhaps, the end of the beginning.
Winston Churchill

CHAPTER 20

Getting Started with Beyond Dilute Solutions at Equilibrium

20.1 INTRODUCTION

In the first five parts of this text, you have been exposed to all of the major types of chemical equilibria in aquatic systems: acid-base, complexation, redox, gas-liquid, and solid-liquid equilibria. You should feel confident in your abilities to develop a conceptual model for the equilibrium chemistry of an aqueous system. In addition, you have the tools to translate the conceptual model into a mathematical model (i.e., into n equations in the n unknown species concentrations) and solve the mathematical model.

So what is left to learn? In this part of the text, you will challenge some of the assumptions made throughout the text. For example, until now we have assumed routinely that activities and concentrations are interchangeable. In Chapter 21, you will learn to identify the conditions under which this assumption is valid, and you will correct for the differences between activity and concentration, if necessary. In addition, you have been limited so far to temperature and pressure conditions for which you have thermodynamic data. In Chapter 21, you will learn how to calculate equilibrium constants at non-standard temperatures and pressures.

Another assumption inherent in all the examples so far in this text is that the system is at equilibrium. Recall that equilibrium was defined in Chapter 3 as the condition where all thermodynamic functions (including species concentrations) do not change with time. In many cases, you can still calculate species concentrations when the system is not at equilibrium; that is, when species concentrations change with time. Calculating species concentrations over time requires an additional set of tools, which are introduced in Chapter 22.

Finally, this text would not be complete without a few integrated case studies to show you how all the tools you have learned can be brought together and used to solve new problems. Several integrated case studies are presented in Chapter 23.

20.2 EXTENSIONS TO NONIDEAL AND NONSTANDARD CONDITIONS

20.2.1 Introduction

Even though you have learned about many types of equilibria, the applications have been limited. Why? Almost all applications so far in this

Key idea: All applications so far in this text have been limited to dilute solutions (where ideal behavior is assumed) at 25°C and 1 atm

ideal solution: a mixture in which species do not interact with other nonreactive species

text have been limited to dilute solutions at 25°C and 1 atm. Clearly, the real world is not always dilute and at 25°C and 1 atm.

Remember that dilute conditions are required because we have *assumed* that the systems behaved ideally. *Ideal behavior* means that species mimic ideal gases and do not interact with any other species. In the systems you have examined, chemical species clearly *do* interact (otherwise, no reactions would occur). However, in *ideal solutions*, species do not interact with other nonreactive species.[†] Ideal behavior is observed only in dilute solutions.

20.2.2 When are activities and concentrations interchangeable?

Try this experiment. Fill a beaker with water and measure its pH. *Carefully* (with proper eye and skin protection) add dilute HCl dropwise until the pH is about 3. At this pH, you can convince yourself from an equilibrium calculation that HCl is nearly completely dissociated to H^+ and Cl^-. Now add some table salt to the water and remeasure the pH. If you keep adding salt, you will find that the pH eventually will increase. Does this make sense? Not really. A charge balance on this system is as follows:

$$[H^+] + [Na^+] = [Cl^-]_{from\ HCl} + [Cl^-]_{from\ NaCl} + [OH^-]$$

Since $[Na^+] = [Cl^-]_{from\ NaCl}$, the charge balance tells you that at low pH, the H^+ concentration should be equal to the chloride concentration from the original HCl regardless of how much salt was added. Thus, $[H^+]$ should be constant and the pH should remain constant.

Thoughtful Pause

Can you see a flaw in this reasoning?

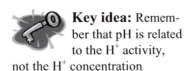

Key idea: Remember that pH is related to the H^+ activity, not the H^+ concentration

Remember that pH is related to the H^+ *activity*, not the H^+ concentration. The H^+ ions are constrained by the added sodium and chloride ions. Thus, the activity of H^+ is reduced and, as a result, the pH increases (since pH = $-\log\{H^+\}$).

In the presence of the added salt, activities and concentrations are *not* interchangeable. Fortunately, models have been developed to quantify the relationship between activity and concentration as a function of added unreactive species. The models will be presented in Chapter 21. The models also will allow you to express equilibrium constants in terms of the *concentrations* of the participating chemical species instead of the *activities* of the participating chemical species.

[†] Formally, ideal solutions are mixtures where the chemical potential of each species increases linearly with the log of the activity of the species.

20.2.3 Effects of temperature

Your everyday experiences tell you that temperature affects the equilibrium position of thermodynamic systems. For example, you know that the solubility of common substances depends on temperature: you can dissolve more sugar in hot coffee than in cold water.[†] Another example in the environment is the solubility of gases in water. Aqueous solubility data for several gases are plotted as a function of temperature in Figure 20.1. Note that all the gases shown are *more* soluble at *lower* temperature, implying that you can heat water to drive off dissolved gases. From Figure 20.1, you can see that the solubility of hydrogen gas apparently is much less sensitive to temperature than the solubility of carbon dioxide.

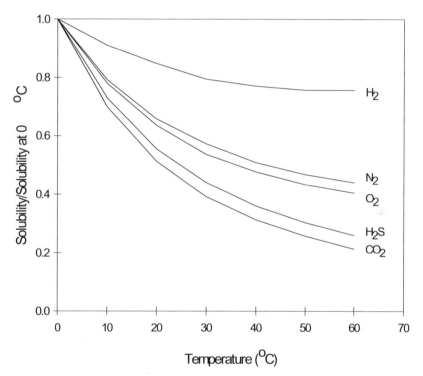

**Figure 20.1: Effect of Temperature on the Solubility
of Several Gases
(calculated from Henry's Law constants found in Metcalf and Eddy, 2003)**

[†] As a quantitative example, glucose is four times more soluble in water at 70°C than in water at 25°C.

Since the solubility of solids and gases is determined by equilibrium constants, it makes sense that such equilibrium constants are dependent on temperature. In fact, temperature affects other equilibrium constants as well. Temperature affects the strength of acids (determined by K_a values), oxidation strength (determined by redox equilibrium constants), and the extent of complexation (determined by complex formation equilibrium constants). In fact, nearly all equilibrium constants are affected in some way by temperature. Why? Remember that equilibrium constants are thermodynamic functions and thermodynamic functions are affected by temperature and pressure.

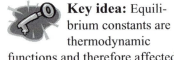 **Key idea:** Equilibrium constants are thermodynamic functions and therefore affected by temperature and pressure

Once again, models have been developed to adjust equilibrium constants for temperature and pressure. These models and their applications to aquatic systems are presented in Chapter 21.

20.3 NONEQUILIBRIUM CONDITIONS

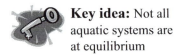 **Key idea:** Not all aquatic systems are at equilibrium

Are all aquatic systems at equilibrium? The answer is emphatically no! As evidence that not all systems are at equilibrium, equilibrium calculations sometimes give answers that do not make physical sense. Refer back to Figure 16.12. In this figure, the pe-pH diagram for the Cl(−I)-Cl(0)-Cl(+I) family is presented. Note that chloride is the only chlorine-containing species in the pe-pH diagram that is thermodynamically stable. This result certainly would be a big surprise to anyone who has used bleach to clean clothes or maintained a swimming pool. We use chlorine species other than chloride all the time in environmental engineering. Is it possible that only chloride is thermodynamically stable in water?

In fact, Figure 16.12 is correct: chloride *is* the only chlorine-containing compound in Figure 16.1 that is thermodynamically stable in water. At equilibrium, HOCl, OCl⁻, and Cl$_2$(aq), will oxidize water to oxygen, as shown in the pe-pH diagram. However, the oxidation of water by Cl(+I) is very slow. Nevertheless, the slow decay of chlorine must be included with the thermodynamic relationships in Figure 16.12 to understand chlorine chemistry.

The slow decay of chlorine is an example of species concentrations that change over time. As stated in Section 20.1, a change in species over time indicates a nonequilibrium situation. Unfortunately, the thermodynamically based tools developed for equilibrium conditions do not help to determine species concentrations that change over time. A different approach is needed. The branch of chemistry devoted to calculating species concentrations as they change over time is called *chemical kinetics*.

Chapter 22 will provide a brief introduction to chemical kinetics. The focus in this chapter will be on reaction *rates* rather than reaction thermodynamics. You will find that certain types of energy profiles in reactions will lead to nonequilibrium conditions. Although the tools in Chapter 22 look different than those in the rest of the text, it still will be necessary to perform mass balances and write reaction stoichiometry. In addition, analogies with the material presented in Chapter 21 will allow you to determine the effects of nonreactive species and temperature on the rate of chemical reactions.

20.4 INTEGRATED CASE STUDIES

This text has taken a traditional approach in presenting you with the different types of chemical equilibria separately. In the real world, all pertinent reactions (acid-base equilibria, redox equilibria, etc.) occur simultaneously. In addition, rapid equilibria occur at the same time as slower, kinetically controlled reactions. Through several integrated case studies in Chapter 23, you will see how the tools and approaches you have learned can be used to analyze more complex systems.

Although the examples in Chapter 23 are fairly complex, it is important to realize that they require the same approaches you have mastered: develop a species list, write the pertinent equilibria, and develop mass balances. Whether your system has three species (H^+, OH^-, and H_2O) or 3,000 species, the techniques you have studied here can be used to quantify and understand speciation.

20.5 PART V ROAD MAP

Part VI of this book discusses nonideal, nonstandard, nonequilibrium, and nontrivial systems. By the conclusion of this part of the text, you should know the effects of nonreactive species, temperature, and pressure equilibria; have a quantitative appreciation for chemical kinetics; and be comfortable with the application of equilibrium and kinetic calculation tools to larger systems.

In Chapter 21, the theory and applications of thermodynamics from Chapter 3 will be extended. The concept of nonideal behavior is introduced, along with a common measure of the contribution of nonreactive species (called *ionic strength*). Quantitative models will allow you to determine the effects of nonreactive species on the activities of individual species and equilibrium constants. In addition, the effects of temperature and pressure on equilibrium constants will be quantified. By the conclusion of Chapter 21, you should be able to determine species concentrations at equilibrium under a variety of thermodynamic conditions.

Chapter 22 provides solution techniques for systems under kinetic control. The energy profiles leading to both equilibrium and nonequilibrium behavior will be discussed. Several common kinetic models and their applications to the environment will be presented. By the conclusion of Chapter 22, you should be able to answer quantitative questions about the chemical species under kinetic control.

Finally, in Chapter 23, you will see the power of the techniques you have mastered thus far. By the conclusion of Chapter 23, you should appreciate and be less anxious about complex aquatic systems.

20.6 SUMMARY

In this chapter, you were reminded of the limitations caused by the assumptions inherent in previous chapters in the text. You learned that nonreactive species (such as Na^+ and Cl^- in the HCl example in Section 20.2.2) can influence the activity of chemical species. In addition, everyday examples with the solubility of solids and gases were used to illustrate the point that temperature (and pressure) affect almost all equilibria.

Spurious equilibrium calculation results were used to demonstrate that some environmental systems are not at equilibrium. Clearly, nonequilibrium tools are needed to determine the concentrations of chemical species in such systems.

Finally, the need for using all pertinent equilibrium types during equilibrium calculations was emphasized. In many cases, *all* the equilibrium types discussed so far in the text must be combined to understand environmental problems. Nonequilibrium (kinetic) tools also may be needed and must be integrated with equilibrium solution methods where appropriate.

SUMMARY OF KEY IDEAS

- All applications so far in this text have been limited to dilute solutions (where ideal behavior is assumed) at 25°C and 1 atm

- Remember that pH is related to the H^+ activity, not the H^+ concentration

- Equilibrium constants are thermodynamic functions and therefore affected by temperature and pressure

- Not all aquatic systems are at equilibrium

CHAPTER 21

Thermodynamics Revisited: The Effects of Ionic Strength, Temperature, and Pressure

21.1 INTRODUCTION

Chapter 3 provided an introduction to the broad field of thermodynamics. The purpose of Chapter 3 was to develop the thermodynamic basis of equilibrium. Among other things, you learned in Chapter 3 that the changes in standard Gibbs free energy of reaction (and thus, the equilibrium constant) are thermodynamic functions. As a result, they are dependent on temperature, pressure, and the number of moles of material in the system. In this chapter, quantitative relationships describing the effects of temperature, pressure, and the number of moles of material in the system on the equilibrium constant will be developed and used.

In Section 21.2, the effects of the medium composition on the equilibrium position of a system will be examined in two steps. First, quantitative effects of the medium composition on the activity of individual chemical species will be developed (Sections 21.2.1 through 21.2.5). Recall that activity is an idealized concentration. Second, the dependency of equilibrium constants (and equilibrium concentrations) on medium composition will be explored in a quantitative fashion in Sections 21.2.6 and 21.2.7.

Section 21.3 will explore the effects of temperature on equilibrium constants. You will learn that equilibrium constants can increase or decrease with temperature, depending on whether heat is evolved or absorbed during the reaction. A model will be developed from thermodynamic principles to allow for the calculation of equilibrium constants at temperatures other than 25°C. As a result, the effects of temperature on the equilibrium composition of a system can be quantified.

Similarly, the effects of pressure on equilibrium constants will be investigated in Section 21.4. You will learn that equilibrium constants are relatively insensitive to pressure. A quantitative model will be developed to allow for the calculation of equilibrium constants and equilibrium concentrations at pressures other than 1 atm.

21.2 EFFECTS OF IONIC STRENGTH

Key idea: Chemical species may behave differently under different environmental conditions, even at the same mass or molar concentration

Key idea: Even if two species do not react with one another, the behavior (i.e., the activity) of each species will be influenced by the other species

21.2.1 Ionic strength

The concept of activity was introduced in Section 2.10. Recall the idea that chemical species may behave differently under different environmental conditions, even at the same mass or molar concentration. One factor that influences the behavior of chemical species is the presence of other species. *Even if two species do not react with one another, the behavior of each species will be influenced by the other species.*

If you want to determine the effects of other species on the activity of a given species, you must first devise a way to quantify the concentration of other species. A first approach to creating a measure of the concentration of other species would be to simply sum the concentrations of all the species. This idea makes some sense. However, the experimental evidence reveals that the *charge* of the species plays a role in determining the effects on its and other species activities.

Thoughtful Pause
Why does charge play a role?

Charge plays an important role because the interactions between species are mainly electrostatic in nature.

A second approach to creating a measure of the concentration of other species, then, would be to sum the product of a species charge and its concentration. This new parameter suffers from a serious problem: negatively charged species would cancel out positively charged species if found at the same concentration (as, for example, sodium and chloride ions in a 0.01 M NaCl solution). In addition, additional experimental evidence suggests that the effects of species on activity are not simply proportional to charge.

ionic strength (*I*): a measure of the concentration of charged species, given by $I = \frac{1}{2}\sum z_i^2 C_i$

The accepted parameter used to describe the quantity of material influencing species activity is the ***ionic strength***. The ionic strength (abbreviated I or μ) is one-half the sum of the product of a species charge squared and its concentration. For a system containing n species:

$$I = \frac{1}{2}\sum_{i=1}^{n} z_i^2 C_i$$

eq. 21.1

Example 21.1: **Calculation of Ionic Strength**

What is the average ionic strength of the Great Lakes and the Dead Sea?

where z_i and C_i are the charge on species i and concentration of species i, respectively. Note that I has the same units as C_i, usually moles/L. Note that by squaring the charge, the effects of cations and anions do not cancel each other out.

Solution:
You can calculate the ionic strength from the data in Table 13.1, assuming that the table lists all the main ions. Converting the data from meq/L to mM, the concentrations for the Great Lakes are:

Ion	C_i (mM)	$z_i^2 C_i$
Na^+	0.28	0.28
K^+	0.03	0.03
Ca^{2+}	0.75	3.00
Mg^{2+}	0.30	1.18
SO_4^{2-}	0.19	0.76
Cl^-	0.37	0.37
HCO_3^-	1.65	1.65
Sum:		7.27

Thus, **the average ionic strength of the Great Lakes is about 3.6 mM**. Similar calculations for the Dead Sea show that **the ionic strength of the Dead Sea is about 8 M**.

The calculation of I is straightforward. For the 0.01 M NaCl solution discussed above, the ionic strength is:

$$I = \tfrac{1}{2}(z_{Na}^2 C_{Na} + z_{Cl}^2 C_{Cl})$$
$$= \tfrac{1}{2}[(+1)^2(0.01 \text{ M}) + (-1)^2(0.01 \text{ M})]$$
$$= 0.01 \text{ M}$$

The ionic strength depends on the ionic charges. For a 0.01 M Na_2SO_4 solution, for example (assuming that sulfate does not react appreciably with water):

$$I = \tfrac{1}{2}(z_{Na}^2 C_{Na} + z_{SO_4}^2 C_{SO_4})$$
$$= \tfrac{1}{2}[(+1)^2(0.02 \text{ M}) + (-2)^2(0.01 \text{ M})]$$
$$= 0.03 \text{ M}$$

Other cases are shown in Example 21.1.

The calculation of ionic strength is simple when the species concentrations are known. Even when the water composition is not known, the ionic strength can be estimated. Butler (1982) reviewed several methods of estimating the ionic strength, summarized here in Table 21.1. In general, I can be estimated from physical measurements such as conductivity and total dissolved solids (see Section 2.4.3 for a review of TDS). Since most natural waters have TDS values in the 50 to 500 mg/L range, expected ionic strengths of fresh waters range from about 0.001 to 0.01 M.

Table 21.1: Correlations Used to Estimate the Ionic Strength

Estimate	Reasoning	Reference
$I = 1.6 \times 10^{-5} \kappa$ $I = 1.4 \times 10^{-5} \kappa$	From surveys	see Snoeyink and Jenkins (1980)
$I = 2.04 \times 10^{-5} (\text{TDS})$	Assumes the same I:TDS ratio as seawater	Butler (1982)
$I = 2.50 \times 10^{-5} (\text{TDS})$	Assumes water equivalent to a monovalent salt of MW = 40 g/mol	Langelier (1936)
$I = \tfrac{1}{2}\Sigma z_i^2 C_i - 2.50 \times 10^{-5} R$	Accounts for known species concentrations	Butler (1982)

κ = conductivity (in μmho/cm = μS/cm; S = siemens)
R = unaccounted for TDS
   = measured TDS − contribution of species of known concentration to TDS

21.2.2 Activity coefficients

Concentration (C_i) and activity (a_i) are related through an activity coefficient, γ_i, as follows:

$$a_i = \gamma_i C_i \quad \text{or:} \ \{i\} = \gamma_i \{i\}$$

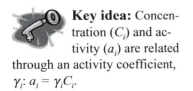

Key idea: Concentration (C_i) and activity (a_i) are related through an activity coefficient, γ_i: $a_i = \gamma_i C_i$.

The activity coefficient is necessary because ions in solution appear to behave differently than expected. In other words, ions exhibit *nonideal* behavior.

How do you know that ions exhibit nonideal behavior? One way that investigators measure the effects of ionic strength on species activities is through the effects of ionic strength on equilibria. In the mathematical form of an equilibrium expression, activities are multiplied or divided by one another. This means that activity coefficients are multiplied or divided by one another. In this way, the products and ratios of activity coefficients can be determined experimentally. The effects of ionic strength on equilibria will be considered in Section 21.2.6. Products and ratios of activity coefficients can be determined experimentally but the activity coefficients of single ions cannot be determined by experiment. However, single-ion activity coefficients are very useful and will be examined in Sections 21.2.3 through 21.2.5.

As with other thermodynamic functions, activities must be defined relative to some standard state. Several standard states are possible. In this text, the ***infinite dilution standard state*** will be used. In this convention, γ_i approaches 1 as I approaches 0. The infinite dilution standard state is equivalent to the idea you have seen many times in this text: if the system is dilute enough, then the concentrations and activities are interchangeable (i.e., $\gamma_i \to 1$).

infinite dilution standard state :
$\gamma_i \to 1$ as $I \to 0$

21.2.3 Single-ion activity coefficients for ions

For ions with activities *not* controlled by equilibria, activity coefficients can only be estimated. Remember that activity coefficients are necessary because ions in solution exhibit nonideal behavior. Peter J.W. Debye (1884-1966) and Erich Hückel (1896-1980) hypothesized in 1923 that the deviation from ideal behavior for ions stemmed from the interactions between an ion and the cloud of counterions surrounding it. Thus, to estimate γ_i, we must go back to the electric double-layer (EDL) theory described in Section 19.6.4. Debye and Hückel assumed that the interaction was only electrostatic in nature. This assumption allowed them to calculate the electrostatic Gibbs free energy from the electrostatic work required to bring a second ion close to the ion of interest. Their theory predicted that the activity coefficient was given by:

$$\ln \gamma_i = \frac{-\kappa (z_i e)^2}{2 \varepsilon k T} \qquad \text{eq. 21.2}$$

In eq. 21.2 (all units in the cgs system):

κ = Debye radius (inversely related to the thickness of the EDL)
e = unit charge on the electron = 4.8032×10^{-10} esu
ϵ = dielectric constant of water
k = Boltzmann's constant = 1.381×10^{-16} erg/°K
T = temperature (°K)

The Debye radius is given by:

$$\kappa = \left(\frac{8\pi N_A e^2 \rho}{1000\epsilon kT} \right)^{1/2} I^{1/2}$$

where N_A = Avogadro's number = 6.022×10^{23} mol^{-1} and ρ = water density. Evaluating κ in water and combining with eq. 21.2:

$$\log_{10}\gamma_i = -Az_i^2 I^{1/2} \qquad\qquad \text{eq. 21.3}$$

In eq. 21.3, $A = 1.82\times10^6 (\epsilon T)^{-3/2}$. At 25°C, ϵ is 78.54 and $A \approx 0.5$.

Equation 21.3 is called the ***Debye-Hückel limiting law***. The limiting law appears to correct for nonideal behavior only at relatively low ionic strengths. Over the years, different modifications of the limiting law have been proposed to increase its ability to describe non-ideal behavior. For example, the theory behind the limiting law assumes that the ions can get infinitely close to one another. One modification involves the recognition that ions have a finite size. This modification results in the replacement of κ in eq. 21.2 with $\kappa/(1 + \kappa a)$, where a is the distance of closest approach of the ions. The parameter a is called the ***ion size parameter***. With this replacement, eq. 21.3 becomes:

$$\log_{10}\gamma_i = -Az_i^2 I^{1/2}/(1 + BI^{1/2})$$

In this equation, B for water is $50.3a/(\epsilon T)^{1/2}$ where a is in Angstroms. At 25°C, $B = 0.33a$ (a in Angstroms). The ion size parameter is listed for several common ions in Table 21.2.

Brønsted suggested that at high ionic strength, $\log_{10}\gamma_i$ should be proportional to I. This led to the following semiempirical modification of the limiting law:

$$\log_{10}\gamma_i = -Az_i^2 I^{1/2}/(1 + BaI^{1/2}) - CI \qquad\qquad \text{eq. 21.4}$$

In eq. 21.4, B and C often are treated as empirical constants, with C usually equal to 0.2 or 0.3. These and other approximations for individual ion activity coefficients are listed in Table 21.3.

Debye-Hückel limiting law: an approximation for the single-ion activity coefficient as a function of ionic strength: $\log_{10}\gamma_i = -Az_i^2 I^{1/2}$, where $A \approx 0.5$ at 25°C

ion size parameter (***a***): a parameter, used in a common extension to the Debye-Hückel limiting law, that accounts for the finite size of ions

21.2.4 Using single-ion activity coefficients

Single-ion activity coefficients can be used to relate the concentration and activity of a chemical species at a given ionic strength. For example, consider a 10^{-3} M NaCl solution. If NaCl dissociates completely, then the *concentrations* of both Na^+ and Cl^- will be 10^{-3} M. What are the *activities* of sodium and chloride ion? Recall that $a_i = \gamma_i C_i$ or $\{i\} = \gamma_i[i]$.

Table 21.2: Values of the Ion Size Parameter *a*
(from Kielland, 1937)

a (Å)	Univalent ions	Divalent ions	Trivalent ions
3	Ag^+, K^+, NH_4^+ Cl^-, ClO_4^-, HS^-, I^-, NO_3^-, OH^-		
4	Na^+ CH_3COO^-, HCO_3^-, $H_2PO_4^-$	HPO_4^{2-}, SO_4^{2-}	PO_4^{3-}
5		Ba^{2+}, Pb^{2+}, Sr^{2+} CO_3^{2-}	
6		Ca^{2+}, Cu^{2+}, Fe^{2+}, Mn^{2+}, Sn^{2+}, Zn^{2+}	
8		Be^{2+}, Mg^{2+}	
9	H^+		Al^{3+}, Ce^{3+}, Fe^{3+}, La^{3+}

Thoughtful Pause

What is the ionic strength of a 10^{-3} M NaCl solution?

You can verify that the ionic strength of a 10^{-3} M NaCl solution is 10^{-3} M. Using the Debye-Hückel limiting law:

$$\log_{10}\gamma_{Na} = \log_{10}\gamma_{Cl} = (-0.5)(1)(0.001)^{\frac{1}{2}} = -0.0158 \text{ and:}$$
$$\gamma_{Na} = \gamma_{Cl} = 0.96$$

Thus, $\{Na^+\} = \{Cl^-\} = 9.6 \times 10^{-4}$ M. For a 10^{-2} M NaCl solution, $I = 0.01$ M, and the Debye-Hückel limiting law no longer applies. Using the other models in Table 12.3, you can verify that the estimates of the activity coefficients for sodium and chloride ions range from 0.89 to 0.90. Note that at higher ionic strength, the single-ion activity coefficient *decreases*: in other words, *the activity of the ions decreases as the ion of interest is constrained by other ions.*

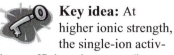 **Key idea:** At higher ionic strength, the single-ion activity coefficient decreases (i.e., the activity of the ions decreases as the ions of interest are constrained by other ions)

Table 21.3: Common Approximations for Individual Ion Activity Coefficients
(after Stumm and Morgan, 1996, and Sun et al., 1980; all constants at 25°C)

Name	Constants[1]	Equation	Comments
Debye-Hückel limiting law	$B = 0$ $C = 0$	$\log_{10}\gamma_i = -0.5z_i^2I^{1/2}$	Valid for $I \lesssim 5\times10^{-3}$ M
Extended limiting law	$B \approx 0.33$ $C = 0$	$\log_{10}\gamma_i =$ $-0.5z_i^2I^{1/2}/(1 + 0.33aI^{1/2})$	Valid for $I \lesssim 0.1$M
Güntelberg[1]	$Ba = 1$ $C = 0$	$\log_{10}\gamma_i =$ $-0.5z_i^2I^{1/2}/(1 + I^{1/2})$	Valid for $I \lesssim 0.1$ M Good for mixtures
Davies[2] ($C = 0.2$)	$Ba = 1$ $C = 0.2$	$\log_{10}\gamma_i =$ $-0.5z_i^2I^{1/2}/(1 + I^{1/2}) - 0.2I$	Valid for $I \lesssim 0.5$ M
Davies[2] ($C = 0.3$)	$Ba = 1$ $C = 0.3$	$\log_{10}\gamma_i =$ $-0.5z_i^2I^{1/2}/(1 + I^{1/2}) - 0.3I$	Valid for $I \lesssim 0.5$ M
Scatchard[3] ($C = 0$)	$Ba = 1.5$ $C = 0$	$\log_{10}\gamma_i =$ $-0.5z_i^2I^{1/2}/(1 + 1.5I^{1/2})$	Valid for $I \lesssim 0.1$ M

Notes: 1. Constants in: $\log_{10}\gamma_i = -Az_i^2I^{1/2}/(1 + BaI^{1/2}) - CI$ (eq. 21.4)
2. $Ba \approx 1$ when $a = 3$. This is valid for certain common ions such as Cl^- and OH^- (see Table 21.2).
3. $Ba \approx 1.5$ when $a = 4$ to 5. This is valid for certain common ions such as Na^+, HCO_3^-, and CO_3^{2-} (see Table 21.2).

At first glance, the calculation of activity coefficients appears to be a daunting task. First, you must select an approximation for the single-ion activity coefficient. As listed in Table 12.2, there are several different models for calculating γ_i. Which model should you use? It is important to recognize that the approximations have ranges of ionic strength over which they are applicable. Thus, use a model valid for the ionic strength of interest. Within their appropriate ion strength ranges, the models give reasonably similar estimates for $I < 0.01$ M. Second, the parameters in the approximations for single-ion activity coefficients are functions of temperature and pressure. Thus, the activity coefficients will vary with temperature and pressure. Fortunately, over conditions of environmental interest, the temperature and pressure dependence is very small and generally can be ignored.

Key idea: Always use single-ion activity coefficient approximations only over the ionic strength range for which they are valid

The single-ion activity coefficients are very useful for the interpretation of analytical measurements make with electrochemical methods. Some electrochemical methods inherently measure species activity rather than concentration. For example, potentiometric methods such as ion-specific electrodes (where the voltage is measured at zero current) measure the *activity* of the target ion, not the ion concentration. Consider the pH electrode. What would be the measured pH of a solution of 0.001 M nitric acid (HNO_3) in 5×10^{-3} M NaCl? Is the pH ($= -\log\{H^+\}$) exactly 3.00? To answer this question, calculate the H^+ concentration and then use the activity coefficient to calculate the H^+ activity.

***Example 21.2:* Use of Single-Ion Activity Coefficients**

In seawater, an ion-specific electrode for copper gives a response of 100 μM Cu^{2+}. What is the concentration of Cu^{2+}?

Solution:
The ion-specific electrode gives the Cu^{2+} activity = 100 μM. The major ions in seawater are Cl⁻ (0.54 M), Na⁺ (0.47 M), Mg^{2+} (0.053 M), SO_4^{2-} (0.028 M), K⁺ (0.01 M), and Ca^{2+} (0.01 M), giving an ionic strength of about 0.7 M. Using the Davies approximation (although it is valid only up to about 0.5 M), $\gamma_{Cu} \approx 0.43$. Thus: $[Cu^{2+}] = \{Cu^{2+}\}/\gamma_{Cu} \approx (100 \text{ μM})/0.43 \approx$ **230 μM**. Note the large effect of the high I.

Thoughtful Pause
What is the H^+ concentration (assuming the nitric acid dissociates completely)?

If the nitric acid dissociates completely, the charge balance tells you that the H^+ concentration will be 0.001 M (charge balance: $[H^+] = [NO_3^-] + [OH^-] \approx [NO_3^-] = 10^{-3}$ M). In 5×10^{-3} M NaCl, you can show that the ionic strength is 5×10^{-3} M. Therefore (using the Davies approximation):

$$\log_{10}\gamma_H = (-0.5)(1)(0.005)^{\frac{1}{2}} = -0.034 \text{ and } \gamma_H = 0.92$$

Thus: $\{H^+\} = \gamma_H[H^+] = (0.92)(0.001) = 9.2 \times 10^{-4}$ M or pH 3.04. In this case, at low I, ionic strength has only a small effect on pH. In 5×10^{-2} M NaCl, the pH becomes 3.10. Another illustration is shown in Example 21.2.

21.2.5 Single-species activity coefficients for uncharged species
Uncharged species also are affected by ions in solution, although not to the extent that ions are affected by other ions. For uncharged species, the Debye-Hückel limiting law does not apply. Why? Recall that the theory behind the limiting law assumes that a cloud of counterions surrounds a central ion.

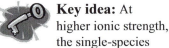

Key idea: At higher ionic strength, the single-species activity coefficients for uncharged species usually increase

Activity coefficients for uncharged species usually are determined empirically. The general form of the approximation for activity coefficients for uncharged species is shown in eq. 21.5:

$$\log_{10}\gamma_i = +kI \text{ (for uncharged species)} \qquad \text{eq. 21.5}$$

Uncharged species typically (but not always) partition *out* of water at higher ionic strengths. Why? At very high ionic strengths, water is associated with the ions making up I. This leaves less water to solubilize the uncharged species. Thus, the activity coefficients usually are greater than 1. This partitioning behavior is called ***salting out*** and the coefficients k sometimes are called ***salting-out coefficients***. For many species, k is between 0.01 and 0.15 (Snoeyink and Jenkins, 1980). This leads to activity coefficients very close to 1 for most fresh waters. Thus, in fresh waters, activity and concentration are very similar for many uncharged species and activity coefficient corrections usually are ignored.

salting out: the tendency of most uncharged species to partition out of water to a larger extent at higher ionic strengths

salting-out coefficient: the parameter k in the common model for the single-species activity coefficient for uncharged species: $\log_{10}\gamma_i = +kI$

21.2.6 Single-species activity coefficients and equilibrium constants

The incorporation of single-species activity coefficients into equilibrium expressions is fairly straightforward. Consider, for example, the dissociation of a monoprotic acid: $HA = A^- + H^+$, with equilibrium constant K. The equilibrium constant is based on species activities:

$$K = \{A^-\}\{H^+\}/\{HA\}$$

Rewriting in terms of concentrations:

$$K = \gamma_A[A^-]\gamma_H[H^+]/\gamma_{HA}[HA] = (\gamma_A\gamma_H/\gamma_{HA})([A^-][H^+]/[HA])$$

We can introduce a new type of equilibrium constant based on concentration, ^cK:

^cK: an equilibrium constant based on concentrations, not activities

$$^cK = [A^-][H^+]/[HA]$$

Thus: $K = (\gamma_A\gamma_H/\gamma_{HA})^cK$. In general, for the equilibrium $aA + bB = cC + dD$:

$$K = (\gamma_C^c\gamma_D^d/\gamma_A^a\gamma_B^b)(^cK)$$

Does ionic strength affect equilibrium constants significantly? The answer, of course, depends on the charge on each species. For example, K_a for acetic acid at 25°C and $I \to 0$ is $10^{-4.7}$. What is ^cK_a for acetic acid in 0.01 M NaCl? In 0.01 M NaCl, $I = 0.01$ M (assuming that acetate, H^+, and OH^-

do not contribute significantly to the ionic strength). Thus: $\gamma_A = \gamma_H = 0.9$ and $\gamma_{HA} \approx 1$ (since it is uncharged: see Section 21.2.5). Therefore:

$$^cK = K/(\gamma_A\gamma_H/\gamma_{HA}) = 10^{-4.7}/(0.9)^2 = 10^{-4.6}$$

Acetic acid becomes a slightly stronger acid in a higher ionic strength environment. For the dissociation of bicarbonate, you can show that in 0.01 M NaCl:

$$^cK = K/(\gamma_{CO_3}\gamma_H/\gamma_{HCO_3}) = (0.9)(10^{-10.3})/\{(0.9)(0.65)\} = 10^{-10.1}$$

In this case, ionic strength has a larger effect on ^cK since one species (carbonate) carries a charge of -2. Another illustration of the effects of ionic strength on equilibria is shown in Example 21.3.

21.2.7 Single-species activity coefficients and equilibrium calculations

It is a simple matter to include activity coefficient corrections in equilibrium calculations. The key to performing equilibrium calculations at $I \neq 0$ is to recognize that *equilibrium expressions are written in terms of activities, but charge and mass balances are written in terms of concentrations*. This statement leads to a good approach for solving systems at $I \neq 0$ (also summarized in Appendix C, Section C.8):

Step 1: Set up the equilibrium system in the usual fashion

Step 2: Using single-species activity coefficients, convert all equilibrium constants to ^cK values at the ionic strength of interest

Step 3: Solve the system for species concentrations

Step 4: If desired, compute species activities from the concentrations calculated in Step 3 and the activity coefficients calculated in step 2

This process will be illustrated for the simplest aquatic system: pure water. What is neutral pH in water with $I = 0.1$ M? To answer this question, set up the system as usual:

Species:
　　H^+, OH^-, and H_2O

Equilibrium:
　　$H_2O = H^+ + OH^-$, $K = K_W$ or:

Key idea: Equilibrium expressions are written in terms of activities, but charge and mass balances are written in terms of concentrations

---

Example 21.3: **Effect of I on Equilibrium Calculations**

How do the equilibrium concentrations and equilibrium pH vary for a closed system containing $CaCO_3(s)$ at $I = 0.01$ M and as $I \to 0$?

Solution:
You have solved this system in section 19.5.3. The equilibrium $[H^+]$ is the $[H^+]$ satisfying the following equation:

$$2(^cK_{s0}/^c\alpha_2)^{1/2} =$$
$$(^c\alpha_1 + 2^c\alpha_2)(^cK_{s0}/^c\alpha_2) +$$
$$^cK_W/[H^+] - [H^+] \quad (*)$$

Here, $^c\alpha_i$ is α_i evaluated by using $^cK_1$ and $^cK_2$.

Now, assuming $\gamma_i = 1$ for all uncharged species:

$$K_{s0} = \{Ca^{2+}\}\{CO_3^{2-}\}, \text{ so:}$$
$$^cK_{s0} = K_{s0}/(\gamma_{Ca}\gamma_{CO_3})$$

$K_1 = \{HCO_3^-\}\{H^+\}/\{H_2CO_3\},$
    so: $^cK_1 = K_1/(\gamma_{HCO_3}\gamma_H)$

$K_2 = \{CO_3^-\}\{H^+\}/\{HCO_3^-\},$ so:
    $^cK_2 = K_2/(\gamma_{CO_3}\gamma_H/\gamma_{HCO_3})$

and:

$K_W = \{H^+\}\{OH^-\},$ so:
    $^cK_W = K_W/(\gamma_H\gamma_{OH})$

At $I = 0.01$ M and using the Davies approximation: $\gamma_H = \gamma_{OH} = \gamma_{HCO_3} = 0.90$ and $\gamma_{Ca} = \gamma_{CO_3} = 0.65$. Thus:

$^cK_{s0} = 10^{-8.48}/[(0.65)(0.65)]$
    $= 10^{-8.11}$
$^cK_1 = 10^{-6.3}/[(0.90)(0.90)]$
    $= 10^{-6.2}$
$^cK_2 =$
    $10^{-10.3}/[(0.90)(0.65)/(0.90)]$
    $= 10^{-10.1},$ and
$^cK_W = 10^{-14}/[(0.90)(0.90)]$
    $= 10^{-13.9}$

Solving (*), **the equilibrium pH is 9.89 as $I \to 0$ and 9.93 at $I = 0.01$ M**. More calcium carbonate dissolves at the higher ionic strength, with $[Ca^{2+}]$ increasing from 1.09×10^{-4} M as $I \to 0$ to 1.46×10^{-4} M at $I = 0.01$ M.

***Example 21.4:* Effect of *I* on Equilibrium Calculations with Uncharged Species**

The dissolved oxygen (DO) concentration in equilibrium with the atmosphere at 1 atm and 25°C is 8.24 mg/L as $I \to 0$.

$K_W = \{H^+\}\{OH^-\}/\{H_2O\} = 10^{-14}$ at 25°C

Mass balance:
    None

Charge balance:
    $[H^+] = [OH^-]$ (note the use of concentrations, not activities)

Other:
    $\{H_2O\} = 1$

Now convert all K to ^cK at $I = 0.1$ M. At $I = 0.1$ M, $\gamma_H = \gamma_{OH} = 0.72$ (Davies approximation). Thus:

$$^cK_W = [H^+][OH^-]$$
$$= \{H^+\}\{OH^-\}/(\gamma_H\gamma_{OH})$$
$$= K_W/(\gamma_H\gamma_{OH})$$
$$= 10^{-14}/(0.72)^2 = 10^{-13.71}$$

(assuming that the activity of water is unity). Neutral pH (i.e., the equilibrium pH of pure water) occurs where the charge balance is satisfied. (Remember that pH is calculated from the H^+ *activity*, **not** the H^+ *concentration*.) This occurs when $[H^+] = [OH^-] = 10^{-6.86}$ M. Thus, at $I = 0.1$ M, the *concentration* of H^+ is $10^{-6.86}$ M. As $I \to 0$, $\gamma_H \to \gamma_{OH} = 1$ and so: $^cK_W = K_W = 10^{-14}$ and neutral pH occurs when $[H^+] = [OH^-] = 10^{-7}$ M. Thus, at $I = 0.1$ M, the *concentration* of H^+ ($= 10^{-6.86}$ M) is different than at $I \to 0$ ($[H^+] = 10^{-7}$ M).

What about pH at $I = 0.1$ M? To answer this question, convert the equilibrium H^+ concentration to activity. If $[H^+] = 10^{-6.86}$ M, then:

$$\{H^+\} = \gamma_H[H^+] = (0.72)(10^{-6.86} \text{ M}) = 10^{-7} \text{ M or pH 7}$$

Thus, neutral pH also occurs at pH 7 when $I = 0.1$ M. In this case, the effects of ionic strength on K_W and the effects of ionic strength on calculating $\{H^+\}$ from $[H^+]$ essentially cancel each other out.

In general, ionic strength effects do not cancel out. Consider a system consisting of 10^{-3} M HAc (acetic acid) and 0.05 M NaCl. To simplify matters, recognize in this case that the contribution of acetate to the ionic strength is negligible. Setting up the system:

Species:
    H^+, OH^-, H_2O, HAc, and Ac^-

Equilibrium:
    $H_2O = H^+ + OH^-$; $K = K_W = 10^{-14}$
    $HAc = Ac^- + H^+$; $K = K_a = 10^{-4.7}$

Find the DO in water with a salinity of 10‰ in equilibrium with the atmosphere.

Solution:
For oxygen: $K_H = \{O_2(aq)\}/P_{O2}$. As $I \to 0$, the value of $[O_2(aq)] = \{O_2(aq)\} = 8.24$ mg/L = 2.58×10^{-4} M implies a Henry's Law constant of $(2.58 \times 10^{-4}$ M)/(0.209 atm) = 1.23×10^{-3} mol/L-atm.

A salinity of 10‰, if made up solely by sodium and chloride ions, corresponds to an I of about 0.17 M (see section 21.2.1). At $I = 0.17$ M, you must calculate:

$$^cK_H = [O_2(aq)]/P_{O_2}$$
$$= \{O_2(aq)\}/(\gamma_{O_2}P_{O_2})$$
$$= K_H/\gamma_{O_2}$$

For dissolved oxygen in NaCl, $\log_{10}\gamma_{O_2} = +0.132I$. At $I = 0.17$ M, $\gamma_{O_2} = 1.05$. Thus:

$$^cK_H = K_H/\gamma_{O_2}$$
$$= (1.23 \times 10^{-3}$$
$$\text{mol/L-atm})/1.05$$
$$= 1.05 \times 10^{-3} \text{ mol/L-atm}$$

The DO concentration is:
$^cK_H P_{O_2} = (1.05 \times 10^{-3}$ mol/L-atm)(0.209 atm) = 2.45×10^{-4} M or **7.83 mg/L**. This is close to the tabulated value of 7.79 mg/L (Metcalf and Eddy, 2003).

Or:

$$K_W = \{H^+\}\{OH^-\}/\{H_2O\}$$
$$K_a = \{H^+\}\{Ac^-\}/\{HAc\}$$

Mass balance:
$$C = [HAc] + [Ac^-] = 0.001 \text{ M}$$
(note the use of concentrations, not activities)

Charge balance:
$$[H^+] = [OH^-] + [Ac^-]$$
(note the use of concentrations, not activities)

Other:
$$\{H_2O\} = 1$$

You can use the standard solution techniques if you express the equilibrium expressions in terms of ^cK values instead of K values. Note that in this example: $^cK_a = K_a/(\gamma_{Ac}\gamma_H/\gamma_{HAc})$ and $^cK_W = [H^+][OH^-] = K_W/(\gamma_H\gamma_{OH})$. At $I = 0.05$ M, $\gamma_H = \gamma_{OH} = \gamma_{Ac} = 0.79$. If $\gamma_{HAc} = 1$, then:

$$^cK_a = 10^{-4.7}/(0.79)^2 = 10^{-4.5} \text{ and:}$$
$$^cK_W = 10^{-14}/(0.79)^2 = 10^{-13.8}$$

You can solve the system by using the usual techniques. The equilibrium concentrations and activities are given in Table 21.4. Note that ionic strength has only a small effect on the equilibrium pH. The change in pH from 3.88 to 3.89 represents a 9% increase in $\{H^+\}$. The acetate *concentration* is increased by about 24% at $I = 0.05$ M compared to $I \to 0$. Ionic strength plays a larger role with multivalent ions, as demonstrated in Example 21.3. Another illustration of the effects of ionic strength on species activities controlled by equilibria is given in Example 21.4.

Table 21.4: Effects of Ionic Strength on the Equilibrium Composition of a 10^{-3} M Acetic Acid Solution at $I = 0.05$ M and as $I \to 0$
(all values except pH in M)

Species	$\{i\} = [i]$ as $I \to 0$	$[i]$ at $I = 0.05$ M	$\{i\}$ at $I = 0.05$ M
pH	3.88		3.89
HAc	8.68×10^{-4}	8.36×10^{-4}	8.36×10^{-4}
Ac$^-$	1.32×10^{-4}	1.64×10^{-4}	1.29×10^{-4}
OH$^-$	7.60×10^{-11}	9.82×10^{-11}	7.76×10^{-11}
H$^+$	1.32×10^{-4}	1.63×10^{-4}	1.29×10^{-4}

21.3 EFFECTS OF TEMPERATURE ON CHEMICAL EQUILIBRIUM

21.3.1 The van't Hoff equation

The goal in this section is to develop an expression for the manner in which equilibrium constants change with temperature. Rather than deriving an expression for K as a function of temperature (T), the job at hand will be much easier if we seek an expression for $\ln K$ as a function of T. Recall from Section 3.9.2 that: $\Delta G^o_{rxn} = -RT\ln K$, where R is the ideal gas constant. Thus: $\ln K = -(1/R)(\Delta G^o_{rxn}/T)$ and:

$$\frac{d \ln K}{dT} = -\frac{1}{R}\frac{d\left(\dfrac{\Delta G^o_{rxn}}{T}\right)}{dT} \qquad \text{eq. 21.6}$$

Equation 21.6 uses full derivatives with respect to temperature and is valid only at constant pressure and ionic strength. You can use the chain rule of differentiation to expand the right-hand side of eq. 21.6 and show that:

$$\frac{d \ln K}{dT} = \frac{\Delta G^o_{rxn} - T\dfrac{d\Delta G^o_{rxn}}{dT}}{RT^2} \qquad \text{eq. 21.7}$$

All you have left to do is evaluate $d\Delta G^o_{rxn}/dT$, plug it into eq. 21.7, and integrate with respect to temperature to get $\ln K$ as a function of T.

The key to evaluating $d\Delta G^o_{rxn}/dT$ comes from eq. 3.7 (repeated here for convenience) as eq. 21.8:

$$dG = VdP - S_{sys}dT \qquad \text{eq. 21.8}$$

At constant pressure ($dP = 0$) for a chemical reaction: $d\Delta G^o_{rxn} = -\Delta S^o_{rxn}dT$ or $d\Delta G^o_{rxn}/dT = -\Delta S^o_{rxn}$. Plugging into eq 21.7:

$$d(\ln K)/dT = (\Delta G^o_{rxn} + T\Delta S^o_{rxn})/(RT^2)$$

Recall from the definition of Gibbs free energy that: $\Delta G^o_{rxn} = \Delta H^o_{rxn} - T\Delta S^o_{rxn}$. Thus:

van't Hoff equation: the relationship showing how equilibrium constants change with temperature: $d(\ln K)/dT = \Delta H^o_{rxn}/RT^2$, where ΔH^o_{rxn} is the standard enthalpy of reaction

$$d(\ln K)/dT = \Delta H^o_{rxn}/RT^2 \qquad \text{eq. 21.9}$$

Equation 21.9 is called the *van't Hoff equation* (after Jacobus Henricus van't Hoff, 1852-1911: see the *Historical Note* at the end of this chapter).

21.3.2 Using and interpreting the van't Hoff equation

Qualitatively, the van't Hoff equation (eq. 21.9) tells you what type of equilibria have equilibrium constants that increase with temperature and which have equilibrium constants that decrease with temperature.

Thoughtful Pause

For what type of equilibria do equilibrium constants increase with temperature?

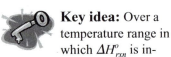

Key idea: Endothermic reactions will show increasing equilibrium constants with temperature, and exothermic reactions will show decreasing equilibrium constants with temperature

Key idea: For many equilibria of environmental interest and over temperatures of 0 to 30°C or so, ΔH^o_{rxn} is fairly independent of temperature

Key idea: Over a temperature range in which ΔH^o_{rxn} is independent of temperature, $\ln(K_2/K_1) = (\Delta H^o_{rxn}/R)(1/T_1 - 1/T_2)$

Recall from Section 3.7.4 that reactions with $\Delta H^o_{rxn} < 0$ are called *exothermic* reactions and produce heat, and those reactions with $\Delta H^o_{rxn} > 0$ are called *endothermic* reactions and absorb heat. Thus, endothermic reactions will show increasing equilibrium constants with temperature, and exothermic reactions will show decreasing equilibrium constants with temperature.

For example, you can use the van't Hoff equation to determine if bicarbonate becomes a stronger or weaker acid at 30°C than at 25°C. For the equilibrium $HCO_3^- = CO_3^{2-} + H^+$:

$$\Delta H^o_{rxn} = \Delta H^o_{f,H} + \Delta H^o_{f,CO_3} - \Delta H^o_{f,HCO_3}$$
$$= 0 + (-677.1 \text{ kJ/mol}) - (-692.0 \text{ kJ/mol})$$
$$= +14.9 \text{ kJ/mol}$$

So, for a positive ΔT (25°C to 30°C), the change in $\ln K$ should be positive. Thus, K_a increases with temperature and bicarbonate becomes a stronger acid at higher temperature.

The van't Hoff equation also can be used quantitatively. However, eq. 21.9 can be integrated only if you know how the standard enthalpy of reaction, ΔH^o_{rxn}, varies with temperature.[†] For many equilibria of environmental interest and over temperatures of interest in water (say, 0 to 30°C or so), ΔH^o_{rxn} *is fairly independent of temperature*. For equilibria where ΔH^o_{rxn} is fairly independent of temperature, eq. 21.9 can be integrated to show:

$$\ln(K_2/K_1) = (\Delta H^o_{rxn}/R)(1/T_1 - 1/T_2) \qquad \text{eq. 21.10}$$

where K_2 and K_1 are the equilibrium constants at temperatures T_2 and T_1 (in °K), respectively. For the bicarbonate example given above and if ΔH^o_{rxn} is independent of temperature over the temperature range of interest:

$$\ln(K_{30}/K_{25}) = [(+14.9 \text{ kJ/mol})/(8.314 \times 10^{-3} \text{ kJ/mol-°K})] \times$$
$$(1/298.16°K - 1/303.16°K)$$
$$= +0.099$$

[†] You may be wondering how ΔH^o_{rxn} can possibly vary with temperature since it is evaluated under standard conditions (usually 25°C). We really are asking how ΔH^o_{rxn} varies with the standard temperature at which it is evaluated.

Thus: $K_{30}/K_{25} = e^{+0.099} = 1.10$. Since $K_{25} = 10^{-6.3}$, then:

$$K_{30} = (1.10)(10^{-6.3}) = 10^{-6.26}$$

Therefore, bicarbonate becomes very slightly more acidic at 30°C compared to 25°C.

21.3.3 Examples of changes in K with temperature

Armed with eqs. 21.9 and 21.10, you can determine the effects of temperature on many equilibrium constants. A few more examples will be presented here to illustrate some important classes of equilibria. In particular, you will see whether the integrated form of the van't Hoff equation matches your preconceived view of the effects of temperature on chemical equilibria.

What is your gut feeling about the effects of temperature on the solubility of solids? Most people know intuitively that many solids are *more* soluble at higher temperatures (but see a very important counterexample in Section 21.3.4). As an example, consider the equilibrium: $AgCl(s) = Ag^+ + Cl^-$. The change in enthalphy for the reaction is:

$$\begin{aligned}\Delta H^o_{rxn} &= \Delta H^o_{f,Ag} + \Delta H^o_{f,Cl} - \Delta H^o_{f,AgCl(s)} \\ &= 105.6 \text{ kJ/mol} + (-167.2 \text{ kJ/mol}) - (-127.1 \text{ kJ/mol}) \\ &= +65.5 \text{ kJ/mol}\end{aligned}$$

Thus, for a positive ΔT, the change in $\ln K$ should be positive. As a result, K_{s0} increases with temperature and silver chloride becomes more soluble at higher temperatures. An increase in temperature from 25°C to 30°C would result in a 1.55-fold increase in K_{s0} and thus a 24% increase in both the $\{Ag^+\}$ and $\{Cl^-\}$ in equilibrium with AgCl(s).

How about the solubility of gases? Again, most people would guess that gases become *less* soluble at higher temperatures. For example, if you warm up a carbonated beverage, it is likely to go flat. You should be able to determine exactly how the solubility of a gas varies with temperature by using the van't Hoff equation.

One gas of environmental interest is oxygen. The Henry's Law equilibrium for oxygen is: $O_2(g) = O_2(aq)$; $K_H = 1.26 \times 10^{-3}$ mol/L-atm at 25°C. For this equilibrium:

$$\begin{aligned}\Delta H^o_{rxn} &= \Delta H^o_{f,O_2(aq)} - \Delta H^o_{f,O_2(g)} \\ &= -11.71 \text{ kJ/mol} - 0 = -11.71 \text{ kJ/mol}\end{aligned}$$

Thus: $\ln(K_{H,T}/K_{H,25}) = (\Delta H^o_{rxn}/R)(1/298.16 - 1/T)$. This form of the equation makes clear that a plot of $\ln K_T$ versus $1/T$ should be a straight line with a slope of $\Delta H^o_{rxn}/R = -1.41 \times 10^3$ °K. Solving for $K_{H,T}$:

$$K_{H,T} = (1.26 \times 10^{-3} \text{ mol/L-atm}) \exp[(1.41 \times 10^3)(1/298.16 - 1/T)]$$

The dissolved oxygen concentration in equilibrium with the atmosphere is $P_{O_2}K_{H,T} = 0.209K_{H,T}$. The dissolved oxygen concentration is plotted over temperature for $T = 0$ to $30^{\circ}C$ in Figure 21.1. Note that oxygen is less soluble at higher temperatures, as anticipated.

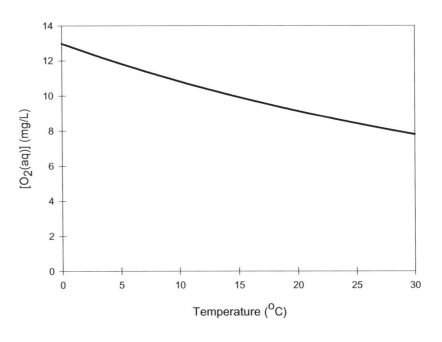

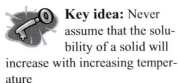

Key idea: Never assume that the solubility of a solid will increase with increasing temperature

Figure 21.1: Effect of Temperature on the Solubility of Oxygen in Equilibrium with the Atmosphere

21.3.4 The strange case of some carbonate and hydroxide solids

Homeowners in areas with hard water sometimes complain of scale formation in their water heaters. Scale is often modeled as calcium carbonate. The homeowners' observations suggest that $CaCO_3(s)$ precipitates at *higher* temperatures. This seems to run counter to the discussion in Section 21.3.2.

In fact, the homeowners are not crazy: scale *is* more likely to form in water heaters. Why? For calcite, you can write: $CaCO_3(s) = Ca^{2+} + CO_3^{2-}$. Here:

$$\Delta H^o_{rxn} = \Delta H^o_{f,Ca} + \Delta H^o_{f,CO_3} - \Delta H^o_{f,CaCO_3(s)}$$
$$= -542.83 \text{ kJ/mol} + (-677.1 \text{ kJ/mol}) - (-1207.4 \text{ kJ/mol})$$
$$= -12.53 \text{ kJ/mol}$$

Note that a *negative* value of ΔH^o_{rxn} means that K_{s0} will *decrease* with an increase in temperature. In other words, calcium carbonate will be *less* soluble at higher temperatures. A water heater at $130^{\circ}F = 54^{\circ}C$ will experience a K_{s0} value about 0.64 times the K_{s0} value at 25°C. Thus, it may be possible for the solubility product to be exceeded at 54°C in the water heater but not at 25°C (see also Problem 21.9).

Similar behavior is observed with some hydroxide solids. For example, many drinking water treatment plants add alum to remove particles from water. It is not uncommon to find that the residual aluminum concentration in the treated water increases in the winter. Since most drinking waters are filtered to remove precipitated aluminum (among other things), the residual aluminum concentration can be estimated by the total soluble aluminum in equilibrium with $Al(OH)_3(s)$. One factor affecting the residual aluminum concentration is the effect of temperature on the solubility of $Al(OH)_3(s)$. To determine the effects of temperature on the solubility of $Al(OH)_3(s)$, you must do a complete analysis as follows:

Species list:
$$H_2O,\ H^+,\ OH^-,\ Al^{3+},\ AlOH^{2+},\ Al(OH)_2^+,\ Al(OH)_3^0,\ Al(OH)_4^-,$$
and $Al(OH)_3(s)$

Equilibria (ΔH_{rxn}^o values in parentheses):

$H_2O = H^+ + OH^-$	$K_W = 10^{-14}$	(+55.9 kJ/mol)
$Al^{3+} + OH^- = AlOH^{2+}$	$K_1 = 10^{+9.0}$	(−7.8 kJ/mol)
$Al^{3+} + 2OH^- = Al(OH)_2^+$	$K_2 = 10^{+17.9}$	(+0.7 kJ/mol)
$Al^{3+} + 3OH^- = Al(OH)_3^0$	$K_3 = 10^{+21.5}$	(−0.8 kJ/mol)
$Al^{3+} + 4OH^- = Al(OH)_4^-$	$K_4 = 10^{+33.3}$	(−46.6 kJ/mol)
$Al(OH)_3(s) = Al^{3+} + 3OH^-$	$K_{s0} = 10^{-32.1}$	(+56.8 kJ/mol)

Mass balance:
$$Al(+III)_T = [Al^{3+}] + [AlOH^{2+}] + [Al(OH)_2^+] + [Al(OH)_3^0] + [Al(OH)_4^-]$$

You can rewrite the mass balance as:

$$Al(+III)_T = [Al^{3+}](1 + K_1[OH^-] + K_2[OH^-]^2 + K_3[OH^-]^3 + K_4[OH^-]^4)$$

If $Al(OH)_3(s)$ is present, then:

$$Al(+III)_T = (K_{s0}/[OH^-]^3)(1 + K_1[OH^-] + K_2[OH^-]^2 + K_3[OH^-]^3 + K_4[OH^-]^4)$$

The equilibrium constants can be evaluated at temperatures other than 25°C by using eq. 21.10 (assuming the ΔH_{rxn}^o are constant over the desired temperature range). For example, K_1 at 5°C is given by:

$$K_{1,5} = K_{1,25}\exp[\,(-7.8\ \text{kJ/mol}/8.314 \times 10^{-3}\ \text{kJ/mol-°K}) \times (1/298.16 - 1/278.16)]$$
$$= 10^{+9.1}$$

The total soluble aluminum lines at 5°C and 25°C are shown in Figure 21.2. Note that aluminum is *more* soluble at the lower temperature below about pH 6.2 but *less* soluble at the lower temperature at pH above about pH 6.2. The pH of minimum solubility has shifted from about 6.3 to about 6.8. Note that the change in the minimum soluble aluminum concentration with temperature is relatively small on the pC-pH diagram but not insignificant. For example, the total aluminum solubility at pH 7 is about 5.5 times larger at 25°C than at 5°C.

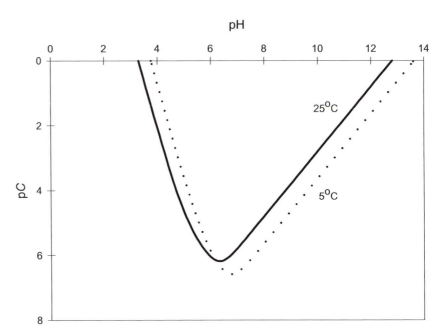

Figure 21.2: Effect of Temperature on the Solubility of Aluminum in Equilibrium with Al(OH)$_3$(s)
(lines are total soluble aluminum = *Al(+III)$_T$*)

The effects of temperature on the equilibrium position can be determined with equilibrium calculation software. With *Nanoql*, the temperature can be varied and the equilibrium concentration of species monitored. Varying temperature requires that the standard enthalpy of formation values are known for all species in the system.

21.4 EFFECTS OF PRESSURE ON EQUILIBRIUM CONSTANTS

21.4.1 Introduction and theory

Pressure exerts very little effect on most equilibrium constants. Significant effects on equilibria involving only dissolved species are expected only with large pressure (> 1000 bars = 987 atm). Such conditions may be found in the ocean. However, in the ocean (even at 1000 bars), the effects

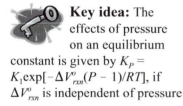

Key idea: The change in ln$K$ with pressure is related to the change in the molar volume of reaction, ΔV^o_{rxn}

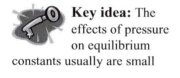

Key idea: The effects of pressure on an equilibrium constant is given by $K_P = K_1\exp[-\Delta V^o_{rxn}(P - 1)/RT]$, if ΔV^o_{rxn} is independent of pressure

of ionic strength and temperature on equilibrium constants can exceed the effects of pressure on equilibrium constants.

To show the effects of pressure on K, reconsider eq. 21.8: $dG = VdP - S_{sys}dT$. This relationship suggests that dG/dP [and thus $d(\ln K)/dP$] should be related to the change in volume in a system at constant temperature. A more detailed analysis reveals the following relationship at constant T:

$$d(\ln K)/dP = -\Delta V^o_{rxn}/RT \quad \text{(at constant temperature)} \qquad \text{eq. 21.11}$$

Here, ΔV^o_{rxn} is the change in the molar volume of reaction. It can be calculated from the standard molar volumes of the reactants and products. The standard molar volume is the contribution of a species (per mole) to the volume of solution under standard conditions. At low pressures, ΔV^o_{rxn} is relatively constant. The integrated form of eq. 21.11 is:

$$K_P = K_1\exp[-\Delta V^o_{rxn}(P - 1)/RT] \qquad \text{eq. 21.12}$$

In eq. 21.12, K_P is the equilibrium constant at pressure P and K_1 is the equilibrium constant at 1 atm.

21.4.2 Examples of the effects of pressure on equilibrium constants

A few examples will demonstrate the small influence of pressure on equilibrium constants. Table 21.5 lists several ΔV^o_{rxn} values and the corresponding effects of pressure on the equilibria. Note that pressure has almost no effect at 2 atm ($K_2/K_1 = K$ at 2 atm divided by K at 1 atm $\approx$ 1). Even at 1,000 atm, the effects of pressure are relatively small. For comparison, the effects of pressure at 1,000 atm on K_W are equivalent to an increase in temperature from 25°C to about 30°C. Except for the ocean, conditions with 30°C are much more likely to be encountered in systems of environmental interest than conditions with 1,000 atm.

Key idea: The effects of pressure on equilibrium constants usually are small

Table 21.5: Effects of Pressure on Selected Equilibrium Constants (ΔV^o_{rxn} values from Byrne and Laurie, 1999)

Equilibrium	ΔV^o_{rxn} (cm^3/mol)[1]	K_2/K_1	K_{1000}/K_1
$H_2O = H^+ + OH^-$	-22.05	0.999	0.406
$H_2CO_3 = HCO_3^- + H^+$	-27.4	0.999	0.327
$HCO_3^- = CO_3^{2-} + H^+$	-28.1	0.999	0.317
$H_2S = HS^- + H^+$	-15.0	0.999	0.542
$NH_4^+ = NH_3 + H^+$	$+6.9$	1.000	1.325
$CaCO_3(s) = Ca^{2+} + CO_3^{2-}$ (calcite)	-36.5^2	0.999	0.225

Notes: 1. At 25°C, 1 atm
2. At 5°C in seawater

Pressure effects in the ocean are not insignificant. As can be seen in Table 21.5, reactions that create charged species have slightly lower K values at higher pressures. As a result, minerals in general are less soluble at higher pressures. The effects of pressure in the ocean on calcite and aragonite solubility significantly influence global carbon dioxide cycling. Since calcite and aragonite are less soluble at higher pressures, more carbonate is tied up as $CaCO_3(s)$ (Bryne and Laurie, 1999).

21.5 SUMMARY

In this chapter, the effects of the medium composition on the equilibrium position of a system were determined. An important measure of the medium composition is the ionic strength: $I = \dfrac{1}{2} \sum_{i=1}^{n} z_i^2 C_i$, where z_i and C_i are the charge on species i and concentration of species i, respectively. The concentration (C_i) and activity (a_i) of a species are related through an activity coefficient, γ_i: $a_i = \gamma_i C_i$. The single-ion activity coefficient can be estimated through several approximations stemming from models of the electric double layer. In general, increasing I decreases the activity of ions Increasing I generally increases the activity of uncharged species slightly. The latter process is called *salting out*. Once the single-species activity coefficients are calculated, the system can be solved at equilibrium. The usual process is followed, except activity coefficients are used to express all equilibrium constants in terms of concentrations rather than activities.

Thermodynamic arguments were used to derive relationships between equilibrium constants and temperature. The dependency of K on temperature is described by the van't Hoff equation. The integrated form of the van't Hoff equation is $\ln(K_2/K_1) = (\Delta H^o_{rxn}/R)(1/T_1 - 1/T_2)$, where K_2 and K_1 are the equilibrium constants at temperatures T_2 and T_1 (in $^\circ$K), respectively. This form of the equation is valid only if ΔH^o_{rxn} is fairly independent of temperature. According to the van't Hoff equation, endothermic reactions will show increasing equilibrium constants with temperature and exothermic reactions will show decreasing equilibrium constants with temperature. With the van't Hoff equation, equilibrium constant values can be calculated at temperatures other than 25°C. Some solids, including $CaCO_3(s)$, show higher solubility at *lower* temperatures.

The effects of pressure on equilibrium constants also was investigated in this chapter. Equilibrium constants were found to vary with pressure as follows: $K_P = K_1 \exp[-\Delta V^o_{rxn}(P - 1)/RT]$, where K_P is the equilibrium constant at pressure P and K_1 is the equilibrium constant at 1 atm. This equation is valid only if ΔV^o_{rxn} is independent of pressure over the pressure range of interest. In general, pressure exerts only a small effect on equilibrium constants.

SUMMARY OF KEY IDEAS

- Chemical species may behave differently under different environmental conditions, even at the same mass or molar concentration

- Even if two species do not react with one another, the behavior (i.e., the activity) of each species will be influenced by the other species

- Concentration (C_i) and activity (a_i) are related through an activity coefficient, γ_i: $a_i = \gamma_i C_i$

- At higher ionic strength, the single-ion activity coefficient decreases (i.e., the activity of the ions decreases as the ions of interest are constrained by other ions)

- Always use single-ion activity coefficient approximations only over the ionic strength range for which they are valid

- At higher ionic strength, the single-species activity coefficients for uncharged species usually increase

- Equilibrium expressions are written in terms of activities, but charge and mass balances are written in terms of concentrations

- Endothermic reactions will show increasing equilibrium constants with temperature and exothermic reactions will show decreasing equilibrium constants with temperature

- For many equilibria of environmental interest and over temperatures of 0 to 30°C or so, ΔH^o_{rxn} is fairly independent of temperature

- Over a temperature range in which ΔH^o_{rxn} is independent of temperature, $\ln(K_2/K_1) = (\Delta H^o_{rxn}/R)(1/T_1 - 1/T_2)$

- Never assume that the solubility of a solid will increase with increasing temperature

- The change in $\ln K$ with pressure is related to the change in the molar volume of reaction, ΔV^o_{rxn}

- The effects of pressure on an equilibrium constant is given by $K_P = K_1 \exp[-\Delta V^o_{rxn}(P - 1)/RT]$, if ΔV^o_{rxn} is independent of pressure

- The effect of pressure on equilibrium constants usually is small

HISTORICAL NOTE: JACOBUS HENRICUS VAN'T HOFF

Jacobus Henricus van't Hoff was one of the most creative chemists of his time, leaving his imprint on nearly all branches of chemistry. In organic chemistry, he created the idea of *asymmetrical carbon atoms* (stereochemistry, in modern terms). The idea evidently came to him during a walk. He published the revolutionary concept in a small pamphlet several months before submitting his doctoral dissertation on cyanoacetic and malonic acids. In physical chemistry, his work on the effects of temperature on the equilibrium position of a system was first published in 1884. The concept was to be expanded by Le Chatelier (see Chapter 4). The next year, van't Hoff was to make a another breakthrough: using osmotic pressure to demonstrate that the ideal gas law can be translated to solutions. This idea, coupled with Arrhenius's dissociation theory, took chemistry in a new direction.

Although van't Hoff received many honors over the years, his greatest recognition was as recipient of the first Nobel Prize in Chemistry in 1910. Imagine that first Nobel ceremony at the Musical Academy in Stockholm, with honorees van't Hoff, x-ray discoverer Wilhelm Conrad von Röntgen, Emil Adolf von Behring (developer of therapies for diphtheria and tuberculosis), and French poet René François Armand (Sully) Prudhomme. Evidently, van't Hoff enjoyed himself in Stockholm. At the banquet following the ceremony, according to an eyewitness: "There were many toasts and a splendid ambience. And in the small hours, two marshals carried the little van't Hoff on a gold chair around the room."

For one man to have developed such far-reaching and innovative ideas seems almost beyond belief. During the Nobel Prize ceremony, van't Hoff's work was described as the most important work in theoretical chemistry in 100 years. In fact, many people now consider van't Hoff, along with fellow laureates Arrhenius and Wilhelm Ostwald, to be the fathers of physical chemistry.

PROBLEMS

Note: Thermodynamic data are listed at the end of the problems.

21.1 What is the ionic strength of a water of the following chemical composition?

30 mg/L Ca^{2+}, 183 mg/L HCO_3^-, 35 mg/L Na^+, 27 mg/L Cl^-, 9 mg/L Fe^{2+}, pH 7.0, 9 mg/L Mg^{2+}, and 15 mg/L SO_4^{2-}

21.2 What is the relationship between I and TDS in a NaCl solution?

21.3 Show that, for a C M solution of a salt M_xY_z (which dissolves to form xC moles/L of M^{z+} and zC moles/L of Y^{x-}), the ionic strength is $\frac{1}{2}xz(x+z)C$.

21.4 What is $-\log[H^+]$ of water in an estuary at pH 8.2 and $I = 0.05$?

21.5 What is neutral pH at 10°C?

21.6 Is $H_2CO_3^*$ a stronger acid at 25°C or at 5°C?

21.7 Compare the K_{s0} values of $MnCO_3(s)$ in fresh water ($I \approx 0$) and in estuarine waters ($I = 0.08$) at 25°C. If each of the waters contained a concentration of CO_3^{2-} equal to 10^{-6} M and $MnCO_3(s)$ controls the Mn solubility, what is the Mn^{2+} *concentration* in each water? What is the Mn^{2+} *activity* in each water?

21.8 Compute the equilibrium constant at 10°C for: $MnS(s,pink) + H^+ = Mn^{2+} + HS^-$
(Hint: first calculate the equilibrium constant at 25°C given that $pK_{a,1}$ for H_2S at 25°C is 7.0.) Is $MnS(s,pink)$ more or less soluble at higher temperatures?

21.9 A treated drinking water has the following characteristics: pH 7.5, alkalinity = 75 mg/L as $CaCO_3$, $[Ca^{+2}]$ = 50 mg/L as Ca, and temperature of the distribution system = 77°F (= 25°C). For this problem, ignore ionic strength effects and assume that the only equilibrium affected by temperature is the precipitation of $CaCO_3(s)$.

A. Will $CaCO_3(s)$ precipitate in the distribution system?

B. Will $CaCO_3(s)$ precipitate in a home water heater (at about 140°F = 60°C)?

C. Find the lowest temperature at which $CaCO_3(s)$ precipitates at equilibrium in the water described above.

D. Will your answer to part B change if you consider the effects of temperature on all the equilibria in the system?

21.10 A chemical reactor containing water is maintained at 50 atm and 25°C. What is K_W in the reactor? What is neutral pH in the reactor? (Hint: $R = 0.082057$ L-atm/mol-°K)

21.11 The value of pK_a for the equilibrium $H_2S = HS^- + H^+$ is 7.0 at $I \to 0$, $T = 25$°C, and $P = 1$ atm. Find the ionic strength (at 25°C and $P = 1$ atm), temperature (at $I \to 0$ and $P = 1$ atm), and pressure (at $I \to 0$ and 25°C) needed to decrease the pK_a (expressed in concentration) to 6.9.

H_f^0 values in kJ/mol are given below (from Stumm and Morgan, 1996). See Table 21.5 for ΔV_{rxn}^o values.

Ca^{2+}	-542.83	CO_3^{2-}	-677.1	H_2S	-39.75	$CaCO_3(s)$	-1207.4
HCO_3^-	-692.0	H^+	0	HS^-	-17.6	$MnS(s)$	-213.8
$H_2CO_3^*$	-699.6	$H_2O(l)$	-285.83	OH^-	-230.0	Mn^{2+}	-220.7

CHAPTER 22

Chemical Kinetics of Aquatic Systems

22.1 INTRODUCTION

So far in this text, you have had great success in using equilibrium models to describe aqueous systems. At equilibrium in a isolated system, species concentrations do not change with time. However, not all chemical systems are at equilibrium. In this chapter, your collection of modeling tools will be extended to allow the investigation of nonequilibrium systems.

In Section 22.2, you will see an illustration of species that are not present at their equilibrium concentrations. You will learn that some nonequilibrium conditions may exist for species that react slowly. The rate of reactions will be related qualitatively to the energy profile as the reaction proceeds from reactants to products.

A more quantitative discussion of reaction rates may be found in Section 22.3. In that section, the relationship between reaction stoichiometry and the number of species participating in the reaction will be presented. The relationships among the reaction rate, rates of change of each species with respect to time, and reaction stoichiometry will be revealed. An experimental measure, the kinetic order, also will be introduced in this section. Finally, an approach to developing expressions for the rates of change of species concentrations with several reaction types will be developed.

In Section 22.4, several simple rate expressions will be introduced. Each rate expression will be integrated to yield equations showing species concentrations over time. The units of the rate constants and time for half completion of the reaction will be presented for the most common rate expressions. More complex systems, including reversible and sequential reactions, are discussed in Section 22.5. For all rate expressions, environmental examples will be provided.

The focus of Section 22.6 is the effects of temperature and ionic strength on rate constants. It is interesting that the common models describing the effects of temperature and ionic strength on rate constants were developed from the models describing the effects of temperature and ionic strength on equilibrium constants. The energy profile of a reaction will be linked to the rate constant in a quantitative way in this section.

22.2 THE NEED FOR CHEMICAL KINETICS

22.2.1 Not all reactions are at equilibrium

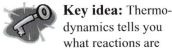

Key idea: Thermodynamics tells you what reactions are possible but not how fast the reactions occur

Chemical equilibrium is a thermodynamic concept. Recall from Chapter 3 that thermodynamics tells you what reactions are *possible*. However, not all possible reactions actually proceed to equilibrium. As an example, consider the following innocent question: do glucose solutions spontaneously combust? To answer this question, write the balanced redox reaction for the oxidation of glucose with oxygen:

$$\text{oxidation half reaction: } (1/6)C_6H_{12}O_6 + H_2O = CO_2(g) + 4H^+ + 4e^-$$

$$\text{reduction half reaction: } O_2(g) + 4H^+ + 4e^- = 2H_2O$$

$$\text{overall redox reaction: } (1/6)C_6H_{12}O_6 + O_2(g) = CO_2(g) + H_2O$$
$$\text{eq. 22.1}$$

The standard Gibbs free energy of reaction for eq. 22.1 is negative. Based on your experience, you would say that the reaction in eq. 22.1 proceeds spontaneously under standard conditions and that glucose is expected to burn in air.

In reality, you are unlikely to observe spontaneous combustion of glucose. Why? In fact, the reaction in eq. 22.1 proceeds *very* slowly. Thermodynamics tells you what reactions are possible *but not how fast the reactions occur*. There is a need for chemical kinetics because some reactions of environmental interest in water are slow and not at equilibrium over the time scale of interest.

22.2.2 Why are some chemical reactions not at equilibrium?

To understand why some chemical systems are not at equilibrium, it is necessary to review why systems equilibrate. Recall from Chapter 3 that the driving force in approaching equilibrium is the minimization of the Gibbs free energy of the system. The example of oxygen partitioning between air and water was discussed quantitatively in Section 3.8.4. Recall that the energy profile was similar to that shown in Figure 22.1. The equilibrium position corresponds to the minimum in free energy. The equilibrium composition is determined by the equilibrium constant, which is in turn calculated from the standard Gibbs free energy of reaction.

Not all reactions have energy profiles similar to Figure 22.1. One important feature of Figure 22.1 is that *all* system compositions have free energies less than the free energy of the reactants. In other words, it takes little or no energy to start the process of converting reactants to the equilibrium position. A system with this characteristic is said to be under

equilibrium (or ***thermodynamic***) ***control***. The species concentrations in a system under equilibrium control are at their equilibrium values. In the case of the energy profile in Figure 22.1, it also takes little or no energy to start the process of converting products to the equilibrium position.

The energy profile in Figure 22.1 is similar to the energy path taken by a skateboarder standing at the lip of a concrete trough. With little energy, the skateboarder can drop into the trough and travel back and forth until reaching the equilibrium position at the bottom of the trough. At the equilibrium position, the potential energy of the skateboarder is minimized.

equilibrium control (thermodynamic control): the situation in which it takes little or no energy to start the process of converting reactants to products and therefore species concentrations are determined by thermodynamics

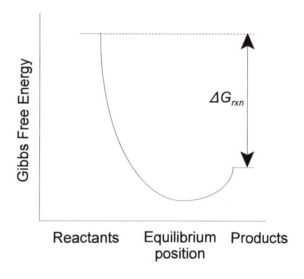

Figure 22.1: Energy Profile for a System Under Equilibrium Control

Another common energy profile is shown in Figure 22.2. It is important to see that the systems in Figures 22.1 and 22.2 have the same ΔG_{rxn}. However, for the system depicted in Figure 22.2, energy must be added before reactants can proceed to the equilibrium position. The system in Figure 22.2 is said to be under ***kinetic control***. In this system, the energy barrier to be overcome ($\Delta G^{\ddagger}$) is more important than the difference in energy between reactants and products (ΔG_{rxn}). This is analogous to a skateboarder who must go over a small hill before entering the trough. The size of the hill (analogous to $\Delta G^{\ddagger}$) will determine the progress of the skateboarder, not the depth of the trough (analogous to ΔG_{rxn}).

kinetic control: the situation in which it takes energy to start the process of converting reactants to products and therefore species concentrations are determined by kinetics

To return to the question at hand, why are all reactions *not* at equilibrium? Some reactions require energy input to overcome an energy barrier. This means that significant time may be required to reach the equilibrium position, where the energy of the system is minimized.

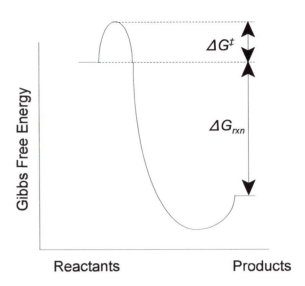

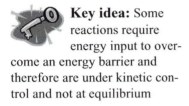 **Key idea:** Some reactions require energy input to overcome an energy barrier and therefore are under kinetic control and not at equilibrium

Figure 22.2: Energy Profile for a System Under Kinetic Control

22.3 REACTION RATES

22.3.1 Mechanisms and molecularity

The field of kinetics is concerned with the change in species concentrations over time. Thus, the *rates* of chemical reactions become important. In Section 22.3.2, you will learn how the rates of change of chemical species are related and how to determine the rates for some reactions by looking at reaction stoichiometry.

Before diving into the details, take a moment to reflect on how chemical reactions occur. Chemical reactions in homogeneous systems take place with chemical species that are close enough to interact. The interaction can take many forms, such as proton transfer, complexation, and electron exchange (redox reactions). If you had a viewing device with sufficient resolution, you could observe the exact ways that chemical species interact. The nature of the interaction is called the ***chemical mechanism***. Kinetics and mechanisms are related. Chemical kinetics provides important information for the elucidation of chemical mechanisms. Also, predicted mechanisms can be used as a starting point to study chemical kinetics.

chemical mechanism: the nature of the interaction between chemical species in a chemical reaction

When the time-dependent nature of chemical reactions is to be emphasized, the reactions are written with a small arrow pointing from reactants to products. An example is the chemical reaction: $A + 2B \rightarrow 3C$. The symbols A, B, and C here represent species, **not** species concentrations. The numbers in front of the species names are stoichiometric coefficients. They represent the number of molecules (or ions) participating in the reaction. Thus, writing the reaction as $A + 2B \rightarrow 3C$ symbolizes that 1 mole of A and 2 moles of B react to form 3 moles of C. The total *molecularity*: the total number of reactant molecules (or ions) participating in the reaction

number of *reactant* molecules (or ions) participating in the reaction is called the ***molecularity*** of the reaction. The reaction A + 2B → 3C has a molecularity of 3 (1 from A plus 2 from B). The terms *unimolecular*, *bimolecular*, and *termolecular* describe reactions with molecularities of 1, 2, and 3, respectively. It is common practice to ignore the role of water in determining the molecularity of reactions in the aqueous phase. For example, the dissociation of acetic acid (HAc) to H^+ and acetate (A^-) usually is written as a unimolecular reaction (HAc → H^+ + A^-), even though you know that the actual mechanism is proton transfer to water (i.e., the bimolecular reaction HAc + H_2O → H_3O^+ + A^-).

22.3.2 Rates and stoichiometry

What is meant by the rate of a chemical reaction? To understand the idea of the rate of a reaction, it is easier to start with the rate of change with time of the concentrations of *individual species*. Thus, we are seeking expressions for, say, $d[A]/dt$ in a system containing species A. Determining the rates of change of individual species can be difficult. However, it is fairly easy with simple systems to calculate how the rate of change of the concentration of one species relates to the rate of change of the concentration of another species.

As an example, suppose you are interested in a chemical system described only by the following chemical reaction: A + B → C. Without experimentation, it is not possible to determine $d[A]/dt$, $d[B]/dt$, or $d[C]/dt$. However, you can determine by inspection how $d[A]/dt$, $d[B]/dt$, or $d[C]/dt$ are related. Suppose you know by experiment that after 5 seconds of reaction, $d[A]/dt = -3×10^{-3}$ mole/s.

Thoughtful Pause

What is $d[B]/dt$ after 5 seconds of reaction?

You know by inspection that $d[B]/dt$ also must be $-3×10^{-3}$ mole/s after 5 seconds of reaction since the reaction stoichiometry tells you that 1 mole of B is consumed for every mole of A consumed. What about $d[C]/dt$? Again by inspection, you know that $d[C]/dt = +3×10^{-3}$ mole/s after 5 seconds of reaction since the reaction stoichiometry dictates that 1 mole of C is *produced* for every mole of A or B consumed.

Now consider the chemical reaction: A + 2B → C. Again, assume you know by experiment that after 5 seconds of reaction, $d[A]/dt = -3×10^{-3}$ mole/s. Now what is $d[B]/dt$ after 5 seconds of reaction? You can see by the reaction stoichiometry that B is consumed twice as fast as A. For each mole of A that is consumed, 2 moles of B are consumed. Thus, $d[B]/dt = -6×10^{-3}$ mole/s after 5 seconds of reaction.

Thoughtful Pause
What about $d[C]/dt$ in this case?

Key idea: In a reaction, the rate of change of each species concentration divided by the species stoichiometry is the same for each species

rate of a chemical reaction: the rate of change of any species in the reaction over time divided by the stoichiometric coefficient of the species

Again by inspection, you know that $d[C]/dt = +3 \times 10^{-3}$ mole/s after 5 seconds of reaction since the reaction stoichiometry dictates that 1 mole of C is *produced* for every mole of A consumed or for every 2 moles of B consumed.

The relationships between the rates of changes of species concentrations can be generalized. To generalize, note that it is useful in chemical kinetics to treat the stoichiometric coefficients of reactants as negative and the stoichiometric coefficients of products as positive. This trick also was used with equilibria. With this sign convention in mind, consider this slightly more general reaction: $v_A A + v_B B \rightarrow v_C C + v_D D$, where v_i is the stoichiometric coefficient for species i. The rates of changes of the individual species are related by:

$$\frac{1}{v_A}\frac{d[A]}{dt} = \frac{1}{v_B}\frac{d[B]}{dt} = \frac{1}{v_C}\frac{d[C]}{dt} = \frac{1}{v_D}\frac{d[D]}{dt}$$

eq. 22.2

Note that the signs work out in eq. 22.2: both v_A and $d[A]/dt$ are negative so $(1/v_A)(d[A]/dt)$ is positive and both v_D and $d[D]/dt$ are positive so $(1/v_D)(d[D]/dt)$ also is positive. For the most general reaction involving n species $A_1, A_2, ..., A_n$, you can write the reaction itself as $\sum_{i=1}^{n} v_i A_i$ and the relationships between the rates of change of species concentrations as $\frac{1}{v_i}\frac{d[A_i]}{dt} = \frac{1}{v_j}\frac{d[A_j]}{dt}$, for all i and $j = 1$ to n.

How can you use these observations? First, relationships between the rates of change of species concentrations will save you time. Once you find an expression for the rate of change of one species concentration with time, you can write immediately the rate of change of all the other species in the reaction. Second, you can use these observations to write an expression for the rate of a chemical reaction. Equation 22.2 defines the *rate of a chemical reaction* as the rate of change of any species in the reaction over time divided by the stoichiometric coefficient of the species:

$$\text{rate} = (1/v_i)(d[A_i]/dt) \text{ for } \textit{any} \text{ species } A_i \text{ in the reaction}$$

Another calculation of reaction rates is given in Example 22.1.

22.3.3 Kinetic order and elementary reactions
At first appearance, it seems that the concepts of molecularity and

---

***Example 22.1:* Reaction Rates**

The solvent methylene chloride (formally, dichloromethane: CH_2Cl_2) reacts with bisulfide to form dithiomethane:

$CH_2Cl_2 + 2HS^- \rightarrow$
$\quad CH_2(SH)_2 + 2Cl^-$

The reaction is being monitored in the lab by measurements of the chloride concentration. What is the reaction rate and rate of methylene chloride disappearance after 2 minutes of reaction if chloride appears at a rate of 10 mmol/min after 2 minutes?

Solution:
The reaction rate is
$(1/v_i)(d[A_i]/dt)$ for any species
A_i. If A_i is chloride, then the
reaction rate after two minutes
is $(\frac{1}{2})(10 \text{ mmol/min}) = $ **5
mmol/min**. If A_i is methylene
chloride, then $(-1/1)(d[A_i]/dt) = $
reaction rate = 5 mmol/min.
Thus, the rate of change of
methylene chloride after two
minutes is $-$**5 mmol/min**.

---

rate expression (*rate law*): a
mathematical expression for the
rate of change of a species
concentration

rate constant: the proportionality constant in the rate expression that relates the rate of
change of a species concentration with a function of species
concentration

reaction order: the apparent
molecularity, calculated as the
sum of the exponents on the
species concentrations in the
rate expression

elementary reactions: reactions
for which the reaction order and
molecularity are identical and
thus the reaction order (and the
rate expression) can be determined from the reaction
stoichiometry

stoichiometry could be combined to determine reaction rates. For example, suppose you know that species A and B react with 1:1 stoichiometry. In this case, the molecularity is 2. If A and B react with 1:1 stoichiometry, then you might expect that the reaction rate ($= -d[A]/dt = -d[B]/dt$) is proportional to the concentrations of both A and B. If the reaction rate is proportional to the concentrations of *both* A and B, then it makes sense that the reaction rate is proportional to the *product* of the reactant concentrations, [A][B]. Thus, you might be willing to accept reaction rates such as: $d[A]/dt = d[B]/dt = -k[A][B]$. In this expression, k is some proportionality constant and the negative sign is used to indicate that A and B are disappearing (so that $d[A]/dt$ and $d[B]/dt$ are both less than zero). Equations such as $d[B]/dt = -k[A][B]$ are called ***rate expressions*** or ***rate laws***. Proportionality constants such as k are called ***rate constants***.[†]

The problem with this approach is that the reaction mechanism must be known before both the molecularity and stoichiometry of the reaction can be determined. In many cases, reaction rates are determined experimentally. If determined experimentally, then you do not really know how many molecules (or ions) react. In other word, you do not know the true molecularity. All you know is *apparently* how many molecules (or ions) react.

This apparent molecularity is called the ***reaction order***. If the experimentally determined rate of change of species A is given by $d[A]/dt = -k[A][B]$, then we say that the reaction is first order in A, first order in B, and second order overall. The reaction order with respect to each species is the *exponent on the concentration of that species* in the rate expression. The overall reaction order is the *sum of the exponents on the species concentrations* in the rate expression.

The difference between the (true) molecularity and apparent molecularity is important. The (true) molecularity is determined by the true mechanism. The reaction order comes from the hypothesized mechanism through experimental data or good guessing. For some (but certainly not all) chemical reactions, the reaction order and molecularity are identical. Reactions for which the reaction order and molecularity are identical are called *elementary reactions*. Another way to think of elementary reactions is that *the reaction order (and thus the rate expression) can be determined from the stoichiometry in an elementary reaction*.

As an example, suppose that you measure the rate at which the concentrations of species A and B change over time. You find that A disappears twice as fast as B appears.

[†] Rate constants have a deeper meaning than just a proportionality constant. The meaning of rate constants is explored more fully in Section 22.6.

Thoughtful Pause
What is your guess of the reaction stoichiometry?

You might guess that the reaction stoichiometry is: $2A \rightarrow B$. If the reaction is not elementary, then you cannot write the rate expression. However, if the reaction is elementary, then the rate expression is given by the stoichiometry.

Thoughtful Pause
What is the rate expression if the reaction is elementary?

If elementary, the rate expression for B is: $d[B]/dt = -k[A]^2$. Using the approach of Section 22.3.2, the rate expression for A is: $\frac{1}{2}d[A]/dt = -k[A]^2$ or $d[A]/dt = -2k[A]^2$.

Similarly, if you know the reaction is elementary, then you can deduce the reaction stoichiometry from the rate expression. Suppose that B is produced and the rate of change of the concentration of B is found to be first order in the concentrations of both A and C.

Thoughtful Pause
What is the reaction stoichiometry if the reaction is elementary?

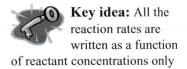 **Key idea:** All the reaction rates are written as a function of reactant concentrations only

The reaction stoichiometry is likely to be that 1 mole of A reacts with 1 mole of C to produce 1 mole of B, or $A + C \rightarrow B$.

Without being stated explicitly, an important convention has been used to write rate expressions in the examples shown in this section. The convention is that *all the reaction rates are written as a function of reactant concentrations only*. In other words, a reaction such as $A + C \rightarrow$ B, if elementary, means that A and C react in some fashion to form B. Thus, the rates of change of A, B, and C all depend only on reactant concentrations (i.e., only on [A] and [C]) and not on [B]. If the rate of change of the concentration of B is found to depend on [B], then a new reaction must be written.

22.3.4 Systems of elementary reactions
In environmental settings, it is not unusual to have many chemical reactions involving the same species occurring simultaneously. Thus, a given species may be found in several or even dozens of reactions.

Regardless of the number of reactions, it is fairly simple to write rate expressions for each species *if the reactions are elementary*. Three basic principles are used to write rate expressions with a system of reactions. First, as discussed in Section 22.3.3, the reaction rates for individual reactions are determined from reactant concentrations only. Second, the reaction rates for individual reactions can be determined directly from the reaction stoichiometry for elementary reactions. Third, rates are additive. Thus, if the rate of change of the concentration of A with respect to time is $-k_1[A][B]$ from one reaction and $+k_2[C]^2$ from another reaction, then the overall rate of change of [A] with time is the sum of the rates:

$$d[A]/dt = -k_1[A][B] + k_2[C]^2$$

The process of writing rate expressions with more complex systems can be broken down into a series of steps. The steps are shown here and summarized in Appendix C (Section C.9). The process will be illustrated with the following set of elementary reactions:

Reaction 1: $A + B \rightarrow C$ k_1
Reaction 2: $C + A \rightarrow B$ k_2
Reaction 3: $2C \rightarrow A$ k_3

We seek to write the rate expressions for the concentrations of A, B, and C.

Step 1: Identify the reaction(s) that include the species of interest as either a reactant or product

For A, the pertinent reactions containing A are reactions 1, 2, and 3; B appears only in Reactions 1 and 2, whereas C appears in all three reactions.

Step 2: Write the rate expressions for each species of interest, using the reactions identified in Step 1

Each rate expression should be the rate constant multiplied by the reactant concentrations, each raised to their stoichiometric powers. Use a negative sign with the rate when the species of interest is a reactant and a positive sign when the species of interest is a product. Remember to correct for the stoichiometry of the species of interest, if the stoichiometry is not one.

For A:
    rate from reaction $1 = -k_1[A][B]$
        (rate < 0 since A is consumed)

Key idea: To write rate expressions with a system of reactions, remember that (1) the reaction rates for individual reactions are determined from reactant concentrations only, (2) the reaction rates for individual reactions can be determined directly from the reaction stoichiometry for elementary reactions, and (3) rates are addi-tive

---

Example 22.2: **Systems of Reactions**

When ozone gas dissolves in water, it decomposes fairly rapidly. One set of pertinent reactions in "pristine" water is:

$O_3 + OH^- \rightarrow HO_2^- + O_2$ k_1
$HO_2^- + O_3 \rightarrow HO_2 + O_3^-$ k_2
$HO_2 \rightarrow H^+ + O_2^-$ k_3
$O_2^- + O_3 \rightarrow O_2 + O_3^-$ k_4
$O_3^- + H_2O \rightarrow OH + OH^- + O_2$ k_5
$OH + O_3 \rightarrow HO_2 + O_2$ k_6

Write an expression for the rate of change in the ozone concen-tration.

Solution:
Although the reaction sequence is complex, simply apply the steps in the text. The reactions

having ozone as a reactant or product are reactions 1, 2, 4, and 6. The rate terms for $d[O_3]/dt$ from reactions 1, 2, 4, and 6 are, respectively, $-k_1[O_3][OH^-]$, $-k_2[HO_2^-][O_3]$, $-k_4[O_2^-][O_3]$, and $-k_6[OH][O_3]$. Note that the rate expressions use reactant concentrations only. Adding the rate expressions for the individual reactions:

$d[O_3]/dt = -(k_1[OH^-] + k_2[HO_2^-] + k_4[O_2^-] - k_6[OH])[O_3]$

rate from reaction 2 $= -k_2[A][C]$
    (rate < 0 since A is consumed)
rate from reaction 3 $= +k_3[C]^2$
    (rate > 0 since A is formed)

For B:
    rate from reaction 1 $= -k_1[A][B]$
        (rate < 0 since B is consumed)
    rate from reaction 2 $= +k_2[A][C]$
        (rate > 0 since B is formed)

For C:
    rate from reaction 1 $= +k_1[A][B]$
        (rate > 0 since C is formed)
    rate from reaction 2 $= -k_2[A][C]$
        (rate < 0 since C is consumed)
    rate from reaction 3 $= -2k_3[C]^2$
        (rate < 0 since C is consumed)

Note that the rate expressions use only reactant concentrations. Also note the factor of 2 in the rate of [C] over time with reaction 3. For that reaction: $\frac{1}{2}d[C]/dt = -k_3[C]^2$, so the rate from reaction 3 is $d[C]/dt = -2k_3[C]^2$.

Step 3: Add up the rates from each reaction to find the overall rate of change of each species concentration

From Step 2:

$$d[A]/dt = -k_1[A][B] - k_2[A][C] + k_3[C]^2 \qquad \text{eq. 22.3}$$
$$d[B]/dt = -k_1[A][B] + k_2[A][C] \qquad \text{eq. 22.4}$$
$$d[C]/dt = +k_1[A][B] - k_2[A][C] - 2k_3[C]^2 \qquad \text{eq. 22.5}$$

If you have experience with solving differential equations, you will recognize that eqs. 22.3-22.5 constitute a set of three coupled differential equations in three unknowns. This set of equations can be solved numerically to find [A], [B], and [C] over time if the rate constants and initial concentrations of A, B, and C are known. For another instance of writing rate expressions, see Example 22.2.

22.4 COMMON RATE EXPRESSIONS

22.4.1 Introduction
A number of simple rate expressions are seen repeatedly in aquatic chemistry. Several common forms will be discussed in this section. For

each form, the rate law will be presented. As indicated in Section 22.3, the rate laws are differential equations and are sometimes called the *differential form of the rate law*. To determine the species concentration over time, the rate expression will be integrated (sometimes called the *integrated form of the rate law*). In addition, environmental examples of each rate law will be presented.

The important features of each rate expression are summarized in Table 22.1. In addition, examples of species concentrations over time with each rate law are shown in Figure 22.3.

22.4.2 Zero-order kinetics

Recall that the kinetic order is the sum of the exponents on species concentrations in the rate law. The zero-order rate expression refers to a rate law in which the exponent for each species is zero. For zero-order disappearance:

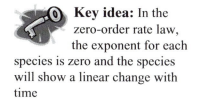

Key idea: In the zero-order rate law, the exponent for each species is zero and the species will show a linear change with time

$$d[A]/dt = -k_0[A]^0 = -k_0$$

(The subscript 0 is used here simply as a reminder that the rate constant is a zero-order rate constant.) Note that a zero-order reaction has a constant reaction rate; that is, $d[A]/dt$ is constant. In addition, zero-order rate constants have units of mol/L-time.

The rate expression can be integrated to show:

$$[A] = [A]_0 - k_0t$$

half-time of the reaction $(t_{1/2})$: the time required to reduce a species concentration to one-half its initial value

where $[A]_0$ is the concentration of A at $t = 0$. Thus, a substance exhibiting zero-order disappearance will show a linear decrease in concentration over time (see Figure 22.3). One way of expressing the rate of a reaction is to determine the time required to reduce a species concentration to one-half is initial value. This is called the ***half-time of the reaction*** (or sometimes, the half-life of species A) and is given the symbol $t_{1/2}$. From the integrated form of the rate expression, the half-time of a zero-order reaction is $[A]_0/(2k_0)$. Note that the half-time increases with increasing $[A]_0$.

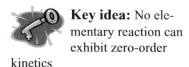

Key idea: No elementary reaction can exhibit zero-order kinetics

Recall that elementary reactions have rate laws that come directly from reaction stiochiometry. Thus, no elementary reaction by itself can exhibit zero-order kinetics. In spite of this conclusion, zero-order rate expressions are observed fairly frequently in environmental systems. Why? Zero-order kinetics can result as the limiting form of more complicated rate laws. One example of this phenomenon is the chemical reaction of adsorbed species (called *heterogeneous catalysis*). Using the symbols from Section 19.6, consider a species P equilibrating with a surface $\equiv$S to form a surface complex $\equiv$S-P with an equilibrium constant K. From eq. 19.10, the adsorbed species concentration is given by: $[\equiv SP] = K[P]S_T/(1 + K[P])$, where S_T is the total site concentration. Now imagine that the adsorbed species reacts on the surface to form a product, P':

$$\equiv SP \rightarrow P' \quad \text{rate constant} = k$$

The Langmuir-Hinshelwood equation (after Irving Langmuir and Cyril Norman Hinshelwood, 1897-1967) states that the rate of change of [P'] is:

$$d[P']/dt = +k[\equiv SP] = kK[P]S_T/(1 + K[P])$$

If $K[P] \gg 1$, then $d[P']/dt = kS_T$. Thus, at the beginning of the adsorption process, when [P] is large (and $K[P]$ may be $\gg 1$), the formation of P' will exhibit zero-order kinetics with this type of catalysis, since:

$$d[P']/dt = kS_T = \text{constant}$$

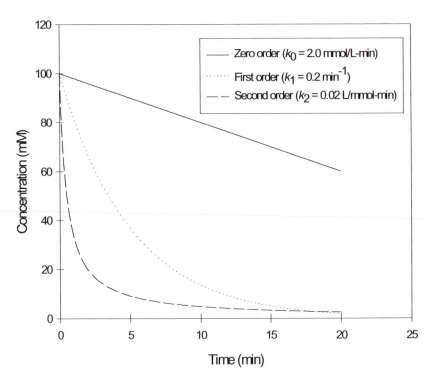

Figure 22.3: Examples of Zero-, First-, and Second-order Kinetics

Another example comes from the transfer of gases between the aqueous phase (where they exist as dissolved gases) and gaseous phase. A common kinetic model takes the following form:

$$d[A(aq)]/dt = k_L a(P_A/K_H - [A(aq)])$$

where k_L is a mass-transfer coefficient, a is the ratio of the interfacial surface area to the liquid volume, P_A is the partial pressure of A in the gas phase, and K_H is the Henry's Law constant. Under conditions in which [A] is small and P_A is constant (e.g., in the early stages of mass transfer from the atmosphere to a small liquid volume):

$$d[A(aq)]/dt \approx k_L a P_A / K_H = \text{constant}$$

Thus, zero-order kinetics are observed.

Table 22.1: Summary of the Characteristics of Simple Kinetic Orders

Order	Elementary Reaction	Differential form	Integrated Form	Units of k^1	$t_{1/2}$
Zero	None	$d[A]/dt = -k_0$	$[A] = [A]_0 - k_0 t$	mol/L-time	$[A]_0/(2k_0)$
First	A → products	$d[A]/dt = -k_1[A]$	$[A] = [A]_0 e^{-k_1 t}$	1/time	$\ln(2)/k_1$
Second²	2A → products	$d[A]/dt = -k_2[A]^2$	$[A] = \dfrac{[A]_0}{1 + [A]_0 k_2 t}$	L/mol-time	$1/([A]_0 k_2)$
	A + B → products	$d[A]/dt = -k_2[A][B]$	$[A] = \dfrac{[B]_0 - [A]_0}{\dfrac{[B]_0}{[A]_0} e^{([B]_0 - [A]_0)k_2 t} - 1}$	L/mol-time	note 3
n^{th} ($n \neq 1$, see note 2)	nA → products	$d[A]/dt = -k_n[A]^n$	$[A] = \left([A]_0^{1-n} - k_n t(1-n)\right)^{\frac{1}{1-n}}$	(L/mol)$^{n-1}$time^{-1} note 4	

Notes: 1. If the concentration units are mol/L.

2. It is traditional to define the rate constants here so that the rate laws are written as $d[A]/dt = -k_2[A]^2$, not $d[A]/dt = -2k_2[A]^2$ and $d[A]/dt = -k_n[A]^n$, not $d[A]/dt = -nk_n[A]^n$.

3. If $[A]_0 \geq 2[B]_0$, then [A] will never be as small as $[A]_0/2$ and the half-time is infinite. Otherwise, $t_{1/2} = \dfrac{\ln\left(2 - \dfrac{[A]_0}{[B]_0}\right)}{([B]_0 - [A]_0)k_2}$; for $[A]_0 < 2[B]_0$.

4. $t_{1/2}$ is given by: $\dfrac{[A]_0^{1-n}\left(2^{n-1} - 1\right)}{(n-1)k_n}$ $(n \neq 1)$

Key idea: In the first-order rate law, the exponent for the species is 1 and the species will show an exponential change with time

---

Example 22.3: **First-Order Kinetics**

The pesticide ethylene dibromide (EDB, formally 1,2-dibromoethane: BrH_2C-CH_2Br) can hydrolyze (react with water) to form the corresponding alcohols. The first-order hydrolysis rate constant is 6×10^{-9} s^{-1} at pH 7 and 25°C (Haag and Mill, 1988). What is the half-time of EDB pH 7 and 25°C?

Most uses of EDB were banned in 1984. If the EDB concentration in a water was 2 µg/L in 1984, when will the EDB concentration fall below the drinking water standard of 0.05 µg/L at pH 7 and 25°C?

Solution:
The reaction is first order in EDB. If the hydrolysis reaction is elementary, then the appropriate rate expression is: $dC/dt = -kC$, where C is the EDB concentration. Integrating: $C = C_0 e^{-kt}$, where C_0 is the initial DBE concentration. Rearranging: $t = (-1/k)\ln(C/C_0)$.

The half-time is $\ln(2)/k = \ln(2)/(6 \times 10^{-9}$ $s^{-1}) =$ **3.7 years**. The time required to reduce the concentration from 1 to 0.05 µg/L is: $t = (-1/k)\ln(C/C_0) = (-1/6 \times 10^{-9}$ $s^{-1})\ln(0.05/2) = 19.5$ years, **or in about 2004**. The hydrolysis reaction is slow.

22.4.3 First-order kinetics

For first-order disappearance, the rate law is:

$$d[A]/dt = -k_1[A]$$

Thus, first-order rate constants have units of 1/time. The rate expression can be integrated to show:

$$[A] = [A]_0 \, e^{-k_1 t}$$

Thus, a substance exhibiting first-order disappearance will show an exponential decrease in concentration over time (see Figure 22.3). Similarly, exponential growth in a population stems from a first-order rate law: $dN/dt = +k_1 N$, where N is the population at time t. The half-time for a first-order process is $\ln(2)/k_1$. First-order kinetics is the only kinetic order where $t_{1/2}$ is independent of the initial concentration.

Many processes in the environment exhibit first-order kinetics. Examples include radioactive decay and exponential growth. A good example of first-order kinetics is Chick's law of disinfection, a common model for disinfection (after Henrietta Chick). Chick's law states that the rate of disinfection is proportional to the number concentration of surviving organisms. Thus: $dN/dt = -k_1 N$, where N is the number of surviving organisms per liter. If Chick's law holds, you would expect an exponential decrease over time in surviving organisms. Another illustration is shown in Example 22.3.

Even more common than first-order elementary reactions is a kinetic form that *appears* as first-order kinetics. Consider the reaction of two species as follows: $A + B \rightarrow$ products. If this is an elementary reaction, then the reaction is second order and the rate expression is given by:

$$d[A]/dt = -k[A][B]$$

Now imagine that $[B]$ is constant. If $[B]$ is relatively constant, then the rate expression can be rewritten as: $d[A]/dt = -k'[A]$, where: $k' = k[B]$. This form of first-order kinetics is called ***pseudo-first-order kinetics***. The pseudo-first-order rate constant, k', is the product of a second-order rate constant and (usually) the concentration of another species. Pseudo-first-order kinetics is very common in environmental systems. For example, reactions with H^+ at constant pH (and ionic strength) generally result in pseudo-first-order kinetics with $k' = k[H^+]$. Pseudo-first-order kinetics also is shown in Example 22.4.

22.4.4 Second-order kinetics

Second-order kinetics can take two very different forms. In the first form, a chemical species reacts with itself to go to products: $2A \rightarrow$ products.

Example 22.4: Pseudo-First-Order Kinetics

The compound ethyl acetate (formally, ethyl ethanoate) is used as a solvent, as an insecticide, and to impart a fruity aroma. It hydrolyzes by three mechanisms: reaction with water (considered first order with rate constant $k_W = 1.5 \times 10^{-10}$ s^{-1}), reaction with H$^+$ (second order, $k_H = 1.1 \times 10^{-4}$ M^{-1}s^{-1}), and reaction with OH$^-$ (second order, $k_{OH} = 1.1 \times 10^{-1}$ M^{-1}s^{-1}). Show how ethyl acetate hydrolysis depends on pH.

Solution:
Since rates are additive, the overall rate expression (if the reactions are elementary) is:

$$dC/dt = -(k_W + k_H[H^+] + k_{OH}[OH^-])C$$

If the pH is constant, this becomes: $dC/dt = -kC$, where: $k = k_W + k_H[H^+] + k_{OH}[OH^-]$. The half-time is $\ln(2)/k$ and is plotted as a function of pH below. Note that the slowest rate (largest $t_{1/2}$) occurs at about pH 5.5.

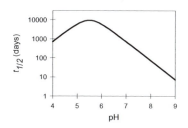

 Key idea: Second-order kinetics typically come from reactions that are second order in one reactant or from reactions that are first order in two reactants

The resulting rate expression is:

$$d[A]/dt = -2k_2[A]^2$$

The integrated form is $[A] = \dfrac{[A]_0}{1 + 2[A]_0 k_2 t}$ and the half-time is $1/(2[A]_0 k_2)$[†].

Note that the half-time decreases with increasing $[A]_0$. The units of the second-order rate constant are L/mol-time. This "second order in A" form of second-order kinetics is relatively rare in environmental systems. It is sometimes used to model particle-particle collisions, where [A] is interpreted as the number of particles per unit volume.

A more common form of second-order kinetics involves the interaction of two chemical species: A + B → products. This might be called the "first order in both A and B" case. In this case:

$$d[A]/dt = -k_2[A][B] \qquad \text{eq. 22.6}$$

Again, note that the units of the second-order rate constant are L/mol-time. Clearly, the differential equation in eq. 22.6 cannot be solved because it is one equation in two unknowns (namely, [A] and [B]). From the stoichiometry of the reaction, it makes sense that the amount of A consumed in the reaction must be equal to the amount of B consumed. Thus:

$$\text{moles/L of A consumed} = [A]_0 - [A]$$
$$= \text{moles/L of B consumed}$$
$$= [B]_0 - [B]$$

So: $[B] = [A] + [B]_0 - [A]_0$. Substituting into eq. 22.6:

$$d[A]/dt = -k_2[A]([A] + [B]_0 - [A]_0) \text{ or:}$$

$$\frac{d[A]}{[A]([A] + [B]_0 - [A]_0)} = -k dt \qquad \text{eq. 22.7}$$

Solving the differential equation in eq. 22.7:

$$[A] = \frac{[B]_0 - [A]_0}{\dfrac{[B]_0}{[A]_0} e^{([B]_0 - [A]_0)k_2 t} - 1} \qquad \text{eq. 22.8}$$

[†] These equations are slightly different than those in Table 22.1, since reaction stoichiometry is included here.

Does eq. 22.8 make sense? One way to decide if expressions such as eq. 22.8 are reasonable is to evaluate them at the extreme values of t. As you can see from eq. 22.8, $[A] = [A]_0$ at $t = 0$, as required. As $t \to \infty$, $[A] \to 0$ if $[B]_0 > [A]_0$. This makes sense: in the presence of excess B, all A will be consumed eventually (Figure 22.4, dashed line). However, $[A] \to [A]_0 - [B]_0$ as $t \to \infty$ if $[A]_0 > [B]_0$. Again, this makes sense: if A is in excess, B will be consumed completely given enough time. If $[B]_0$ moles/L of B are consumed, then $[B]_0$ moles/L of A will be consumed and $[A] \to [A]_0 - [B]_0$ (Figure 22.4, dotted line). Equation 22.8 is not valid if $[A]_0 = [B]_0$. If $[A]_0 = [B]_0$, then eq. 22.7 becomes $d[A]/dt = -k_2[A]^2$ and the "second order in A" solution holds (Figure 22.4, solid line). Another illustration of second-order kinetics is shown in Example 22.5.

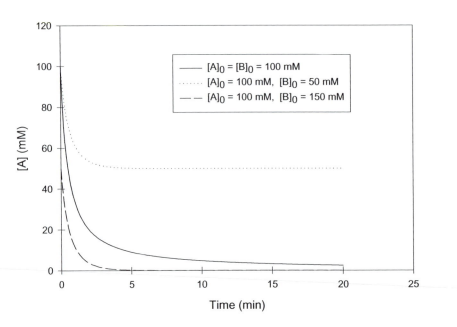

Figure 22.4: Effects of Initial Concentrations on the Concentration Profiles in Second-Order Reactions
($k_2 = 0.02$ L/mmol-min)

22.5 MORE COMPLEX KINETIC FORMS

22.5.1 Introduction

To analyze more complex kinetic systems of elementary reactions, write the set of differential equations and solve them with the appropriate initial conditions. The required initial conditions typically are the initial concentrations of all species. *Writing* the set of differential equations is fairly easy, using the approach shown in Section 22.3.4. *Solving* the differential equations can be tedious. Numerical solution methods frequently are needed. In this section, some representative examples of more complex systems are presented and solved.

22.5.2 Kinetics of reversible reactions

Most of this book has focused on reversible reactions at equilibrium. Reversible reactions also can be analyzed to determine how the species change over time as equilibrium is approached. Without doing the mathematics, you *know* there must be a link between kinetics and equilibrium.

Thoughtful Pause

What is the link between the kinetic and equilibrium approaches to a system with reversible reactions?

The kinetic equations for a reversible system must give the equilibrium concentrations as t → ∞.

To demonstrate this point, consider a simple reversible system consisting of two reactions:

$$A \rightarrow B \ k_f$$
$$B \rightarrow A \ k_r$$

This system is equivalent to: $A \rightleftharpoons B$. Both k_f and k_r are first-order rate constants. The subscripts used in this example are a reminder that the rate constants refer to the forward (reactant-to-product) and reverse (product-to-reactant) reactions that make up the reversible system. Following the procedure in Section 22.3.4, you can write the rate expressions for $d[A]/dt$ and $d[B]/dt$ (you may wish to try this on your own before going on):

$$d[A]/dt = -k_f[A] + k_r[B]$$
$$d[B]/dt = -k_f[B] + k_r[A]$$

eq. 22.9

A mass balance reveals that $[A] + [B]$ must equal the total initial concentration added. For this example, consider the addition of A only ($[B]_0 = 0$), so: $[A] + [B] = [A]_0$ or: $[B] = [A]_0 - [A]$. Substituting into eq. 22.9:

$$\begin{aligned} d[A]/dt &= -k_f[A] + k_r[B] \\ &= -k_f[A] + k_r([A]_0 - [A]) \\ &= -(k_f + k_r)[A] + k_r[A]_0 \end{aligned}$$

Solving with $[A] = [A]_0$ at $t = 0$:

Example 22.5: **Second-Order Kinetics**

Dissolved ozone reacts with toluene with a second-order rate constant of 14 $M^{-1}s^{-1}$ (Staehelin and Hoigné, 1983). If the initial dissolved ozone and toluene concentrations are 2 mg/L and 10 mg/L, respectively, what will the toluene and ozone concentrations be after 20 minutes?

Solution:
The initial ozone and toluene concentrations are $(2 \times 10^{-3}$ g/L)/(48 g/mol) = 4.17×10^{-5} M and $(1 \times 10^{-2}$ g/L)/(92 g/mol) = 1.09×10^{-4} M, respectively. Plugging these values into eq. 22.8 (species A corresponding to toluene) with $k = 14$ $M^{-1}s^{-1}$ and $t = 1200$ s yields a toluene concentration of 8.57×10^{-5} M or **7.9 mg/L**. Assigning A to ozone yields a dissolved ozone concentration of 1.87×10^{-5} M or **0.9 mg/L**.

Note: by 4400 seconds, the ozone has decreased to about 1% of its initial value and the toluene concentration is approaching its limiting value of $[toluene]_0 - [ozone]_0 = 6.7 \times 10^{-5}$ M or about 6.2 mg/L.

 Key idea: The kinetic equations for a reversible system must give the equilibrium concentrations as $t \rightarrow \infty$

$$[A] = [A]_0 \left(\frac{k_r}{k_f + k_r} \right) \left[1 + \frac{k_f}{k_r} e^{-(k_f + k_r)t} \right] \qquad \text{eq. 22.10}$$

Similarly: $[B] = [A]_0 - [A] = [A]_0 \left(\frac{k_f}{k_f + k_r} \right) \left[1 - e^{-(k_f + k_r)t} \right]$.

There are several interesting features of the time-dependent solution to the reversible system (eq. 22.10). First, you can compare the time-dependent solution as $t \rightarrow \infty$ with the equilibrium solution. At equilibrium, $[B]/[A] = K$, where K is the equilibrium constant for A = B. At $t \rightarrow \infty$ in eq. 22.10, $[A] \rightarrow [A]_0 k_r/(k_f + k_r)$ and $[B] \rightarrow [A]_0 k_f/(k_f + k_r)$. Thus: $[B]/[A] \rightarrow k_f/k_r$. This must mean that $k_f/k_r = K$. The kinetic calculations yield a remarkable result: *an equilibrium constant can be expressed as the ratio of the forward and reverse rate constants.*

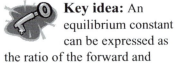 **Key idea:** An equilibrium constant can be expressed as the ratio of the forward and reverse rate constants

In fact, you can show that $k_f/k_r = K$ much more easily. At equilibrium, the species concentrations do not change with time. From eq. 22.9: $d[A]/dt = -k_f[A] + k_r[B] = 0$ at equilibrium. Thus: $[B]/[A] = k_f/k_r$ at equilibrium. Since $[B]/[A] = K$ at equilibrium, then, as before, $k_f/k_r = K$.

A second interesting feature of eq. 22.10 is the kinetics of the *approach* to equilibrium.

Thoughtful Pause

What is the apparent kinetic order and rate constant for the approach to equilibrium?

Based on eq. 22.10, the approach to equilibrium appears to be first order, with a first-order rate constant equal to $k_f + k_r$. Thus, the *slower* of the forward and reverse reactions determines the apparent reaction rate constant.

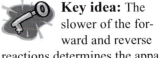 **Key idea:** The slower of the forward and reverse reactions determines the apparent reaction rate constant in the approach to equilibrium

You can use the discussion in this section to decide if elementary reactions can be considered at equilibrium over the time frame of interest. We almost always think of acid-base reactions as being at equilibrium. Why? The usual explanation is that proton-transfer reactions are fast. In fact, the transfer of a proton from H_3O^+ to a conjugate base of the form A^- is exceedingly fast. As an example, the second-order rate constant for $OH^- + H_3O^+ \rightarrow 2H_2O$ is about $10^{+11.1}$ L/mol-s. This is the fastest known reaction in water.[†] Even for acetic acid, the proton-transfer reactions are very fast:

[†] This reaction is so fast that the rate of the reaction is limited only by the diffusion of the reactants through the media (water). Reactions such as this are called *diffusion-controlled* reactions. The theoretical upper limit for second-order rate constants of diffusion-controlled reactions is in the $10^9 - 10^{11}$ L/mol-s range.

$$HAc + H_2O \rightarrow Ac^- + H_3O^+ \qquad k_{2,f} = 10^{+5.9} \text{ L/mol-s}$$
$$Ac^- + H_3O^+ \rightarrow HAc + H_2O \qquad k_{2,r} = 10^{+10.7} \text{ L/mol-s}$$

***Example 22.6:* Approach to Equilibrium**

In wastewater that has been disinfected by chlorination, ammonia reacts with hypochlorous acid to form monochloramine (NH_2Cl) as follows:

$$HOCl + NH_3 \rightarrow NH_2Cl + H_2O$$
$$k_f = 4.2 \times 10^6 \text{ L/mol-s}$$
$$\text{at } 25°C$$

Monochloramine also can hydrolyze:

$$NH_2Cl + H_2O \rightarrow HOCl + NH_3$$
$$k_r = 2.1 \times 10^{-5} \text{ s}^{-1} \text{ at } 25°C$$

Find the HOCl and monochloramine concentrations over time if 0.1 mg/L as Cl_2 of chlorine is added to 10 mg/L as N of ammonia.

Solution:
The initial ammonia and chlorine concentrations are (0.01 g/L)/(14 g/mol) = 7.1×10^{-4} M and (0.0001 g/L)/(70.9 g/mol) = 1.4×10^{-6} M, respectively. Since the ammonia is in great excess, assume that $[NH_3]$ is constant. You can calculate a pseudo-first-order rate constant = $k_f' = k_f[NH_3]$ = (4.2×10⁶ L/mol-s)(7.1×10^{-4} mol/L) = 3.0×10^3 s⁻¹. Plugging into eq. 22.10 with $[A]_0$ = initial chlorine concentration = 1.4×10^{-6} M and $[B]_0$ = initial monochloramine concentration = 0, you will obtain the plot shown below.

If sodium acetate (NaA) is added to water at a constant pH of 7.0, what is the half-time of the conversion of Ac^- to HAc? The analysis in this section has been conducted with first-order reactions approaching equilibrium. Therefore, you need to convert the second-order rate constants into pseudo-first-order rate constants. For the forward reaction: $k_f = k_{2,f}[H_2O] = (10^{+5.9}$ L/mol-s)(55.56 M) = $10^{+7.6}$ s⁻¹. For the reverse reaction: $k_r = k_{2,r}[H_3O^+] = (10^{+10.7}$ L/mol-s)(10^{-7} M) = $10^{+3.7}$ s⁻¹. The half-time is $\ln(2)/(k_f + k_r) = \ln(2)/(10^{+7.6}$ s⁻¹ $+ 10^{+3.7}$ s⁻¹) = 1.6×10^{-8} s or 16 nanoseconds. The equilibrium assumption for fast reactions clearly is justified at reasonable times. Another example of the approach to equilibrium is shown in Example 22.6.

22.5.3 Kinetics of sequential reactions

In the environment, it is not uncommon for the products of one reaction to be reactants in a second reaction and subsequently converted into new products. The overall reaction sequence is A → B → C, or: A → B; $k_{A,1}$ and B → C; $k_{B,1}$. If the two reactions are elementary, then the corresponding rates are (you may wish to try writing the rate expressions before proceeding):

$$d[A]/dt = -k_{A,1}[A]$$
$$d[B]/dt = +k_{A,1}[A] - k_{B,1}[B]$$
$$d[C]/dt = +k_{B,1}[B]$$

Solving for [A] first, then [B], and finally [C], you can show that for $[B]_0 = [C]_0 = 0$:

$$[A] = [A]_0\exp(-k_{A,1}t)$$
$$[B] = k_{A,1}[A]_0[\exp(-k_{A,1}t) - \exp(-k_{B,1}t)]/(k_{B,1} - k_{A,1})$$
$$[C] = [A]_0 - [A] - [B]$$

An example is shown in Figure 22.5. Note that [A] decreases exponentially, whereas [B] shows a maximum and [C] exhibits a sigmoidal (*S*-shaped) growth. You can show (Problem 22.2) that the maximum value of [B] and the time at which the maximum value of [B] occurs depend on both rate constants.

If $k_{B,1}$ is much greater than $k_{A,1}$, then [B] will be very small at all times and [C] will increase without a lag. Under these conditions, we say that A → B is the ***rate-determining step*** in the sequence A → B → C. In other words, $k_{A,1}$ will determine the concentrations of the species. Under such conditions, it is easy to miss B, since the concentration profiles are consistent with A → C (see Figure 22.6). Another set of sequential chemical reactions is shown in Example 22.7.

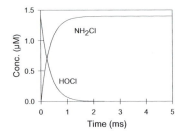

The reaction is very fast, with a half-time of about 0.25 milliseconds.

---

rate-determining step: the slowest step in a reaction sequence and thus the step that sets the overall rate of reaction (also called the rate-limiting step)

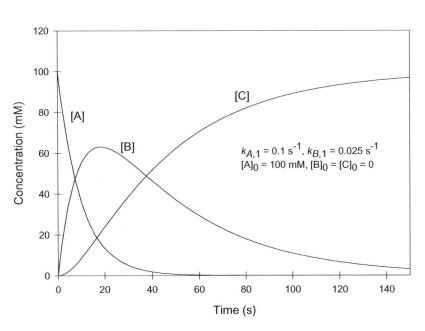

Figure 22.5: Sequential First-Order Reactions
(A → B ; $k_{A,1}$ and B → C ; $k_{B,1}$)

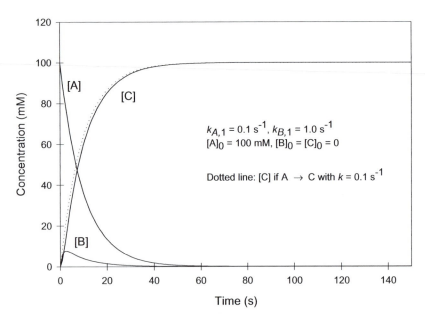

Figure 22.6: Sequential Reaction Sequence ($k_{B,1}/k_{A,1} = 10$)
Compared with A → C

22.6 EFFECTS OF TEMPERATURE AND IONIC STRENGTH ON REACTION KINETICS

22.6.1 Effects of temperature

There are many approaches to determining or modeling the effects of temperature on rate constants. All of the most common approaches can be summarized in the following dependency (Espenson, 1981):

$$k = (\text{constant})T^n\exp(-\text{energy term}/RT) \qquad \text{eq. 22.11}$$

In eq. 22.11, T is the temperature in °K and R is the ideal gas constant. The values of the constant n and the energy term for two models will be discussed.

The Swedish chemist Svante August Arrhenius[†] (1859-1927) reasoned that rate constants should change with temperature in the same fashion that equilibrium constants change with temperature. His based this idea on the fact that an equilibrium constant can be thought of as the rate of two rate constants (see Section 22.5.2). Relating eq. 22.11 with the van't Hoff equation (Section 21.3.1), he found $n = 0$. In Arrhenius's approach, the constant is given the symbol A (also called the ***preexponential factor***) and the energy term is given the symbol E_a (also called the ***activation energy*** or the ***Arrhenius activation energy***. Note that A has the same units as the rate constant. Thus, eq. 22.11 becomes:

$$k = A\exp(-E_a/RT) \qquad \text{eq. 22.12}$$

This is called the *Arrhenius equation*. Although the Arrhenius equation is empirical, the activation energy, E_a, is related to the energy required to overcome the energy "hill" ($\Delta G^{\ddagger}$ in Figure 22.2). The linearized form of the Arrhenius equation is:

$$\ln k = \ln A - (E_a/R)(1/T)$$

In other words, the $\ln k$ should be linearly related to $1/T$.

Another approach was adopted by Henry Eyring (1901-1981) and is called ***transition-state theory***. Transition-state theory assumes that all chemical reactions proceed through a complex called the *activated complex*. Reactants are considered to be in equilibrium with the activated complex. Thus, according to transition-state theory, the reaction A + B → C is more properly written as: A + B = AB‡ and AB‡ → C. The superscript "$\ddagger$" denotes the properties of the activated complex. Thus, $\Delta G^{\ddagger}$ in Figure 22.2 refers to the free energy of reaction for the activated complex.

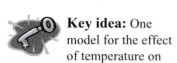

Key idea: One model for the effect of temperature on rate constants is the Arrhenius equation:
$k = A\exp(-E_a/RT)$

preexponential factor: the parameter in the Arrhenius equation that multiplies the exponential term

activation energy (Arrhenius activation energy): the energy term in the Arrhenius equation

transition-state theory: a reaction theory based on the assumption that the reactants equilibrate with an activated complex, which subsequently decomposes to form products

[†] Arrhenius also is known to the environmental science and engineering community as the first person to recognize (in 1896) that increased fossil fuel consumption may lead to global warming.

Example 22.7: **Sequential Reactions**

Thiocyanate (SCN^-) is used in a number of industrial processes and is a by-product of coke manufacturing. It can be removed from water by oxidation with ozone. The oxidation of thiocyanate produces cyanide, which is subsequently oxidized by ozone. The pertinent reactions are:

$$SCN^- + 2O_3 + 2OH^- \rightarrow$$
$$CN^- + products; \; k_1$$
$$CN^- + O_3 \rightarrow$$
$$CNO^- + products; \; k_2$$

The species CNO^- is called cyanate. At pH 7, Tuan and Jensen (1991) found that $k_1 = 5\times10^4$ $L^2/mmol^2$-min and $k_2 = 2$ $L/mmol$-min. If the dissolved ozone concentration is fixed at 0.002 mM, how do the thiocyanate, cyanide, and cyanate concentrations change over time? Assume that the initial thiocyanate concentration is 8 mM and the initial cyanide and cyanate concentrations are zero.

Solution:
The equations in the text for sequential first-order reactions are valid if pseudo-first-order rate constants are calculated. Use the equations with A = thiocyanate, B = cyanide, C = cyanate, $k_1 = (5\times10^4$ $L^2/mmol^2$-min)$(0.002$ mM$)^2 = 0.2$ min^{-1}, and $k_2 = (2$ $L/mmol$-min$)(0.002$ mM$) = 4\times10^{-3}$ min^{-1}. The concentration profiles are shown in the plot below.

In transition-state theory, the reaction rate is assumed to be proportional to the concentration of the activated complex, $[AB^\ddagger]$. If the equilibrium constant for $A + B = AB^\ddagger$ is $K^\ddagger$ and the rate is proportional to $[AB^\ddagger] = K^\ddagger[A][B]$, then the second-order rate constant should be proportional to $K^\ddagger$. The proportionality constant is equal to $\kappa R/Nh$, where $\kappa =$ transmission coefficient (usually taken as 1), $N =$ Avogadro's constant, and $h =$ Planck's constant. Expressing the equilibrium constant in terms of the corresponding standard Gibbs free energy, you can show that the values in eq. 22.11 are: $n = 1$, constant $= (\kappa R/Nh)\exp(-\Delta S^{o\ddagger})$, and the energy term $= \Delta H^{o\ddagger}$. Here: $\Delta S^{o\ddagger}$ is the standard entropy of reaction for $A + B = AB^\ddagger$, and $\Delta H^{o\ddagger}$ is the standard enthalpy of reaction for $A + B = AB^\ddagger$.

In practice, the Arrhenius equation and transition-state theory give similar fits to plots of $\ln k$ versus $1/T$. As an example of the use of the Arrhenius equation, we can explore the science behind the observation that many biochemical reactions double in rate for a 10°C increase in temperature.

Thoughtful Pause

Using the Arrhenius equation, what can you say about the activation energy if a rate constant doubles with a 10°C increase in temperature?

Evaluating the Arrhenius equation at two temperatures, T_1 and T_2, yields the following:

$$k_{T_1}/k_{T_2} = \exp[(-E_a/R)(1/T_1 - 1/T_2)]$$

Rearranging:

$$E_a = -R\ln(k_{T_1}/k_{T_2})/(1/T_1 - 1/T_2)$$

Over temperatures of interest with aquatic systems (0 to 30°C), the term $1/T_1 - 1/T_2$ varies only slightly for a constant value of $T_1 - T_2$. For $T_1 = 10°C$ and $T_2 = 20°C$, $1/T_1 - 1/T_2 = 1.20\times10^{-4}$ °K^{-1}. In this case $k_{T_1}/k_{T_2} = \frac{1}{2}$, so:

$$
\begin{aligned}
E_a &= -R\ln(k_{T_1}/k_{T_2})/(1/T_1 - 1/T_2) \\
&= (-8.314\times10^{-3} \; kJ/mol\text{-}°K)\ln(\tfrac{1}{2})/(1.20\times10^{-4} \; 1/°K) \\
&\approx 48 \; kJ/mol
\end{aligned}
$$

In other words, reactions with activation energies of about 48 kJ/mol will experience about a doubling in their rate constants with a 10°C increase in temperature. Another illustration showing the effect of temperature on rate constants is shown in Example 22.8.

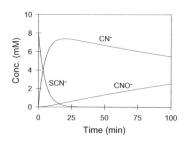

Note that the cyanide concentration remains high because the k_2 process is slow relative to the k_1 process.

***Example 22.8:* Effect of Temperature**

For the 1,2-dibromoethane case in Example 22.3, the value of the preexponential factor (A) is about $10^{10.5}$ s^{-1} and E_a is about 105 kJ/mol. If the hydrolysis rate constant is 6×10^{-9} s^{-1} at 25°C (and pH 7), what is the half-time at an average groundwater temperature of 10°C?

Solution:
From the text:
$$k_{T_1}/k_{T_2} = \exp[(-E_a/R)(1/T_1 - 1/T_2)]$$

If $T_1 = 10°C$ ($= 283.16°K$) and $T_2 = 25°C$ ($= 298.16°K$), then:

$$k_{10}/k_{25} = \exp[(-105$$
$$\text{kJ/mol/8.314}\times10^{-3}\text{ kJ/mol-}$$
$$°K)\ (1/283.16°K -$$
$$1/298.16°K)] = 0.106$$

Thus: $k_{10}/k_{25} = 0.106$ and $k_{10} = (0.106)(6\times10^{-9}\text{ s}^{-1}) = 6.4\times10^{-10}$ s^{-1}. The halftime at 10°C is $\ln(2)/k = \ln(2)/(6.4\times10^{-10}\text{ s}^{-1}) =$ **34.6 years**.

Another common way of correcting k for temperature in environmental engineering is given by:

$$k_T = k_{20}\theta^{T-20} \qquad \text{eq. 22.13}$$

where k_T = rate constant value at temperature T (in °C), k_{20} = rate constant value at 20°C, and θ = constant near 1.0 that differs in value for each type of reaction. From the Arrhenius equation:

$$k_{T_1}/k_{T_2} = \exp[(-E_a/R)(1/T_1 - 1/T_2)] = \exp[(-E_a/R)(T_2 - T_1)/T_1T_2]$$

Thus:

$$\frac{k_{T_1}}{k_{T_2}} = \left(e^{-\frac{E_a}{RT_1T_2}}\right)^{T_2-T_1} \qquad \text{eq. 22.14}$$

If $T_2 = 20°C$ and T_1 is expressed in °C, then, comparing eqs. 22.13 and 22.14: $\theta = \exp[-E_a/(RT_1T_2)]$. (Note: $T_2 - T_1$ is the same, whether temperatures are expressed in °K or °C.) Now for temperatures between 0 and 30°C, T_1T_2 does not change much: $T_1T_2 = 8.59\times10^4$ °K^2 at $T_1 = T_2 = 20°C$. Reported values of θ for biological treatment processes range from 1.00 to 1.10 (Metcalf and Eddy, 2003). Values of θ between 1.00 and 1.10 correspond to E_a between about 0 and 69 kJ/mol. If $\theta = 1.07$, then the rate constant would double when the temperature is increased from 20°C to 30°C.

22.6.2 Effects of ionic strength

The effects of ionic strength can best be seen by revisiting the transition-state theory introduced in Section 22.6.1. Recall that in transition-state theory, the reaction rate is assumed to be proportional to the *concentration* of the activated complex. Thus, the second-order rate constant is proportional to $K^{\ddagger}$ written as *concentrations*, not activities. As a result, you can use the ionic strength corrections for equilibrium constants (developed in Section 21.2.6) to describe the effect of ionic strength on rate constants. The result is that the rate constant for the reaction A + B → products depends on ionic strength in the following fashion:

$$k_I = k_{I-0}(\gamma_A\gamma_B/\gamma_{AB\ddagger})$$

where k_I is the rate constant at ionic strength I, k_{I-0} is the rate constant as ionic strength approaches zero, and γ_i is the activity coefficient of species i. The activity coefficients can be calculated from one of the single-ion activity coefficient approximations (see Table 21.3).

As an example, consider the work by Weil and Morris (1949) on the reaction between hypochlorous acid and ammonia to form monochlor-

Note that the relatively large E_a value means that the rate constant (and thus the half-time) is sensitive to temperature.

amine, NH_2Cl (see also Example 22.6). At the time of their work, it was not known if the chlorine-containing reactant was $HOCl$ or OCl^- or if the nitrogen-containing reactant was NH_3 or NH_4^+. Four reactions are possible:

$$HOCl + NH_3 \rightarrow NH_2Cl + H_2O \qquad \text{eq. 22.15a}$$
$$OCl^- + NH_3 \rightarrow NH_2Cl + H_2O \qquad \text{eq. 22.15b}$$
$$HOCl + NH_4^+ \rightarrow NH_2Cl + H_2O, \text{ or} \qquad \text{eq. 22.15c}$$
$$OCl^- + NH_4^+ \rightarrow NH_2Cl + H_2O \qquad \text{eq. 22.15d}$$

Weil and Morris attacked the problem in two ways. First, they varied the pH and observed the reaction rate. The four mechanisms in eq. 22.15a-d have somewhat different pH dependencies. If, for example, the mechanism in eq. 22.13a is correct, then the reaction rate should be proportional to $[HOCl][NH_3] = \alpha_{0,Cl}\alpha_{1,N}Cl_T N_T$, where the alpha values are for the acid-dissociation reactions of $HOCl$ and NH_4^+ ($pK_a = 7.5$ and 9.3, respectively), $Cl_T = [HOCl] + [OCl^-]$, and $N_T = [NH_3] + [NH_4^+]$. The reaction rate showed a maximum at a pH equal to about the average of the pK_a values. This dependency of the rate on pH allowed for the elimination of the mechanism in eq. 22.15b (which would show an increased rate with increasing pH) and the mechanism in eq. 22.15c (which would show an increased rate with decreasing pH). However, the effects of pH on the rate could not differentiate between the mechanisms in eqs. 22.15a and 22.15d.

Thoughtful Pause

How could you differentiate between the mechanisms in eqs. 22.15a and 22.15d?

The rate constants from the reactions in eqs. 22.15a and 22.15d should show very different dependencies on ionic strength. If the mechanism is as shown in eq. 22.15a, then the rate constant should be relatively independent of I (assuming that the activated complex is uncharged and recalling that the activity coefficients for uncharged species are near 1.0). However, if the mechanism is as shown in eq. 22.15d, then the rate constant should *decrease* significantly with increasing I. The data showed that the measured rate constant was nearly independent of ionic strength between $I = 0.01$ and 0.21 M. Thus, Weil and Morris concluded that the formation of monochloramine occurs between uncharged molecules ($HOCl$ and NH_3), as in eq. 22.13a.

22.7 SUMMARY

In this chapter, you added several kinetic modeling tools to your repertoire to allow you to calculate species concentrations when the system is controlled by kinetics rather than equilibrium. Nonequilibrium conditions may exist for species that react slowly. The rate of a reaction is related to the energy "hill" that must be overcome before reactants can be converted into products.

The number of species participating in the reaction, called the *molecularity*, is given by the reaction stoichiometry. Another useful characteristic of a chemical reaction is its reaction rate. The reaction rate is calculated by the rate of change of any species with respect to time divided by the species stoichiometry in the reaction. Thus, the rates of reactions of all species in a reaction are related through their stoichiometry. The expression describing the rate of change of a species is called the rate expression or rate law. The molecularity determined through experiments is called the *reaction order*. For certain reactions, called *elementary reactions*, the reaction order and molecularity are identical, and thus the reaction order (and the rate expression) can be determined from the reaction stoichiometry. The rate expressions for elementary reactions are a function of only the reactant concentrations, not the product concentrations.

These observations about individual reactions can be applied to a system of elementary reactions. The overall rate of change of a species is the sum of the rate expressions from each reaction in the system in which the species participates. As in all systems, rates of change of conservative substances are additive.

Simple rate expressions, such as zero-, first-, and second-order kinetics, can be integrated easily to give equations showing species concentrations over time. Integration also is possible for simple systems of reactions, such as reversible and sequential reaction schemes. Reversible reactions are controlled by the slowest of the forward and backward reactions. In sequential reaction systems, products may be degraded to form new products.

The common models describing the effects of temperature and ionic strength on *rate constants* were inspired by models describing the effects of temperature and ionic strength on *equilibrium constants*. Temperature effects often are described by the Arrhenius equation and transition-state theory. In the Arrhenius equation, the log of the rate constant is linear with the reciprocal of the temperature (in °K). Ionic strength effects can be handled by single-ion activity coefficient approximations.

SUMMARY OF KEY IDEAS

- Thermodynamics tells you what reactions are possible but not how fast the reactions occur

- Some reactions require energy input to overcome an energy barrier and therefore are under kinetic control and not at equilibrium

- In a reaction, the rate of change of each species concentration divided by the species stoichiometry is the same for each species

- All the reaction rates are written as a function of reactant concentrations only

- To write rate expressions wi th a system of reactions, remember that (1) the reaction rates for individual reactions are determined from reactant concentrations only, (2) the reaction rates for individual reactions can be determined directly from the reaction stoichiometry for elementary reactions, and (3) rates are additive

- In the zero-order rate law, the exponent for each species is zero and the species will show a linear change with time

- No elementary reaction can exhibit zero-order kinetics

- In the first-order rate law, the exponent for the species is one and the species will show an exponential change with time

- Second-order kinetics typically come from reactions that are second order in one reactant or from reactions that are first order in two reactants

- The kinetic equations for a reversible system must give the equilibrium concentrations as $t \rightarrow \infty$

- An equilibrium constant can be expressed as the ratio of the forward and reverse rate constants

- The slower of the forward and reverse reactions determines the apparent reaction rate constant in the approach to equilibrium

- One model for the effect of temperature on rate constants is the Arrhenius equation: $k = A\exp(-E_a/RT)$

PROBLEMS

22.1 A catalyst is a substance that increases the rate of a chemical reaction without being consumed. Catalysts do not change the equilibrium position of a reaction if the reaction is allowed to proceed to equilibrium. What characteristic energy of the system is affected by a catalyst? (Hint: refer to Figure 22.1)

22.2 For the reactions shown in Section 22.5.3, show that the maximum value of [B] ($= [B]_{max}$) and the time at which the maximum value of [B] occurs ($= t_{max}$) are given by:

$$[B]_{max} = [A]_0 \left(\frac{k_{B,1}}{k_{A,1}} \right)^{\frac{k_{B,1}}{k_{A,1}-k_{B,1}}} \text{ and } t_{max} = [\ln(k_{B,1}/k_{A,1})]/(k_{B,1} - k_{A,1})$$

22.3 It is useful to determine rate constants from experimental data (i.e., from values of C versus t).

A. How would you find a zero-order rate constant from experimental data?

B. You suspect a reaction is first order because the species concentration looks vaguely like an exponential decrease over time. How can you find the best-fit first-order rate constant? Can you linearize the data [that is, find a function of C so that $f(C) = mt + b$]?

C. How can you find the best-fit second-order rate constant? Can you linearize the data [that is, find a new function of C so that $f(C) = mt + b$]?

D. It is a truism in chemical kinetics that it is difficult to determine the kinetic order of a reaction by using the methods you developed in parts A, B, and C if the kinetic data collected are only for the first one-half of the reaction (i.e., only for $C > \frac{1}{2}C_0$ if C disappears). Create some "perfect" data for zero-, first-, and second-order kinetics for $C > \frac{1}{2}C_0$. Fit the data to zero-, first-, and second-order models. Is it difficult to determine the kinetic order under these conditions? (In other words, is it difficult to decide which model fits the data best when the data set is limited?)

22.4 Several systems of reactions yield an exponential decrease in the concentrations of certain species over time, as with a simple first-order reaction.

A. For a reversible system, the reactant concentration decreases exponentially over time initially. Is its initial rate of decrease identical to that which would be observed if the reverse reaction did not occur? Why? (The initial rate is $d[A]/dt$ at $t = 0$.)

B. For a reversible reaction approaching equilibrium, show that the time required to reach a concentration halfway between the initial concentration and the equilibrium concentration is $\ln(2)/(k_f + k_r)$.

C. For a sequential reaction system, the first reactant concentration decreases exponentially over time. Is its rate of decrease at all times identical to that which would be observed if the second reaction did not occur? Why?

22.5 Derive the equations for the concentrations of A, B, and C in the sequential reactions discussed in Section 22.5.

22.6 For very reactive intermediate species in sequential reactions, we sometimes assume that the concentration of the intermediate species is small and relatively constant over time. This is called the *steady-*

state assumption. A conservative approach to the steady-state assumption is to assume that $d[B]/dt = 0$ (in the sequence: $A \rightarrow B \rightarrow C$).

A. How does your analysis of the sequential system change if you apply the steady-state assumption to species B? (Hint: The mass balance equation is $[A] + [B] + [C] = [A]_0$. Differentiate the mass balance equation with respect to time.)

B. In the ozone example (Example 22.2), write the rate expressions for all the species in the system (except H^+, OH^-, H_2O, and O_2: assume their concentrations to be fixed). Apply the steady-state assumption to HO_2^-, HO_2, O_2^-, O_3^-, and OH. Show that with the steady-state assumptions: $d[O_3]/dt = -3k_1[OH^-][O_3]$.

22.7 For the monochloramine example in Example 22.6, the Arrhenius parameters for the forward reaction are $A = 6.6 \times 10^8$ L/mol-s and $E_a/R = 1510°K$. The Arrhenius parameters for the reverse reaction are $A = 1.38 \times 10^8$ s^{-1} and $E_a/R = 8800°K$ (values from Morris and Isaac, 1983). Plot the forward and reverse rate constants over temperature from 0 to 30°C.

22.8 In the example in Section 22.6.2, the lack of dependency of the rate constant on ionic strength was used to infer a reaction mechanism. What would the expected change in the rate constant have been from $I \rightarrow 0$ to $I = 0.01$ if the reaction mechanism in eq. 22.15d was the true reaction mechanism?

CHAPTER 23

Putting It All Together: Integrated Case Studies in Aquatic Chemistry

23.1 INTRODUCTION

In this capstone chapter of the text, you will have a chance to apply your knowledge of aquatic chemistry to several challenging problems. The case studies in this chapter were selected to integrate the major tools from the text for the solution of environmentally interesting problems.

The first case study involves complexation and precipitation chemistry in an industrial setting. Equilibrium concentrations of chemical species in a metal plating bath will be calculated. The reaction conditions used in practice necessitate temperature and ionic strength corrections.

The second case study will allow you to integrate equilibrium and kinetics to understand the rate of oxidation of ferrous iron by oxygen in natural waters. Fast reactions will be considered to be at equilibrium. Thus, rate expressions will be developed that contain both rate constants and equilibrium constants.

The third case study is a rather involved analysis of mercury chemistry in natural waters. In addition to dissolved phase and precipitation chemistry, the complexation of mercury by surfaces will be included.

23.2 INTEGRATED CASE STUDY 1: METAL FINISHING

23.2.1 Background

In metal plating, thin coats of metals are applied to metal or plastic parts to enhance the appearance and physical characteristics of the parts. In the typical plating process, dissolved metal is reduced and plated out as zero-valent metal. It is desirable to maintain high concentrations of dissolved metal in solution to increase the efficiency of the process. High dissolved metal concentrations usually are maintained in the plating bath by using acidic conditions or adding a strongly binding ligand such as cyanide.

Thoughtful Pause
Why does maintaining low pH or high ligand concentrations help to keep the metals in solution?

Recall that many metals precipitate as hydroxide solids. Thus, low pH conditions or the presence of other ligands generally increase metal solubility. Copper is one of the most common metals to be plated. In a copper strike bath, very high cyanide concentrations are used to plate out a thin copper coating. Usually, copper or other metals are used to form the final finish on top of the copper strike. Typical chemical additions to the copper strike bath are (from Lowenheim, 1978): $CuCN(s) = 15$ g/L, $NaCN(s) = 28$ g/L, and $Na_2CO_3(s) = 15$ g/L. The strike bath commonly is maintained at 50°C-63°C. The typical pH is 10-10.5.

The sodium carbonate is added for pH control. As carbonates (carbonate ion and carbonate-containing species) build up in the bath, it is necessary to remove them. One common process is to refrigerate the entire bath and precipitate carbonates as sodium carbonate decahydrate ($Na_2CO_3 \cdot 10H_2O$).

23.2.2 Analysis goals

The typical conditions of the plating bath are designed to maintain nearly all the copper as soluble cyanide complexes. Copper is added in the form of $CuCN(s)$. Although $CuCN(s)$ is not very soluble, it is supposed to be more soluble in the presence of excess cyanide.

Thoughtful Pause

Why should excess cyanide aid in the solubilization of $CuCN(s)$?

Excess cyanide should help form higher concentrations of copper-cyanide complexes and thus increase the solubility of $CuCN(s)$. In addition to enhancing the solubility of $CuCN(s)$, the $NaCN(s)$ dose is supposed to be sufficient to form higher copper-cyanide complexes, $Cu(CN)_3^{2-}$ and $Cu(CN)_4^{3-}$. Finally, during carbonate freeze-out, the temperature of the refrigeration unit is supposed to be sufficiently low to precipitate carbonates but not copper. The specific goals of the analysis are to answer the following questions:

- Is $CuCN(s)$ soluble in the strike bath?
- Is nearly all the copper soluble under standard operating conditions?
- Is most of the soluble copper present as $Cu(CN)_3^{2-}$ and $Cu(CN)_4^{3-}$?
- Will carbonate precipitate at lower temperatures (say, 0°C) without precipitating copper?

23.2.3 Analysis

For this analysis, assume the system is at equilibrium and open to the atmosphere. Although the system is complicated, use the same steps you have followed throughout this text. Start with a species list. A potential list of species is given in Table 23.1. You know that carbon dioxide will be present, but you do not know if the solid phases exist at equilibrium. Note that all the copper-containing species in the species list are in the +I oxidation state. This is the usual state of affairs in a copper-cyanide bath. Table 23.1 contains some unusual species. We normally do not include sodium-bicarbonate or sodium-carbonate complexes such as $NaHCO_3^0$ and $NaCO_3^-$. These species are included here because of the large added sodium concentration.

The second step in the analysis process is to write the pertinent equilibria. A list of equilibria may be found in Table 23.2. Most of the thermodynamic data (ΔH^o_{rxn} and log K values at 25°C and $I \to 0$) were obtained from Martell and Smith (1974). (As usual, the equilibrium list was used to generate the species list in Table 23.1.) Other ΔH^o_{rxn} values were calculated from the ΔH^o_f data in Stumm and Morgan (1996).

The log K values at 25°C and $I \to 0$ were manipulated to allow you to draw conclusions at strike bath conditions (about 57°C and $I = 2$ M) and refrigerated bath conditions (about 0°C and $I = 2$ M). Ionic strength corrections were made by using the Davies approximation, with parameters adjusted for temperature (with the caveat that $I = 2$ M is outside of the range of the Davies approximation). The van't Hoff equation was used to correct equilibrium constants for temperature. Because of the missing standard enthalpy of formation data, equilibria 6, 11, 12, and 13 were not adjusted for temperature with the van't Hoff equation.

Table 23.1: Species List for the Metal Plating Case Study

Total Mass	Species	Amount Added (M)
Cu	soluble: Cu^+, $Cu(CN)_2^-$, $Cu(CN)_3^{2-}$, $Cu(CN)_4^{3-}$ insoluble: $CuCN(s)$, $Cu_2O(s)$	0.17
CN	soluble: HCN, CN^-, $Cu(CN)_2^-$, $Cu(CN)_3^{2-}$, $Cu(CN)_4^{3-}$ insoluble: $CuCN(s)$	0.74
C_T	soluble: $H_2CO_3^*$, HCO_3^-, CO_3^{2-}, $NaHCO_3^0$, $NaCO_3^-$ insoluble: $CO_2(g)$, $Na_2CO_3(s)$	see note
Na	soluble: Na^+, $NaHCO_3^0$, $NaCO_3^-$ insoluble: $Na_2CO_3(s)$	0.85

Note: Although 0.14 M of total carbonate was added, the system is open to the atmosphere.

The temperature dependence of the K_{s0} value for $Na_2CO_3(s)$ is a little more complicated. Above about 34°C, sodium carbonate is found mainly as sodium carbonate monohydrate ($Na_2CO_3 \cdot H_2O$) (see FMC, 2001). The monohydrate is fairly soluble in water. From about 0°C to about 32°C, sodium carbonate is present mainly as sodium carbonate decahydrate ($Na_2CO_3 \cdot 10H_2O$). The decahydrate is fairly *insoluble* in water. Therefore, the K_{s0} values for $Na_2CO_3(s)$ in Table 23.2 are for different species, as indicated in the table.

The third step in the analysis process is writing the mass balances. The appropriate mass balances are:

$$Cu(+I)_T = [Cu^+] + [Cu(CN)_2^-] + [Cu(CN)_3^{2-}] + \\ [Cu(CN)_4^{3-}] \quad \text{eq. 23.1}$$

$$CN_T = [HCN] + [CN^-] + 2[Cu(CN)_2^-] + 3[Cu(CN)_3^{2-}] + \\ 4[Cu(CN)_4^{3-}] \quad \text{eq. 23.2}$$

$$C_T = [H_2CO_3^*] + [HCO_3^-] + [CO_3^{2-}] + [NaHCO_3^0] + \\ [NaCO_3^-]$$

$$Na_T = [Na^+] + [NaHCO_3^0] + [NaCO_3^-] \quad \text{eq. 23.3}$$

At this point, the values of $Cu(+I)_T$, CN_T, C_T, and Na_T are unknown.

Thoughtful Pause

Why are the values of $Cu(+I)_T$, CN_T, C_T, and Na_T unknown?

The values of the total masses are unknown because some portion of each may precipitate (and exchange with the atmosphere, in the case of C_T).

The system was solved first at 57°C and $I = 2$ M. A pH of 10.25 was assumed. To solve the system, a spreadsheet was set up with a cell for each species containing the equation for its concentration as a function of the concentrations of Cu^{2+}, CN^-, and Na^+; the fixed species concentrations [H^+, OH^-, and $CO_2(g)$]; and equilibrium constants. As a starting point, it was assumed that no solids precipitated at 57°C (the expected situation in a plating bath). With the use of Excel's *Solver*, the concentrations (actually, pC) of Cu^{2+}, CN^-, and Na^+ were fitted to satisfy the objective function with constraints. The objective function was to set the copper total mass to 0.17 M (eq. 23.1), with the mass balances of cyanide and sodium as constraints (eqs. 23.2 and 23.3). The results indicated that CuCN(s) would precipitate under these conditions ([Cu^+][CN^-]/$K_6 > 1$). If CuCN(s) precipitates, then the same amount of copper and cyanide should be lost to solids formation [since the stoichiometry of copper and cyanide in CuCN(s) is 1:1]. Thus, the calculations were repeated by assuming that CuCN(s) precipitates. The objective function was changed to set the difference between the calculated loss of copper [$Cu(+I)_T$ - sum of copper-containing species] and the calculated loss of cyanide [CN_T - sum of cyanide-containing species] equal to zero. The only constraint was on the sodium mass balance.

Table 23.2: Equilibria for the Metal Plating Case Study
[see text for explanation of the temperature effects on the
K_{s0} value for $Na_2CO_3(s)$]

No.	Equilibrium	ΔH^o_{rxn} (kJ/mol)	log K (temp. in °C, I in M)			
			25, 0	25, 2	57, 2	0, 2
1	$Cu^+ + 2CN^- = Cu(CN)_2^-$	−121.6	16.3	16.1	14.1	17.6
2	$Cu^+ + 3CN^- = Cu(CN)_3^{2-}$	−168.0	21.6	21.6	18.7	23.7
3	$Cu^+ + 4CN^- = Cu(CN)_4^{3-}$	−214.9	23.1	23.5	19.8	26.2
4	$\frac{1}{2}Cu_2O(s) + \frac{1}{2}H_2O = Cu^+ + OH^-$	−6.3	−14.7	−14.5	−14.6	−14.4
5	$HCN = CN^- + H^+$	+54	−9.0	−8.8	−7.9	−9.5
6	$CuCN(s) = Cu^+ + CN^-$ −19.3[1]	N/A	−19.5	−19.3	−19.1[1]	
7	$H_2O = H^+ + OH^-$	+79.9	−14.0	−13.8	−12.5	−14.8
8	$CO_2(g) + H_2O = H_2CO_3*$	+8.4	−1.5	−1.5	−1.5	−1.5
9	$H_2CO_3* = HCO_3^- + H^+$	+36.4	−6.3	−6.1	−5.5	−6.6
10	$HCO_3^- = CO_3^{2-} + H^+$	+58.9	−10.3	−9.9	−9.8	−10.0
11	$Na^+ + HCO_3^- = NaHCO_3^0$	N/A	−0.25	−0.44	−0.42[1]	−0.45[1]
12	$Na^+ + CO_3^{2-} = NaCO_3^-$	N/A	1.27	0.90	0.93[1]	0.86[1]
13	$Na_2CO_3(s) = 2Na^+ + CO_3^{2-}$	N/A	40.9[2]	41.6[2]	101.[2]	1.3[3]

Notes: 1. Temperature correction in activity coefficient only
2. For monohydrate salt - see text
3. For decahydrate salt - see text
N/A = not available

Upon recalculation, it appeared that the plating conditions were reasonable. Most of the added CuCN(s) dissolved: the total soluble copper was 62% of the added copper. As required, the soluble copper was found to be mainly $Cu(CN)_4^{3-}$ (78%). The addition of excess cyanide resulted in CN^- being the cyanide-containing species of highest concentration (41% of the total soluble cyanide). It appears, as expected, that $Na_2CO_3(s)$ would not precipitate at 57°C and $I = 2$ M. The primary form of sodium was $NaCO_3^-$ (92%), and carbonate was found mainly as carbonate ion. The total soluble carbonate concentration was very large (3.2 M). Thus, carbonate removal may be necessary.

The calculations were repeated with the equilibrium constants adjusted to 0°C and $I = 2$ M. Again, as a starting point, it was assumed that no solids would precipitate. The calculated concentrations revealed that $Cu_2O(s)$ and $CuCN(s)$ would not precipitate, but $Na_2CO_3(s)$ would precipitate (calculated value of $K_{13}/[Na^+]^2[CO_3^{2-}] = 1.01$). Thus, lowering the temperature significantly could indeed be used to freeze out carbonates. The species $Cu(CN)_4^{3-}$ accounted for 95% of the total copper and 87% of the total cyanide at 0°C. Soluble sodium was primarily found as Na^+ (60%) and $NaCO_3^-$ (39%). Most of the soluble carbonate was found as carbonate and bicarbonate ion.

The equilibrium calculations summarized in this case study shed light on the speciation of copper and cyanide at equilibrium in the plating bath. The speciation information was used to answer the questions posed in Section 23.2.1 and could be used to optimize both the plating process further. Knowledge of the species concentrations also could be used to develop and optimize treatment processes for removing cyanide and residual copper from plating waste streams. The calculations indicate that lowering the bath temperature should result in the freezing out of carbonates under equilibrium conditions.

23.3 CASE STUDY 2: OXIDATION OF FERROUS IRON BY OXYGEN

23.3.1 Background

The oxidation of ferrous iron, Fe(+II), by oxygen to form ferric iron, Fe(+III), is an important process in the environment. In natural systems, the oxidation of ferrous iron is a critical step in the production of very acidic waters from abandoned mines. These streams, called *acid mine drainage*, can be extremely acidic, with pH values less than 1. The acidic conditions are created when the sulfur in pyrite, FeS(s), is oxidized by atmospheric oxygen, with the subsequent release of ferrous iron. Ferrous iron is oxidized by oxygen (often catalyzed by bacteria) to ferric iron. The oxidation of Fe(+II) to Fe(+III) is the rate-determining step in the dissolution of pyrite and thus in the generation of acid mine drainage (Singer and Stumm, 1970).

The oxidation of ferrous iron by oxygen also is important in engineered systems. Ferrous iron is fairly soluble in water, but ferric iron is much less soluble (see Figure 23.1). If ferric iron precipitates in the home, then reddish-brown staining of laundry and cooking utensils may occur. Iron precipitates can clog pipes and water heaters and encourage the growth of iron-reducing bacteria. One solution to iron precipitation in potable water is to oxidize ferrous iron in the drinking water treatment plant and remove Fe(+III) precipitates by filtration. The least expensive oxidant is atmospheric oxygen. Thus, the rate of ferrous iron oxidation is important because it will drive the sizing of basins in drinking water treatment facilities where ferrous iron is to be oxidized.

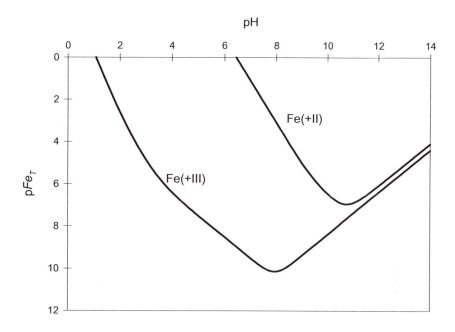

Figure 23.1: Comparison of the Solubility of Fe(+II) and Fe(+III)
[in equilibrium with $Fe(OH)_2(s)$, $\log K_{s0} = 15.1$, and FeOOH, $\log K_{s0} = 38.8$, respectively; assumes no dissolved carbonate species and the formation of $Fe_2OH_2^{4+}$ is ignored]

23.3.2 Analysis goals

Oxidation experiments have revealed the following information about the rate of oxidation of ferrous iron by oxygen:

- The rate is slow at pH values below 8
- The rate increases by a factor of 100 for a one-unit increase in pH
- At fixed oxygen partial pressure and pH, the initial disappearance of Fe(+II) is first order in Fe(+II)

It is desirable to develop a kinetic scheme consistent with the observations listed above. For applications to natural waters, it is of interest to explore the effects of ligands on the oxidation of ferrous iron by oxygen.

23.3.3 Analysis

The second and third observations suggest second-order dependence on $[OH^-]$ and first-order dependence on $Fe(+II)_T$. A possible rate law consistent with the kinetic data is:

$$dFe(+II)_T/dt = -k[Fe(+II)_T][OH^-]^2P_{O_2}$$

The rate constant k is known to be 8.0×10^{13} L^2/atm-mol^2-min. At $P_{O_2} = 0.2$ atm (atmospheric conditions), the pseudo-first-order rate constant for ferrous iron oxidation is: $k' = k[OH^-]^2P_{O_2} = 1.6 \times 10^{-15}/[H^+]^2$ min^{-1}. Thus, you would expect that a plot of $\log(k')$ (with k' in min^{-1}) against pH would have a slope of 2.0 and an intercept of $\log(1.6 \times 10^{-15}) = -14.8$. Literature data collected between pH 5 and 8 are consistent with this rate constant and rate law (Figure 23.2).

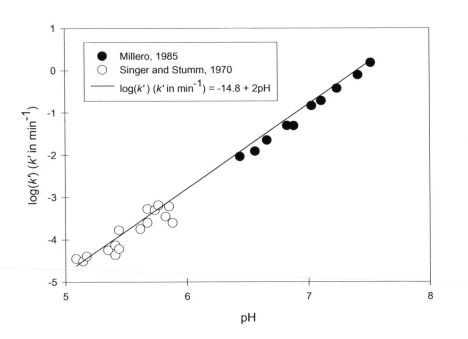

Figure 23.2: Dependency of the Pseudo-First-Order Rate Constant for Ferrous Iron Oxidation by Oxygen on pH (pH 5 to 8)

The inclusion of data collected at pH < 5 reveals a different story (see Figure 23.3). The plot of $\log(k')$ against pH appears to have a slope of about 2 between about pH 5 and 8, a slope of about 1 between about pH 3 and 5, and a slope of about zero below pH 3. In other words, the dependency of k' on $[OH^-]$ appears to change from $[OH^-]^2$ to $[OH^-]^1$ to $[OH^-]^0$ as the pH decreases.

Thoughtful Pause
Can you think of a reason for the change in the dependency of k' on [OH$^-$] as the pH decreases?

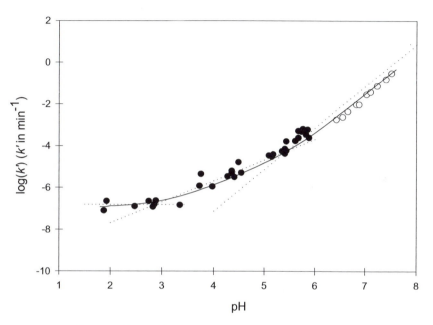

Figure 23.3: Model Prediction for Ferrous Iron Oxidation by Oxygen
[data symbols from Figure 23.2; solid line is eq. 23.5; dotted lines (left to right) represent slopes of 0, 1, and 2]

One explanation is that dissolved oxygen reacts at different rates with different ferrous iron species. The overall reaction [Fe(+II) + O$_2$(aq) → products] should be separated into three reactions:

$$Fe^{2+} + O_2(aq) \rightarrow \text{products} \qquad k_1$$
$$FeOH^+ + O_2(aq) \rightarrow \text{products} \qquad k_2$$
$$Fe(OH)_2^0 + O_2(aq) \rightarrow \text{products} \quad k_3$$

The rate law becomes:

$$dFe(+II)_T/dt = -\{k_1[Fe^{2+}] + k_2[FeOH^+] + k_3[Fe(OH)_2^0]\}P_{O_2} \qquad \text{eq. 23.4}$$

Now, recall from Section 22.5.2 that acid-base reactions are very fast. Thus, you might expect that the reactions between Fe^{2+} and OH$^-$ to form

$FeOH^+$ and $Fe(OH)_2^0$ are very fast and the hydroxy ferrous iron complexes may be in equilibrium with OH^-. The appropriate mass balance and equilibrium expressions are:

$$Fe(+II)_T = [Fe^{2+}] + [FeOH^+] + [Fe(OH)_2^0]$$
$$Fe^{2+} + OH^- = FeOH^+ \qquad K_1 = 10^{4.5}$$
$$Fe^{2+} + 2OH^- = Fe(OH)_2^0 \qquad K_2 = 10^{7.4}$$

Solving:

$$[Fe^{2+}] = Fe(+II)_T/(1 + K_1[OH^-] + K_2[OH^-]^2)$$
$$[FeOH^+] = K_1 Fe(+II)_T[OH^-]/(1 + K_1[OH^-] + K_2[OH^-]^2), \text{ and:}$$
$$[Fe(OH)_2^0] = K_2 Fe(+II)_T[OH^-]^2/(1 + K_1[OH^-] + K_2[OH^-]^2)$$

Substituting into the rate expression (eq. 23.4), you can show that:

$$k' = (k_1 + k_2 K_1[OH^-] + k_3 K_2[OH^-]^2)P_{O_2}/(1 + K_1[OH^-] + K_2[OH^-]^2) \qquad \text{eq. 23.5}$$

The values of the rate constants are (Wehrli, 1990): $k_1 = 1.0 \times 10^{-8}\ atm^{-1}s^{-1}$, $k_2 = 3.2 \times 10^{-2}\ atm^{-1}s^{-1}$, and $k_3 = 1.0 \times 10^4\ atm^{-1}s^{-1}$. The calculated values of $\log(k')$ are shown by the solid line in Figure 23.3. The model appears to describe the data very well.

The revised rate law suggests that complexation of Fe(+II) with ligands other than hydroxide ion may slow down the rate of Fe(+II) oxidation. The effects of a ligand (e.g., chloride) can be quantified with the model. In the presence of chloride, the mass balance becomes:

$$Fe(+II)_T = [Fe^{2+}] + [FeOH^+] + [Fe(OH)_2^0] + [FeCl^+]$$

An additional equilibrium is required: $Fe^{2+} + Cl^- = FeCl^+$; K_3. Thus:

$$k_{Cl}' = (k_1 + k_2 K_1[OH^-] + k_3 K_2[OH^-]^2)P_{O_2}/(1 + K_1[OH^-] + K_2[OH^-]^2 + K_3[Cl^-])$$

Here, k_{Cl}' is the pseudo-first-order rate constant for ferrous iron oxidation by oxygen in the presence of chloride. If $FeCl^+$ reacts very slowly with dissolved oxygen compared to the hydroxy complexes, then the ratio of the rate in the presence of chloride to the rate in the absence of chloride is:

$$k_{Cl}'/k' = (1 + K_1[OH^-] + K_2[OH^-]^2)/(1 + K_1[OH^-] + K_2[OH^-]^2 + K_3[Cl^-])$$

For pH values less than about 9, the term $1 + K_1[OH^-] + K_2[OH^-]^2$ is about equal to 1. Thus, for pH less than about 9:

$$k_{Cl}'/k' \approx 1/(1 + K_3[Cl^-])$$

If chloride is present in great excess over ferrous iron, then $[Cl^-] \approx Cl_T =$ constant and:

$$k_{Cl}'/k' \approx 1/(1 + K_3 Cl_T) \qquad\qquad \text{eq. 23.6}$$

To test this model for the effects of chloride on the rate of oxidation, use the data of Millero and Izaguirre (1989) shown in Figure 23.4. The value of K_3 was estimated by fitting eq. 23.6 to the data for $Cl_T = 4$ to 6 M. The fitted value of K_3 is 1.2, similar to literature values for K_3 in marine waters (see Millero and Izaguirre, 1989).

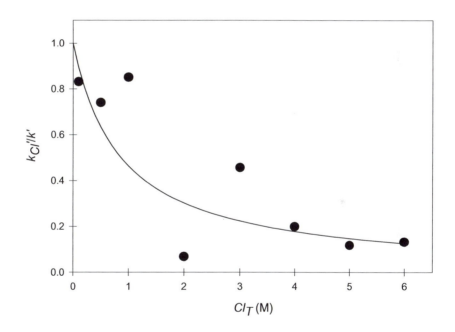

**Figure 23.4: Effect of Chloride on the Rate Constant
for the Oxidation of Fe(+II) by Oxygen**
**(symbols are data from Millero and Izaguirre, 1989; line was obtained by
best fit of eq. 23.6 to data for $Cl_T = 4$ to 6 M)**

23.4 INTEGRATED CASE STUDY 3: INORGANIC MERCURY CHEMISTRY IN NATURAL WATERS

22.4.1 Background

Mercury is an important pollutant because of its well-known adverse health effects. Dissolved mercury forms complexes with many common ligands in the environment, including hydroxide, chloride, bicarbonate, carbonate, and nitrate ions. Mercury can form environmentally important organic complexes as well, including methyl- and dimethylmercury (see the Part I case study). The focus in this case study is on the inorganic chemistry of mercury; soluble organic complexes will not be considered here. Mercury also can precipitate as hydroxide and carbonate solids.

Many cations, including Hg^{2+} and its cationic complexes, can adsorb onto negatively charged surfaces. A common surface in the environment is formed at the interface of water and *hydrous ferric oxides* (sometimes abbreviated *HFO*). As discussed in Section 19.6, charged surfaces often exhibit acid-base chemistry. The uncharged surface species on HFO (denoted by the symbol $\equiv$S-OH) is amphoteric: it can both accept a proton from water and donate a proton to water.

A detailed model of inorganic Hg(+II) chemistry must take into account complexation, precipitation, and adsorption. Such a model was developed by Tiffreau and colleagues (1995). The approach summarized in this case study is based on their work.

23.4.2 Analysis goals

The overall analysis goal is to determine the speciation of mercury-containing species as a function of pH in a typical natural water containing HFO. In particular, determination of the adsorption edge for mercury on HFO is desired.

23.4.3 Analysis

The analysis approach is similar to the metal plating case study in Section 23.2.3. A list of potential species is shown in Table 23.3. The system has 41 species (including H^+, OH^-, and H_2O). The analysis will be conducted for a water containing the total masses shown in Table 23.3. The amorphous HFO concentration used represents 50 mg/L of HFO containing 0.011 mol Fe/g HFO and 0.205 mol sites/mol Fe (Tiffreau et al., 1995). The low total mercury concentration was selected to minimize mercury precipitation. The pertinent equilibria are summarized in Table 23.4.

In the calculations, the system was assumed to be in equilibrium with the atmosphere and no corrections were made for the effects of surface charge. (See Tiffreau et al., 1995, for the influences of surface charge on adsorption in this system.) Similar to the first case study, a spreadsheet was set up with a cell for each species containing the equation for its concentration as a function of the concentrations of Hg^{2+}, Cl^-, $\equiv$S-OH, and NO_3^-; fixed species concentrations [H^+, OH^-, and $CO_2(g)$]; and equilibrium constants. As a starting point, it was assumed that no solids would precipitate. With Excel's *Solver*, the concentrations (or pC) of Hg^{2+}, Cl^-, $\equiv$S-OH, and NO_3^- were fitted to satisfy the objective function with constraints. The objective function was to minimize the square of the difference between the mercury total mass and 1.84×10^{-7} M, with the mass balances of chloride, total surface, and nitrate as constraints (see Table 23.3).

The results indicated that no solids would precipitate in the pH range of 4 to 12 under these conditions.

Table 23.3: Species List for the Mercury Case Study

Total Mass	Species	Amount Added (M)
Hg(+II)	soluble: Hg^{2+}, $HgOH^+$, $Hg(OH)_2^0$, $Hg(OH)_3^-$, $Hg(OH)_4^{2-}$, Hg_2OH^{3+}, $Hg_3(OH)_3^{3+}$, $HgCl^+$, $HgCl_2^0$, $HgCl_3^-$, $HgCl_4^{2-}$, $HgOHCl^0$, $HgOHCl_2^-$, $HgOHCl_3^{2-}$, $Hg(OH)_2Cl^-$, $Hg(OH)_3Cl^{2-}$, $Hg(OH)_2Cl_2^{2-}$, $HgNO_3^+$, $Hg(NO_3)_2^0$, $HgCO_3^0$, $Hg(CO_3)_2^{2-}$, $HgHCO_3^+$, $Hg(OH)CO_3^-$ insoluble: $HgCO_3(s)$, $Hg(OH)_2(s)$, $2HgO \cdot HgCO_3(s)$ surface: $\equiv S\text{-}OHg^+$, $\equiv S\text{-}OHgOH$, $\equiv S\text{-}OHgCl$	1.84×10^{-7}
Cl	soluble: Cl^-, $HgCl^+$, $HgCl_2^0$, $HgCl_3^-$, $HgCl_4^{2-}$, $HgOHCl^0$, $HgOHCl_3^{2-}$, $Hg(OH)_2Cl^-$, $Hg(OH)_3Cl^{2-}$, $Hg(OH)_2Cl_2^{2-}$ insoluble: None surface: $\equiv S\text{-}OHgCl$	5.64×10^{-4} (20 mg/L)
C_T	soluble: $H_2CO_3^*$, HCO_3^-, CO_3^{2-}, $HgCO_3^0$, $Hg(CO_3)_2^{2-}$, $HgHCO_3^+$, $Hg(OH)CO_3^-$ insoluble: $CO_2(g)$, $HgCO_3(s)$, $2HgO \cdot HgCO_3(s)$ surface: None	see note
NO_3	soluble: NO_3^-, $HgNO_3^+$, $Hg(NO_3)_2^0$ insoluble: None surface: None	7.86×10^{-5} (1.1 mg/L as N)
$\equiv S$	soluble: None insoluble: None surface: $\equiv S\text{-}OH_2^+$, $\equiv S\text{-}OH$, $\equiv S\text{-}O^-$, $\equiv S\text{-}OHg^+$, $\equiv S\text{-}OHgOH$, $\equiv S\text{-}OHgCl$	1.15×10^{-4} (see text)

Note: The system is open to the atmosphere.

The absorption edge for mercury on HFO is shown in Figure 23.5. Mercury apparently does not adsorb very much at low and high pH values. Between pH 6.5 and 9.0, over 90% of the mercury is adsorbed.

Table 23.4: Equilibria for the Mercury Chemistry Case Study

Equilibrium	K	Reference
Hg(+II) Hydroxo Complexes		
$Hg^{2+} + OH^- = HgOH^+$	$10^{10.6}$	S&M
$Hg^{2+} + 2OH^- = Hg(OH)_2^0$	$10^{21.8}$	S&M
$Hg^{2+} + 3OH^- = Hg(OH)_3^-$	$10^{20.9}$	S&M
$Hg^{2+} + 4OH^- = Hg(OH)_4^{2-}$	$10^{19.2}$	T
$2Hg^{2+} + OH^- = Hg_2OH^{3+}$	$10^{10.7}$	T
$3Hg^{2+} + 3OH^- = Hg_3(OH)_3^{3+}$	$10^{35.6}$	T
Hg(+II) Chloro Complexes		
$Hg^{2+} + Cl^- = HgCl^+$	$10^{7.2}$	S&M
$Hg^{2+} + 2Cl^- = HgCl_2^0$	$10^{14.0}$	S&M
$Hg^{2+} + 3Cl^- = HgCl_3^-$	$10^{15.1}$	S&M
$Hg^{2+} + 4Cl^- = HgCl_4^{2-}$	$10^{15.4}$	S&M
Hg(+II) Chloro-Hydroxo Mixed-Ligand Complexes		
$Hg^{2+} + OH^- + Cl^- = HgOHCl^0$	$10^{18.1}$	T
$Hg^{2+} + OH^- + 2Cl^- = HgOHCl_2^-$	$10^{17.5}$	T
$Hg^{2+} + OH^- + 3Cl^- = HgOHCl_3^{2-}$	$10^{17.1}$	T
$Hg^{2+} + 2OH^- + Cl^- = Hg(OH)_2Cl^-$	$10^{19.4}$	T
$Hg^{2+} + 3OH^- + Cl^- = Hg(OH)_3Cl^{2-}$	$10^{18.9}$	T
$Hg^{2+} + 2OH^- + 2Cl^- = Hg(OH)_2Cl_2^{2-}$	$10^{17.9}$	T
Other Hg(+II) Complexes		
$Hg^{2+} + NO_3^- = HgNO_3^+$	$10^{0.39}$	T
$Hg^{2+} + 2NO_3^- = Hg(NO_3)_2^0$	$10^{0.42}$	T
$Hg^{2+} + CO_3^{2-} = HgCO_3^0$	$10^{12.1}$	T
$Hg^{2+} + 2CO_3^{2-} = Hg(CO_3)_2^{2-}$	$10^{15.5}$	T
$Hg^{2+} + HCO_3^- = HgHCO_3^+$	$10^{6.0}$	T
$Hg^{2+} + OH^- + CO_3^{2-} = Hg(OH)CO_3^-$	$10^{19.2}$	T
Hg(+II) Solids		
$HgCO_3(s) = Hg^{2+} + CO_3^{2-}$	$10^{-16.1}$	S&M
$Hg(OH)_2(s) = Hg^{2+} + 2OH^-$	$10^{-25.4}$	S&M
$2HgO \cdot HgCO_3(s) + 6H^+ =$		
$\quad 3Hg^{2+} + CO_2(g) + 3H_2O$	$10^{7.0}$	T

Table 23.4 Continued

Equilibrium	K	Reference
Surface Equilibria		
$\equiv\!S\text{-}OH_2^+ = \equiv\!S\text{-}OH + H^+$	$10^{-7.29}$	T
$\equiv\!S\text{-}OH = \equiv\!S\text{-}O^- + H^+$	$10^{-8.93}$	T
$\equiv\!S\text{-}OH + Hg^{2+} = \equiv\!S\text{-}OHg^+ + H^+$	$10^{6.9}$	T
$\equiv\!S\text{-}OH + Hg^{2+} + H_2O = \equiv\!S\text{-}OHgOH + 2H^+$	$10^{-0.9}$	T
$\equiv\!S\text{-}OH + Hg^{2+} + Cl^- = \equiv\!S\text{-}OHgCl + H^+$	$10^{9.8}$	T
Other Equilibria		
$H_2O = H^+ + OH^-$	10^{-14}	S&M
$CO_2(g) + H_2O = H_2CO_3^*$	$10^{-1.5}$	S&M
$H_2CO_3^* = HCO_3^- + H^+$	$10^{-6.3}$	S&M
$HCO_3^- = CO_3^{2-} + H^+$	$10^{-10.3}$	S&M

Abbreviations: S&M = Stumm and Morgan, 1996
                          T = Tiffreau et al., 1995

Thoughtful Pause

Why does mercury only adsorb onto this HFO in the near neutral pH range?

To explain the effects of pH on adsorption, it is necessary to look at the speciation of mercury and the surface species. Figures 23.6 and 23.7 show the pC-pH diagrams for, respectively, the predominant mercury-containing and surface species. At low pH, most of the mercury is complexed with chloride (since the ligands OH^- and $\equiv\!S\text{-}O^-$ are in low concentration). Most of the surface is protonated since the metal H^+ is at much higher concentration than mercury species. Nearer to neutral pH, the concentration of the ligand $\equiv\!S\text{-}O^-$ increases and outcompetes chloride and hydroxide for mercury. Note that mercury-containing surface complexes account for a large amount of the mercury (Figures 23.5 and 23.6) but represent only a small fraction of the total surface concentration (Figure 23.7). At higher pH values, the carbonate concentration increases significantly, because of the high solubility of carbon dioxide at high pH. Most of the mercury is complexed with carbonate, whereas the surface is mainly uncomplexed. This case study demonstrates that knowledge of individual species concentrations greatly increases your understanding of the behavior of pollutants in the environment.

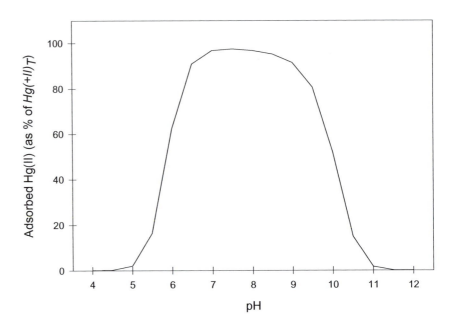

**Figure 23.5: Adsorption Edge for Mercury on a Hypothetical
Hydrous Ferric Oxide**

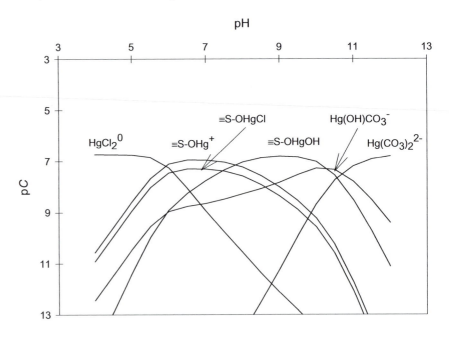

**Figure 23.6: pC-pH Diagram Showing the Major
Mercury-Containing Species**

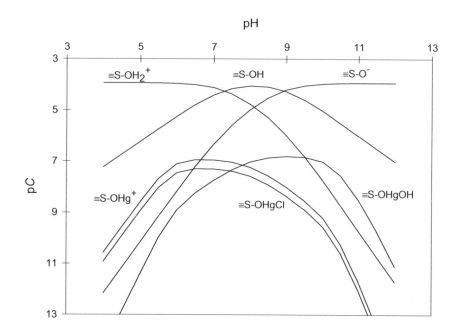

Figure 23.7: p*C*-pH Diagram Showing the Major Surface Species

23.5 SUMMARY

The case studies in this chapter show how your knowledge of aquatic chemistry can be applied to a wide variety of challenging problems. When your aquatic chemistry odyssey began, the analysis techniques in this chapter were probably far out of reach. Now, at the end of the text, you should feel comfortable with the conceptualization of complex problems and trust that you have the tools for their solution. It is hoped that your newly developed intuition and tools will allow you to face aquatic chemistry problems with confidence and some small element of pleasure.

APPENDIX A

Background Information

A.1 INTRODUCTION

It is assumed that the reader has some background in chemistry and mathematics prior to using this text. A chemistry background equivalent to freshman chemistry is required. Good skills in the manipulation of algebraic equations are highly desirable. A few basic concepts in chemistry, mathematics, and the use of spreadsheet programs are reviewed in this appendix.

A.2 CHEMICAL PRINCIPLES

A.2.1 Chemical species

ion: an atom or molecule carrying a net electric charge

cation: a positively-charged ion

anion: a negatively-charged ion

radicals: molecules containing atoms with unpaired electrons

chemical species: the collection of atoms, molecules, ions, and radicals in a chemical system

ionic equilibria: chemical equilibria involving molecules and/or ions

Atoms share electrons to form molecules. Molecules (from the diminutive of the Latin *moles* mass) are the smallest division of a substance that retains the properties of the substance. Charged molecules are called *ions* (from the Greek *ienai* to go, since ions migrate to electrodes). *Cations* are positively charged and *anions* are negatively charged (from, respectively, *katienai* to go down and *anienai* to go up: to differentiate the movement of ions in an electric field). Charges are indicated by a superscript at the end of the chemical formula of the ion, as is SO_4^{2-} for sulfate ion. Formally, the names of ions include the word "ion". For example, HCO_3^- is called "bicarbonate ion", not "bicarbonate". In this text, the word "ion" frequently will be omitted. Molecules containing atoms with unpaired electrons are called *radicals*. The atoms, molecules, ions, and radicals are referred to as *chemical species*.

This text is concerned with chemical equilibria involving molecules and/or ions, sometimes called *ionic equilibria*. Radical chemistry can be important in some environmental systems, but is outside the scope of this text. The concentration of chemical species will be indicated by square brackets: $[NO_3^-]$ = nitrate concentration.

A.2.2 Species types

You will work with four types of chemical species defined by their phase: dissolved species, pure liquids, pure solids, and gases. Each type will be indicated by a term following the species name in parentheses. The species type labels are summarized in Table A.1. The dissolved species type will

537

be indicated by the symbol (aq). For example, dissolved chloride ion will be written as: Cl^-(aq). Since most of the species you shall work with in this text are dissolved, the (aq) label frequently will be omitted. It will be used explicitly when the type of species is ambiguous.

Pure liquids will be indicated with the symbol (l). In most of the text, the only pure liquid you will encounter is water. Thus, the (l) symbol usually will be omitted and only used to emphasize water as a pure phase. The symbols H_2O(l) and H_2O will be used interchangeably to indicate pure water.

Key idea: Dissolved species, pure liquids, pure solids, and gases will be labeled (aq), (l), (s), and (g), respectively

Pure solids will be indicated by the symbol (s), as in $CaCO_3$(s). The (s) symbol always will be used with pure solids. A species name lacking the symbol (s) is a dissolved species. For example, $CaSO_4$(s) indicates the pure solid calcium sulfate, while $CaSO_4$ (or less ambiguously, $CaSO_4$(aq) or $CaSO_4^0$) represents the dissolved calcium sulfate complex. Solid phases with which are not pure solids will be indicated by the symbol (am) for amorphous.

Gases will be indicated by the symbol (g), as in CO_2(g). A species name lacking the symbol (g) is a dissolved species, not a gas. For example, O_2(g) indicates gaseous oxygen, while O_2 (or less ambiguously, O_2(aq)) represents dissolved oxygen.

Table A.1: Labels for Species Types

Species Type	Formal Label	Common Label
Dissolved species	(aq)	no label
Pure liquids	(l)	no label
Solids		
Pure solids	(s)	(s)
Other solids	(am)	(am)
Gases	(g)	(g)

A.2.3 Chemical reactions

reactants: reacting species

products: species formed through the reaction of reactants

Several common symbols and abbreviations will be used throughout this book. Reactions involve the interactions of some chemical species (called *reactants*) to form other chemical species (called *products*). Three symbols will be used to emphasize the direction of the reaction. The symbol → will be used to emphasize the progression of the reaction from left to right, as in: H_2O → H^+ + OH^-. In the reaction, the common approach of writing reactants on the left side of the reaction is observed.

Similarly, The symbol ← will be used to emphasize the progression of the reaction from right to left, as in: H^+ + OH^- ← H_2O. Note in this case, the reactant (H_2O) is written on the right side.

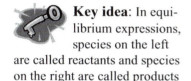

Key idea: In equilibrium expressions, species on the left are called reactants and species on the right are called products

The symbol $=$ will be used to indicate the state of chemical equilibrium. The equals sign signals that reactions have occurred both to the right and left until the species concentrations no longer change with time. For example, writing the equilibrium expression $H_2O = H^+ + OH^-$ indicates that the concentrations of the three species H_2O, H^+, and OH^- do not change with time.

Since the equals sign indicates that reactions occur both to the right and left, you may wonder which species should be called reactants and which should be called products. Indeed, at equilibrium, all reacting species are both reactants and products. By convention, the species on the left of the equilibrium expression are called reactants and species on the right of the equilibrium expression are called products.

oxidation state (oxidation number): the effective charge on an atom needed to account for the overall charge on the ion or molecule

A.2.4 Oxidation state

The *oxidation state* (or *oxidation number*) refers to the effective charge on an atom in a molecule or ion. Oxidation states usually are indicated by signed Roman numerals (e.g., $+I$, $-III$). To determine the oxidation state, you must have a starting point. For aquatic systems, it is usually assumed that the oxidation state of H is $+I$ and the oxidation state of O is $-II$. This allows the calculation of the oxidation state of other atoms in a molecule or ion. The sum of the oxidation states (weighted by the number of atoms of each type in the chemical species) must equal the charge on the species. In mathematical form, you can write:

$$\sum_{i=1}^{n} v_i (OS)_i = z$$

where: v_i = stoichiometric coefficient of atom i in the chemical species, $(OS)_i$ = oxidation state of atom i, z = overall charge on the chemical species, and n = the number of different elements in the species.

The steps in calculating the oxidation state are as follows (repeated in Appendix C, Section C.1.1):

Step 1: Assume the oxidation state of H is $+1$ ($+I$) and the oxidation state of O is -2 ($-II$)

Step 2: Calculate the oxidation state of other atoms by
$$\sum_{i=1}^{n} v_i (OS)_i = z$$
where: v_i = stoichiometric coefficient of atom i in the chemical species, $(OS)_i$ = oxidation state of atom i, z = overall charge on the chemical species, and n = the number of different elements in the species.

An example will make the calculation of oxidation state more clear. Consider the oxidation state of sulfur in sulfate (SO_4^{2-}). Here, $n = 2$ (S and

O), $z = -2$, $v_S = 1$ (one sulfur atom in sulfate), and $v_O = 4$ (four oxygen atoms in sulfate). If you assume that the oxidation state of O is $-II$ (or -2), then: $(1)(OS)_S + (4)(-2) = -2$. Solving, the oxidation state of sulfur, OS_S, is $+6$ or $+VI$.

A.3 MATHEMATICAL PRINCIPLES

In this text, a fluency with the algebraic manipulations of equations will be assumed. A knowledge of calculus is useful for understanding how thermodynamic properties change with other thermodynamic properties (Chapter 3), but not essential. In terms of nomenclature, $\log(x)$ will mean $\log_{10}(x)$ and $\ln(x)$ will mean $\log_e(x)$. Recall the meaning of the summation and product signs: $\sum_{i=1}^{3} a_i = a_1 + a_2 + a_3$ and $\prod_{i=1}^{4} a_i = a_1 a_2 a_3 a_4$. Also remember the properties of the log function: $\log(ab) = \log(a) + \log(b)$ and $\log(a^b) = b\log(a)$.

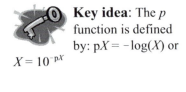

Key idea: The p function is defined by: $pX = -\log(X)$ or $X = 10^{-pX}$

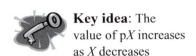

Key idea: The value of pX increases as X decreases

You shall use another function in this text: the p function. The p function is defined by: $pX = -\log(X)$. Thus: $X = 10^{-pX}$. The p function can be applied to constants. For example, if $K = 0.001 = 10^{-3}$, then $pK = 3$. In addition, the p function can be applied to species concentrations. For example, $pNH_3 = -\log([NH_3])$. There are two important aspects of the p function to remember. First, *the value of pX increases as X decreases*. As an example, $pX = 7$ if $X = 10^{-7}$, but pX increases to 9 if X decreases to 10^{-9}. Second, when applying the p function, you will sometimes ignore the usual rule that you cannot take the logarithm of a number with units. As will be developed in Chapter 2, concentrations in aqueous systems usually are written in units of moles per L ($= \text{mol/L}$). We write: $[Cl^-] = 10^{-3}$ mol/L and $pCl = 3$ without difficulty.

A.4 COMPUTER SKILLS

A.4.1 Solving nonlinear equations with spreadsheets

It is assumed that you are familiar with the use of standard spreadsheet programs (i.e., Microsoft *Excel* or Corel *QuattroPro*). It is expected that you can enter and copy formulas in their spreadsheet software of choice. The examples in this text will use the *Excel* format.

It is useful to know how to solve nonlinear equations with spreadsheets. For example, let us find x which satisfies: $5x = x^2 + 6$. There are three ways to use a spreadsheet program to solve this equation. For each method, it is useful to rewrite the equation with zero on one side: $x^2 - 5x + 6 = 0$. In the first method, you guess x and iterate until $x^2 - 5x + 6 = 0$. To accomplish this, enter any number (say **1**) in cell A1 and enter the formula = **A1^2−5*A1+6** in cell A1. If the value of cell A1 is **1**, then the value of cell B1 will be **2**. Now manually change the value of cell A1 until the value of cell B1 is sufficiently close to zero. This method also can be implemented easily on a programmable calculator.

In the second method, use a nonlinear optimization approach such as the *Excel* Solver function. Fill cells A1 and B1 as describe above. Now access Solver (*Tools•Solver*). If Solver is not shown, you will have to use the distribution CD for Excel and install it. Set the target cell to **B1** (the cell containing the formula you wish to solve), select the "Value of:" option in "Equal To:" and type **0** in the box, and set the "By Changing Cells:" to **A1** (the cell containing the value you is to change). Now click "Solve". The approximate answer (2.00...) will appear in cell A1.

In the third method, we make use of the auto-calculation feature. In *Excel*, enable the auto-calculation mode (select *Tools•Solver*, select the Calculation tab, select the "Automatic" option, and check the "Iteration" box). Now solve the original equation for x: $x = (x^2 + 6)/5$. Type the right hand side in cell A1: $= (A1*A1+6)/5$. Excel will interpret this a command to find a number which equals one-fifth of itself squared plus six. (If you try this and get a circular reference error, make sure the "Iteration" box is checked as describe above.)

A.4.2 Pitfalls with using spreadsheets to solve nonlinear equations

None of these methods are foolproof. For example, each methods breaks down when multiple roots are present. In the example above, each method returns an answer of 2, when actually both 2 and 3 are roots of the equation. Care should be taken when multiple roots are expected. In such cases, it is useful to plot the function against the independent variable and look for roots (i.e., values of the independent variable where the function equals zero).

Each method may fail if the function and/or cell to be varied are extremely small. This is often the case in equilibrium calculations. In such instances, it is useful to multiply the function by a large number so that its value is reasonable. For example, if the equation to be set equal to zero is $x^2 - 5 \times 10^{-8} x$ (with x in cell A1), one might enter $= (A1^2-5e-8*A1)*1e10$. In addition, it is useful to transform the variable to be changed so that its values are reasonably large. For example, one may choose to vary $\log(x)$ rather than x. In chemical equilibrium problems, it may be useful to vary p(concentration) rather than the concentration.

The third method may fail if the formula entered gives an error with the variable value of zero. For example, to solve the equation $x = 1/x$ (which has roots 1 and -1), one would enter into cell A1: $=1/A1$. This will give a division by zero error, since Excel begins the calculation with the value of A1 equal to zero. One possible workaround is to tweak the formula to avoid division by zero when $x = 0$. In the example above, one could enter the formula as: $=1/(a1+1e-20)$[†].

[†] Thanks to John E. Van Benschoten at the University at Buffalo for this suggestion.

SUMMARY OF KEY IDEAS

- Dissolved species, pure liquids, pure solids, and gases will be labeled (aq), (l), (s), and (g), respectively

- In equilibrium expressions, species on the left are called reactants and species on the right are called products

- The p function is defined by: $pX = -\log(X)$ or $X = 10^{-pX}$

- The value of pX increases as X decreases

Equilibrium and Steady State

B.1 INTRODUCTION

In Chapter 3, equilibrium was defined as the state where all thermodynamic functions are invariant with time *in a system that does not exchange mass or energy with its surroundings*. If the system does exchange mass with its surroundings, then a different approach is required. In fact, a different name is given to the state where all thermodynamic functions are invariant with time in a system that does exchange mass or energy with its surroundings. This state is called *steady state*. At first glance, the steady state and the equilibrium state appear very similar. In this appendix, the differences between steady-state solutions and equilibrium solutions will be explored.

To analyze a system which exchanges mass with its surroundings, it is necessary to have a basic understanding of chemical equilibria and kinetics. The material in this appendix is best understood after completing Chapters 3 and 22 of this text.

B.2 STEADY-STATE ANALYSIS

B.2.1 Problem statement

Consider a system receiving and discharging water so that the volume of the system is constant. Constant system volume means that the flow of water into the system equals the flow of water out of the system (assuming constant water density). This is a good model for a lake of constant volume. Imagine that the influent stream brings with it water with a constant molar concentration $[A]_{in}$ for species A. Use the symbol $[A]$ for the concentration of A inside the system. If the system is well-mixed, then $[A]$ is the same everywhere in the system. This means that the effluent concentration also will be equal to $[A]$.

B.2.2 Mole balance

A mole balance on species A reveals:

$$\text{accumulation rate inside the system} =$$
$$\text{rate of mass into the system} - \text{rate of mass}$$
$$\text{out of the system} + \text{rate of change of}$$
$$\text{mass inside the system}$$

eq. B.1

Each term in the mole balance has units of mole/time. The accumulation rate is given by $Vd[A]/dt$. The rates of mass flow into and out of the system are $Q[A]_{in}$ and $Q[A]$, respectively, where Q = flow in = flow out. To determine the rate of change of mass inside the system, assume that A is transformed into another species B. Two reactions are of interest:

$$A \rightarrow B \qquad k = k_f$$
$$B \rightarrow A \qquad k = k_r$$

Assuming the reactions are elementary (Section 22.3.3), the techniques of Section 22.3.4 can be used to write the rate of change of $[A]$ inside the system:

$$\text{rate of change of } [A] \text{ inside the system} = -k_f[A] + k_r[B]$$

The rate of the moles of A is equal to V(rate of change of $[A]$ inside the system), where V is the system volume.

Substituting into the mole balance (eq. B.1):

$$Vd[A]/dt = Q[A]_{in} - Q[A] + (-k_f[A] + k_r[B])V, \text{ or:}$$

$$d[A]/dt = ([A]_{in} - [A])/\theta - k_f[A] + k_r[B]$$

where: θ = hydraulic residence time = V/Q. Similarly, for B (assuming it is not present in the influent stream):

$$d[B]/dt = -[B]/\theta + k_f[A] - k_r[B] \qquad \text{eq. B.2}$$

We seek the steady-state solution; that is, the solution when the species concentrations do not change with time. At steady state: $d[A]/dt = d[B]/dt = 0$. The steady state concentrations are denoted $[A]_{ss}$ and $[B]_{ss}$. From eq. B.2:

$$0 = -[B]_{ss}/\theta + k_f[A]_{ss} - k_r[B]_{ss}$$

Solving:

$$[B]_{ss}/[A]_{ss} = k_f\theta/(k_r\theta + 1) \qquad \text{eq. B.3}$$

B.3 COMPARISON OF STEADY STATE AND EQUILIBRIUM SOLUTIONS

B.3.1 Equilibrium solution
The ratio of the species concentrations at steady state is given by eq. B.3. What is the ratio of the species concentrations at equilibrium?

At equilibrium (with no mass exchange with the surroundings):

$$A = B \qquad \text{equilibrium constant} = K$$

Thus:

$$[B]_{eq}/[A]_{eq} = K \qquad\qquad \text{eq. B.4}$$

B.3.2 Impact of mass exchange with surroundings

Recall from Section 22.5.2 that for a system at equilibrium with no inputs from its surroundings: $K = k_f/k_r$. Thus, eq. B.3 becomes:

$$[B]_{ss}/[A]_{ss} = K/(1 + 1/k_r\theta) \qquad\qquad \text{eq. B.5}$$

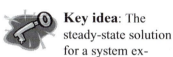

Key idea: The steady-state solution for a system exchanging mass with its surroundings and the equilibrium solution for a system that does not exchange mass with its surrounding are different

Comparing eqs. B.4 and B.5, it is clear that *the steady-state solution for a system exchanging mass with its surroundings and the equilibrium solution for a system that does not exchange mass with its surrounding are different*.

B.3.3 Environmental examples

It is interesting to note that the relationship between the steady-state solution and equilibrium solution depends on k_r but not on k_f. Note that the steady state and equilibrium solutions approach each other as $k_r\theta$ gets large. The solutions will be within 1% of each other when $k_r\theta > 100$.

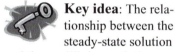

Key idea: The relationship between the steady-state solution and the equilibrium solution depends on the hydraulic residence time and rate constant(s)

Consider two examples from Section 22.5.2. In the case of monochloramine synthesis (Example 22.6), $k_r = 2.1 \times 10^{-5}\,\text{s}^{-1}$. Thus, for the steady-state and equilibrium solutions to be within 1% of each other requires $\theta > 100/k_r = 100/(2.1 \times 10^{-5}\,\text{s}^{-1})$ or about 55 days. At hydraulic residence times of interest in engineered systems, the steady-state and equilibrium concentrations clearly will be different and an equilibrium approach may be inappropriate.

On the other hand, consider the example of acetic acid hydrolysis, where $k_r = 10^{+3.7}\,\text{s}^{-1}$ at pH 7. In this case, for the steady-state and equilibrium solutions to be within 1% of each other requires $\theta > 100/k_r = 100/(10^{+3.7}\,\text{s}^{-1})$ or about 20 milliseconds. For this chemistry, the steady-state and equilibrium concentrations are likely to be very similar and an equilibrium approach may be appropriate even if mass is exchanged with the surroundings.

SUMMARY OF KEY IDEAS

- The steady-state solution for a system exchanging mass with its surroundings and the equilibrium solution for a system that does not exchange mass with its surrounding are different

- The relationship between the steady-state solution and the equilibrium solution depends on the hydraulic residence time and rate constant(s)

Summary of Procedures

In this text, several step-by-step procedures were outlined. The procedures are summarized here for the convenience of the reader.

C.1 OXIDATION STATE AND BALANCING REACTIONS

C.1.1 Calculating oxidation states (section A.2.4)

Step 1: Assume the oxidation state of H is +1 (+I) and the oxidation state of O is −2 (−II)

Step 2: Calculate the oxidation state of other atoms by $\sum_{i=1}^{n} v_i (OS)_i = z$

where: v_i = stoichiometric coefficient of atom i in the chemical species, $(OS)_i$ = oxidation state of atom i, z = overall charge on the chemical species, and n = the number of different elements in the species.

C.1.2 Balancing chemical reactions (sections 4.7.1 and 16.3.1)

Step 1: Write the known reactants on the left and known products on the right

Step 2: Adjust stoichiometric coefficients to balance all elements (except H and O)

Step 3: Add water (H_2O) to balance the element O

Step 4: Add H^+ to balance the element H

Step 5: Add electrons (e^-) to balance the charge

Note: For basic conditions, add or subtract the reaction $H_2O = H^+ + OH^-$ to eliminate H^+ and express the reaction in terms of OH^-

C.1.3 Balancing overall redox reactions (section 16.3.2)

Step 1: Balance the half reactions separately

Step 2: Add the half reactions to eliminate free electrons

C.2 SETTING UP CHEMICAL EQUILIBRIUM SYSTEMS (SECTION 6.4)

Step 1: Define the chemical system
  Goal: Decide if an open or closed model is to be used
  Tool: Match the model to the system to be modeled

Step 2: Generate a species list
  Goal: Create list of *n* unknowns (equilibrium concentrations of *n* species)
  Tools: Identify the starting materials
            Always include H_2O as a starting material
          Identify the initial hydrolysis products
                Look for the formation of simple ions (see Table 6.1)
                Look for H^+ transfer
                Always include H_2O, H^+, and OH^-, usually ignore $N_2(aq)$
          Enumerate subsequent hydrolysis products
          Enumerate other reaction products
          Generate a species list diagram (see Figure 6.1)

Step 3: Define the constraints on species concentrations
  Goal: Create list of *n* equations
  Tools: Equilibrium constraints
                Always include the self-ionization of water
          Mass balance constraints
                Identify components
                Find the total concentration of each component
                Associate each soluble species with a component (exceptions: H_2O, H^+, and OH^-)
                Determine the stoichiometry of each soluble species with respect to its component(s)
          Electroneutrality constraint (charge balance)
          Other constraints
                Activity of pure liquids = 1
                Activity of pure solids = 1 (if solid exists)
                Activity of solids = 0 (if solid dissolves completely)

C.3 ALGEBRAIC SOLUTION TECHNIQUES

C.3.1 Method of substitution (section 7.3)

Step 1: Simplify the system by setting the concentrations of all pure liquids and solids equal to unity and all solids that dissolve completely equal to zero

Step 2: Express each species concentration in terms of one species concentration

Step 3: Substitute the equations derived in Step 2 into the unused equation in the Step 1

Step 4: Insert known equilibrium constants and total masses and solve

Step 5: Back substitute to calculate the equilibrium concentration of each species

C.3.2 Method of approximation (section 7.4)

Step 1: Simplify the system by setting the concentrations of all pure liquids and solids equal to unity and all solids that dissolve completely equal to zero

Step 2: Make approximations in the additive equations

Step 3: Write one equation in one unknown

Step 4: Insert known equilibrium constants and total masses and solve

Step 5: Back substitute to calculate the equilibrium concentration of each species

Step 6: Check the assumptions and original equations: iterate if necessary

C.4 GRAPHICAL SOLUTIONS

C.4.1 General pC-pH diagram (section 8.3.3)

Step 1: Set up the pC-pH diagram
Set up the chemical system and prepare a pC-pH diagram with lines representing the H^+ and OH^- concentrations

Step 2: Simplify

Simplify the system by setting the concentrations of all pure liquids and solids equal to unity and all solids that dissolve completely equal to zero

Step 3: Express concentrations as function of $[H^+]$

Express each species concentration in terms of $[H^+]$ using equilibria and mass balance(s)

Step 4: Plot

Plot the functions derived in Step 3 on the pC-pH diagram

Step 5: Find the equilibrium pH

To find the equilibria pH (and equilibrium concentrations of all plotted species), find the pH where the charge balance is satisfied

Step 6: Check assumptions

If assumptions were made in Step 5 (i.e., if ions were assumed to be of negligible concentration), check all assumptions and iterate if the original assumptions are invalid

C.4.2 Shortcut graphical method for a monoprotic acid (section 8.4.2)

Step 1: Locate the system point

Prepare a pC-pH diagram with lines representing the H^+ and OH^- concentrations and locate the system point (where pH = pK_a and pC = pA_T).

Step 2: Draw species concentration lines away from the system point

Draw the [HA] and [A$^-$] lines at least 1.5 pH unit away from the system point by noting that: (1) *below* the pK_a, pHA $\approx$ pA_T and pA increases by one log unit for every unit decrease in pH below the pK_a, and (2) *above* the pK_a, pA $\approx$ pA_T and pHA increases by one log unit for every unit increase in pH above the pK_a. Make sure that the [HA] and [A$^-$] lines, if extended, would both go through the system point.

Step 3: Curve the lines

Curve the [HA] and [A$^-$] lines so that they intersect about 0.3 log units below the system point.

Step 4: Find the equilibrium pH

To find the equilibria pH (and equilibrium concentrations of all plotted species), find the pH where the charge balance is satisfied.

Check any assumptions made in the charge balance and iterate if necessary.

C.4.3 Graphical shortcut method for an *n*-protic acid (section 11.4.3)

Step 1: Locate system points

Prepare a pC-pH diagram with lines representing the H^+ and OH^- concentrations. Locate the *n* system points. These occur where pH = $pK_{a,1}$ and $pC = pA_T$, pH = $pK_{a,2}$ and $pC = pA_T$, ..., and pH = $pK_{a,n}$ and $pC = pA_T$.

Step 2: Draw species lines

Draw the lines for the p(concentration) of each species at least 1.5 pH units away from each system point. The p(concentration) line of species H_iA has the following slopes: $i - n$ at pH values less than the first system point, $i - n + j$ between the jth and $j + 1^{th}$ system point, and $-i$ at pH values greater than the final (n^{th}) system point. (Recall that a negative slope means that the pC decreases with increasing pH and thus the line is upward sloping to the right.) Make sure that the lines, if extended, would both go through the system points.

Step 3: Make species lines intersect below the system points

Curve the lines so that they intersect about 0.3 log units below the system points.

Step 4: Find the equilibria pH

To find the equilibria pH (and equilibrium concentrations of all plotted species), find the pH where the charge balance or proton condition is satisfied. Check any assumptions made in the charge balance and iterate if necessary.

C.4.4 pe-pH diagrams (section 16.7.5)

Step 1: Make a species list, including species from all reasonable oxidation states

Step 2: Carefully consider which chemical processes should be included

It is useful to make a list of how each species in the species list could be related to every other species.

Step 3: List the equilibria describing the chemical processes listed in Step 2

Tables of redox and acid/base equilibria (see Appendix D and Table 16.3) provide the following possible equilibria for inclusion in the chemical model.

Step 4: Write the expressions for the K values in terms of species activities
Set the activity of water to unity and set the ratio of activities of redox pairs equal to one.

Step 5: Plot the lines generated in Step 4 on a pe-pH diagram

Step 6: Erase line segments in the pe-pH diagram that are thermodynamically impossible
Blindly plotting the equations in Step 5 usually does not result in a clear delineation of the pe-pH regions where each species predominates. This step requires you to examine each line to determine over what pe and pH range (if any) it makes sense.

C.5 COMPUTER SOLUTIONS: TABLEAU METHOD (SECTION 9.6)

Step 1: Make a list of species and equilibria

Step 2: Select the components
The usual approach is to include H_2O, H^+, and the most dissociated forms of each species family as components.

Step 3: Draw the initial tableau
Make one column for each component and one column labeled "log K". Make rows for each species and rows at the bottom of the tableau for each starting material.

Step 4: Enter the component stoichiometric coefficients from equilibria
Rearrange the equilibria to express each species as a product of the components raised to a power. For example, the self-ionization of water can be rearranged to: $[OH^-] = K_w[H_2O]^1[H^+]^{-1}$. Enter the coefficients for each component on the row for that species.

Step 5: Enter the component stoichiometric coefficients from mass balances
Express the elemental composition of each starting material as a linear combination of the components. For example: $NaOH = (Na^+)_1(H_2O)_1(H^+)_{-1}$. Enter the coefficients for each component on the row for that starting material.

C.6 ACID-BASE TITRATIONS

C.6.1 Monoprotic acids (section 12.3.3)

Step 1: pC-pH Diagram
Prepare a pC-pH diagram for a C M HA solution

Step 2: Zeroth and first equivalence points
Apply the proton conditions for an HA and NaA solution to the pC-pH diagram to find the pH at $f = 0$ and $f = 1$, respectively.

Step 3: Half-equivalence point
The half equivalence point occurs at about pH = pK_a. Using the pC-pH diagram, check to make sure that $[H^+]$ and $[OH^-]$ are much smaller than [HA] and $[A^-]$.

Step 4: "Connect the Dots"
Draw a smooth curve through the three points determined in the previous steps. Make sure the slope is steepest at $f = 1$.

C.6.2 Diprotic acids (section 12.4.2)

Step 1: pC-pH diagram
Prepare a pC-pH diagram for a C M H_2A solution

Step 2: Zeroth, first, and second equivalence points
Apply the proton conditions for H_2A, NaHA, and Na_2A solutions to the pC-pH diagram to find the pH at $f = 0$, 1, and 2, respectively.

Step 3: Half-equivalence points
As a first guess, half-equivalence points are *estimated* to occur at about pH = $pK_{a,1}$ and pH = $pK_{a,2}$. Using the pC-pH diagram, check the assumptions: (1) At $f = \frac{1}{2}$, check that $[H^+]$ and $[OH^-]$ can be ignored and $[A^{2-}] \ll [HA^-]$, and (2) At $f = 3/2$, check that $[H^+]$ and $[OH^-]$ can be ignored and $[H_2A] \ll [HA^-]$.

Step 4: "Connect the Dots"
Draw a smooth curve through the three points determined in the previous steps. Make sure the slope is steepest at $f = 1$.

C.7 COMPLEXATION

Step 1: Solve for each species concentration as a function of the uncomplexed (i.e, aquo complexed) metal and uncomplexed ligand

Step 2: Substitute the expressions from Step 1 into the mass balances

Step 3: Rework the mass balances to obtain one equation in one species concentration

C.8 IONIC STRENGTH EFFECTS (SECTION 21.2.7)

Step 1: Set up the equilibrium system in the usual fashion

Step 2: Using single-species activity coefficients, convert all equilibrium constants to ^cK values at the ionic strength of interest
In general, for the equilibrium $a\mathrm{A} + b\mathrm{B} = c\mathrm{C} + d\mathrm{D}$, $K = (\gamma_C^c \gamma_D^d / \gamma_A^a \gamma_B^b)(^cK)$

Step 3: Solve the system for species concentrations

Step 4: If desired, compute species activities from the concentrations calculated in Step 3 and the activity coefficients calculated in step 2

C.9 CHEMICAL KINETICS (SECTION 22.3.4)

Step 1: Identify the reaction(s) which include the species of interest as either a reactant or product

Step 2: Write the rate expressions for each species of interest using the reactions identified in Step 1

Step 3: Add up the rates from each reaction to find the overall rate of change of each species concentration

Selected Equilibrium Constants

This appendix contains tables of equilibrium constants. The equilibria selected for tabulation represent common chemical equilibria in the environment. The lists are not meant to be exhaustive.

The equilibrium constants were obtained primarily from the thorough database edited by Arthur E. Martell and Robert M. Smith and published by the National Institute of Standards and Technology (NIST, 2001). This reference should be consulted for citations of the original literature.

The following types of equilibrium constants are presented in this appendix:

acid dissociation constants:
first row of Table D.1 for inorganic acids and first row of Table D.2 for organic acids

base dissociation constants:
can be calculated from acid dissociation constants by $pK_b = 14 - pK_a$

complex formation constants:
Table D.1 for inorganic complexes and Table D.2 for organic complexes

redox equilibria:
Table 16.3, repeated for convenience as Table D.5

Henry's Law constants:
Table D.4

vapor pressures:
Table D.4

K_{s0} and other solid-liquid equilibrium constants:
Table D.3

Table D.1: Formation Constants for Selected Inorganic Complexes and Acid Dissociation Constants for Selected Inorganic Acids

Tabulated values are $\log K$ values for the equilibria as written in the second column. Note that equilibria with H^+ are written as acid dissociation reactions, while equilibria with all other metals are written as complex formation reactions. The acid dissociation reactions are listed with $\log K_{a,1}$ as the topmost value.

Metal (M)	Equilibrium	Inorganic Ligand (L)							
		OH^-	CO_3^{2-}	SO_4^{2-}	Cl^-	F^-	NH_3	PO_4^{3-}	CN^-
H^+	$H_3L = H_2L + H$							-2.222	
	$H_2L = HL + H$		-6.352					-7.179	
	$HL = L + H$	-13.997	-10.329	-1.99		-3.17	-9.244	-12.375	-9.21
Ag^+	$M + L = ML$	2.0		1.3	3.31		3.31		
	$M + 2L = ML_2$	3.99			5.25		7.22		20.48
	$M + 3L = ML_3$				5.2				21.7
Al^{3+}	$M + L = ML$	9.00		3.89		7.00			
	$M + HL = MHL$							$6.12^{I=0.1}$	
	$M + 2L = ML_2$	17.9				12.6			
	$M + 3L = ML_3$	25.2				16.7			
	$M + 4L = ML_4$	33.3				19.4			
	$M + 5L = ML_5$								
	$M + 6L = ML_6$								
	$2M + L = M_2L$							$16.7^{I=0.1}$	
	$2M + 2L = M_2L_2$	20.3							
	$3M + 4L = M_3L_4$	42.1							
Ba^{2+}	$M + L = ML$	0.64	2.71	2.13			-0.1		
	$M + HL = MHL$		0.98						
Ca^{2+}	$M + L = ML$	1.30	3.20	2.36	$0.2^{I=0.1}$	$0.63^{I=1}$	0.2		
	$M + HL = MHL$		1.27					2.66	
	$M + H_2L = MH_2L$							1.35	
	$2M + L = M_2L$						-0.1		
Cd^{2+}	$M + L = ML$	3.9	$3.5^{I=0.1}$	2.37	1.98	1.2	2.55		6.01
	$M + HL = MHL$		$0.9^{I=3}$					$2.91^{I=0.1}$	
	$M + H_2L = MH_2L$							$0.76^{I=3}$	
	$M + 2L = ML_2$	7.7			2.6		4.56		11.12
	$M + 3L = ML_3$	$10.3^{I=3}$			2.4		5.90		15.65
	$M + 4L = ML_4$	8.7					6.72		17.92
	$2M + L = M_2L$	4.6							
	$4M + 4L = M_4L_4$	23.2							
Cr^{3+}	$M + L = ML$	10.34	$1.45^{I=1}$		$0.5^{I=1}$	5.2		$2.57^{I=0.1}$	
	$M + HL = MHL$								
	$M + 2L = ML_2$	$17.2^{I=1}$				$7.7^{I=0.5}$			
	$M + 3L = ML_3$					$10.1^{I=0.5}$			

Metal (M)	Equilibrium	Inorganic Ligand (L)							
		OH^-	CO_3^{2-}	SO_4^{2-}	Cl^-	F^-	NH_3	PO_4^{3-}	CN^-
Cu^{2+}	$M + L = ML$	6.5	6.77	2.36	0.2	1.8	4.02		
	$M + HL = MHL$		1.8					$3.2^{I=0.1}$	
	$M + H_2L = MH_2L$							$0.64^{I=3}$	
	$M + 2L = ML_2$	11.8	10.2				7.40		
	$M + 3L = ML_3$	$14.5^{I=1}$					10.2		
	$M + 4L = ML_4$	$15.6^{I=1}$					12.3		
	$2M + 2L = M_2L_2$	17.4							
	$4M + 4L = M_4L_4$	35.2							
Fe^{2+}	$M + L = ML$	4.6		(2.39)	−0.2	$0.8^{I=1}$	1.40		
	$M + HL = MHL$		1.10					$2.46^{I=3}$	
	$M + H_2L = MH_2L$							$0.55^{I=3}$	
	$M + 2L = ML_2$	(7.5)					2.25		
	$M + 3L = ML_3$	13					2.68		
	$M + 4L = ML_4$	10					2.75		
	$M + 6L = ML_6$								35.4
Fe^{3+}	$M + L = ML$	11.81		4.05	1.48	6.04			
	$M + HL = MHL$							$8.30^{I=0.5}$	
	$M + H_2L = MH_2L$							$3.47^{I=0.5}$	
	$M + 2L = ML_2$	23.4			2.13	$9.12^{I=0.5}$			
	$M + 3L = ML_3$					$12.0^{I=0.5}$			
	$M + 4L = ML_4$	34.4							
	$M + 6L = ML_6$								43.6
	$2M + 2L = M_2L_2$	25.14							
	$3M + 4L = M_3L_4$	49.7							
Hg^{2+}	$M + L = ML$	10.60	$(11.0)^{I=0.5}$	$1.34^{I=0.5}$	7.3	$1.03^{I=0.5}$	$8.8^{I=2,T=22}$	$9.5^{I=3}$	17.00
	$M + HL = MHL$		$5.48^{I=0.5}$					$8.8^{I=3}$	
	$M + 2L = ML_2$	21.8	$14.5^{I=0.5}$	$2.4^{I=0.5}$	14.0		$17.3^{I=2}$		32.75
	$M + 3L = ML_3$	20.9			15.0		$18.3^{I=2}$		36.31
	$M + 4L = ML_4$				15.6		$19.0^{I=0.5}$		38.97
	$2M + L = M_2L$	10.7							
	$3M + 3L = M_3L_3$	35.6							
K^+	$M + L = ML$	$0.0^{I=0.1}$			−0.5	$-1.2^{I=1}$		(1.43)	
	$M + HL = MHL$							0.88	
	$M + H_2L = MH_2L$							(0.3)	
Li^+	$M + L = ML$	0.36		0.64			−0.7	$0.73^{I=0.1}$	
Mg^{2+}	$M + L = ML$	2.6	2.92	2.26	0.6	2.05	0.2		
	$M + HL = MHL$		1.01					2.80	
	$M + H_2L = MH_2L$							$0.16^{I=3}$	

Metal (M)	Equilibrium	Inorganic Ligand (L)							
		OH^-	CO_3^{2-}	SO_4^{2-}	Cl^-	F^-	NH_3	PO_4^{3-}	CN^-
Mn^{2+}	$M + L = ML$	3.4	4.7	2.25	$-0.2^{I=1}$	1.6	0.84		$1.9^{I=1}$
	$M + HL = MHL$		1.30						
	$M + 2L = ML_2$						1.25		$3.36^{I=1}$
	$M + 3L = ML_3$						1.38		
	$M + 4L = ML_4$	7.7					1.24		
	$2M + L = M_2L$	6.8							
	$2M + 3L = M_2L_3$	18.1							
Na^+	$M + L = ML$	0.1	1.27	0.73	-0.5	-0.2		(1.43)	
	$M + HL = MHL$		-0.25					1.07	
	$M + H_2L = MH_2L$							0.3	
Ni^{2+}	$M + L = ML$	4.1	$3.57^{I=0.7}$	2.30		1.4	2.72		
	$M + HL = MHL$		$1.59^{I=0.7}$					$2.10^{I=0.1}$	
	$M + H_2L = MH_2L$							$0.5^{I=0.1}$	
	$M + 2L = ML_2$	9					4.88		
	$M + 3L = ML_3$	12					6.54		
	$M + 4L = ML_4$						7.67		30.2
	$M + 5L = ML_5$						8.33		
	$M + 6L = ML_6$						8.30		
	$4M + 4L = M_4L_4$	28.3							
Pb^{2+}	$M + L = ML$	6.4	$5.4^{I=0.5}$		1.55	$1.44^{I=1}$	1.55		
	$M + HL = MHL$		$1.91^{I=0.5}$		2.2			3.1	
	$M + H_2L = MH_2L$				1.8			1.5	
	$M + 2L = ML_2$	10.9	$8.86^{I=0.5}$			$2.53^{I=1}$			
	$M + 3L = ML_3$	13.9							
	$2M + L = M_2L$	7.6							
	$3M + 4L = M_3L_4$	32.1							
	$4M + 4L = M_4L_4$	36.0							
	$6M + 8L = M_6L_8$	68.4							
Sr^{2+}	$M + L = ML$	0.82	2.81	2.30	$0.2^{I=1}$	$0.14^{I=1}$	0.0		
	$M = HL = MHL$		1.21					$1.64^{I=0.1}$	
Zn^{2+}	$M + L = ML$	5.0	4.76	2.34	0.4	1.3	2.21		
	$M = HL = MHL$		1.5					$2.4^{I=0.1}$	
	$M + H_2L = MH_2L$							$0.37^{I=3}$	
	$M + 2L = ML_2$	10.2	7.3		0.6		4.50		11.07
	$M + 3L = ML_3$	13.9					6.86		16.05
	$M + 4L = ML_4$	15.5					8.89		19.62

Source: NIST, 2001

$\log K$ values are at 25°C and $I \to 0$, unless otherwise indicated

Values in parentheses carry higher uncertainty

Table D.2: Formation Constants for Selected Organic Complexes and Acid Dissociation Constants for Selected Organic Acids

Tabulated values are $\log K$ values for the equilibria as written in the second column. Note that equilibria with H^+ are written as acid dissociation reactions, while equilibria with all other metals are written as complex formation reactions. The acid dissociation reactions are listed with $\log K_{a,1}$ as the topmost value.

Metal (M)	Equilibrium	Organic Ligand (L)				
		NTA^{3-}	$EDTA^{4-}$	$Citrate^{3-}$	Glycinate	Acetate
H^+	$H_4L = H_3L + H$		$-2.00^{I=0.1}$			
	$H_3L = H_2L + H$	-2.0	$-2.69^{I=0.1}$	$-3.05^{I=0.1}$		
	$H_2L = HL + H$	$-2.94^{T=20}$	-6.273	$-4.30^{I=0.1}$		
	$HL = L + H$	$-10.334^{T=20}$	-10.948	$-5.73^{I=0.1}$	-2.350 (acid) -9.778 (amino)	-4.757
Ag^+	$M + L = ML$	$4.85^{I=0.1}$	$7.20^{I=0.1}$		$3.20^{I=0.1}$	0.73
	$M + 2L = ML_2$				$6.63^{I=0.1}$	
Al^{3+}	$M + L = ML$	$11.4^{I=0.1}$	$16.4^{I=0.1}$		$5.91^{I=0.1}$	2.75
	$M + 2L = ML_2$					4.6
Ba^{2+}	$M + L = ML$	$4.81^{I=0.1}$	$7.88^{I=0.1}$			
Ca^{2+}	$M + L = ML$	$6.3^{I=0.1}$	$10.65^{I=0.1}$		$1.09^{I=0.1}$	1.26
	$M + 2L = ML_2$	$8.81^{I=0.1}$				
Cd^{2+}	$M + L = ML$	$9.76^{I=0.1}$	$16.5^{I=0.1}$	$3.10^{I=0.1}$	$4.25^{I=0.1}$	1.92
	$M + 2L = ML_2$	$14.47^{I=0.1}$			$7.77^{I=0.1}$	$1.91^{I=0.5}$
	$M + 3L = ML_3$				$10.0^{I=0.1}$	$2.18^{I=0.5}$
Cr^{3+}	$M + L = ML$		$(24.3)^{I=0.1,T=20}$			$4.63^{I=0.5}$
	$M + 2L = ML_2$					$7.08^{I=0.5}$
	$M + 3L = ML_3$					$9.6^{I=0.5}$
Cu^{2+}	$M + L = ML$	$12.7^{I=0.1}$	$18.78^{I=0.1}$	$5.20^{I=0.1}$	$8.19^{I=0.1}$	2.21
	$M + 2L = ML_2$	$17.4^{I=0.1}$			$15.1^{I=0.1}$	3.4
	$M + 3L = ML_3$					$3.3^{I=0.1}$
Fe^{2+}	$M + L = ML$	$8.9^{I=0.1}$	$14.30^{I=0.1}$		$3.73^{I=1}$	
	$M + 2L = ML_2$	$11.98^{I=0.1}$			$6.65^{I=1}$	
	$M + 3L = ML_3$				$8.87^{I=1}$	
Fe^{3+}	$M + L = ML$	$15.9.3^{I=0.1}$	$25.1^{I=0.1}$		$(8.57)^{I=0.5}$	$3.6^{I=0.1}$
	$M + 2L = ML_2$	$24.0^{I=0.1}$				$6.5^{I=0.1}$

Metal (M)	Equilibrium	Organic Ligand (L)				
		NTA^{3-}	$EDTA^{4-}$	$Citrate^{3-}$	Glycinate	Acetate
Hg^{2+}	$M + L = ML$	$14.3^{I=0.1}$	$21.5^{I=0.1}$		$10.3^{I=0.5, T=20}$	4.3
	$M + 2L = ML_2$				$19.2^{I=0.5, T=20}$	$8.45^{I=3}$
K^+	$M + L = ML$	$0.5^{I=0.5}$	$0.8^{I=0.1}$			-0.27
Li^+	$M + L = ML$	$2.45^{I=0.1}$	$2.95^{I=0.1}$			0.28
Mg^{2+}	$M + L = ML$	$5.50^{I=0.1}$	$8.79^{I=0.1}$	$2.51^{I=0.1}$	$1.66^{I=0.1}$	
	$M + 2L = ML_2$				$2.26^{I=3}$	
Mn^{2+}	$M + L = ML$	$7.27^{I=0.1}$	$13.89^{I=0.1}$	$2.62^{I=0.1}$	$2.80^{I=0.1}$	$0.80^{I=0.1}$
	$M + 2L = ML_2$	$10.44^{I=0.1}$			$4.8^{I=0.1}$	
Na^+	$M + L = ML$	$1.2^{I=0.1}$	$1.86^{I=0.1}$		(-0.6)	-0.12
Ni^{2+}	$M + L = ML$	$11.51^{I=0.1}$	$18.4^{I=0.1}$		$5.74^{I=0.1}$	1.44
	$M + 2L = ML_2$	$16.321^{I=0.1}$			$10.58^{I=0.1}$	2.40
	$M + 3L = ML_3$				$14.1^{I=0.1}$	
Pb^{2+}	$M + L = ML$	$11.48^{I=0.1}$	$18.0^{I=0.1}$		$4.72^{I=0.1}$	2.58
	$M + 2L = ML_2$	$12.8^{I=0.1, T=20}$			$7.7^{I=0.1}$	4.02
	$M + 3L = ML_3$					$3.59^{I=3}$
Sr^{2+}	$M + L = ML$	$4.99^{I=0.1}$	$8.72^{I=0.1}$	$2.09^{I=0.1}$	$0.6^{I=0.1}$	1.12
Zn^{2+}	$M + L = ML$	$10.45^{I=0.1}$	$16.5^{I=0.1}$		$4.96^{I=0.1}$	1.57
	$M + 2L = ML_2$	$14.24^{I=0.1}$			$9.19^{I=0.1}$	$1.1^{I=0.5}$
	$M + 3L = ML_3$				$11.6^{I=0.1}$	$1.57^{I=3}$

Source: NIST, 2001

$\log K$ values are at 25°C and $I \to 0$, unless otherwise indicated

Values in parentheses carry higher uncertainty

Table D.3: Solubility Constants for Selected Solids

Metal	Main Ligand	Solid (solubility equilibrium included if K_s is not K_{s0})	log K_s
Ag(+I)	CO_3	$Ag_2CO_3(s)$	-11.09
	SO_4	$Ag_2SO_4(s)$	-4.82
	PO_4	$Ag_3PO_4(s)$	-17.59
	Cl	$AgCl(s)$	-9.75
	S	$Ag_2S(s)$ $(Ag_2S(s) + 2H^+ = 2Ag^{2+} + H_2S)$	-29.2
	CN	$AgCN(s)$	-15.74
		$Ag_2(CN)_2(s)$ $(Ag_2(CN)_2(s) = 2Ag^+ + 2CN^-)$	-10.9
Al(+III)	OH	$Al(OH)_3(s,\alpha)$	-33.7
Ba(+II)	OH	$Ba(OH)_2(H_2O)_8(s)$	-3.6
	CO_3	$BaCO_3(s,Witherite)$	-8.57
	SO_4	$BaSO_4(s,Barite)$	-9.98
	PO_4	$BaHPO_4(s)$ $(BaHPO_4(s) = Ba^{2+} + HPO_4^{2-})$	-7.40
	S	$BaS(s)$ $(BaS(s) + 2H^+ = Ba^{2+} + H_2S)$	23.2
Ca(+II)	OH	$Ca(OH)_2(s)$	-5.19
	CO_3	$CaCO_3(s,Calcite)$	-8.48
		$CaCO_3(s,Aragonite)$	-8.30
		$CaCO_3(s,Vaterite)$	-7.91
		$CaCO_3(s,Monohydocalsite)$	-7.60
		$CaCO_3(H_2O)_6(s)$	-7.46
	SO_4	$CaSO_4(H_2O)_2(s,Gypsum)$	-4.61
	PO_4	$Ca(HPO_4)(H_2O)_2(s)$ $(Ca(HPO_4)(H_2O)_2(s) = Ca^{2+} + HPO_4^{2-} + 2H_2O)$	-5.73
		$Ca_3(PO_4)_2(s, \beta)$	-28.92
		$Ca_4H(PO_4)_3(H_2O)_3(s)$ $(Ca_4H(PO_4)_3(H_2O)_3(s) = Ca^{2+} + H^+ + 3PO_4^{2-} + 3H_2O)$	-47.08
		$Ca_5OH(PO_4)_3(H_2O)(s)$ $(Ca_5OH(PO_4)_3(H_2O)(s) = Ca^{2+} + OH^- + 3PO_4^{2-} + H_2O)$	-58.33
	F	$CaF_2(s)$	-10.50
	S	$CaS(s)$ $(CaS(s) + 2H^+ = Ca^{2+} + H_2S)$	18.2
Cd(+II)	OH	$Cd(OH)_2(\beta)$	-14.35
	CO_3	$CdCO_3(s,Otavite)$	-12.1
	PO_4	$Cd_5H_2(PO_4)_4(H_2O)_4(s)$ $(Cd_5H_2(PO_4)_4(H_2O)_4(s) + 2H^+ = 5Cd^{2+} + 4HPO_4^{2-} + 4H_2O)$	$-25.4^{I=3}$
	S	$CdS(s)$ $(CdS(s) + 2H^+ = Cd^{2+} + H_2S)$	-7.0
Cr(+III)	OH	$Cr(OH)_3(s)$	$-29.8^{I=0.1}$
Cu(+I)	OH	$CuO_{0.5}(s)$ $(CuO_{0.5} + \frac{1}{2}H_2O = Cu^+ + OH^-)$	-14.7

Metal	Main Ligand	Solid (solubility equilibrium included if K_s is not K_{s0})	log K_s
Cu(+II)	OH	$Cu(OH)_2(s)$	-19.32
		$CuO(OH)_2(s)$	-20.35
	CO_3	$Cu_2(OH)_2CO_3(s,Malachite)$	-33.3
		$Cu_3(OH)_2(CO_3)_2(s,Azurite)$	-44.9
		$CuCO_3(s)$	-11.5
	SO_4	$Cu(OH)_{3/2}(SO_4)_{1/4}(s)$	-17.19
	S	$CuS(s)$ $(CuS(s) + 2H^+ = Cu^{2+} + H_2S)$	-15.2
Fe(+II)	OH	$Fe(OH)_2(s)$	-14.43
	CO_3	$FeCO_3(s)$	-10.8
	PO_4	$Fe_3(PO_4)_2(H_2O)_8(s,Vivianite)$	-37.76
	S	$FeS(s)$ $(FeS(s) + 2H^+ = Fe^{2+} + H_2S)$	3.0
Fe(+III)	OH	$Fe(OH)_3(s)$	-38.8
		$FeOOH(s,\alpha)$ $(FeOOH(s) + H_2O = Fe^{3+} + 3OH^-)$	-41.5
		$FeO_{1.5}(s,\alpha)$ $(FeO_{1.5}(s) + 3/2H_2O = Fe^{3+} + 3OH^-)$	-42.7
	PO_4	$Fe(PO_4)(H_2O)_2(s)$	-26.4
Hg(+I)	CO_3	$Hg_2CO_3(s)$	-16.05
Hg(+II)	OH	$HgO(s,red)$ $(HgO(s,red) + H_2O = Hg^{2+} + 2OH^-)$	-25.44
	CO_3	$Hg_3O_2CO_3(s)$ $(Hg_3O_2CO_3(s) + 4H^+ = 3Hg^{2+} + CO_3^{2-} + 2H_2O)$	-11.1
	PO_4	$Hg(HPO_4)(s)$ $(Hg(HPO_4)(s) = Hg^{2+} + HPO_4^{2-})$	$-13.1^{I=3}$
		$Hg_3(PO_4)_2(s)$ $(Hg_3(PO_4)_2(s) + 2H^+ = 3Hg^{2+} + 2HPO_4^{2-})$	$-24.6^{I=3}$
		$Hg_3(OH)_3(PO_4)(s)$ $(Hg_3(OH)_3(PO_4)(s) + 4H^+ = 3Hg^{2+} + HPO_4^{2-} + 3H_2O)$	$-9.4^{I=3}$
	S	$HgS(s,black)$ $(HgS(s,black) + 2H^+ = Hg^{2+} + H_2S)$	-31.7
		$HgS(s,red)$ $(HgS(s,red) + 2H^+ = Hg^{2+} + H_2S)$	-32.1
Mg(+II)	OH	$Mg(OH)_2(s,active)$	-9.2
		$Mg(OH)_2(s,Brucite)$	-11.15
	CO_3	$MgCO_3(H_2O)_5(s)$	-4.54
		$MgCO_3(H_2O)_3(s)$	-4.67
		$MgCO_3(s,Mangesite)$	-7.46
	PO_4	$Mg(HPO_4)(H_2O)_3(s)$ $(Mg(HPO_4)(H_2O)_3(s) = Mg^{2+} + HPO_4^{2-} + 3H_2O)$	-5.80
	F	$MgF_2(s)$	-8.13
	S	$MgS(s)$ $(MgS(s) + 2H^+ = Mg^{2+} + H_2S)$	24.7
Mn(+II)	OH	$Mn(OH)_2(s)$	-12.8
	CO_3	$MnCO_3(s)$	-10.5
		$MnCO_3(s,Rhodochrosite)$	-11.0
	S	$MnS(s,pink)$ $(MnS(s,pink) + 2H^+ = Mn^{2+} + H_2S)$	10.0
		$MnS(s,green)$ $(MnS(s,green) + 2H^+ = Mn^{2+} + H_2S)$	7.0
Ni(+II)	OH	$Ni(OH)_2(s)$	-15.2
	CO_3	$NiCO_3(s)$	-11.2
	S	$NiS(s,\alpha)$ $(NiS(s,\alpha) + 2H^+ = Ni^{2+} + H_2S)$	1.5
		$NiS(s,\beta)$ $(NiS(s,\beta) + 2H^+ = Ni^{2+} + H_2S)$	-4.0
		$NiS(s,\gamma)$ $(NiS(s,\gamma) + 2H^+ = Ni^{2+} + H_2S)$	-5.7

Metal	Main Ligand	Solid (solubility equilibrium included if K_s is not K_{s0})	log K_s
Pb(+II)	OH	$PbO_{1/2}OH(s)$ ($PbO_{1/2}OH(s) + \frac{1}{2}H_2O = Pb^{2+} + 2OH^-$)	-14.9
		$PbO(s,yellow)$ ($PbO(s,yellow) + H_2O = Pb^{2+} + 2OH^-$)	-15.1
		$PbO(s,red)$ ($PbO(s,red) + H_2O = Pb^{2+} + 2OH^-$)	-15.3
	CO_3	$PbCO_3(s,Cerussite)$	-13.2
	SO_4	$PbSO_4(s)$	-7.79
	PO_4	$PbHPO_4(s)$ ($PbHPO_4(s) = Pb^{2+} + HPO_4^{2-}$)	-11.43
	Cl	$PbCl_2(s)$	-4.78
	F	$PbF_2(s)$	-7.44
	S	$PbS(s)$ ($PbS(s) + 2H^+ = Pb^{2+} + H_2S$)	-7.9
Sr(+II)	CO_3	$SrCO_3(s,Strontianite)$	-9.27
	SO_4	$SrSO_4(s,Selestite)$	-6.62
	PO_4	$SrHPO_4(s)$ ($SrHPO_4(s) = Sr^{2+} + HPO_4^{2-}$)	-6.92
	F	$SrF_2(s)$	-8.58
	S	$SrS(s)$ ($SrS(s) + 2H^+ = Sr^{2+} + H_2S$)	20.9
Zn(+II)	OH	$Zn(OH)_2(s,am)$	-15.52
		$Zn(OH)_2(s,\beta_1)$	-16.24
		$Zn(OH)_2(s,\beta_2)$	-16.20
		$Zn(OH)_2(s,\gamma)$	-16.26
		$Zn(OH)_2(s,\delta)$	-16.15
		$Zn(OH)_2(s,\epsilon)$	-16.46
		$ZnO(s)$ ($ZnO(s,red) + H_2O = Zn^{2+} + 2OH^-$)	-16.66
	CO_3	$ZnCO_3(s)$	-10.8
	PO_4	$Zn_3(PO_4)_2(H_2O)_4(s)$	-35.42
	S	$ZnS(s,\alpha)$ ($ZnS(s,\alpha) + 2H^+ = Zn^{2+} + H_2S$)	-3.8
		$ZnS(s,\beta)$ ($ZnS(s,\beta) + 2H^+ = Zn^{2+} + H_2S$)	-1.6
	CN	$Zn(CN)_2(s)$	-15.5

Source: NIST, 2001

log$K$ values are at 25°C and $I \to 0$, unless otherwise indicated

Table D.4: Henry's Law Constants (K_H) and Vapor Pressures (P_i^o) for Selected Species of Environmental Interest

Species	$\log K_H$ (K_H in mol/L-atm at 25°C)	$\log P_i^o$ (P_i^o in atm at 25°C; omitted for gases)
Inorganics		
Oxygen	−2.92	—
Nitrogen	−3.21	—
Ozone	−2.05	—
Carbon dioxide	−1.49	—
Hydrogen sulfide	−1.01	—
Chlorine	−1.07	—
Sulfur dioxide	0.116	—
Ammonia	1.76	—
Organics		
Methane	−2.82	—
Chloroform	−0.60	−0.59
Benzene	−0.74	−0.90
Toluene	−0.83	−1.42
Phenol	3.39	−3.59
Benzo[a]pyrene	2.92	−11.14

Sources:

Inorganics: Calculated from Montgomery, 1985
Organics: Schwarzenbach et al., 1993

Table D.5: Redox Equilibria for Selected Species of Environmental Interest[1]

Reaction	E^o (V)	$\log K$	pe^o
$Zn^{2+} + 2e^- = Zn(s)$	-0.763	-25.8	-12.9
$Fe^{2+} + 2e^- = Fe(s)$	-0.409	-13.8	-6.92
$2H^+ + 2e^- = H_2(g)$	$\equiv 0$	0.0	0.0
$S(s) + 2H^+ + 2e^- = H_2S$	$+0.14^2$	4.8	2.4
$Cu^{2+} + e^- = Cu^+$	$+0.158$	2.7	2.7
$Cu^{2+} + 2e^- = Cu(s)$	$+0.340$	11.5	5.7
$SO_4^{2-} + 8H^+ + 6e^- = S(s) + 4H_2O$	$+0.36^2$	36.2	6.03
$Fe^{3+} + e^- = Fe^{2+}$	$+0.770$	13.0	13.0
$NO_3^- + 2H^+ + 2e^- = NO_2^- + H_2O$	$+0.84^2$	28.3	14.15
$O_2(g) + 4H^+ + 4e^- = 2H_2O$	$+1.229$	83.1	20.78
$MnO_2(s) + 4H^+ + 2e^- = Mn^{2+} + 2H_2O$	$+1.21$	40.9	20.5
$Cl_2(g) + 2e^- = 2Cl^-$	$+1.36$	46.0	23.0
$O_3(g) + 2H^+ + 2e^- = O_2(g) + H_2O$	$+2.09$	70.7	35.3

Note: 1. All at 25°C; E^o data are from Bard and Faulkner (1980), with $\log K$ and pe^o values calculated.
2. The pe^o value is from Stumm and Morgan (1996), with E^o and $\log K$ calculated.

Animations and Example Computer Files

E.1 ANIMATIONS

The CD accompanying this text contains several *animations*. Animations are Excel files that illustrate key points in the text by plottting user controlled data before your eyes.

The animations are found on the CD in at the *Animations* folder. To run an animation, you must have Excel installed on the computer you are using. Open the *Animations* folder and double click the animation you wish to run. The animations are macros written in Visual Basic for Applications. It is likely that Excel will warn you about the dangers of running macros. If a warning screen appears, you must click "Enable Macros" to run the animations. Only click "Enable Macros" if the macro is from a trusted source.

E.2 VARIATION OF THE EQUILIBRIUM pH OF A MONOPROTIC ACID SOLUTION WITH THE TOTAL ACID CONCENTRATION AND K_a

E.2.1 Background

The purpose of this animation is to show the effects of the total acid concentration (A_T) or K_a on the equilibrium pH of a monprotic acid solution. The animation is discussed in Section 8.4.2.

The animation file is located on the accompanying CD at **\Animations\AT and Ka animation.xls**. To run the animation, double click the file name on the CD. Make sure that the "Animate" worksheet is selected. The opening screen of the animation is shown in Figure E.1.

E.2.2 Running the animation

The animation allows you to vary pA_T or pK_a over the range of your choice. The resulting pC-pH diagrams are plotted, along with a plot of the equilibrium pH as a function of the parameter varied.

To observe the effects of pA_T on the pC-pH diagram and equilibirum pH, enter the range of pA_T values you wish to consider in the "maximum" and "minimum" cells (C2 and C3, respectively). Enter the step size in cell C4. For example, minimum and maximum pA_T values of 3 and 5 with a step size of 0.5 will give you pC-pH diagrams and equilibrium pH values at $pA_T = 3.0, 3.5, 4.0, 4.5$, and 5.0. Set the number in cell H2 to the desired value pK_a value.

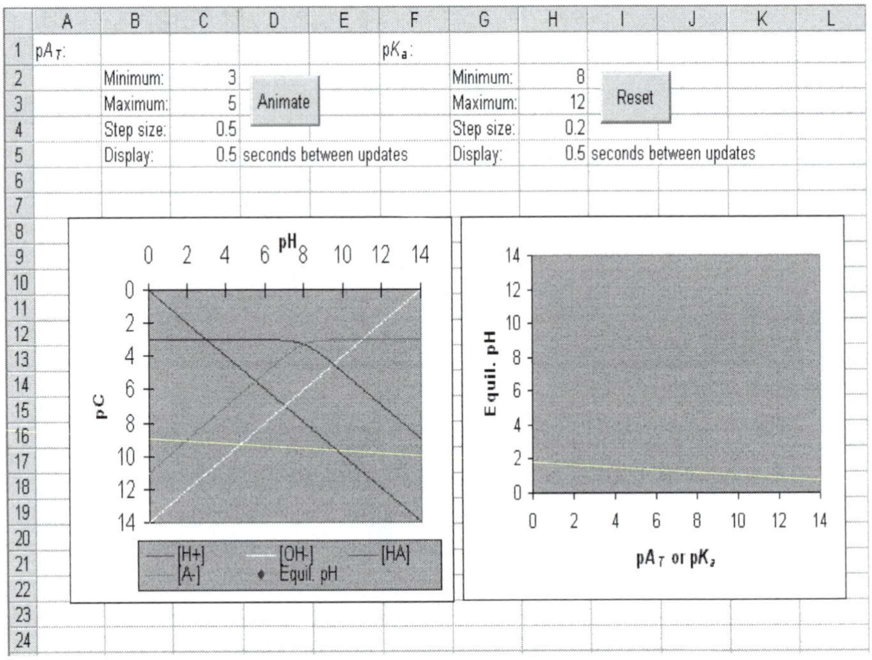

Figure E.1: Opening Screen in the A_T and K_a Animation

To run the animation, make sure that the button near the pA_T section (near column D) reads "Animate". If the button reads "Reset", press the button to clear the previous graphs. Press the "Animate" button and watch the lines in the pC-pH diagram adjust to each new value of pA_T. A red dot on the pC-pH diagram indicates the equilibrium pH. The equilibrium pH also is plotted against pA_T in the leftmost graph. If the animation is too fast or too slow for your taste, you can adjust the animation speed in cell C5. A speed of about 0.5 seconds per step is a good place to start.

The effects of pK_a can be observed by using the data entry cells in column H and the button near column I. Set the number in cell C2 to the desired value pA_T value. Again, press the button to clear the previous graphs if the button reads "Reset".

E.2.3 Using the animation

This animation can be used to explore the behavior of monoprotic acid systems. Use it to see that the charge balance shifts from $[H^+] \approx [A^-]$ at high values of A_T (low pA_T) to $[H^+] \approx [OH^-]$ at low values of A_T (high pA_T). Use it to investigate the behavior of very strong acids (i.e., acids with low or negative pK_a values). Is it true that the equilibrium pH of very strong acid solutions is about equal to pA_T?

E.3 HOW TO DRAW A pC-pH DIAGRAM

E.3.1 Background

The purpose of this animation is to illustrate the steps involved in drawing a pC-pH diagram. The animation is discussed in Section 8.4.3.

The animation file is located on the accompanying CD at **\Animations\pC pH animation.xls**. To run the animation, double click the file name on the CD. Make sure that the "Animate" worksheet is selected. The opening screen of the animation is shown in Figure E.2.

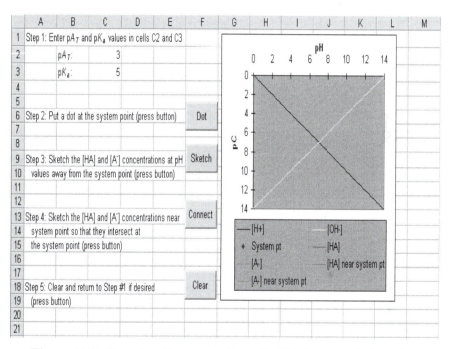

Figure E.2: Opening Screen in the pC-pH Diagram Animation

E.3.2 Running the animation

The animation allows you to draw a pC-pH diagram for a monoprotic acid with your choice of pA_T and pK_a values. The new points and lines in the pC-pH diagram are drawn for you as you go through each step outlined in Section 8.4.3.

To begin the simulation, enter the pA_T and pK_a values of your choice in cells C2 and C3, respectively. Click the "Dot" button to plot the system point on the pC-pH diagram. Next, click the "Sketch" button to sketch the [HA] and [A$^-$] lines at pH values away from the system point. Click the "connect" button to draw the [HA] and [A$^-$] lines so that they intersect at the system point. The diagram can be reset by clicking the "Clear" button. Before you reset the graph, you may wish to practice locating the point where the charge balance is satisfied.

E.3.3 Using the animation

This animation can be used to practice the short-cut method for drawing pC-pH diagrams. It is also an excellent tool for practicing the skill of locating the equilibrium pH (i.e., the pH where the charge balance is satisfied).

E.4 EQUILIBRIUM PH DURING THE TITRATION OF A MONOPROTIC ACID WITH A STRONG BASE

E.4.1 Background

The purpose of this animation is to illustrate the steps involved in drawing a pC-pH diagram. The animation is discussed in Section 12.2.3.

The animation file is located on the accompanying CD at **\Animations\Titration animation.xls**. To run the animation, double click the file name on the CD. Make sure that the "Animate" worksheet is selected. The opening screen of the animation is shown in Figure E.3.

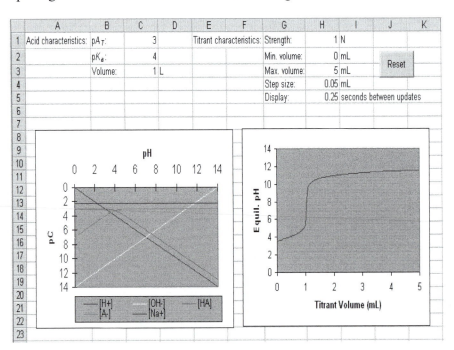

Figure E.3: Opening Screnn for the Titration Animation

E.4.2 Running the animation

The animation allows you to draw a pC-pH diagram and a titration curve at each step in the titration of a monoprotic acid with a strong acid. You can change the characteristics of the acid and titrant and replot the pC-pH diagram and a titration curve.

To begin the simulation, enter the monoprotic acid characteristics (pA_T value, pK_a value, and acid volume in L) in cells C1 through C3, respectively. Enter the titrant characteristics (strength in N, minimum volume in mL, maximum volume in mL, and titration increment [step size] in mL) in cells H1 through H4, respectively.

To run the animation, make sure that the button near column J reads "Animate". If the button reads "Reset", press the button to clear the previous graphs. Press the "Animate" button and watch the [Na^+] line in the pC-pH diagram adjust to each new titrant volume. The equilibrium pH also is plotted against titrant volume in the titration curve (leftmost graph). If the animation is too fast or too slow for your taste, you can adjust the animation speed in cell H5. A speed of about 0.25 seconds per titration step is a good place to start.

E.4.3 Using the animation

This animation can be used to explore the titration behavior of monoprotic acid systems. Use it to see how the shape of the titration curve varies with both the acid and titrant characteristics. Is it true that the titration curve for the titration of a very strong acid solution with a concentrated basic titrant is very sharp at the first equivalence point?

E.5 OTHER COMPUTER FILES

A list of spreadsheet files on the CD is given in Table E.1.

Table E.1: Spreadsheet Files on the Accompanying CD

Purpose	File Location on CD	Text Section
Ch. 9 acetic acid problem	**/Chapter 9/Acetic acid.xls**	Section 9.3.2
NTA-Ca(+II)-Cu(+II)-Fe(+II) system	**/Chapter 15/NTA system.xls**	Section 15.5.4

Nanoql chemical system files are listed in Table E.2. More information about *Nanoql* may be found in Appendix F.

Table E.2: *Nanoql* Chemical System Files on the Accompanying CD

Purpose	File Location on CD	Text Section
Ch. 9 acetic acid problem	**/Chapter 9/Acetic acid.nqs**	Section 9.5.2
Part II case study	**/Chapter 9/Case study.nqs**	Section 9.8
10^{-3} M H_2CO_3* solution	**/Chapter 11/Carbonic acid.nqs**	Section 11.4.2
Titration of a 10^{-3} M acetic acid solution with a strong base	**/Chapter 12/Acetic acid titration.nqs**	Section 12.2.3
Hg(+II)-chloride system	**/Chapter 15/Hg chloride.nqs**	Section 15.4.4
NTA-Ca(+II)-Cu(+II)-Fe(+II) system	**/Chapter 15/NTA.nqs**	Section 15.5.4
$H_2S(g)$ system	**/Chapter 18/H2S fixed.nqs**	Section 18.3.3
$Zn(OH)_2(s)$ system	**/Chapter 19/zinc.nqs**	Section 19.3.3

Nanoql

F.1 INSTALLATION

To install *Nanoql*, locate and run the setup file on the accompanying CD (file location: **\Nanoql\setup.exe**). During installation, you will be given a choice of where you would like the *Nanoql* files installed. Select a location and click the large button to begin installation.

F.2 USING AND MANIPULATING FILES IN *NANOQL*

F.2.1 Introduction

The basic operations in *Nanoql* were described in Section 9.5.2. Material from that section is elaborated upon here. *Nanoql* uses two types of files: a thermodynamic database (called a Possible Species file and having the extension nps) and a file for your system (called a *Nanoql* Chemical System and having the extension nqs). It is likely that the only Possible Species file you will ever use is the file **Nanoql Possible Species.nps**. This file is unpacked into the installation directory upon installation of the software. *It is highly recommended that you make a backup copy of this file.*

F.2.2 Opening and using files

You can open files using the *Open* command in the Main Menu or by clicking the *Open* button. *Nanoql* will prompt you for the Possible Species file you wish to use. It is possible to create you own Possible Species file, but it is sufficient to use the file **Nanoql Possible Species.nps** for all the uses of *Nanoql* discussed in this text

You can examine the opened file using the *View•Chemical System* menu item. The arrow controls near the bottom of the screen allow you to navigate through the chemical system. On the first screen, basic system characteristics are indicated (whether the electroneutrality constraint is used, temperature, etc.). On the second screen, system species are listed. Species are either dissolved or fixed. A fixed species has a fixed concentration. For example, you may want to work at a fixed pH or the concentration of an unreactive species such as Na^+ might be fixed. (The current version of *Nanoql* also supports variable species. Variable species are solids that may be allowed to dissolve completely or precipitate.) The next screen shows the equilibria, followed by mass balances on the fourth screen. The fifth screen shows User Defined Functions (see Section 9.5.2). From this view screen, the chemical system also can be edited (see Section E.2.3).

To calculate equilibrium concentrations, press the Σ button (or select *Solve* under the *Calculate* menu). *Nanoql* will prompt you for initial guesses for each species. To avoid the tedium of entering each value separately, select *Helper*. *Helper* will assign initial guesses for species concentration. *Be careful using Helper*: you may want to use *Helper* and then check the values of the fixed species concentrations to make sure they are hat you want. Concentration guesses can be very crude. Press *Ok* and *Nanoql* will calculate the species concentrations. The output can be viewed or saved to a data file.

You will make frequent use of the *Vary* menu. A number of system characteristics can be varied, including a species concentration, equilibrium constant, total mass, temperature, and ionic strength. The example of varying $[H^+]$ to draw a pC-pH diagram was discussed in section 9.5.2. Another example of varying a species concentration is an acid-base titration, where, say, the concentration of Na^+ could be varied. Variation results can be viewed in a table, plotted, or saved as a data file.

F.2.3 Entering a new chemical system

The basic process for entering a new chemical system was outlined in Section 9.5.2. To create a new chemical system, click on the *New* button on the right side of the toolbar or select *New* under the *File* menu. You will be prompted to open a Possible Species file (.nps file). Usually, you will choose the default file.

There are five steps for entering a new system. In Step 1, choose the system characteristics. The most important choice is whether or not to use the electroneutrality condition. In most cases, the electroneutrality condition is used. It is not used when, for example, the pH is fixed by buffers not included in the chemical system. This may occur if you are modeling only part of a larger system. Press *Next*.

In Step 2, you will choose the species. You can do this by clicking on the species individually in the top table or clicking on the components in the top list box to select all species in a given component. You can click on the selected species (bottom grid) to delete them, view their properties, or change their species type.

In Steps 3 and 4, equilibria and mass balances are entered. You can click on species in the top table to add the information manually or click on *Helper* to let *Nanoql* assist you with the process. Be sure to enter the equilibrium constant and total mass. If sufficient thermodynamic data are available in the database, *Nanoql* can calculate the equilibrium constant.

Remember from Section 9.5.2 that numbers in *Nanoql* can be entered several different ways. Enter $10^{-3.5} = 3.16 \times 10^{-4}$ as **3.16e-4** or **p3.5** or **l-3.5**. Here, **p** means that the number that follows is -1 times $\log_{10}$ of the value to be entered and **l** (lowercase letter ell) means that the number that follows is $\log_{10}$ of the value to be entered. Always press the *Enter* key after typing in a number.

The fifth step is optional. You can enter User Defined Functions (UDFs). A UDF is a linear combinations of species concentrations. The UDFs can be tracked at equilibrium or during a variation exercise. For example, you could enter the charge balance to see if it really is zero at equilibrium. Other common UDFs are the alkalinity equation and subsets of mass balances (e.g., the sum of the concentrations of complexed or free metal species). After entering a system, be sure to save it.

F.3 TIPS ON USING *NANOQL*

This section lists several tips for using *Nanoql* to solve specific types of chemical equilibrium problems

F.3.1 Acid-base and titrations

Use the Components list box to select entire families of species rather than selecting them individually. You can use the check boxes to avoid adding solid or gaseous species automatically. In titrations, you typically start with the titrant concentration equal to zero. If you use *Helper* to set initial concentration guesses, be sure to manually reset the C_B or C_A species to zero.

F.3.2 Complexation and redox equilibria

Again, use the Components list box to select entire families of species. Be sure to enter a mass balance equation for each component.

F.3.3 Open systems

Nanoql is well-suited to performing equilibrium calculations when the gas-phase concentration is fixed and the solid exists. In these cases, be sure to make the gas-phase species or solid species a fixed species. If you use *Helper* to set initial concentration guesses, be sure to manually reset the gas-phase species [e.g., $CO_2(g)$] to its known value.

With some open systems, the total masses of some species are not known. For example, if $CaCO_3(s)$ partially dissolves in water, you do not know the total Ca(+II) concentration or C_T. However, you do know that $Ca(+II)_T = C_T$ [if $CaCO_3(s)$ is the only source of calcium and carbonate]. Thus, you can enter the following mass balance equation (assuming that Ca^{2+} is the only soluble calcium species of interest):

$$[Ca^{2+}] - [H_2CO_3] - [HCO_3^-] - [CO_3^{2-}] = 0$$

Use negative coefficients in this artificial mass balance to force $Ca(+II)_T = C_T$.

REFERENCES

American Public Health Association, American Water Works Association, and Water Environment Federation. **Standard Methods for the Examination of Water and Wastewater**. 20th ed., APHA, AWWA, WER, Washington, DC, 1998.

Bard, A.J. and L.R. Faulkner. **Electrochemical Methods: Fundamentals and Applications**. John Wiley & Sons, Inc., New York, NY, 1980.

Benefield, L.D. and Morgan, J.S. Chemical Precipitation. In: **Water Quality and Treatment**. 4th ed., F.W. Pontius (ed.), McGraw-Hill, Inc., New York, NY, pp. 641-708, 1990.

Byrne, R.H. and S.H. Laurie. Influence of Pressure on Chemical Equilibria in Aqueous Systems - with Particular Reference to Seawater (Technical Report). *Pure Appl. Chem.*, **71**(5), 871-890, 1999.

Butler, J.N. **Carbon Dioxide Equilibria and Their Applications**. Addison-Wesley Publ. Co., Reading, MA, 1982.

Butler, J.N. **Ionic Equilibrium: Solubility and pH Calculations**. John Wiley & Sons, Inc., New York, NY, 1998.

Deffeyes, K.S. Carbonate Equilibria: A Graphic and Algebraic Approach. *Limnol. Oceanogr.*, **10**(3), 412-426, 1965.

Eisenberg, D. and D. Crothers. **Physical Chemistry with Applications to the Life Sciences**. Benjamin/Cummings Publ. Co., Inc., Menlo Park, CA, 1979.

Espenson, J.H. **Chemical Kinetics and Reaction Mechanisms**. Mc-Graw Hill Book Co., New York, NY, 1981.

Fyfe, W.S. From Molecules to Planetary Environments: Understanding Global Change. In: **Aquatic Surface Chemistry: Chemical Processes at the Particle-Water Interface**. W. Stumm (Ed.), pp. 495-508, 1983.

Haag, W.R., and T. Mill. Some Reactions of Naturally Occurring Nucleophiles with Haloalkanes in Water. *Environ. Toxicol. Chem.*, **7**, 917-924, 1988.

Hendrix, P.F., J.A. Hamala, C.L. Langner, and H.P. Kollig. Effects of Chlorendic Acid, A Priority Toxic Substance, on Laboratory Aquatic Ecosystems. *Chemosphere*, **12**(7/8), 1083-1099, 1983.

Jackson, G.A. and J.J. Morgan. Trace Metal-Chelator Interactions and Phytoplankton Growth in Seawater Media: Theoretical Analysis and Comparison with Reported Observations. *Limnol. Oceanogr.*, 23, 268-282, 1978.

Jensen, J.N. and J.D. Johnson. Interferences by Monochloramine and Organic Chloramines in Free Available Chlorine Methods. 2. DPD. *Environ. Sci. Technol.*, **24**(7), 985-990, 1990.

Kielland, J. Individual Activity Coefficient of Ions in Aqueous Solutions. *J. Amer. Chem. Soc.*, **59**, 1675-1678, 1937.

Kling, G.W., W.C. Evans, G. Tanyileke, M. Kusakabe, and Y. Yoshida. The Nyos-Monoun Degassing Program: Preliminary Report of the U.S.-OFDA Technical Project, October-November, 1999

Kling, G.W., W.C. Evans, M.L. Tuttle, and G. Tanyileke. Degassing of Lake Nyos, *Nature*, **368**, 405-406, 1994.

Langelier, W.F. The Analytical Control of Anti–Corrosion Water Treatment. *J. Amer. Water Works Assoc.*, **28**(10), 1500-1521, 1936.

Le Chatelier, H.L. *Annales des Mines*, **13** (2), 157, 1888, as found in: **http://www.woodrow.org/teachers/chemistry/institutes/1992/LeChatelier.html**

Lowenheim, F.A. **Electroplating**. McGraw-Hill Book Co., New York, NY, 1978.

Luck, W.A.P. The Importance of Cooperativity for the Properties of Liquid Water. *J. Mol. Struct.*, **448**(2-3), 131-142, 1998.

March, J. **Advanced Organic Chemistry: Reactions, Mechanisms, and Structure**. 3rd ed., John Wiley & Sons, Inc., New York, NY, 1985.

Martell, A.E. and R.M. Smith. **Critical Stability Constants**. Plenum Press, New York, NY, 1974.

Metcalf and Eddy. **Wastewater Engineering: Treatment and Reuse**. 4th ed., McGraw-Hill Book Co., New York, NY, 2003.

Millero, F.J. and M. Izaguirre. Effect of Ionic Strength and Ionic Interactions on the Oxidation of Iron(III). *J. Solution Chem.*, **18**(6), 585-599, 1989.

Morel, F.M.M. and J.G. Hering. **Principles and Applications of Aquatic Chemistry**. John Wiley & Sons, Inc., New York, NY, 1993.

Morgan, J.J. and A.T. Stone. Kinetics of Chemical Processes of Importance in Lacustrine Environments. In: **Chemical Processes in Lakes**, W. Stumm (ed.), John Wiley and Sons, New York, NY, pp. 389-426, 1985.

Morris, J.C. and R.A. Isaac. A Critical Review of Kinetic and Thermodynamic Constants for the Aqueous Chlorine-Ammonia System. In: **Water Chlorination: Environmental Impact and Health Effects**, Vol. 4,

Book 1, R.L. Jolley, W.A. Brungs, J.A. Cotruvo, R.B Cummings, J.S. Mattice, and V.A. Jacobs (Eds.), Ann Arbor Science Publ., Ann Arbor, MI, 1983.

NRC (Committee on the Toxicological Effects of Methylmercury, Board on Environmental Studies and Toxicology, National Research Council). **Toxicological Effects of Methylmercury**. National Academy Press, Washington, DC, 2000.

Pauling, L. **General Chemistry**. Dover Publ., Mineola, NY, 1970.

Pauling, L. **College Chemistry**, 3rd ed., Freeman, San Francisco, CA, 1964.

Plechanov, N., B. Josefsson, D. Dyrssen, and K. Lundquist. Investigations on Humic Substances in Natural Water. In: **Aquatic and Terrestrial Humic Materials**. R.F. Christman and E.T. Gjessing (Eds). Ann Arbor Science, Ann Arbor, MI, pp. 387-405, 1983.

Sawyer, C.N. and P.L. McCarty. **Chemistry for Environmental Engineers**. 3rd ed., McGraw-Hill Book Co., New York, NY, 1978.

Sillén, L.G. How have sea water and air got their present compositions? *Chemistry in Britain*, **3**, 291, 1967.

Sillén, L.G. Graphical Presentation of Equilibrium Data. In: **Treatise on Analytical Chemistry**. I.M. Kolthoff and P.J. Elving (Eds.), Interscience Encyclopedia, New York, NY, pp. 277-317, 1959.

Singer, P.C. and W. Stumm. Acid Mine Drainage: The Rate Determining Step. *Science*, **167**, 1121-1123, 1970.

Snoeyink, V.L. and D. Jenkins. **Water Chemistry**. John Wiley & Sons, Inc., New York, NY, 1980.

Staehelin, J. and J. Hoigné. Decomposition of Ozone in Water in the Presence of Organic Solutes Acting as Promoters and Inhibitors of Radical Chain Reactions. *Environ. Sci. Technol.*, **19**(12), 1206-1213, 1983.

Stumm, W. and J.J. Morgan. **Aquatic Chemistry: Chemical Equilibria and Rates in Natural Waters**. 3rd ed., John Wiley & Sons, Inc., New York, NY, 1996.

Sun, M.S., D.K. Harriss, and V.R. Magnuson. Activity Corrections for Ionic Equilibriums in Aqueous Solutions. *Can. J. Chem.*, **58**(12), 1253-1257, 1980.

Thurman E.M. and R.L Malcolm. Structural Study of Humic Substances: New Approaches and Methods. **In**: **Aquatic and Terrestrial Humic Materials**. R.F. Christman and E.T. Gjessing (Eds), Ann Arbor Science, Ann Arbor, MI, pp. 1-23, 1983.

Tiffreau, C., J. Lützenkirchen, and P. Behra. Modeling the Adsoprtion of Mercury(II) on (Hydr)oxides. I. Amorphous Iron Oxide and α-Quartz. *J. Colloid Interface Sci.*, **172**, 82-93, 1995.

Tuan, Y.-J. and J.N. Jensen. Chemical Oxidation of Thiocyanate by Ozone. *Ozone Sci. Eng.*, **15**(4), 343-360, 1993.

Van Slyke, D.D., *J. Biol Chem.*, **52**, 525, 1922.

Wehrli, B. Redox Reactions of Metals at Mineral Surfaces. In: **Aquatic Chemical Kinetics: Reaction Rates of Processes in Natural Waters**. W. Stumm (ed.), John Wiley & Sons, Inc., New York, NY, 1990.

Weil, I. and J.C. Morris. Kinetic Studies on the Chloramines. 1. The Rate of Formation of Monochloramine, *N*-Chloromethylamine and *N*-Chlorodimethylamine. *J. Amer. Chem. Soc.*, **71**, 1664-1671, 1949.

Wetzel, R. **Limnology**. Saunders Publ., Philadelphia, PA, 1983.

Index

G

g function, 240
Gibbs free energy, 33, 44–45, 47, 48, 50, 52, 54, 55, 56, 57, 61, 67, 74, 75, 79, 175, 176, 208, 346, 349, 352, 354, 449
Gibbsite (α-Al(OH)$_3$(s)), 419
Gibbs Phase Rule, 37
Gram molecular weight, 16
Graphical solutions
    general pC-pH diagram, 549–550
    graphical shortcut method, 551
    pe-pH diagrams, 551–552
    shortcut graphical method, 550–551
Guoy-Chapman diffuse layer model, 449

H

Half-equivalence point, 236, 245
Half reaction, 338
    balancing, 339–341
Half-time of the reaction ($t_{1/2}$), 501
Hardness, 434
Hemocyanin, 312
Hemoglobin, 312
Henry's Law, 395, 398, 401, 402, 405, 406, 410
Henry's Law constant, 396, 397, 564
Hess's Law of Heat Summation, 47, 49
Homogeneous system, 193, 194, 200
Hydration, 9
Hydrazine (NH$_2$NH$_2$), 346
Hydrogen bonding, 8, 9
Hydrogen bonds, 8, 195
Hydrogen cyanide (HCN), 5
Hydrolysis reactions, 3
Hydrolyzing dissolved gases, 401
Hydrostatic pressure, 411
Hydroxo complexes, 310, 311, 315–316

I

Ideal gas constant, 54
Ideal gas law, 25, 54
Ideal gas mixture, 54
Ideal solution, 462
Ideal substances, 391–393
Immersion deposits, 332
Individual chemical species, 4
Infinite dilution standard state, 470
Intensive equivalents, extensive properties, 47–48

Intensive property, 35, 36
Intrinsic equilibrium constants, 449
Ion, 537
Ionic bond, 299
Ionic equilibria, 537
Ionic strength, 468
Ionic strength effects, 554
Ionic theory, 204
Ion product of water (K_W), 104
Ion size parameter (a), 471
Isotherm, 448

K

K_{s0}, 418
Kinetic control, 493

L

Langmuir isotherm, 446–448
Le Chatelier's principle, 69–70, 84–85
Lewis acid, 307
Lewis base, 307
Ligand, 306
Linear independence, 73–74, 76
Linearly independent equilibria, 76
Log concentration diagram, 142, 143

M

Manipulating equilibria, 77–79, 80–82
Mass balance, 160, 161, 165, 172, 173, 175, 176, 177, 183–185, 188, 239, 240, 324, 398
Mass balance constraints, 122
Mass balance equation, 93–94, 127, 128, 211, 313
Mass concentration units, 18–21
Master variables, 10
    pe, 7. *See also* pe
    pH, 7. *See also* pH
Mathematical equations, 68
Mercury (Hg), 10
Metal, 306
Metal buffers, 330
Metal finishing, 519–524
Method of approximation, 126–133
Method of substitution, 123
Methylmercury (CH$_3$Hg$^+$), 10, 11, 61, 82–83
Methyl orange endpoint, 276
MINEQL, 171, 176, 177, 178